3rd Edition

Firefighter's Handbook on Wildland Firefighting

Strategy, Tactics and Safety

William C. Teie
Illustrations by Dave A. Hubert

Deer Valley Press
Rescue, California
January 2005

Produced and published by:
DEER VALLEY PRESS
5125 Deer Valley Road
Rescue, California 95672
(530) 676-7401
www.deervalleypress.com

Library of Congress Control Number: 2004096725
ISBN 1-931301-16-6

First edition printed in June 1994
Second edition printed in February 2001
Third edition printed in January 2005

Printed in China

10 9 8 7

Dedication

To all firefighters and the ones who love them.

About the Author

William C. Teie retired from the California Department of Forestry and Fire Protection (CDF) after a successful 34-year career. He worked up through the ranks from seasonal firefighter to Deputy Director for Fire Protection. In this position, he was responsible for all of the fire protection programs within CDF.

Chief Teie was very active in the California fire service. He was a member of the Joint Apprentice Committee, the California State Board of Fire Service, and the Office of Emergency Services' Fire and Rescue Advisory Committee. He was also on the Board of Directors of the California Fire Chiefs Association, and was President in 1986-87. He is also a member of the Institute of Fire Engineers.

He is the author of the ***Firefighter's Handbook on Wildland Firefighting, Fire Officer's Handbook on Wildland Firefighting, Wildland Firefighting Fundamentals,*** the series of ***Fire in the West*** reports, and has developed several training and operational aids for the firefighter. In 2004, two new books were published in South Africa; the ***Fire Manager's Handbook on Veld and Forest Fires*** and the ***Veld and Forest Firefighting Fundamentals***.

South Africa 2002

About the Artist

Dave Hubert has been in the fire service for 27 years. Dave started as a firefighter with CDF, and when he retired, he was a captain with the Orange County Fire Authority. Dave has served as training officer, fire prevention inspector, and fire defense coordinator (specializing in community fire defense programs). Dave is a true practitioner who knows the reality of the job. He was President of the Orange County Firemen's Association in 1980 and was a Deputy Director and the staff artist for the California State Firefighters Association.

Dave is dedicated to his family, and enjoys fire service artwork and working with his horse-drawn steam fire engine with his wife, Barbara, and their dog, "Blaze."

Acknowledgments

Writing a technical book is not a solitary undertaking. It involves many people with many talents. I have had much help and support in the development of this book. There are many people that I want to extend my appreciation and thanks to. Most of all, I want to thank my wife, Linda, for putting up with me being "locked up in the office" for days on end. She has been very supportive, and without that support, this edition of the ***Firefighter's Handbook on Wildland Firefighting*** would have never happened.

There are five very special people that have added so much to this work: Dave Hubert for his wonderful artwork; Brian Weatherford and Tim Murphy for their technical editing, making sure the information is correct and proper; Matt Thau who gathered material and allowed himself to be photographed for pictures in the sections on safety and use and care of hand tools; and, Judy Craig was great in her role as a "non-fire type" that reviewed, edited and polished the handbook from a teaching point of view. Special thanks also has to be given to Ronny Coleman, an old friend, for his kind words in the Foreword, and to Don Artley for the Introduction. The last person to look at the manuscript was Anne Holms, the proofreader. Again, she did an outstanding job of catching all of my numerous mistakes.

One of the things that has made this handbook such a success has been the graphics. Dave's art made all the difference, but a new dimension has been added to this edition—color. By printing the handbook in full color, we have been able to add hundreds of photographs. A lot of the pictures were drawn from thousands of slides available from the National Wildfire Coordinating Group as aids to their training packages, or from their Image Portal. The individuals that took each of these pictures should be recognized, but in the case of thousands of individual 35-mm slides, there was no way of identifying them. If you see a picture you took, please contact me so that you can receive proper recognition. When the photographer is known, their name appears to the left of the picture.

I wish to also acknowledge several individuals who provided special access to their vast library of pictures. They are: Lisa Boyd, Kari Greer, John Hawkins, Steve Huntington, Wes Schultz, Kurt Taylor, Ed Waggoner, and Karen Wattenmaker.

In this edition of the handbook, Jake is introduced as "a wise, old training captain," someone who doesn't say much, but when he does, you should stop and listen. A lot of the "Jake-isms" came out of the author's background, but some of the best came from three real experts in wildland firefighting: Tim Murphy, retired Chief of Aviation and Fire Management, Montana DNRC; Joe Stutler, retired USDA Forest Service; and Ed Waggoner, retired CDF and now Chief of Operations for Northtree Fire International.

Special thanks also needs to be given to Peri Poloni of Knockout Design for the outstanding cover design. The front cover is a montage of pictures taken by Cheryl Prestia (the firefighter on a training burn at the CDF Academy), Kurt Taylor (the helicopter on the Old Fire, CA 2003) and Kari Greer (the background fire somewhere in the west).

Finally, thanks go to my dedicated friends in CDF, from whom I learned so much as we worked together in our chosen career in the fire service.

If you get nothing else out of this book, learn that fighting wildland fires is very dangerous. You must know and understand the basics and meaning behind the ten standard firefighting orders. Fighting fire in a safe way must be automatic, like riding a bike.

Safety is not a peripheral consideration; it is the primary consideration. Your understanding of the material in this book may save not only your life, but the lives of your fellow firefighters and the public.

W.C.T.

Foreword

I started off my career in the fire service as a member of the US Forest Service. I then served time with the National Park Service. The bulk of my career was spent working for cities. Forty-four years later, I look back upon my career and marvel at all of the changes in both the complexity of the fire problem, and the methods we use to combat them. Yet at the same time, I recognize that much of the task of fighting a fire remains physical and very basic.

In the last couple of decades, we have seen the wildland fire become this century's version of the conflagration. I recently sat on the State of California's Governor's Blue Ribbon Commission to look at the lessons we are learning from the most recent of these tragedies. A lesson that is continually reinforced is that quality training and education for our firefighters are the most important elements in assuring that they are able to complete their tasks safely, efficiently and effectively.

This new edition of Chief Teie's text is an important part of creating a well-rounded and highly competent fire suppression force to deal with these major fires. Moreover, it is becoming very important that all firefighters have an understanding of this topic. Knowledge of wildland fire is just as important to a structural firefighter as it is to the wildland firefighter. Massive mutual aid mobilizations have resulted in even the most urbanized areas responding their firefighters into harm's way in the wildland environment. Recent statistics have indicated that a firefighter stands a greater chance of suffering major injury or loss of life on a wildland fire than almost any other location. This is a topic that should be studied by everyone, everywhere who is responding on a piece of fire apparatus.

Chief Teie's approach to preparing this textbook is also worthy of note. This is not a textbook of theory in itself. It is founded on a common sense and real-world application of principles that actually work when facing these kinds of fires. Moreover, Chief Teie has continued to supplement his original text with more and more detail that makes the reader a well-informed person prior to being exposed on an emergency.

Granted, the wildland fire often involves only firefighters from wildland agencies. They will benefit immensely from this book. In my opinion, what makes this book most valuable, however, is that it provides a comprehensive knowledge base for all firefighters; both wildland and structural. Mother nature has demonstrated that she does not discriminate when it comes to the color of the fire trucks, the type of uniform or the name of an agency when it comes to creating a hostile and dangerous workplace on a fire. This book should be a basic text for anyone who has the responsibility of dealing with wildland fire.

Ronny J. Coleman, Chief
California State Fire Marshal (retired)

Preface

This book is written for the firefighter who needs a basic grasp of wildland firefighting. This includes those who have primary wildland responsibilities, and those who may intermittently be called into a wildland fire situation, such as through a mutual aid response.

These words were written to introduce the first edition of this very successful handbook in 1994. They still hold true today, with the writing of this, the third edition. This edition brings a whole new dimension to the ***Firefighter's Handbook on Wildland Firefighting***. It brings full color pictures; pictures that tell a story of how to go about combating a wildland fire safely and effectively.

This edition also introduces "Jake" to the reader. Jake's role is to stress the important points from a working knowledge of the subject...points that only can come from experience on the fireline. Jake is not one person, but a collection of "old-time" firefighters who have, over time, said something that struck a chord, and was worth writing down.

I began my 34-year career with CDF as a seasonal firefighter right out of high school. In those days, there were four books that helped a "newbe" learn about wildland firefighting. They were: ***Forest Fire Fighting Fundamentals***, ***Water vs. Fire***, ***Fireman's Guide*** and ***Principles of Forest Fire Management***. The first three books were developed by the USDA Forest Service; the last one was developed by my old organization, the California Department of Forestry and Fire Protection. What I have attempted to do is to incorporate the material from these four very well-written books into one, and update the material while I was at it.

Throughout most of those challenging and rewarding years, the desire to write a book on wildland firefighting remained an ultimate goal. When I retired (or more accurately changed careers) in February of 1993, the time and opportunity finally presented themselves. Since then I have been privileged to have been involved in many projects that have kept me close to the wildland firefighting community.

The increased attention to firefighter safety and the wildland/urban interface problem—as disastrously exemplified by the fires in Southern California and the South Canyon, the Thirty-mile, and other fires where good people died—has compelled me to produce this new, completely revised third edition.

The primary objective of this handbook remains the same: to present updated information on wildland firefighting in a user-friendly format. ***It is hoped that this book will lead the firefighter to acquire new skills, learn to predict fire behavior, and carry out informed actions with an emphasis on safety.***

If you get nothing else out of this book, learn that fighting wildland fires is very dangerous. You must know and understand the basics and the meaning behind the Ten Standard Firefighting Orders. Fighting fire in a safe way must be automatic, like riding a bike.

Safety is not a peripheral consideration; it is the primary consideration. Your understanding of the material in this book may save not only your life, but the lives of your fellow firefighters and the public.

The material in this handbook is intended to be an informational presentation of some of the issues involved in wildland firefighting. It is not intended to be used in lieu of the rules and regulations of your department or other authorities. The application of any of the recommendations in this handbook should be modified as appropriate by particular circumstances and your own best judgement.

CDF 1989

William C. Teie

Deputy Director for Fire Protection (retired)
California Department of Forestry and Fire Protection

Contents

Acknowledgments v
Foreword vii
Preface ix
Introduction xiii

1 **Fireline Safety 1**

2 **Fire Weather 57**

3 **Topography and Fuels 103**

4 **Wildland Fire Behavior 131**

5 **Fire Extinguishment Methods 167**

6 **Initial Attack Strategy and Tactics 201**

7 **Use of Firefighting Resources 237**

8 **Wildland/Urban Firefighting Strategy and Tactics 289**

9 **Incident Command System 321**

10 **Firefighting Realities 361**

11 **Firefighting Situations 389**

Appendices:
- **Map Reading 417**
- **Fire Prevention 449**
- **Hazardous Materials Recognition 465**
- **Use and Care of Hand Tools 473**
- **Fire Ground Communications 495**
- **Incident Organizer 507**

Glossary 513
Index 525

Introduction

Safe, effective wildland firefighting requires a thorough understanding of many complicated principles and procedures. It also requires great courage and personal commitment. Since you are interested, or are already in the fire service, you have demonstrated that you have the courage and commitment. This handbook strives to teach you the science, principles and skills necessary to enhance your effectiveness as a wildland firefighter.

Safety is your most important rule.

The ***Firefighter's Handbook on Wildland Firefighting, Strategy, Tactics and Safety,*** is designed to be a stand-alone text. But, in a effort to make the learning experience even more meaningful and complete, there are several other aids to help you. There is a ***Study Guide*** that provides checks on your comprehension, a set of ***Flash Cards*** that will aid you in retaining information, and PowerPoint presentations for the instructors.

When Jake talks, listen and learn!

In this edition of the handbook, Chief Teie introduces Jake, "your training officer." Like most "old-timers," when he says something, you should listen. He is an old hand at wildland firefighting, and what he "says" is important. He may be repeating something you have just read...repetition is good for retention. But, most of the time, he will give you his spin on a situation. When Jake speaks...listen.

Chief Teie is also introducing some special highlighting. Since many of you may want to work for—or are working for a wildland firefighting agency—you want to know if studying this text will help you meet the state or federal requirements for basic wildland firefighting. When you see material that is highlighted in yellow, it is important and ties directly back to an objective required in the courses developed by working teams of the National Wildfire Coordinating Group, ***Firefighter Training (S-130)***, ***Introduction to Wildland Fire Behavior (S-190),*** and the ***Incident Command System (I-100 and 200).***

Understanding what is presented in the handbook will help you meet the requirements outlined in S-130, S-190 and I-100/200.

To become a competent wildland firefighter, you can't just read this handbook, and then say "I am one," and get a tattoo! You need to tie this information into courses taught by experienced instructors, and then, most importantly...***get some actual field experience***! In the long run, hands-on practice is where you really start learning. There is an old saying, "Education is what you have leftover after you have forgotten everything you've learned."

Jake says, "If all you do is read this book, don't get your tattoo yet! You ain't a firefighter until you get some experience!"

Reading and studying this handbook is just the beginning of the beginning of your journey towards becoming an experienced wildland firefighter. Read the text; reinforce by reading the red notes in the margins; study the pictures and other graphics. They are not there for their looks, but each of them should tell a story. Learn from what Jake says. It is my earnest hope that this book will assist you in gaining knowledge and skills which will enhance professional growth and personal satisfaction with your performance as a wildland firefighter.

Studying this handbook is just the beginning of your career as a wildland firefighter.

Donald K. Artley
State Forester for Montana (retired)
Past Chair, National Wildfire Coordinating Group

1

Fireline Safety

It is no mistake that fireline safety is the first chapter in the handbook, as safety is the foundation that all other firefighting principles are built on. Always remember that safety is first. Safety is not a class, a certificate or something you can touch. It is an attitude! Catch it!

The issue of safety must not be delegated to the safety officer. It is every firefighter's responsibility to conduct him or herself in a safe manner. You have to have an attitude of safety. You have to want to do it right and perform safely. If you or one of your fellow firefighters gets hurt, you become a liability; you are no longer a firefighting asset contributing to the firefighting effort.

*Jake says,
"Safety is an attitude. Catch it! "*

Firefighting is physically demanding and punishing work. Besides being a very hazardous occupation, it will demand your highest physical and mental efficiency, and a sustained expenditure of energy. You must be prepared, and take every precaution to prevent injury to yourself and others.

Firefighting will demand your highest physical and mental efficiency.

Physical Fitness and Health Maintenance

Physical fitness goes hand-in-hand with productivity and safety. You will be fighting fire when it is hot, dry, dirty and windy. If you can't keep up, you will slow down the firefight. Don't think that all of your fire assignments will

*Jake says,
"Safety is no accident. To fight fire safely you will have to work at it."*

be of short duration. This isn't a nine-to-five day job. You may be on the fireline for a couple of weeks, asking your body and mind to work hard and long. You will be as tired as you have ever been. The best way to combat fatigue and achieve high performance is to follow a physical fitness program.

Fit firefighters are more tolerant of heat.

Fitness has two aspects; aerobic and muscular:

- ***Aerobic fitness*** is a measure of the maximum amount of oxygen that you can take into your body and transport to the muscles. Oxygen intake is the primary factor that regulates work capacity, because working muscles need a continuous supply of oxygen to perform vigorous work for lengthy periods of time. The more efficient your oxygen delivery system, the better you can do the tough job of firefighting.

- ***Muscular fitness*** includes both strength and muscular endurance. Aerobic fitness and muscular fitness together are essential parts of your work capacity. Fit firefighters are more tolerant of heat. They acclimate faster, and work with lower heart rates and body temperatures.

Take physical conditioning seriously.

The best physical fitness program is one that balances aerobic conditioning and muscular training, and that starts well before the beginning of fire season.

The following fitness levels (Figure 1.1) are based on field studies of firefighters performing firefighting tasks. The levels represent the fitness needed to perform tasks safely for extended

Recommended fitness levels for firefighters

Time for 1.5-mile run	Chinups	Situps (total in 60 seconds)	Pushups (total in 60 seconds)	Bench press (lbs.)	Leg press (lbs.)	Curl (lbs).
11:40 minutes	4 to 7 (depends on your weight)	30	20	120	350	50

Figure 1.1 Recommended fitness levels for firefighters. This is a good start. If your department has different guidelines, be sure to follow them. The key is that you be physically fit and ready to fight fire.

periods, with the reserve ability to respond to emergencies. Set up a daily schedule of physical training. Do it as a team if you can (Figure 1.2).

Figure 1.2 *Physical fitness is very important for every firefighter. Set up a routine and follow it. Do it as a team if you can.*

Fatigue

Fatigue is a less visible threat to you than the fire itself. However, without enough sleep and rest, even the most fit firefighter will tire after several long shifts in heat and smoke. Tired people make mistakes. On a wildland fire, mistakes often mean accidents and injury.

Work and Rest

Sleep is a prime factor in controlling fatigue. Most fit firefighters can work hard for 24 to 36 hours with short sleep or rest breaks. After that, without adequate sleep or rest, you will succumb to fatigue. ***To perform well in firefighting, you should average one hour of sleep for every two hours of work.***

Fatigue is a less visible threat to you than the fire itself.

Your pulse rate is a good way to gauge fatigue. Your pulse should recover to less than 110 beats per minute if your break is long enough. Your wake-up pulse can signal potential problems. If it is 10 percent or more above normal, it can mean fatigue, dehydration, or even an impending illness. Ample quality sleep is vital.

One hour sleep for every two worked.

Heat Stress and Hydration

Heat becomes a problem when humidity, air temperature and radiant heat combine with physical exertion to raise body temperature above safe limits. There are three forms of heat stress. The mildest is heat cramps. Heat stress can progress to heat exhaustion and heat stroke if you don't stop, get in the shade, and begin drinking fluids. ***Heat stroke is a medical emergency. Delayed treatment can mean brain damage and even death.*** See Figure 1.3 for a listing of the symptoms and treatment for heat cramps, heat exhaustion, dehydration exhaustion and heat stroke.

Jake says, "Even if you drink a lot of water, you may still be dehydrated. Drink more water than you think you need."

Sweat is your main defense against heat stress.

RECOGNITION AND TREATMENT OF HEAT-RELATED PROBLEMS

	Heat cramps	Heat exhaustion	Dehydration exhaustion	Heat stroke
Symptoms	Painful muscle cramps.	Weakness, unstable gait or extreme fatigue; wet, clammy skin; headache; nausea or collapse.	Weight loss and excessive fatigue.	Hot, often dry skin; high body temperature (106 degrees or higher); mental confusion, delirium, collapse, loss of consciousness and/or convulsions.
Treatment	Drink lightly salted water, fruit juice or athletic drinks.	Same as heat cramps, plus resting in shade.	Increase fluid intake and provide rest until body weight is restored.	Cool the body immediately, by immersing or soaking clothing in cool water; fan to promote cooling; lower temperature to at most 102 degrees; treat for shock; seek medical attention.

Figure 1.3 Recognition and proper treatment of heat-stress-related problems is very important. Drinking enough water is your best method of prevention.

Drink a lot of water.

Sweat is your main defense against heat stress. As sweat evaporates, it cools you, unless the humidity is high. If the water lost in sweat isn't replaced, your body's heat controls break down. Your temperature can climb dangerously. If you stop sweating, evaluate your condition and take appropriate actions.

Fluid replacement is vital. During hard work in the heat, it's common to lose 1 to 2 quarts of sweat an hour. To combat dehydration, you should drink water before going on shift, drink often while working, and continue to replace fluids once off-shift. Everyone on the fireline needs to understand the importance of drinking water often. Replacing 12 or more quarts of fluid a day isn't always easy, but it must be done.

Stay away from caffeinated, carbonated and "diet" drinks. They take water out of your body. Drink water, juices or non-caffeinated sport drinks. Juices and sport drinks contain energy-restoring glucose. When on shift, drink once every hour. Don't wait until you feel thirsty. If you wait until you're thirsty to replace lost body fluids, your body is already exhibiting signs of dehydration.

Pace yourself!

Pace yourself. During breaks, try to get away from the heat. Check your pulse. Heat stress is unlikely if your pulse rate is under 100 beats per minute after 1 minute of rest.

Too much exposure to CO causes headaches, fatigue and drowsiness.

Smoke and Carbon Monoxide Exposure

Unlike work/rest cycles and heat stress, which are more controllable, smoke and carbon monoxide (CO) present a greater danger for the firefighter. Heavy smoke and CO are often present on the typical wildland fire. Some exposure is unavoidable. Your objective is to limit exposure.

High concentrations of smoke particles can irritate mucous membranes and cause allergic and asthmatic reactions in some people. But little or no health risk is likely for healthy firefighters, when the exposure is short.

Smoke and CO make a tough job tougher.

CO is a tasteless and odorless gas. It doesn't "advertise" its presence, but it is always found in the heaviest concentrations of smoke. When CO enters your body, it begins to replace oxygen in the red blood cells. This reduces your blood's capacity to resupply the cells of your body with oxygen. Too much exposure to CO causes headaches, fatigue and drowsiness.

Smoke and CO make a tough job tougher. They reduce work capacity and can impair performance and decision making. They accelerate the onset of fatigue. The solution to this problem is limiting your exposure. Rest in smoke-free areas whenever possible.

Food and Nutrition

Nutritious food is not only a morale booster, but more importantly, it fuels the muscles for hard work. Remember, you may burn 300 to 600 calories an hour and between 5,000 to 6,000 calories during a long shift on the fireline. This will have to be replaced if you are going to avoid fatigue. To replace 6,000 calories, you are going to have to eat a lot of good, healthy food.

Firefighting is a "dirty business," so clean up before you hit the chow line.

The key to a good diet is a balanced one that includes food in all of the main food groups: carbohydrates (grains, vegetables, cereals, fresh fruit, potatoes, etc.); fat; protein (rice, beans, fish, poultry, etc.); fiber; calcium; vitamins and minerals; and sodium.

You should stay away from alcohol, too much red meat, white bread, etc. Eat lots of vegetables, fruits, and drink a lot of water.

Personal Hygiene

Firefighting is dirty work, and it is up to you to attend to personal hygiene. It is not a badge of honor to show up in the feeding line looking like you have just been dragged in by the dog. Wash your hands whenever you can, especially before eating. Inattention to personal hygiene can directly lead to sickness and/or the spread of contagious illnesses. When you can, shower and change your socks.

You are responsible to keep your PPE in serviceable condition.

Personal Protective Equipment

If you are not properly dressed and equipped, you have no business fighting fire (Figure 1.4). You must protect yourself, as best you can, from the dangers of wildland firefighting. Wear only approved safety clothing (check with your agency), or wear cotton material. Do not wear synthetic materials; they will melt when heated and increase the likelihood of major injury.

If you are not "dressed for the occasion," stay at home.

To best protect yourself, you must have appropriate Personal Protective Equipment (PPE); you have to wear it, and ensure that it is properly maintained. It may sound strange that anyone needs to be reminded of this basic requirement, but there have been cases where protective equipment has been left on the engine or back at the station. You, and you alone, are responsible and accountable for your PPE.

Flame-Resistant Shirts and Trousers

The most common materials used in the production of firefighting safety clothing are Nomex® and Kevlar®. If your agency provides safety clothing, ensure that it is clean, without holes, tears, gas and oil stains, and is ready for use. ***The best policy is not to get on the engine until you are properly dressed.*** That way you will never find yourself unprepared to immediately fight fire when you arrive "at scene."

The trousers should be cuffless and loose fitting. They should be worn over the boots. The shirt should be long sleeved, secured at the collar and wrists, and always kept tucked in. Do not paint names on the shirt, because they will tend to absorb heat. You don't want your department logo "burned" into your back. It has happened!

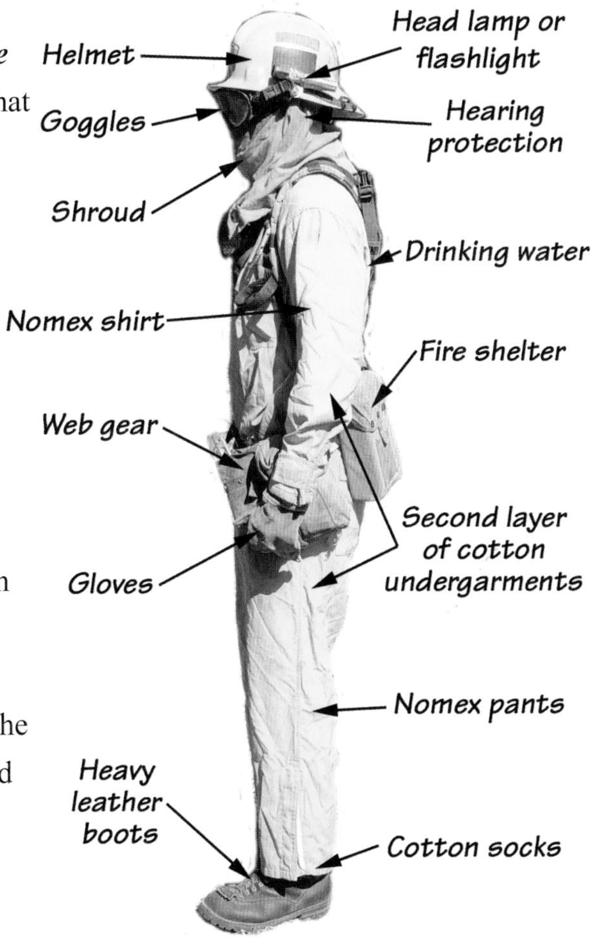

Figure 1.4 The most commonly accepted personal protective equipment (PPE) looks like this.

Don't get on the engine until you are ready to fight fire.

The Importance of a Second Layer

Experience has shown that a single layer of protective clothing is not enough if a firefighter is exposed to extreme heat. There are numerous cases that graphically show the burn lines starting where the undergarments end (Figure 1.5). Firefighting safety gear is designed not to burn or melt, but it can and does transfer heat to the skin.

Jake says, "DO NOT use synthetic materials. They can melt to your skin when heated, which increases your likelihood of serious injury. Think how this will affect your personal undergarments. Wear cotton!"

Second layers are hot, but worth it!

The solution is to wear a second layer of protective clothing. The most practical second layer is a firefighter's undergarments. ***These garments should be made of cotton (again, no synthetics or blends) and must cover the whole body.*** This means a long-sleeved tee shirt and "long johns." Your first impression is that this will be too hot and uncomfortable. But, this layer of clothing will "wick" moisture away from the body and help cool it. ***Your station uniform can be used as the second layer ONLY if it is made of a material that will not melt to your skin.***

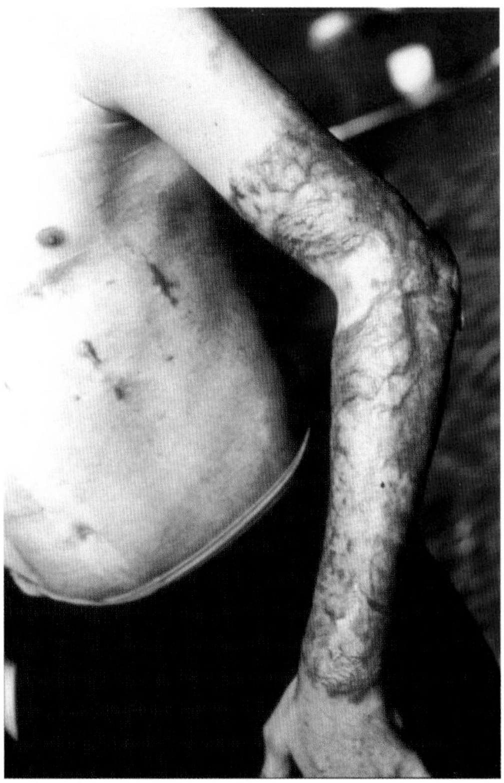

CDF

Figure 1.5 This is graphic proof of the value of a second layer under the PPE. You can see where the cotton tee shirt ended and the gloves began. Yes, it is hotter to wear, but it is worth it.

Boots and Socks

If your feet hurt, you hurt all over.

You must have appropriate footwear. Your boots should be made of leather or another durable material. They should be a lace-up type, with at least eight-inch tops and heavy lugged soles. Hard leather toes provide adequate protection. Boots with steel toes or protection plates and puncture-proof soles will "hold" heat longer, and can cause unnecessary burns.

Cotton, wool or part-wool socks are recommended. Some people prefer wearing a lighter sock when hiking long distances. Extra socks are a must.

Helmet or Hard Hat

There are many types and styles of nonmetal hard hats that can be used for wildland firefighting (Figure 1.6). The best type is one that provides protection from falling objects and is lightweight. Structural firefighting helmets can be worn, but they are usually heavier and can be tiring if worn for long periods of time. Never begin firefighting without appropriate head protection. Use the chin strip.

USDA Forest Service

Figure 1.6 This is the style of hard hat worn by most wildland firefighters. This one saved a firefighter's life by absorbing most of the blow from a falling snag.

Gloves

Gloves should be made of leather, fit well and be long enough so that a gap does not exist between the shirt sleeve and the glove (Figure 1.7). Gloves specifically made for firefighting have a "gauntlet" attached to the glove that protects the wrist area. The only time that firefighters should not be wearing gloves is when they are "feeling for hotspots" while overhauling or mopping-up.

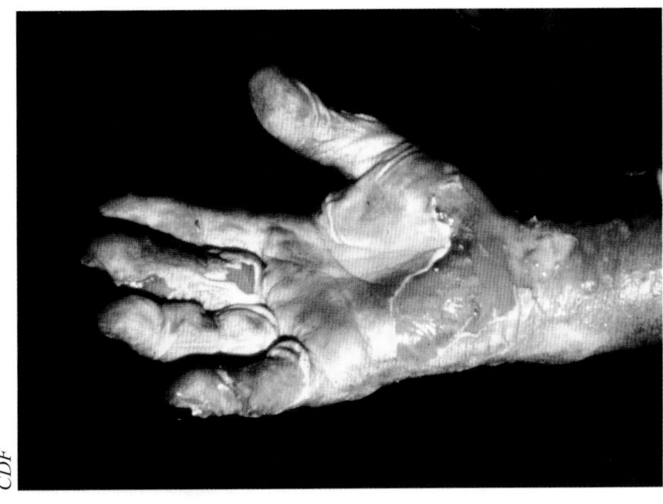

CDF

Figure 1.7 Gloves would not have prevented the accident, but they would have prevented these terrible injuries. A picture like this got the author to wear gloves when he was a firefighter.

Goggles

Protecting your eyes is very important; they are very vulnerable. You should be provided a good set of OSHA approved goggles. They should fit well and easily accommodate eyeglasses, if necessary. Keep them clean and always have them with you.

Hood or Shroud

A hood or shroud is constructed to keep the heat off your ears, neck and face. A hood is pulled over your face; a shroud is attached to your helmet. They are usually made of Nomex® or Kevlar®. When wearing a hood or shroud, your body may not be able to cool itself as well; remove hoods during rest breaks or mop-up.

Hoods or shrouds will protect your ears and neck. Remove them during rest breaks.

Fire Shelter

The individual fire shelter is a critical piece of required firefighting safety gear. Fire shelters have saved hundreds of lives. The shelter protects a trapped firefighter by reflecting radiant heat and trapping air (Figure 1.8). However, don't take more risks "because you have your shelter." The shelter is a "last resort" effort to survive. The key to proper use of the shelter...don't wait until it is too late to deploy it (Figure 1.9).

Fire shelters have saved hundreds of lives.

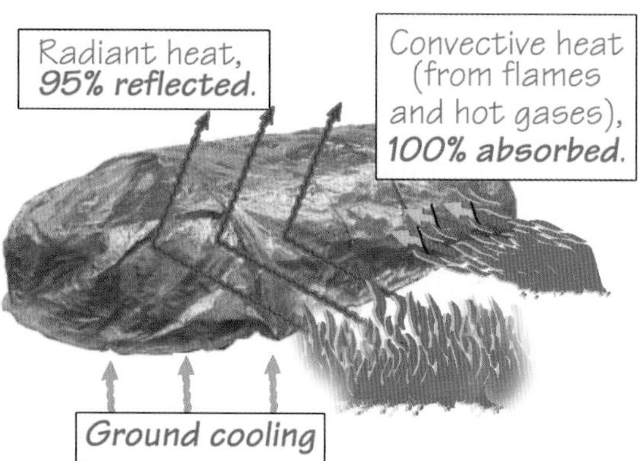

Figure 1.8 The fire shelter design has evolved over the years. They are designed to protect you from radiant heat when escape is not possible.

The cab of an engine can be used for protection...understand the limitations.

If entrapment seems likely, attempt to escape, but only if you are certain you will make it to safety. If there is doubt that you can escape, deploy your shelter. ***When it is time to deploy your shelter, there is no time to read the instructions.*** Practice deploying your shelter as often as you practice laying hose or laddering a building. Always train wearing gloves, helmet and backpack, if you use one. Practice removing your shelter from the carrying case while moving. Even practice deploying the shelter while lying on the ground. You must be able to

If you are going to work during the night, you will need a head lamp and extra set of batteries.

Figure 1.9 They waited too late to deploy their shelters. The firefighter operating the chain saw never heard the order to move and deploy. These two firefighters are in real trouble.

successfully deploy your fire shelter in 25 seconds or less.

When it is time to deploy your shelter, pick a site that is clear of flammable vegetation and away from heavy fuels and flammable equipment. It is absolutely necessary for the fire shelter to be held down on the ground before the fire front arrives (Figures 1.10 to 1.12). With prolonged exposure, temperatures can reach over 150 degrees. But you can survive such temperatures...dry saunas often reach 190 degrees. Breathe through your mouth, stay calm, and stay in your shelter.

Your most important actions are:

- ***Carrying a fire shelter is not an excuse to take risks on the fireline.*** Leaving the shelter on the engine is not an option. There have been cases where people used the shelter carrying case to hold other things. Don't be that dumb!

- ***Your highest priority is to avoid entrapment. If entrapment is imminent, escape if you can.*** The shelter is a last-resort option.

- ***During an escape or entrapment, protect your lungs and airway at any cost.*** Your lungs and airway are very delicate and do not tolerate a lot of heat. While you are in the shelter, keep your face close to the ground where air is cooler.

- ***Drop your gear as soon as you realize your escape may be compromised. Take your fire shelter and your tool, drop all dangerous flammable objects and any items that may slow your escape.*** On the South Canyon Fire, some of those who died did so with all of their gear. Drop it; get away from it, especially items like fusees and packs that can burn.

USFS, MTDC

Figure 1.10 Pick a site that is clear of flammable vegetation and away from heavy fuels.

USFS, MTDC

Figure 1.11 Open the shelter and enter it...you can do it from a standing position or from the ground.

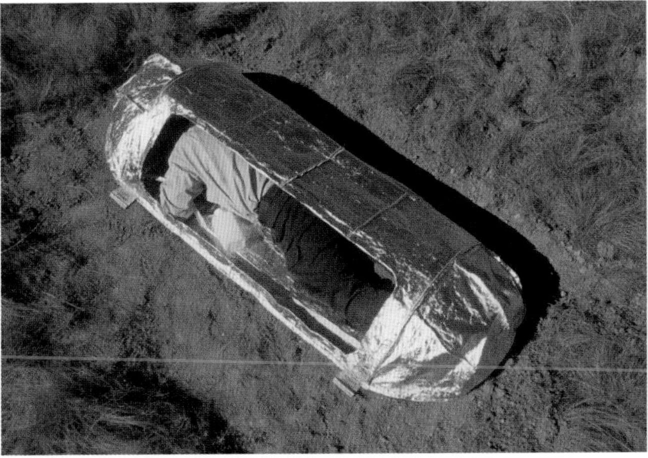

USFS, MTDC

Figure 1.12 Just you and your water bottle belong inside the shelter. Note in the cutaway shelter how the firefighter is lying face-down, holding the shelter down by the handles and flooring.

- ***If you are entrapped, get on the ground before the fire arrives.*** Don't wait to deploy. There are many documented cases where people did wait and died.

- ***Deploy your shelter where flames will be least likely.*** Help the shelter protect you. Get away from large accumulations of fuel. If there is a road, use it. Also, anticipate where the heat is going to come from, and try to protect yourself from it. Get rid of your tool after you use it to clear a spot for the shelter.

Stay away from heavy fuels.

- ***Once you are inside your shelter, stay there. Conditions outside the shelter will be far worse than those inside.*** Moving from one protected place to another, exposing yourself to the full fury of the heat is a bad idea. Stay put.

- ***If you have deployed your shelter with someone else, communicate with him or her. Survivors have reported that talking with your fellow firefighters can help you through your ordeal.*** You are scared. There is considerable comfort in sharing your thoughts.

If someone is in a shelter near you, talk to them.

- ***Train with your fire shelter as though survival is at stake. It may be!*** Train, committing the steps to memory (Figure 1.13). Practice under as real conditions as you can. If you have a smoke ejector, create a little wind to simulate extreme conditions. It will be a lot harder to deploy it...but more real. Some train with real fire...it isn't recommended.

Figure 1.13 When it is time to deploy is no time to read the instructions or practice. Practice deploying your shelter. Your life may depend on it.

Train with your shelter as if your life depended on it. It may!

Another issue that may come up in training is, "Do I attempt to use the cab of the engine for protection, or only use the shelters." It all depends on the situation. The key is, if you cannot escape, seek refuge in the best place possible. If you are in light fuels and the cab of the engine is readily available, use the cab for protection. In heavier fuels, where the duration of the intense heat will be longer, the cab may not be the best place. When the plastics and rubber in the cab begin to heat, they release some real bad gases...that may drive you from the engine. The worst thing is to transition from one protected area to another through an unprotected area. That is how a lot of firefighters have died. In heavy fuels, use the engine as a shield, but deploy on the ground away from where the fire is coming.

Water

A vital part of your PPE is some form of water carrying device. Water is so important.

Always take water with you.

Chaps

If you will be working around or operating a chain saw, you will be required to wear chain saw chaps.

Turnout Gear

If you are going to be operating a chain saw, wear chaps.

If turnout gear (coat, pants and boots) is all you have, you should wear it. You will be limited by this bulky gear, but don't fight fire without protective safety gear. If this is all you have, work with your chief to obtain proper wildland firefighting safety clothing. You should also remember that wildland firefighting gear is not a replacement for turnout gear. Do not attempt to fight a structure fire with the lighter weight wildland gear.

Self-Contained Breathing Apparatus

Normally, self-contained breathing apparatus (SCBA) is not used during wildland firefighting. But, if you have it available, you should know how to use it, and should use it if conditions warrant. Remember, SCBA equipment is good only for a limited time, usually less than 30 minutes.

Respirator

A respirator is an air-filtering device that is worn over the nose and mouth. Full-face respirators also provide eye protection. Respirators have been used in the industry for years. You will begin to see them more and more in wildland firefighting. Respirators are NOT a substitute for SCBAs.

Respirators are not SCBAs.

Respirators filter airborne particles and, depending on the filter media, can filter some chemicals. The firefighters wearing respirators may feel more comfortable in smoky conditions because their lungs are not being invaded by airborne particles, but may still be exposing their lungs to other dangerous chemicals. Wear the respirator when exposed to heavy smoke, but return to fresh air as soon as practical.

Experience has shown that wearing a respirator is not very comfortable. The firefighter will be able to remain calm even under the most smoky of conditions, but will sweat profusely under the mask.

Hearing Protection

Carry and use ear protection.

If you will be working near or with power equipment (portable pumps, chain saws, engine pump panels, etc.), you should have hearing protection. Ear muffs or plugs are just as important as gas and oil for this type of equipment. All firefighters should have ear plugs in their pockets.

Head Lamp

If you will be working the fireline at night, be sure you have a proper head lamp and an extra set of batteries.

Personal Gear for Extended Deployment

There are other items that are not necessarily safety gear, but you will want to have available to you if you are away from your home station for an extended period. Some of the more important ones are: a tent, a sleeping bag, extra clothing, toothbrush and paste, comb, bandana, hair bands to secure long hair (hair is flammable), foot powder, extra boot laces, matches, watch, rain gear, toilet paper, small towel, prescription medications, etc. Put these and any other necessary personal hygiene items in a bag and have it available at a moment's notice. Follow your agency's guidelines regarding personal items. Realize that air travel may present some limitations or restrictions.

There will be times you are going to be away for days. Be prepared.

Ten Standard Wildland Firefighting Safety Orders

The Chief of the Forest Service, Richard E. McArdle, issued the Standard Fire Fighting Orders to USDA Forest Service employees on June 28, 1957. A task force, established by the Chief, developed these orders to study ways they could strengthen efforts to prevent firefighting fatalities. This task force studied 16 fires on which 79 firefighters had been killed. These basic safety orders have been required training ever since they were first issued.

Learn, understand and follow the 10 standard firefighting orders.

The Ten Standard Firefighting Orders address safety issues that are not meant to be compromised. They are intended to be absolute! Some say that you can't fight fire without "breaking" one or two of them. But, which ones? When? There used to be signs along the highways of this country that read, "You may only need your seat belt once, but which once?" The same applies to the Ten Standard Firefighting Orders. How do you determine when you can ignore one or more of the Ten Standard Firefighting Orders without disastrous results?

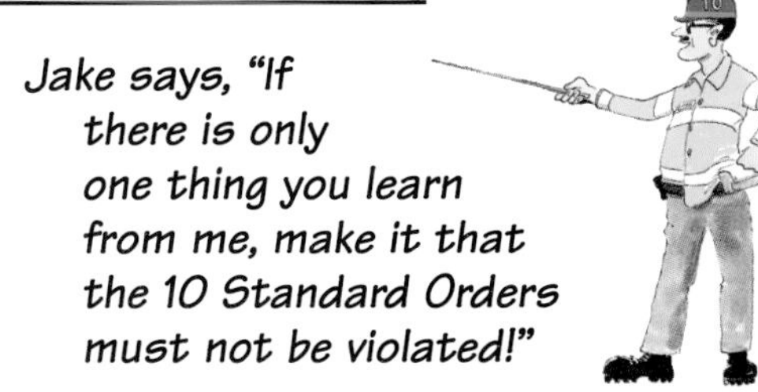

Jake says, "If there is only one thing you learn from me, make it that the 10 Standard Orders must not be violated!"

Studies have shown that we, as a society, don't like to follow orders. The two most commonly violated laws are probably speeding and jay walking. We find it very easy to violate these rules; some firefighters apparently find it very easy to violate the Ten Standard Firefighting Orders. There are a number of reasons why we routinely violate these rules. Following are three of the most common:

- ***Ignorance*** (or lack of training) - When you have not been adequately trained to recognize potential problems and how to respond appropriately.

Jake says, "When can-do, turns to make-do, it often turns to do-do!"

- ***Machismo*** (or an excessive "can do" attitude) - When you feel you are immortal and nothing can hurt you! You are overconfident and feel you can extricate yourself from any situation.

- ***Apathy*** (or lack of empowerment) - When you think someone else will take care of you, or it isn't your job to question violations.

Know your limitations and work within your level of experience.

You must understand how ignorance, egotism and apathy will affect you and those around you. Fighting fire is a lot like driving a car. When you get into your car, you usually have some idea where you are going and the route you will take. In firefighting, this is called your objectives and your plan. You use maps in driving, and you should use maps in firefighting. As you drive down the road, you watch for hazards and signs that indicate when to stop, slow for a curve, how fast you are allowed to go, and detours. In firefighting, you look for changes in weather, fuels, or topography. Your observations may cause you to change your route (strategy) or how you are driving (tactics).

Firefighting is like driving a car. Train to ensure that your skills become automatic!

Then there is the Vehicle Code that gives you the rules of the road. In firefighting, these are called the Ten Standard Firefighting Orders. As a driver, you are required to take a test in order to secure a driver's license. A booklet is provided so that you can study for the test. You also have manuals and tests in firefighting.

The main difference between driving your car and fighting a fire is the development of conscious and subconscious skills. You usually drive on a daily basis, learning and practicing all the time. You allow your subconscious to do a lot of the work. When was the last time you consciously thought about putting your foot on the brake to slow down? Your brain detects the need to slow, but your subconscious moves your foot and applies the brakes. When is the last time you spotted a danger sign and took the time to pull out your copy of the Vehicle Code or DMV guidebook to look up what you are supposed to do? You would be injured or dead before you found the answer. Yet, we rely on printed copies of the Ten Standard Firefighting Orders and the 18 Situations that Shout "Watch Out!" to bail us out on a wildland fire. So, the real difference between driving and firefighting is that we practice driving daily, while most of us do not train for firefighting in such a way that we "train" the subconscious to do some of the thinking for us.

Practice makes perfect. It also makes responses instinctive.

You train to learn skills. You train to ensure that the skills become automatic. You train to develop a team. There is a saying, "Practice makes perfect." A more practical saying for a firefighter should be, ***"Practice makes responses instinctive."*** You also need experience so that you can recognize the subtle changes in weather or fire behavior that tell you what is going to happen.

Let's take each of the prioritized orders and dissect them in an attempt to learn their meaning, purpose, and importance to every firefighter. Consider them your rules of engagement.

1 ORDERS

Keep informed on fire weather conditions and forecasts.

Weather is very influential on the intensity of fire behavior and how it spreads. Of the three primary weather factors that impact a fire (***temperature***, ***relative humidity*** and ***wind***), wind is the most influential.

Jake says, "A change in weather can get you in trouble very quickly."

There are two types of weather information:

- ***Tactical weather information*** is weather that is taken and recorded at the scene of the fire. It includes the temperature, relative humidity, and wind speed and direction that you measure and record at regular intervals during the life of the fire. Remember, the primary purpose of recording weather is to enable you to detect changes, and you can only detect changes if you have data to compare with. When you record the weather readings, you should also make notes as to the state of the fire behavior. Note the flame length, color of the smoke, the ease or difficulty of control, and local wind patterns. This will allow you to start developing an understanding of fire behavior under differing weather conditions.

Weather and fire behavior are interrelated.

You may want to use off-site weather information to supplement your own weather readings. There may be a lookout that takes weather readings. Ask for the readings and record them. If you expect a weather change, like one that may be tied to the passage of a cold front, get weather readings from a station some distance (10 to 20 miles) away, so that you can get a "jump" on the pending change.

Strategic weather is the general weather for the area.

- ***Strategic weather information*** is the information you receive from a weather forecast. It is of a general nature that provides weather trends (e.g., temperature up 7 degrees, relative humidity down 5 to 10 percent, etc.). You can use this data to predict the expected tactical weather information. That, in turn, will aid in the development of your fire behavior predictions.

To comply with this Order, you must take the weather on site, and ask for a forecast. You should attempt to track several weather factors:

Do you expect a change in the weather...what will that mean?

- Take the weather on the fireline. Record the temperature, relative humidity, wind speed and direction, cloud type and extent of cover.

- Assess fire behavior, taking special note of the flame length, smoke color, spotting, and rate of spread.

- Check the actual weather factors against the latest forecast. Obtain off-site weather information when necessary. Note any differences and analyze what impact this will have on your fire.

2 ORDERS

Know what your fire is doing at all times.

You can learn a lot about the fire status by listening to the radio as you respond. Some of this information will have to come from your supervisor, or it may have to be gathered by you during initial size-up. You should know the direction and rate of spread; the location of existing fireline, breaks in fuels or natural barriers; type and status of the fuels; fire behavior and problems with spotting.

If you can't determine this information by personal observation, send someone out to get it (a scout) or ask someone (supervisor, air attack, or lookout).

Some of the factors that you should track are:

- Location of fire perimeter
- How fast it is spreading
- Direction of spread
- Fuel cover (present and potential)
- Fire behavior
- Location of fuel or natural breaks
- Spotting activity

If you do not know what the fire is doing, you are in violation of a fire order.

If you can not see the fire yourself, post a competent lookout who can.

If you do not have current information on fire status, you are out of compliance with this Order, and you will not be able to develop a safe, effective plan of attack.

3 ORDERS

Base all actions on current and expected behavior of the fire.

The fire and associated fire behavior is always changing. If you can't predict what the fire is going to do, how can you develop a meaningful plan to attack it? You must continuously evaluate the fire behavior and detect subtle changes. Is the fire doing what you thought it would? If not, why? Some of the factors that you must ***continuously review*** are:

- ***Weather*** - Is the weather what was predicted? Has it changed? Do you expect it to change? How will these changes impact the fire and the firefighters? You can anticipate the fire will burn hotter if it is above an inversion layer during the morning, and within a thermal belt or mid-slope zones during the night.

This is not a one-time order. To comply you must monitor fire behavior constantly.

- ***Topography*** - Is the fire moving into areas with different topography? What will this mean, and how will you have to adjust your plan to accommodate these changes?

- ***Fuels*** - Are the fuels of such a nature that they will support fire behavior that will make your present plan invalid? Is the fire moving into heavier fuels, and can you prevent it?

- ***Time of day*** - Is the heat of the day still ahead of you? If so, what can you expect to happen in a couple of hours? What changes in temperature, relative humidity, and winds can you expect later in the day/night, etc.? Fires usually burn hotter and more intense between 1000 to 1700 hours, when temperatures are higher, relativity humidity is lower and the winds normally increase.

Three things can happen: the fire behavior stays the same; it lessens; or it gets worse!

Have a plan for all three.

Make an effort to plot your fire behavior predictions on a map. Then check to see "how you did" in estimating the rate of spread. As you are plotting things on the map, be sure to indicate time frames as well. Keep track of when the fire reached each location, when it spotted, when you implemented specific suppression activities, etc. That way you will be able to spot gaps in your plan.

4 Identify escape routes and safety zones and make them known.

ORDERS

A ***safety zone*** is an area that can be used by firefighters seeking refuge from an unexpected change in fire behavior or burning conditions. A safety zone must be void of fuels, such as in "the black" (as long as all fuels are burned). Other examples are roads, rock slides or outcroppings, green areas (lawns, meadows), or wet areas (lakes, swamps, wide creeks). The key is that it is an area that will protect the firefighters from injury. If the area requires the deployment of fire shelters to be safe, it does not technically qualify as a safety zone (Figure 1.14).

Figure 1.14 A safety zone is a place you can safely take refuge. By definition, you should not have to deploy a shelter to be safe. But, it will not normally be this safe!

Jake says, "If you don't have a safety zone and an easy way to get there, simply move off the line."

You may have to construct safety zones. You can do this by burning-out an area, or clearing a zone with equipment or hand tools. The terrain or fuels may require you to construct several safety zones. The type of fuel, terrain, expected fire behavior, and the number of people the site will have to accommodate, dictate how big the safety zone should be. Remember, safety zones should be so carefully selected and properly prepared that the use of a fire shelter should not be necessary in order to survive.

A safety zone does not require the deployment of shelters to make it safe.

The distance between safety zones is dictated by how much warning you can expect to get if conditions change, how fast you think conditions could change, and how fast personnel can be notified and moved.

Jake says, "Before safety zones and escape routes can be effective, you must know where they are and when you have to move to them...SITUATIONAL AWARENESS!"

An ***escape route*** is the way you get personnel from where you are working (and in danger) to a ***safety zone*** (Figure 1.15). Escape routes must allow you to move quickly and safely. Very seldom will a route that takes you uphill qualify as an acceptable escape route. Have two routes selected. That way, if one route is cut off, you still have a way to get to the safety zone. Make sure all escape routes are clearly marked.

This Order should stop more operations than it has in the past. Safety zones, and the routes to them, are your first line of protection. If you don't feel comfortable with the location or size of the safety zone, or the route(s) you have to use to get to it, ***speak up!*** If you are not comfortable with the safety zones and/or escape routes, you are not in compliance with this Order. Strategy and tactics can be modified to ensure safe retreat.

Figure 1.15 One is running without a clue where he is going and the other just doesn't have a clue! Don't put yourself in this type of situation.

Don't panic.

Have more than one route to safety.

5 ORDERS

Post lookouts when there is possible danger.

If you can't see what is happening around you, you must establish a mechanism that will provide timely, accurate, and meaningful information. You may be able to rely on aircraft or a nearby lookout, but in most cases, experienced firefighters will be assigned as on-site lookouts. The lookouts must be knowledgeable in fire behavior so they know what to look for, and understand the significance of any changes. They must have a way to communicate with others to provide a warning or to obtain changes in assignment. Most of all, the lookout must be able to recognize hazardous situations.

Lookouts can take the surprise out of a situation.

Lookouts are a way to take the surprise out of a changing situation. If you don't have a way of keeping informed of changing conditions, you are not in compliance with this Order.

Lookouts are your eyes!

6 ORDERS

Be alert. Keep calm. Think clearly. Act decisively.

The key is to understand and avoid what may cause you to be less alert, to get overexcited, or to become mentally disorganized.

Physical and mental fatigue, the stress of the situation, and heat and carbon monoxide will all "gang up on you" during a fire fight. Fatigue will begin to take its toll when you don't eat properly or get regular rest. Heat stress will begin to affect you if you don't drink enough water. Carbon monoxide (a natural product of combustion) can build up in your system to the point where it will reduce your ability to think clearly.

There may be a lot going on, but stay calm...think.

If you are going to be able to effectively deal with the situation, you should:

- Maintain self control.
- Ensure that you get enough to eat and drink, and get adequate rest.
- Develop contingency plans.
- Monitor the situation closely.
- Seek information and advice from others.
- Take regular breaks from the action.

7 Maintain prompt communications with your forces, your supervisor, and adjoining forces.

You must maintain communications with your crew members, your supervisor, lookouts, and adjoining forces. The purpose of this Fire Order is to ensure that you can report or receive changes in instructions; warnings of changing conditions or impending danger; changes in status; or progress reports.

If you do not have proper communications, how can your report or receive information on dangerous situations?

If you are going to rely on a radio as your communications link, make sure that you have tested the link, have an extra set of batteries, and have established check-in times. You should also have a contingency plan in case the radio link doesn't work.

If you are not linked with those above or below you in the chain of command, you are not in compliance with this Order.

8 Give clear instructions and be sure they are understood.

If fire personnel don't know what is expected of them, how can they effectively contribute to the overall plan of attack? Be clear and concise in your communications. Ask to have instructions repeated if you do not completely understand them. If you are the one receiving the instructions, make sure that you know what they mean.

Repeat the instructions if you have to.

This Order is directed at all fire personnel. EVERYONE on the fire is giving or receiving orders. Ensure that instructions are given to, and understood by all, then follow-up to ensure compliance.

All too often, the communications within the crew is the weakest link.

9 Maintain control of your forces at all times.

The best way to remain in control is to ensure that your instructions are clear, concise, and understood; that you maintain communications; that you know the location of crew members; and that you know the status of the fire, and when and how you will move your forces to safety. As the situation worsens, you may want to concentrate your personnel in one area.

Anticipate the unexpected! Do some "what ifs?"

The key to maintaining control and to be in compliance with this Order is being prepared to react quickly and effectively to the unexpected.

Fight fire aggressively, having provided for safety first.

This order is a general statement that says that you should take meaningful and bold actions to fight the fire, but do so safely.

> *Jake says, "Being aggressive does not mean you can fight fire ignoring safety. You can be both aggressive and safe."*

If you can't ensure that you and your force are capable of fighting the fire on your terms, stop and reevaluate. Ask yourself, "Do I have the forces, skills, and tools necessary to safely control this fire?" ***If not, change your plan.***

What safety considerations need to be taken into account ***prior*** to fighting fire aggressively?

If you don't attack the fire, it usually will not go out right away. Get in there and do what you are told to do...just do it safely.

- Take action only after thoroughly scouting the area.
- Receive and understand instructions and assignments.
- Ensure lookouts alert personnel to danger.
- Maintain good communications.
- Ensure that strategy and tactics do not endanger firefighters.
- Know where your escape routes and safety zones are.

Do not consider review of the Ten Standard Firefighting Orders as something you do only once when you first arrive at the fire. You must constantly review them. Reevaluate the fire situation to detect changes and/or the unexpected, and be prepared to react quickly to provide safety.

Jake says, "These orders are your rules of engagement...do not violate them, period!"

The Ten Standard Firefighting Orders deal with fire behavior concerns, safety precautions, and operations...all of which are summed up in SAFETY FIRST. If you cannot comply with one or more of the Orders, or cannot mitigate the issue, move back and develop a new plan. This is not a failure on your part to do the job. This is just good thinking.

Situations that Shout "Watch Out!"

Each of the 18 situations listed below should raise a warning flag when you see it. Know and understand the risk and the ways to eliminate or mitigate the hazard. It is not enough to just identify the situation; you must know how to change the conditions so that the danger is eliminated.

If you find yourself in one of these situations, Watch Out!

1 Fire not scouted and sized-up.

You are entering a situation where you have no information on the fire, what it is doing and what it may do. DANGER! If there is an air attack assigned to the fire, get information from it. If this isn't available, have an experienced person scout the fire, using a safe route of travel. Get some timely and reliable information before committing to a strategy of attack. You have to identify the hazards and know what you are up against.

Know and understand the risk.

Moving into unfamiliar country without enough information is dangerous.

2 In country not seen in daylight.

At night it is easy to get lost and disoriented. Be careful of cliffs and steep terrain. Work as close to the fireline as possible. Do not attempt to work without headlamps. Keep in close contact with supervisor and crew. You need to scout the fire to identify hazards, escape routes and safety zones, and evaluate the overall safety of the planned suppression action. If you can't scout it, talk to someone who has seen the country during the daylight hours.

It is dark; you haven't seen the country during the day.

3 Safety zones and escape routes not identified.

Don't proceed unless you know where the safety zones are and how to get there. Contact your supervisor, now! This is especially critical if working away from the fire's edge. If on the fire's edge, you may be able to use "the black" or burned-over area as a safety zone. If you continue to fight fire and have not designated escape routes and a safety zone, you are violating a safety order.

No safety zones or escape routes.

4

Unfamiliar with weather and local factors influencing fire behavior.

Know what the fire is doing. Monitor the wind, humidity, and temperature; if they change, be alert. Obtain a weather forecast and a fire behavior prediction, if available. Knowing the weather will help you predict the fire behavior. You must be more alert in strange country than in areas you know.

You don't know what the weather is going to do.

5

Uninformed on strategy, tactics, and hazards.

If you don't know the plan, contact your supervisor as soon as you can. Get briefed on the strategy, tactics and any special hazards. You are not going to be very effective, let alone safe, if you don't know the plan. Be sure to ask about contingencies, in case something unexpected happens. Know how to react to something before it happens.

Stay informed.

6

Instructions and assignments not clear.

Not sure what you are to do? Do not assume anything. Better find out! Contact your supervisor and clarify your assignment. Be sure you receive information on who is in charge, hazards, assignment, strategy and tactics, escape routes and safety zones, etc. Make sure your crew understands what you told them. Have them repeat it.

What is the plan?

7

No communications link with crew members or supervisor.

If you can't communicate with your crew or supervisor, you can't get vital information or pass on information to them about changing conditions, injuries, blowups, etc. You must maintain communications! Find them, now.

You don't have a means to talk to anyone.

8

Constructing line without a safe anchor point.

If you have not anchored your fireline, you run a real risk of the fire going around your line and outflanking you. You must anchor your line. Tie it to a road, creek, lake, or something that will hinder the

No anchor point.

fire moving around your line. If you can't adequately anchor your fireline, keep a close eye on that part of the fire.

9

Building fireline downhill with fire below.

Very special care should be taken. This approach is very dangerous for several reasons: the potential for rapid uphill runs. If your only escape route is uphill, the fire will more than likely beat you; and, even though you may have anchored your line on the ridge, it is meaningless if your are below that point. It can be done, with special precautions (see Chapter 5 Fire Extinguishment Methods), but this is a last-resort situation. Avoid this situation whenever you can. This is a "three-star" watchout. If you are attempting a downhill attack, be very careful.

The fire is below us!!!

10

Attempting frontal assault on fire.

The head of the fire is usually the hottest part. It will be smoky, and visibility will be limited. Unless the fire is small, or you are with a sizable force, this approach will usually not be very effective. This is a "watch out" because: you may not have a good anchor point; the fire may outflank you; the heat and smoke may be very intense; and, the fire may be moving very quickly and not give you time to be adequately set up. If this is the selected strategy, be sure you have safety zones and escape routes established. Even if the unexpected occurs, keep in very close communication with your supervisor and the forces in the area, and stay very alert.

The fire is heading right toward us.

11

Unburned fuel between you and the fire.

This is very dangerous, especially if the fire is burning in your direction. This is also dangerous if the fire is not presently burning in your direction, because the situation may change and catch you by surprise. Post a lookout, someone to warn you when the fire is getting closer. Be prepared to vacate the area by taking a preplanned escape route to a designated safety area.

There is a lot of "green" between us and the fire.

12 Cannot see main fire; not in contact with someone who can.

Where is the fire?

If you can't see the fire, you cannot know what it is doing. Do not let yourself get into this situation. Contact your supervisor, and advise of your situation. Find a place where a lookout can see the fire. You need to know if you or your crew are in danger and if you should be moving to the safety zone.

13 On a hillside where rolling material can ignite fuel below.

Rocks and burning material are rolling down the hill.

The danger here is not just a fire starting below your position; be on the alert for rolling rocks and logs. It is not uncommon for a fire to free logs and rocks, allowing them to roll downhill. They can do this with little or no warning. Use trenching to hold the rolling material. Establish safety zones and escape routes. Any time you are working on a mid-slope, you should be very careful and vigilant.

14 Weather becoming hotter and drier.

Is it getting hotter?

Conditions are changing! And the fire behavior will not be far behind. Hotter weather brings increased fuel combustion and fire behavior, and the chance for blowups and spotting. Check your position and the safety zones. Be prepared to move on a moment's notice. Watch for spot fires.

15 Wind increases and/or changes direction.

The wind just changed direction.

Wind is one of the most influential factors in fire behavior. It can quickly change the rate and direction of the spread of the fire. This condition may bring more spotting; an increase in fire spread and intensity; and the potential for a reburn. Check for the passage of a front (look for any indicator cloud formations). The shift of wind may be as much as 180 degrees; a flank may become the head.

16 Getting frequent spot fires across fireline.

The fire conditions are changing. Humidity and fuel moisture is lower. Wind may be carrying burning embers for some distance. Anticipate a change in fire behavior. Heavy spotting can also cause extreme fire behavior and area ignition. Be alert if you begin to get spot fires below you. Spot fires make your fighting more complex and dangerous.

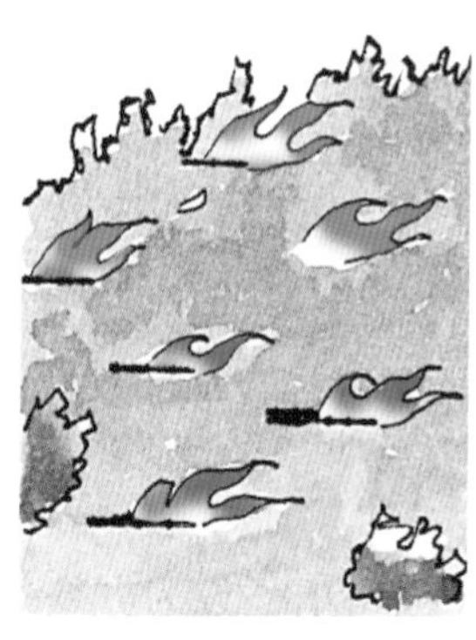

The fire is beginning to spot.

17 Terrain and fuels make escape to safety zones difficult.

Some types of terrain can dramatically increase the time it will take you to escape an advancing fire. If the terrain is rough and fuel cover thick, you will have difficulty moving toward a safety zone. You will have to post lookouts to give you adequate time to move. Improve your escape routes; clear brush or do what it takes to reduce the time it will take to get to your safety zone.

The fuels are really thick; this is steep country.

18 Taking a nap near the fireline.

You should sleep in shifts, and with the rest of your crew. Never sleep in the "green," on the fireline, or away from the crew. Using a bulldozer for an alarm clock is risky business. You also run the risk of being hit by a falling tree or snag. If you must rest, do so in the black or an established safety zone, and only with the permission of your supervisor. Post a lookout.

I'm beat. I'm going to take a nap.

Common Denominators of Fatality or Injury Fires

There are several common denominators on fires that involve a firefighter fatality or "near miss." They are: small fires; innocent appearance; light fuels; uphill runs; and turbulence. These fires do not appear to be a problem, but a sudden change in some factor causes erratic fire behavior (Figure 1.16). Why do tragedies or "near misses" occur under these conditions?

Small fires innocent in appearance!

Light fuels!

Uphill runs!

Air turbulence!

Firefighters tend to be complacent or "lower their guard" on small, innocent-looking fires, or a relatively inactive line. If not keenly watching for signs of change in the weather or burning

Common denominators on fatal fires

Denominator	Why?
Most of the incidents occurred on relatively small fires or isolated sectors of large fires. Most of the fires were innocent in appearance prior to the flareup or blowups In some cases tragedies occurred in the mop-up stage.	Firefighters underestimated the potential. Firefighters did not recognize subtle changes in weather conditions or fire behavior.
Flareups generally occurred in deceptively light fuels.	Firefighters underestimated the extreme rates of spread and heat possible in light fuels.
Accidents occurred in chimneys, gullies, or on steep slopes.	Fires and "super-heated" gas move up steep slopes and canyons with surprising speed.
There is a shift in wind direction or in wind speed.	There have been times that a wind event has been predicted and the impact of such an event not appreciated. There are more times that an unpredicted and unexpected shift in wind occurs. This may be associated with the passage of a cold front.
Suppression tools such as helicopters or airtankers can adversely modify fire behavior. Winds created by wing vortex can change the wind conditions on the ground.	A low pass by a helicopter or airtanker can cause the fire to flare up or spot across the fireline.

Figure 1.16 These are the common denominators for fatal or near-miss fires.

conditions, firefighters can be caught without warning...and without an escape route.

NEVER underestimate the speed of a fire in light fuels.

Fire intensity can change quicker in light fuels than in heavy fuels. The light fuels are more responsive to changes in relative humidity. If this is tied to an increase in wind or a change in topography, extreme fire behavior is probable.

A fire spreading uphill resembles a fire spreading with a strong wind. The rate of spread will usually increase as the slope increases. Not only are the flames closer to the fuel, but the movement of heated air (convection) is more likely to carry firebrands and start spot fires. A fire burning on level ground (up to 5 percent slope) will spread twice as fast when it reaches a 30 percent slope. The rate of spread will double again as the slope reaches 55 percent.

Jake says, "If slope and wind are aligned in the same direction...this is a big watch out!."

Normally, you would think that grass fires can move the quickest. Not necessarily true. Wildland fires in any fuel can move quicker than you think, especially if the fuels are dry and you have a wind. Sometimes, all you need is very dry fuels (Figure 1.17).

Figure 1.17 This sequence of photographs was taken when the Old Fire crossed Highway 18 on October 28, 2003. The time from the first to the last shot is 54 seconds. Note that the terrain is relatively flat and the smoke is going straight up. Within an hour the fire destroyed over 300 homes just east of Lake Arrowhead, CA.

Turbulence created by whirlwinds, microbursts, or vortexes of low-flying aircraft can suddenly fan an inactive fire back to life.

The shape of the land also has an effect on fire behavior. Box canyons, narrow canyons, and gulches tend to act like the chimney of a stove. As radiation and convection increase, spot fires are more likely, as if a damper were opened in a chimney. Isolated extreme fire behavior is possible with even minor changes in

Jake says, "In all of the cases of a fire fatality, the fire simply moved faster than the firefighters could."

Fire can move quicker than you can imagine!

topography. ***The firefighter should always be alert and watch for light fuels, wind shifts, mid-slopes, steep slopes and chimneys. "Look up, look down, look around." Don't ever underestimate the dangers involved with wildland firefighting.***

LACES

LACES: Lookouts, Awareness, Communications, Escape routes and Safety zones.

The acronym LCES was developed by Paul Gleason, a highly experienced and respected fire specialist employed by the USDA Forest Service. He was concerned that trying to remember the Ten Standard Firefighting Orders, 18 Situations that Shout "Watch Out!," the Common Denominators to Fatal Fires, etc., overloads the firefighter. His goal was to provide a simple way to help firefighters remember some key elements to survival. LCES stands for *lookouts, communications, escape routes,* and *safety zones.* An "A" has been added to LCES for ***awareness***, because unless you are aware, all else may fail. Think about these key safety elements *every time you tie your boot LACES!*

The four primary elements of ***LACES***; ***lookouts***, ***communications***, ***escape routes*** and ***safety zones*** are the operational components of the basic rules of engagement, within the Ten Standard Firefighting Orders. Their combined goal is hazard control on the fireline; more specifically, entrapment avoidance.

The "A" has been added for a reason...because if you are not AWARE of what is happening around you, you get hurt.

- ***Lookouts*** - The lookout is the eyes of the crew and its leader. Lookouts should be in a position from which they can see the fireline, the fire itself, and the crews that are working the line. They should be able to recognize and anticipate dangerous situations, and must report changes immediately. The size and complexity of the fire may require more than one lookout. They need to be experienced, able to recognize dangerous situations.

"Look up, look down, look around."

- ***Awareness*** - Lookouts should watch for changes in the fire's location and behavior. They should know the plan, so that they can relate it to what they see the crews accomplishing and what the fire is doing. Lookouts should also track the weather by taking readings at regular intervals. They should watch the sky for telltale signs of change. Finally, you must be able to understand what the lookouts are trying to report, or posting lookouts will serve no purpose. ***Remember, however, that situation awareness is the responsibility of everyone on the fire.***

- ***Communications*** - The fire officer, crew leaders, and lookout(s) should have a quick, reliable, and tested way to communicate with others. This may be by direct radio contact, or through a lookout or other relay point. If you plan on using the radio system, have an alternative way to communicate in case the radios fail for any reason. Establish regular reporting times. The communications link down to the individual firefighter may be by word of mouth. It can be very noisy on the fireline, so as the noise gets louder, the distance between individual firefighters will have to be shortened to ensure adequate communication.

Jake says, "Move to safety before you have to. This is common sense."

- ***Escape Routes*** - Note that routes is plural. Have at least two planned routes of escape. If your primary route is cut off, know what you are going to do. Every person on the fireline must know the plan, and what is expected of them. Everyone must also know what is to trigger a move to the safety zone. It is important to remember that, as the crews tire, they will not be able to retreat as quickly. You may have to shorten the distance between the work site and the safety zone, or provide an escape route that is "faster." Escape routes should not be measured in feet or chains, but in minutes and seconds. Make a conservative estimate of the time you will need to get the crew to safety. Use that as the guide as to how many seconds of travel time you can tolerate. If you are constructing an indirect line, the establishment of two escape routes may be a challenge.

Have a plan of escape., and make sure everyone knows it.

- ***Safety Zones*** - Safety zones are places of refuge, places you can be assured of your safety. Their size is dictated by the fuel, terrain, weather conditions, and worst-case fire behavior. The ***use of fire shelters should not be necessary***; however, that isn't to say you can't use them. Some commonly used safety zones include "the black" or burned area, natural features like green meadows or ridge-top meadows that can be burned, clearings constructed as part of line construction, clear cut blocks, etc. Safety zones should NOT be downwind from the fire; in chimneys, saddles or narrow canyons; require steep uphill escape routes; or be located near heavy fuel concentrations. The time to get to the safety zone is also critical.

Stay calm.

Sometimes, escape routes and safety zones can be compromised and rendered useless. This may require you to deploy your shelter at what is called a ***deployment site***. Do not confuse this with a safety zone...it isn't one. ***This is a last resort location***.

LACES tries to provide for a "safety net" if something goes wrong. It focuses on the important elements of the Orders. The elements of LACES should be as commonplace in the firefighter's vocabulary as fire shelter, helmet, gloves, water and shovel. The elements of LACES are interconnected and interdependent, like a chain. The LACES chain is only as strong as its weakest link.

Like a pilot, always have a way to go.

But wait, let's really look at what LACES means, infers and requires. LACES is much more than having a lookout, a means to talk with him or her, and a way to get to safety. If properly understood, here is what LACES really means:

The Lookout - This person has to be able to see the fire and where the crews in his or her area of responsibility are working. The lookout has to be alert, thinking clearly, knowledgeable and experienced. Specifically, the lookout has to:

There is much more to LACES than five words!

- Know the location of the escape routes and the safety zones, so they can properly direct the crews.

- Be experienced enough to properly evaluate the present and potential fire behavior. The lookout has to anticipate what is happening and why. The lookout has to report any changes in fire behavior.

- Take weather readings periodically, record them and report any changes or trends.

- Be briefed on the strategy and tactics that are to be followed and used. The lookout has to know how the "game is going to be played," so that he/she may be able to detect any conflict between the plan and what is really happening on the ground. There may be cases where a strategy or tactic is in conflict with fire behavior, terrain, or fuels, etc. If there appears to be a problem, this has to be communicated.

The lookout is both eyes and ears for those on the fireline.

- Ensure that as the fire and fireline construction proceed, the view is not compromised. If a lookout can no longer see the active front of the fire or the crews that are working...you don't have a lookout. Before this happens, have plans to move to another site or add another lookout to provide full visual coverage.

- Handle other communication tasks, such as a relay person. There may also be times when the lookout can provide some logistical functions, such as ordering lunches, water, fire hose, etc.

The lookout needs to see the broader picture.

- Look at the big picture. See how the small part of the fire you may be watching fits with the rest of the fire, the broader weather picture, etc. Don't just focus on one spot. If you do, you may miss something very important.

It should be noted, that the lookout does not always have to be a person perched on a hill watching what is going on. A crew supervisor, one who is not actively cutting fireline, may be able to meet all of the requirements of a lookout by watching what is happening, and staying mobile. The key is that as soon as the crew supervisor doesn't meet the requirements of a lookout, one is established quickly.

Communications - Remember, the role of the lookout is to see, track and record, interpret, anticipate and report. If the report isn't made, all the other stuff is meaningless. The lookout has to have a reliable method of communications and use it. Specifically, the lookout has to:

If you can't tell everyone what you see, you fail.

- Be able to contact line supervision and crew personnel quickly. This means that the channel of communication is one that everyone is monitoring, not just checking in on every so often.

- Ensure that if the radio is the selected method of communications, that an alternative system is available, and a full set of extra batteries is available. The lookout also has to know the communications system, so that if problems arise, he/she can easily establish an alternative method.

- Provide regular status checks to ensure that the communications system is working; to ensure that the fireline personnel know someone is looking out for them; and to ensure that no changes in strategies, tactics and location have taken place. Some of the items that need to be communicated, are:

- Known and potential safety concerns
- 10 Standard Orders, Watch Outs! and LACES
- Topography and fuels
- Fire behavior
- Weather, present and changes
- Strategy and tactics
- Duration of assignment
- Radio frequencies and other contact information
- Contingency plans for potential events (medivac, weather changes, etc.)
- Incident organization
- Job assignments and duties
- Gut feelings about the situation, the assignment or individuals

A good lookout is very busy.

- Monitor the radio to pick up on what is happening elsewhere: weather forecasts; information being passed on by others or between other lookouts; and on the latest scoop on fire behavior or personnel performance. The lookout is also the ears of the fireline personnel.

Escape Routes - By orders, you need more than one way to exit the scene and move to safety. These routes have to be identified, and everyone on the fireline needs to know where they are and where they lead. But, it isn't just that simple. Here are some specifics you need to know about escape routes:

- There is more to an escape route than just saying, "we go thataway." It has to be a path that hopefully doesn't go uphill or have other obstacles that slow the retreat. They also can't be miles in length. They have to be passageways that allow firefighters ways to move quickly from a point of potential danger to a point of safety.

Flag it!

- A fire usually gives signals that it is changing. There are more frequent spot fires, trees are torching, or the general feel is more tense. Or, the weather may give you a clue that things are about to change. There may have been a change in wind speed or direction, the humidity just took a drop, or the clouds are changing shape. If you are looking, you should pick up on some of these indicators.

- When to holler, "Fire!" You need to establish "trigger points" that dictate when to activate a plan. A "when, then..." plan is often used. For example: When something exceeds something, then everyone moves to safety. This may be when the relative humidity drops to a specific percentage. Or, it may be, when the fire reaches a certain ridge, then everyone moves. The key is that the lookout(s) and fireline personnel all know what the trigger points are, what they are to do when they are reached, who is going to track them, and who is going to sound the alarm.

Think beyond two escape routes.

There has to be some margin of safety in the timing of the move to safety. Ensure that if you have to move, you have the time you need. Do not wait until just before the fire hits.

- Know the preferred escape route, but think about alternatives, just in case. Think now, when you have time, and before someone says, "Run," you say, "We can't go that way!" and the situation goes from dangerous to deadly.

Safety Zones - By order, you are required to have safety zones where you can seek protection from the fire. A properly designated safety zone should not require the deployment of a fire shelter. This is actually a tough requirement if you are working in tall timber and the potential flame lengths could be measured in the hundreds of feet. Under these conditions, the size of the safety zone would have to be huge. With potentials like that, you have to ask yourself, do I really need to fight fire in that place at that time? But, if you are there, you need to have a safety zone. Here are some specifics about safety zones:

Sometimes it will be tough finding a proper safety zone. Don't be afraid to say you can't... and change your tactics.

- They have to be large enough to protect firefighters under worse than predicted fire behavior. They have to be scouted by experienced people, to ensure they will work in an emergency.

- One of the most reliable safety zones is "the black," as long as there is no canopy that can support a reburn. In some conditions, you can carry "the black" with you as you work. Any time you are in "the black," watch for weakened trees, rolling rocks, etc.

- As work progresses along the line, new safety zones will have to be identified and announced to the crew and lookouts. With new safety zones, new escape routes will also need to be designated.

The new LACES or the original LCES...you still have to know all that stuff!

Note in this detailed discussion, awareness was not specifically addressed. If you follow all of the items listed under the four primary points of LACES, awareness is covered within them.

Just in this one chapter, you have read about lookouts, communications, escape routes and safety zones in at least four places. Enough is enough...but no. These are key points that only repetition is going to help with it "sinking in" and being recorded in your subconscious.

Now, if that isn't enough, there are some other dangers along the fireline you need to know about.

Fireline Hazards

There is more than just fire that can harm you on the fireline.

The fire isn't the only danger on the fireline. There are sharp things, rolling things, falling things, slippery things, and things that bite. All can be dangerous by themselves.

Fire and Other Hot Things

Most of the emphasis of this handbook is how to deal with the dangers of the fire, itself. Along with the dangers of the fire, there is the potential to receive burns from dropping a foot into a burned-out tree stump, burning a hand while "feeling out" for fire during mop-up, etc. One must be very careful when working along a hot fireline (Figure 1.18).

Northtree Fire International

Figure 1.18 Stump holes are very dangerous. Not only will the tree eventually fall, but the stump and most of the larger roots will be consumed.

In some parts of the west, the mountains are so steep, it is said, "you can lean against them." In any terrain, the potential for falls is always present. You must watch your step and ensure that your footing is secure.

Wildland firefighting also is usually conducted when it is hot and dry. These two factors can have a real impact on your body.

When you sweat, you must replace the water that is lost. As mentioned earlier, you must drink a lot of water to keep your body properly hydrated.

Smoke and Dust

Smoke, "products of combustion," and dust will be present along the fireline. If you can, work out of the smoke. Research has shown that a firefighter is exposed to the most toxic smoke during mop-up, when the fire is smoldering. So, take precautions not to spend extended periods of time working in high concentrations of smoke and steam.

Keep your distance; don't be hurt by your fellow firefighter.

Smoke and dust also impair visibility, so one must be very careful when driving in such conditions.

Walking the Fireline

If you are a wildland firefighter, you will be doing a lot of walking. If your footwear is not properly fitted, maintained and broken-in, you will be "out of service" in short order. Here are some tips for you to follow while walking and working the fireline:

- ***Pace and route*** - The crew supervisor will set the pace and path to follow. No going your own way! The fireline is not a place to be lost. If you do get split-up, move into "the black" for safety and sit still, someone will eventually find you.

- ***Safe distance*** - In most cases, stay at least 10 feet apart when walking and working (Figure 1.19). This keeps you from interfering with the other members, keeps you away from their use of tools, and away from branches that may whip back and strike you. If you must move through firefighters along the fireline, announce your intention by shouting, "coming through."

- ***Darkness*** - If it is dark, things can become much more dangerous. You can't see as far or "pick up on" hazards that are lurking in the shadows. Use some form of lighting to mitigate the situation.

Figure 1.19 Keep your distance when working with others close by. Walk and work a minimum of 10 feet apart. And, if you are swinging things, double the spacing.

- ***Hazards*** - Some of the specific hazards that you may encounter on the fireline are: falling trees, tree tops and limbs from the activities of other firefighters or their tools (bulldozers, aircraft, etc.); leaning trees or snags; whipping branches; rolling rocks and/or logs; equipment like bulldozers, ATVs, etc.; and unstable footing.

White ash means heat.

- ***Stream or canal crossings*** - If you have to cross a stream or canal, your footing on slippery rocks and logs will be your biggest hazard. Avoid high log crossings, because one slip may lead to a long fall. If the water is swift, face upstream and use a stick for balance. It is best if you remove your boots and loosen your pack, so if you do fall, the pack will come off easily.

Stump holes can be very deep, and very hot.

- ***Stump holes*** - Holes as the result of a blown over tree should be avoided. Also, if you are in an area that has burned, stay away for stump holes burned-out by the fire. They can hold heat for long periods of time, and are deeper than you think. Beware of white colored ash. This is an indicator of a lot of heat. Test an area of white ash with a stick before working it.

- ***Local hazards*** - Know what is in your surroundings. Is poison oak present; are there bears or other animals you should know about; are there poisonous snakes or insects; is the water polluted? But, most of all, know where the fire is and what it is doing.

Follow your supervisor's instructions, and inform him or her of any hazards you notice.

USFS, MTDC

Figure 1.20 Snags like this one can fall at any time, especially when weakened by fire.

Snags, Trees with Fire in Them, and "Widow Makers"

A snag is a dead tree just waiting for its time to fall (Figure 1.20). A "widow maker" is a dead limb hanging in a tree that may fall when the tree is cut, or when dislodged by wind or a "drop" from an aircraft. They make little or no noise when they fall, thus they are considered silent killers. There are some indicators you can use to spot a potentially dangerous tree. The indicators are:

- Lots of downed trees in an area. You may be in a stand of lodgepole pine or an area where some of the trees have died. All of these trees can fall at any time, for a number of reasons. Watch out for:

- Leaning trees.
- Dead or broken tops and/or limbs hanging in the trees.
- Trees that have lost needles, bark or limbs.
- Conks, basal scars, cat faces, numerous downed limbs, ants or numerous woodpecker holes. These are signs of a dead or dying tree.
- Stump holes burning in the area.
- Smoke or fire at the base or in the top of a tree. This is again a sign of a dead or weakened tree (Figure 1.21).

Figure 1.21 When this snag fell, it hit a firefighter, severely injuring him.

There are four situations that will worsen a snag or dead tree problem. They are:

- You are working in the forest and a strong wind is blowing.
- You are working on steep slopes.
- You are working at night.
- You are working in a diseased or bug-kill area.

Give snags, especially those with fire in them, wide berths. Construct a line well away from them. Estimate their height and give them a 50 percent margin on the uphill and side slopes, and a 150 percent margin on the downhill side. Flag them like any other hazard. Get them on the ground as soon as you can. If you are not experienced enough to do it, get someone who is. Snags with fire in them are very dangerous, both as falling objects and sources of spot fires. Don't underestimate the dangers of working around snags (Figure 1.22). If the area cannot be made safe before entering, avoid it.

Jake says, "Snags kill, and they don't make a lot of noise doing it."

Figure 1.22 Snags are one of the biggest killers of wildland firefighters. Be especially careful of dead trees with fire in them. They don't make a sound when they begin to fall.

Trees with fire in them are also deadly, especially those with heavy lateral (overhanging) limbs. These limbs can break loose from the main trunk, especially when weakened by a fire burning in the interior of the tree. Approach very carefully, and have the tree taken down, if it is considered unsafe.

"Widow makers" are a little more difficult to detect and protect against. Be especially careful when felling a tree or snag. This will dislodge them.

NEVER sit down and rest in an area where there are snags, burning trees or "widow makers" (Figure 1.23).

Rolling Rocks or Logs

When fighting fire in steep, rugged terrain, rolling rocks and logs are always a danger (Figure 1.24). They can be dislodged by the fire burning away material that is supporting them, by firefighters or equipment working above you, or by a misplaced hose stream. Bulldozers are notorious for producing lots of rolling material. ***Never work below a bulldozer.*** When working in steep terrain, post lookouts, and maintain escape routes.

Figure 1.23 Pick your spot to take a break carefully. Dead trees and gravity are a deadly combination.

Sharp Tools

When working on the fireline, always maintain good footing and ensure that the handle of the tool is not slippery. If hit by a retardant drop, clean the handle of your tool. The following are rules for using fireline hand tools:

- Carry hand tools at the balance point of the handle, and on the downhill side.

- Sharp tools should have guards on the cutting edge when not in use.

- Keep tools sharp; dull tools are dangerous.
- When not in use, place tools where they will not be tripped over or hurt anyone.
- Don't use a tool that has a loose head. Keep handles free from splinters and tight in the tool's head.
- Walk and work at least 10 feet from other firefighters.
- Use tools only for their intended purpose.

Figure 1.24 Rocks can be dislodged by bulldozers working above you or by the fire itself burning fuels that are supporting the rocks.

Chain saws are one of the most dangerous cutting tools on the fireline. Special training is necessary for their operation and use. You don't want just anyone cutting trees down on the line. Only trained and experienced people should be allowed to operate a chain saw. Proper safety equipment and a lookout are musts for saw operations.

Chain saws are not only very dangerous, they are very noisy.

Noisy Tools

You should be concerned with noisy tools for two reasons: If you don't protect your ears, the noise may damage your hearing. The noise may also prevent you from hearing a warning of something dangerous. So, when working around noisy chain saws or heavy equipment, wear ear protection and ensure that you have a way to be warned of pending danger.

Vehicle Hazards

As a new firefighter, you may not be operating heavy equipment, but you should know about the potential dangers machines present. Bulldozers and other vehicles are always a danger. They are not very forgiving and the human body is no match for tons of steel. So keep your distance and be sure the operator knows where you are.

Many firefighter has been killed while in vehicles.

Mechanical things also break, especially when put under the rigorous strain of wildland firefighting. Ropes snap, loads shift, brakes fail and unskilled, tired or inattentive operators make mistakes. Don't assume none of these things can happen when you are around equipment. Keep a heads-up, and leave a way to move out of the way.

If you will be using some form of vehicle to move to and from the fireline, here are some simple rules to follow:

"Arrive Alive!"

Always drive defensively.
Reducing response vehicle speed can prevent rollovers.
Red traffic signals and stop signs mean complete STOP.
Insist that vehicle occupants use seat belts.
Verify vehicle occupants are seated and belted.
Evaluate road surface and weather conditions.

Abide by federal and state motor vehicle laws.
Lengthy response distances require frequent rest stops.
Initiate standard vehicle backing operating procedures.
Value occupant and public safety versus time and speed.
Enter dangerous curves and intersections cautiously.

Eric Kurtz

- ***Loading*** - Follow the instructions of the person in charge of loading; use steps and hand-holds; use lights at night; sit in the seats provided, not on the running boards or tail gates; keep the tools in a safe place; and do not overload the vehicle. When approaching a helicopter, have the chin strap fastened on your helmet, or hold it in your hands. Rotor wash has sucked many a helmet up into the rotors, causing thousands of dollars of damage. Also, never approach a helicopter from the rear or from ground higher than the helicopter. Being hit by a rotor blade will ruin your day.

- ***Riding*** - No horse play; keep arms and legs inside the vehicle; do not throw stuff from the vehicle; no smoking; and wear the seat belts, eye protection or life preserver, when applicable. If in an aircraft or boat, don't move around unnecessarily. You know about "rocking the boat."

- ***Unloading*** - Stay seated until the vehicle has stopped; unload one at a time, or as instructed; use the steps or designated exit; and move away from the vehicle as directed.

If in a crew, follow directions so things don't get all balled up.

Retardant Drops and Other Aircraft Hazards

Figure 1.25 If aircraft are working a fire, keep alert to where and when they will be dropping retardant. It might be exciting to be dropped on, but it is dangerous. Avoid it if you can.

It is best that firefighting personnel not be in the airtanker drop zone, however, it does happen. If you are working in and around aircraft, be very careful (Figure 1.25). Retardant is "slippery when wet," so watch your footing. It is also heavy, and when traveling at over one hundred miles an hour as it leaves the airtanker, it can kick up rocks, break the tops out of trees, and knock firefighters off of ridge tops.

In the past, bucket drops from lighter helicopters weren't too much of a danger. But, the bigger Type I helicopters can drop thousands of gallons of water, and this can be quite dangerous. Use the same precautions when working with helicopters as you do when working with airtankers.

Sling loads being delivered by helicopters can also be hazardous. They are heavy, and in a lot of cases, the helicopter pilots don't have a very good view of the load below them. So don't totally rely on them seeing you, simply move out of the area to a safe location.

Helicopters, especially the real big ones, have considerable rotor wash. This high velocity air can cause the fire to spread across lines, and has been a contributing factor in some fatal fires. This air can also break limbs and tops out of trees. This is just another hazard you must be aware of when aircraft and their associated activities are present on the fireline.

Not only does retardant make things slippery, it sometimes knocks things out of trees.

If you are required to be around a helicopter on the ground, follow the instructions very carefully. ***Never*** approach an operating helicopter without the pilot seeing you and acknowledging your approach; ***never*** approach an operating helicopter from the rear. Tail rotors kill!

Snakes, Ticks, and Other Bad Things

We all know about snakes and their bites, but now there are ticks that carry Rocky Mountain spotted fever and Lyme disease.

Ticks, snakes and other wild things are always around you in the woods. You are on their turf.

There is often poison oak (sumac) and ivy. The water in streams sometimes contains giardia, a parasite which causes dysentery.

Be aware of the dangers in your area. If you get bitten by a snake or bug, notify your supervisor immediately. Stay away from poison oak (sumac) and ivy as much as you can; if exposed, bathe as soon as you can and put on clean clothing. Don't drink from a stream or lake, only from your canteen or another known potable water source.

Bears, moose, and other wild animals also pose a threat to you as a wildland firefighter. You are on their turf and they may feel threatened. Avoid "close encounters" with big wildland critters.

Power Lines

Downed power lines are killers (Figure 1.26). ***Consider any power line that is on the ground as one that is charged with electricity!*** Keep everyone away from downed lines. Flag the area to warn others and restrict access. Notify your supervisor and any other firefighting personnel of the situation. Only when the line is cut from the pole by a power company representative can you consider the line safe.

Figure 1.26 Electricity is nothing to fool with. If power lines are down, or involved with your fire, be very careful. Power lines thought to be de-energized have killed firefighters.

It is also dangerous to spray water directly on power lines. If a pole is on fire, or if downed power lines are near a source of heat you want to cool, use only a spray-fog. A straight stream of water may have enough conductivity to deliver you quite a shock.

Another danger when fighting fires near high-tension power lines is the potential for dense smoke to allow the power to go to ground (Figure 1.27). One major component of smoke is carbon. Carbon is a good conductor of electricity, thus if the smoke is thick enough, the electrical energy will find a path to ground. This can be spectacular, and very dangerous. ***If a fire is moving under power lines, stay clear; do not assume the lines have been de-energized.***

Here are some basic firefighting tactics and safety rules to follow when fighting fire near high-voltage power lines:

- If fire is within 100 feet of the power lines, assume that the smoke is thick enough to short the power grid. Abandon any direct attack and stay at least 100 feet from the outermost wire.

Kari Greer

Figure 1.27 Heavy smoke will allow high-voltage electricity to go to ground. This can be very dangerous. Stay away from poles and tower structures, and from under power lines when the fire will be burning under them.

- If the main fire is more than 100 feet from the power lines and there are small spot fires closer to the wires, you can continue your attack. Small spot fires do not generate enough smoke to short out the power grid.

Stay out from under high-voltage power lines.

- Never work within 35 feet of a high-voltage tower or pole structure.

- When using water near power lines, do not spray water toward the power lines, or raise the stream above your height. Do not spray water into heavy smoke or at burning trees within 100 feet of the power lines.

Heavy smoke can short out the power lines.

- Do not use airtanker or helicopter drops near power lines. Retardant could short the lines, or contaminate the insulators.

- Sometimes pipe lines and wire fences parallel or cross under power lines. Keep your distance from them. If a line were to drop on them, they would become energized.

- If fighting fire near high-voltage power lines, have a power company representative come to the scene to assist you.

Jake says this about downed power lines, "You will only step on one in your life!"

Wildland/Urban Interface Hazards

When homes are present, the whole dynamics of a wildland fire changes. From a firefighting point of view, resources are diverted from line construction to structure protection. This presents a new set of hazards you must be aware of and concern yourself with. Some of them are:

When homes are involved, all the rules change, and some new hazards are present.

- ***Hazardous materials*** - Building materials, when they burn, release dangerous gases. Swimming pool chemicals, junk cars and garbage, when burning, also can be very dangerous.

- ***Propane tanks*** - Liquid petroleum gas tanks (LPG) are like bombs if exposed to a lot of heat. Keep them cool if you can. If they are venting and on fire, let it burn, but move some distance away. If you are not specifically trained and equipped to deal with a burning LPG tank, let someone who is do the work.

Lightning not only starts a lot of fires, it can kill.

- ***Traffic*** - If you have a fire moving into any well populated area, people may be leaving on their own, or are under orders to evacuate. You will have to move into the area using the same roads they are wanting to leave on. Narrow and/or congested roads are often a problem, so be alert.

- ***Panicked public*** - Nothing scares an untrained person like a fire does, especially if their home is threatened. If the evacuating public is scared, they can pose a real problem. Your first priority is the protection of life. If you can help move these people from harms way, it is your duty to do so.

Kari Greer

Figure 1.28 Lightning struck this tree and started the fire. If you had been within 20 feet of this tree when it was struck, the incident would have been dispatched as a "medical aid with fire." Or worse, a coroner's case!

Lightning

Lightning has killed firefighters. It doesn't happen often, but it does happen. If a storm is in your area, move away from tall trees that may act as lightning rod (Figure 1.28). If you have a vehicle, stay in it until the danger passes. If you are out in the open, sit down away from high points of terrain and wait it out.

Human Hazards

You will learn more about human limitations in the next section about human factors, but here is a brief list of some of the human hazards you should be concerned with:

Now we are trying to find out why we make mistakes.

- ***Attitude*** - If you or some of the people you are fighting fire with have an "attitude," it could adversely impact you. Firefighting is a full time job, and if your "head isn't screwed on straight," you have a problem that could endanger others.

If you are not physically conditioned, you will not be able to keep up.

- ***Physical conditioning*** - If you are not in proper physical shape, you may not be able to stay up with the crew and pack your weight of the load. Instead of being an asset to the team, you become a liability, a burden and distraction.

- ***Training levels*** - Know your limitations. If you are asked to do something you haven't been trained to do, don't attempt it. Speak up.

- ***Experience, or lack of it*** - Reading this handbook gives you zero experience. Experience is gained by watching and doing it. You will know after reading this handbook that fires usually burn uphill faster than downhill. But, unless you have seen how fast this can happen, your lack of experience will be a danger to you. Again, know your limitations and work within them.

- ***Fatigue*** - It is a fact that when you are tired, your ability to process things is diminished. Just know this is true, and be more careful when you are tired (Figure 1.29).

- ***Critical stress*** - When you see something that really disturbs you, like injured victims of an accident, your subconscious mind can begin to protect you from the pain and horror. Soldiers in battle and firefighters who have been exposed to dramatic situations all suffer from critical stress. Just be aware that if you have been exposed to something like this, you may want to talk to someone about what you are feeling. It isn't bad you may feel stressed, it is only bad if you don't deal with it properly.

Figure 1.29 There are times when you just want to give up. Nothing is going well and you want to go home. Bad days will happen, but people are counting on you to pull your own weight. Firefighting is dangerous, and fatigue can make it worse. Don't let the "can-do" attitude get you into losing situations. Backing off is not a failure...it is a win for common sense.

- ***Viral infections, colds, and influenza*** - Another hazard that you may confront is illness. When large groups of people are working and living in proximity, they will spread what they have to others. If you are tired and cold, you may be more susceptible. One of the best ways to prevent the spread of communicable diseases is to wash your hands.

- ***Sanitation and food handling*** - Washing your hands is also a primary way to help prevent infection from low-quality or improper sanitation facilities and food handling.

Human Factors

Don't be afraid to say you made a mistake!

In recent years, more and more attention has been given to the strengths and weaknesses of human performance skills on the fireline. In some cases, accidents on the fireline happen because something broke, was worn out, or was lost. But most of the accidents on the fireline are due to human error. ***Human factors awareness focuses not on what the firefighter did wrong, but why!***

For years, when a firefighter "screwed up" and someone was injured or died, we tried to determine what went wrong and added another "watch out." We didn't take into consideration that there might be a reason the firefighter made the mistake. Now we ask, "Why?"

When you get tired, you are not as effective.

Firefighting is a dangerous, high-risk business. People become firefighters partly because of the thrill of the fight, the teamwork and the satisfaction when the fire is out. Safety was discussed during training, but we all went out and did what we thought had to be done. If you were properly trained and had a lot of experience, safety was a matter of "common sense." You knew what was safe and what wasn't. You automatically did what was safe for you and your crew. Understand your human limitations. Don't let the "can-do" attitude get you into losing situations. Backing off is not a failure...it is a win for common sense.

Jake says, "We don't hear as well as we speak!"

Another problem you will have to deal with is communication. Even if you see something important, you may be reluctant to report it, or you may not really understand an instruction you have just been given.

Whether you are a new firefighter on your first fire, or an old-timer with years of experience, communication is important. Here are some reasons why communication is important:

- Communication is required for you to receive orders and instructions.
- Communication is required for you to be aware of any new hazards and safety concerns.
- Communication is required for you to make others aware of what you see in a situation.
- Communication will play a major role in how you will learn about the business of firefighting.
- Communication can only be effective if the listener hears and understands what has just been said by the one talking. If a message is not heard, is has not been received. If a message is not understood, it too, is not received. It is the responsibility of the sender to be heard and understood.

We are easily distracted.

To be sure that an order or information has been received, you can ask for the information to be repeated back to you. The key is that you have been heard or understand what you have been told. Clear and concise instructions are vital, especially if your level of experience is low. If you are unclear about the order, ask questions.

Why is it that someone "just isn't listening?" There are some real common barriers to good listening:

Practice making sure people understand what you are saying.

- ***Preconceived opinions*** – "I know what he is saying, and I know the answer."
- ***Distractions*** – You are simply thinking about something else or watching an airtanker circling overhead. Or, wondering what is for lunch; or what movie you will go see during your next days off.
- ***Filtering information*** – "It really isn't that bad."
- ***Not listening*** – but starting to answer before the question or instructions have been completed.
- ***Having an attitude*** either towards the sender or the message he/she is sending. This barrier of having an "attitude" about someone has been the root of several very disastrous

> accidents. Separate your personal or emotional "hang-ups" from your responsibility. ***Every firefighter is responsible for open, effective communication.*** If you ever feel that attitude is negatively affecting communication, speak up! You may be preventing a dangerous situation.

Another barrier to clear communications is that some words or phrases have different meanings. The fire service is full of special phrases or codes. This is one reason why radio codes are not being used as much as in the past. With a greater level of cooperation between agencies, the codes were a liability…a lot of the same codes had different meanings. The key is that if you hear a word or phrase that you do not understand…ask a question. It means you are listening...learning.

There are five basic communication responsibilities:

Don't be afraid to ask questions.

- ***Briefings*** – The passing on of general information.
- ***Debriefing*** – After an incident or event you ask questions from those involved to learn what actually happened; what went right and what went wrong. You learn from debriefing.
- ***Warnings*** – Information about hazards is passed on.
- ***Acknowledge messages*** – This is where you say you understand the information or order.
- ***Questions*** – I do not understand what you are telling me. You ask for clarification.

Only fools don't ask questions.

The last of these communication responsibilities is the toughest. You will be reluctant to ask for clarification, especially if you are just beginning your career as a firefighter. You must ask questions if you are not sure of the message. If you don't ask questions, how else will you learn?

If you are on the fireline and have been just been given an order, here is a list of four questions that you should have the answers to:

- What task am I to perform?
- What are the known hazards?
- Where do I go to be safe?
- How do I get to this safe place?

The fireline is not one of the safest places to work, so what can you do to make it safer? Here are a couple of key things to remember:

- Any hazard or potential hazard should be reported to your immediate supervisor or coworkers.
- If the area is unsafe to work in or around, it should be flagged so others are aware of any dangers.
- There are two options for an unsafe area: make the area safe before entering; or avoid the area if it cannot be made safe.

How can you make a plan if you don't know what is happening around you?

Situational Awareness

One of the main reasons firefighters get hurt is that they really don't know what is happening around them. Their "situational awareness" is lacking. What is situational awareness? It is the:

- Gathering of information by observation or having it reported (communicated) to you. It is on the basis of this information that you will make decisions. This process of observing is not a one-time thing you do when you are sizing-up a fire, but a constant and continuous cycle. Just as a soldier in combat is always looking around, you as a firefighter must always be looking around and sensing what is really happening and how what you see may impact fire behavior, you, and your fellow firefighters.

Be another set of eyes on the fireline.

Until you see "fire in action," you cannot really begin to learn what it can and will do under different conditions. You are one more set of eyes watching what is happening. You may see something no one else does, and your observation may be the difference between life and death. There are many accident reviews of fatal accidents where it was determined that someone saw something and did not report it…and it would have made a difference. Don't be one of those! You are part of a team, and you will play a vital part of that team by keeping vigilant and not hesitating to report what you see. If you have situational awareness, you will be able to detect problems before they can hurt you...or others.

The situation is constantly changing...and it waits for no one.

Hazard assessment is one the primary reasons for situation awareness. Using the 18 Situations that Shout Watch Out!, the Common Denominators of Fatal Fires, and other bits of information you have learned, all will help you assess the situation.

Jake says, "Attitude and safety are linked."

Then you review your rules of engagement, the 10 Standard Firefighting Orders, to ensure that you are in compliance. As mentioned several times, and it will be mentioned several more times in the following chapters, you constantly reassess. Why? Because things are constantly changing:

- ***Time of day*** - Fires burn more actively in the afternoon.
- ***Weather*** - As some people say, "Wait a minute and the weather will change." As the weather changes, the fire behavior changes.
- ***Location of the fire*** - As the fire moves, the topography and fuels being burned, or those about to be burned may be different.

Here are some factors that will hinder your attempts at situation awareness:

Most accidents involve human error.

- ***Inexperience*** – You don't understand the significance of what you are looking at.
- ***Stress*** – You have other things on your mind. They may be related to work or your home life.
- ***Fatigue*** – You are flat tired. You have been up for hours and you are just not thinking straight.
- ***Attitude*** – It is the boss' job to keep track of what is happening, or other attitude barriers that compromise awareness.

If you are distracted, you usually think about other things…you are in another world. If you are less focused, you will not notice key events or hazards. You will not make good decisions, and more information may just confuse you.

Team Work

Firefighting is a team effort. Teams are made up of many individuals, but a team is what it takes to fight fire safely and successfully. There may be stars on your favorite professional team, but it is teamwork that takes teams to championships.

If a team doesn't train and work together...it isn't a team.

A team is a grouping of people who work together, each pulling his or her weight, taking advantage of individual strengths, and compensating for weaknesses. A team understands the mission and how to accomplish it as a cohesive group. Team members support each other and have a pride in what they do. Most of all, they trust that the team and its leadership will work in a safe manner, allowing all members to return safely home and be able to say, "job well done."

Jake says, "There is no 'I' in team."

Nothing will kill a team spirit faster than:

- ***Poor communications*** - Individuals don't pass on vital information. Or, it may involve team members who do not have a working knowledge of English.

- ***Cliques developing*** - Little groupings of individuals focus on the importance of their grouping rather than on the team as a whole. This breakdown of a team has contributed to the deaths of firefighters.

- ***Poor work ethics*** - If someone isn't pulling their own weight, the group will begin to break down. Bad apples need to be pulled into the team or let go.

- ***Lack of respect*** - If team members don't like one another or the team leader, the team will never pull together.

Teams can be grumpy too.

- ***Complacency or lack of desire to improve*** - A team is always trying to learn and do better. If players don't have the desire, they will never come together as a team.

- ***Blaming others*** - If team members start to blame other members for a team failure, the team will begin to break down. A team works together, compensating for individual weaknesses.

- ***Selfishness*** - If individuals put their welfare above the team, the team will not function as a cohesive unit.

Figure 1.30 It looks like this firefighter has had a bad day and has something to report. No doubt we can learn something from his experiences. Learn from all the fire experiences you can...the good, the bad and even the ugly!

A properly functioning team, with many eyes to spot hazards, hands to do the work, and brains to do the thinking can do more work collectively than the sum of all the work they could do individually. It is said that a team is "only as good as its weakest link," but if someone's strength helps the weakest on the team, the whole level of performance comes up a notch. A team is group of people who train and work together with a shared vision, common goals, and a cohesive sense of purpose. If a team does not train and work together, it is just a group of people, not a team.

After Action Fire Reviews

One of the best ways to learn is from one's actions—whether they were right or wrong—is a post fire review. Most firefighters feel that if an agency conducts a fire review, they are looking for someone to blame. This shouldn't be the case. We should consider all fire incidents as learning experiences, and spend just as much time learning from the fires where events went well as the ones where things didn't go so well (Figure 1.30).

Learn from what you have done...the good and the bad.

There are times when good plans and intentions "tank." Yes, sometimes people screw up. But in most cases, "it's the fire's fault;" it just didn't read the plan...it did its own thing and someone got hurt.

Make it a habit to really analyze your responses and actions, and learn from them. Admit when you make an error, but also pat yourself on the back when things go well. Every fire is a learning experience. You will never stop learning. Use your experiences to teach others. Leading by example and good story telling is a part of what makes a good firefighter.

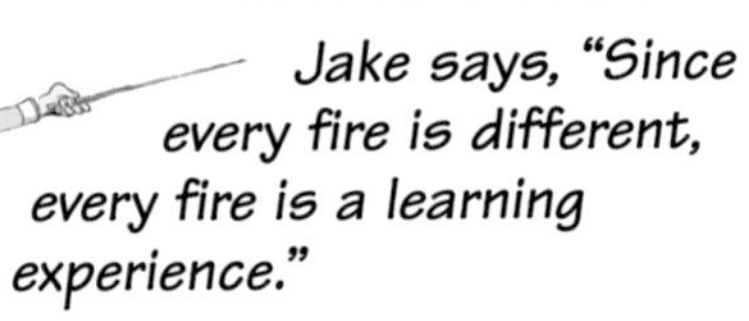

Jake says, "Since every fire is different, every fire is a learning experience."

(***Note***: The reader should "check out" this web site for up-to-date Lessons Learned.
http://www.wildfirelessons.net/Lib_IncdtRevws.html)

2

Fire Weather

It isn't by mistake that fire weather concerns are the first of the Ten Standard Fire Fighting Orders. Weather is the primary driving force behind the changes in fire behavior. Although current and predicted weather forecasts should be available to you, becoming familiar with the underlying dynamics of weather can help you better understand the potential impact of weather on wildland fire behavior.

Weather changes kill firefighters!

Weather changes, even subtle ones, are killing or injuring wildland firefighters each year. If you don't know and understand the actual and predicted (forecasted) weather parameters, you will not be able to predict what a fire is going to do, and will have difficulty developing sound, effective suppression strategies. You might as well fight fire with your eyes closed. More importantly, you need to closely monitor the weather, and know when and why changing conditions require different tactics to protect yourself, or assure the safety of fellow firefighters.

Jake says, "Weather is the primary factor affecting fire behavior. Learn to recognize weather indicators and what they mean."

The earth's atmosphere is very dynamic. Today's weather and tomorrow's predicted weather are but snapshots of constantly changing atmospheric conditions. These conditions are influenced by wind, relative humidity, temperature and rain, etc. Understanding these factors, and how a change in one or more of them can affect the "big picture," is necessary before you can successfully manage a fire.

One given in life is that the weather will change.

Have you ever seen a picture of the earth (Figure 2.1) taken from space and noticed that everything was in motion? The clouds move and circle around constantly. This also happens above "your piece of fireline" to a lesser extent. Undoubtedly, changes in weather will take place. It is your job to detect these changes and to develop a sense of what is going to happen, and when. As noted earlier, for crew safety and for your own, you must be able to observe and respond appropriately to the weather. The weather factors that can affect the start and spread of wildland fires are: ***wind***, ***relative humidity***, ***temperature*** and ***precipitation***."

NASA

Figure 2.1 A picture of earth from space. Visualize the clouds traveling in waves, moving across the land and sea.

All of the weather that will impact your fire is within 10 miles of the surface of the earth.

The Big Weather Picture

We take for granted the air we breathe. But, what is air? Why does weather change, and how does this impact the decisions of the firefighter? Imagine the atmosphere as a moving "ocean" of air surrounding earth, with currents, waves and depth. It contains the oxygen we breathe and the carbon dioxide used by plants. It protects us from the harmful rays of the sun, and keeps us warm at night. Yet while it is doing this, the atmosphere is constantly moving and changing.

The atmosphere consists of several layers. The layer of atmosphere closest to the ground is called the ***troposphere***. All weather phenomena (e.g., thunderstorms, wind, humidity, etc.), as we know them, occur within this 5- to 10-mile thick layer. The air within the troposphere contains these elements:

- 78% nitrogen
- 21% oxygen
- 0.9% argon
- 0.03% carbon dioxide
- dust and salt particles
- smoke
- volcanic ash
- pollutants
- water vapor

Jake says, "Remember that the weather factors that can affect the start and spread of wildland fires are: wind, relative humidity, temperature and precipitation."

From a wildland firefighting perspective, the two most important components of the atmosphere are oxygen and water vapor (Figure 2.2). Oxygen is essential for combustion, and water vapor causes clouds, rain and humidity. Typical weather factors firefighters face are wind, moisture (both air moisture and days since rain), cloud cover, air stability and temperature. Remember, it is not just today's weather that will influence the fire; the weather of the last several days will also have an effect.

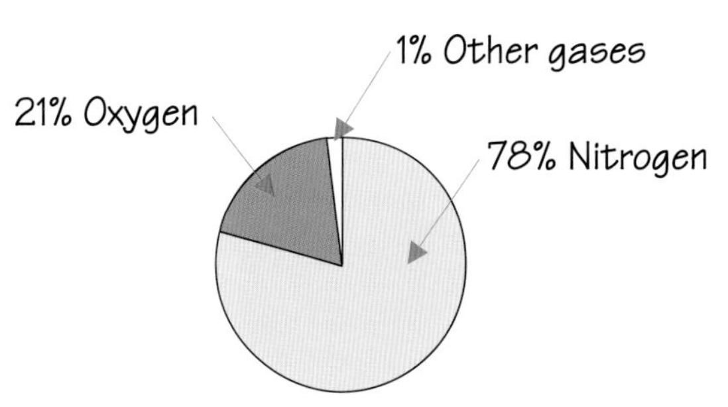

Figure 2.2 The makeup of the atmosphere.

21% of the atmosphere is oxygen.

Atmospheric Pressure

Even air weighs something, and its weight can be measured. A one inch square column of air the full depth of the atmosphere weighs about 14.7 pounds (Figure 2.3). This is equal to a 29.92-inch column of mercury, and is considered the normal or standard atmospheric pressure at sea level. Most barometers measure atmospheric pressure in "inches of mercury." About one-half of the weight of the atmosphere is within 3.5 miles of the earth's surface.

ATMOSPHERIC PRESSURE

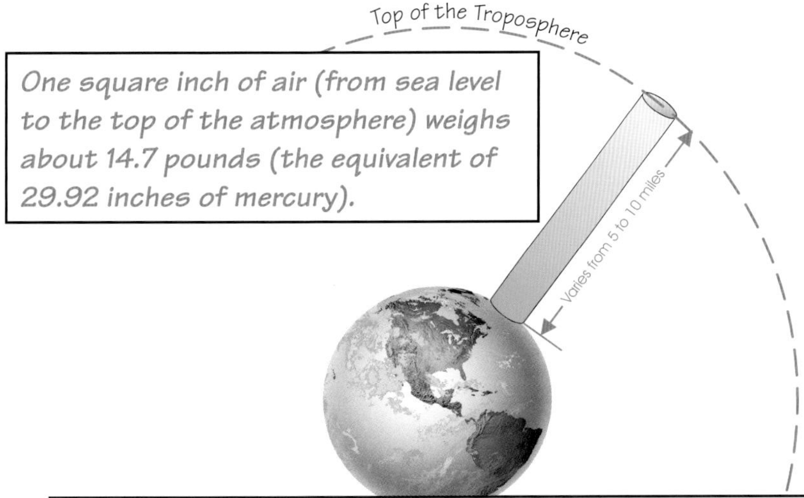

Top of the Troposphere
One square inch of air (from sea level to the top of the atmosphere) weighs about 14.7 pounds (the equivalent of 29.92 inches of mercury).
Varies from 5 to 10 miles

ATMOSPHERIC PRESSURE - the weight of air in a vertical column, in pounds per square inch (psi) or inches of mercury. One half of this weight is within the first 3.5 miles of the surface.

Figure 2.3 The weight of air is 14.7 pounds per square inch. Atmospheric pressure varies with elevation. The higher the elevation, the lower the atmospheric pressure.

Atmospheric pressure varies with elevation.

Atmospheric pressure varies with elevation above the earth's surface. The higher the elevation, the lower the atmospheric pressure. One example of the significance of this concept can be found in aviation. Pilots determine the elevation of their aircraft above sea level with an altimeter. Although the pilot is reading "elevation," the altimeter is actually sensing atmospheric pressure.

Atmospheric pressure also varies in relationship to weather phenomena called "highs" and "lows" which can be measured with a barometer. Knowing the barometric pressure at specific locations can help weather forecasters track high- and low-pressure areas, and the changes that can be associated with their movements and strengths.

The Weather Engine

The power of the sun is the ultimate source of all weather.

The sun is the ultimate source of weather. The conduit for this action is solar radiation energy. Solar radiation strikes the earth on a 24-hour cycle. Most solar radiation strikes earth during the day and is radiated back out at night. However, some of the heat is retained by the earth and its oceans, minimizing wide swings in earth's temperatures.

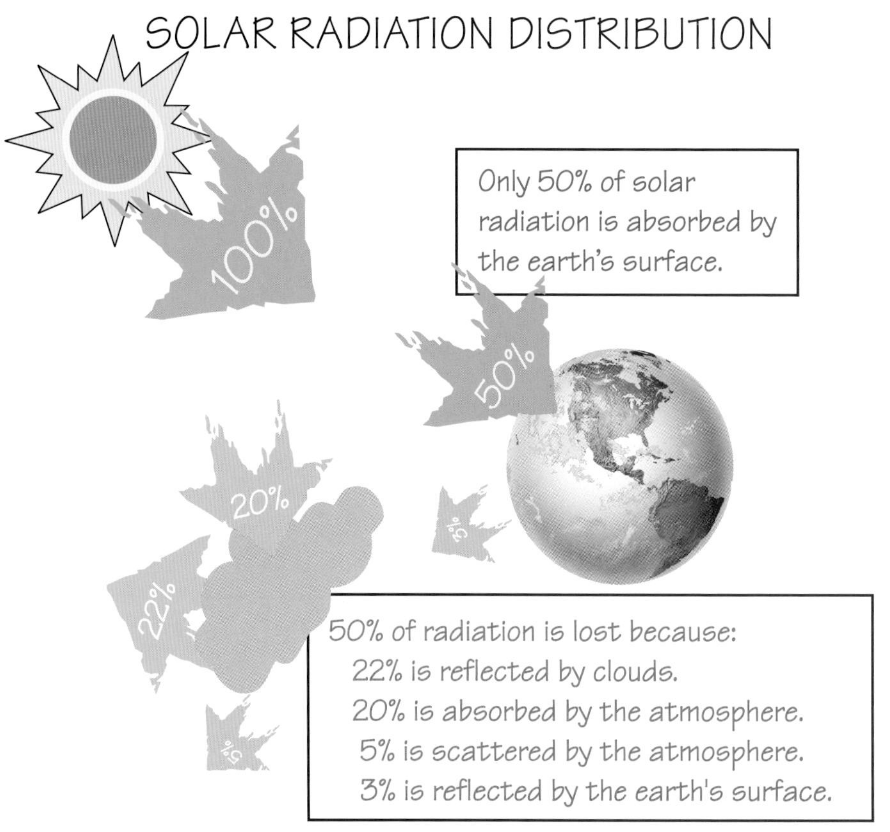

SOLAR RADIATION DISTRIBUTION
100%
Only 50% of solar radiation is absorbed by the earth's surface.
50%
20%
3%
22%
5%
50% of radiation is lost because:
22% is reflected by clouds.
20% is absorbed by the atmosphere.
5% is scattered by the atmosphere.
3% is reflected by the earth's surface.

Figure 2.4 The amount of solar radiation that strikes the earth is related to the density of cloud cover.

The amount of heat energy that reaches the earth varies depending on location, atmospheric conditions (e.g., clouds, moisture, pollution, etc.) and surface orientation or aspect (e.g., south, north). On average, 50% of the sun's energy reaches the earth's surface, 20% is absorbed by the atmosphere (clouds, water vapor or pollutants), and 30% of the energy is reflected back into space. To maintain a favorable climate, the daily outgoing radiation must equal the incoming solar radiation (Figure 2.4). This process is called heat balance. Without it, the earth would become progressively warmer or cooler. However, not all parts of the earth maintain a balance (Figure 2.5). The polar caps lose more heat than they gain, and the equatorial region gains more heat than it loses. This imbalance partially drives the dynamics and movement of what we call weather fronts.

The force that drives the earth's "weather engine" is the increase in the surface temperature from site-to-site and day-to-day. The four primary factors that influence these temperature changes are:

Increased surface temperatures are the primary source of weather.

- *The amount of moisture and/or pollution in the air* - These elements absorb and/or reflect solar radiation. The presence and thickness of clouds, water vapor, and smoke reduces the amount of solar radiation heat that reaches and leaves the earth's surface.

- *The angle between the surface of the earth and the sun* - If the sun is directly above and at right angles to the earth, more energy will reach the surface. As the sun moves through its daily arc, or is in the southern or northern latitudes, the angle is greater, thus the radiation is reduced. Latitude, slope, aspect, elevation, and shape of the terrain can also significantly change the earth's temperature. The changing angle of the sun, influenced by the earth's axis, is what causes the change in seasons. Seasonal changes in the earth's axis help balance overall heating by providing a greater distribution of heat (Figure 2.6).

- *The lag time between when solar radiation strikes the earth and when the heat is radiated back out into space* - This time varies both throughout the day and throughout the year. The greatest amount of solar radiation reaches the earth around noon, but the

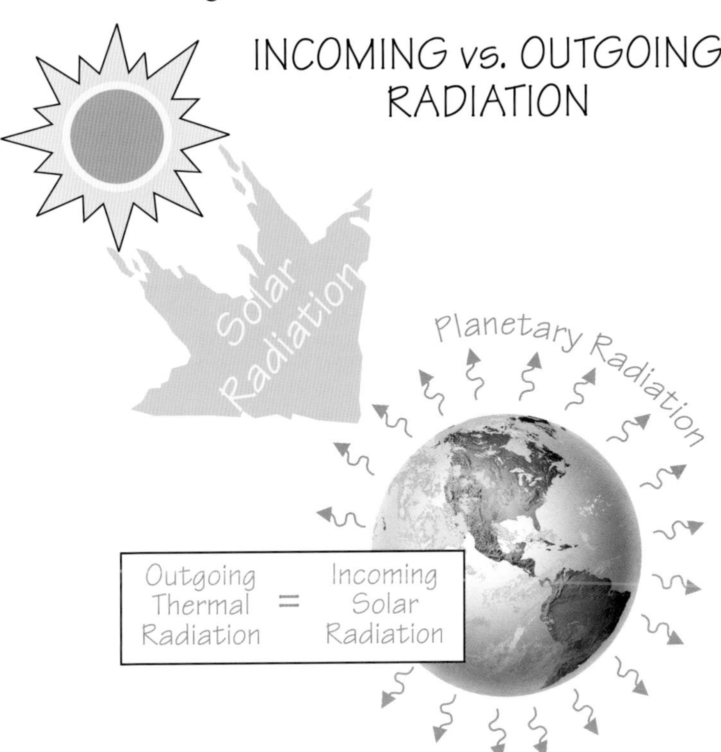

Figure 2.5 The earth gains heat during the day and loses it at night. This gain/loss is what keeps the earth's overall temperature in balance.

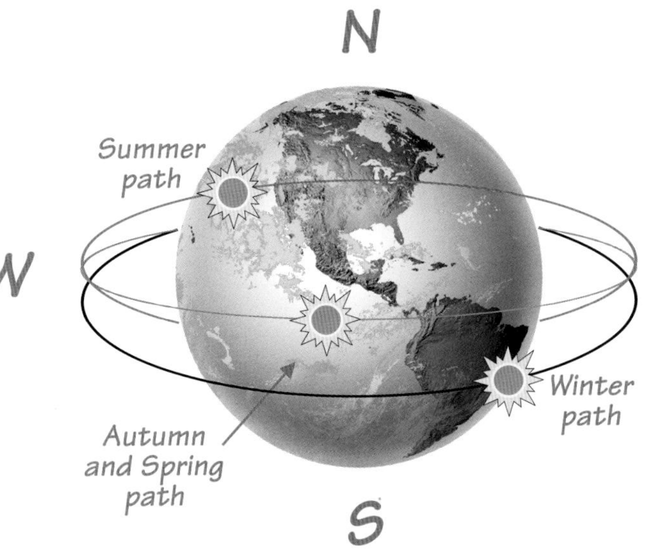

Figure 2.6 The angle at which solar radiation strikes the surface of the earth influences the level of heating. The axis on which the earth rotates helps even-out the amount of solar radiation that hits the earth.

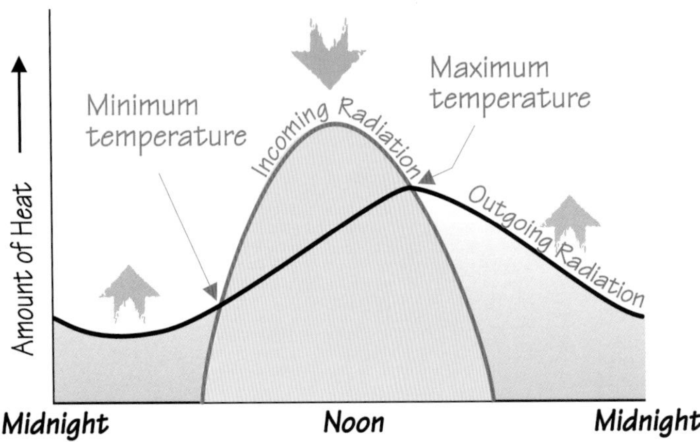

Figure 2.7 When the incoming solar radiation exceeds the level of outgoing radiation, the temperature of the atmosphere will rise. The hottest part of a day will occur after the period during which the earth receives its greatest level of solar radiation.

highest temperatures usually occur between 1400 and 1600 hours. Likewise, the most solar radiation reaches the earth's northern hemisphere in June, but the hottest part of the year is in July. The lag time in both daily and seasonal temperatures can be explained by the following: The earth loses heat continuously through radiation. However, when the incoming energy from the sun exceeds the outgoing energy from the earth (e.g., around noon each day, and during summer), the temperature continues to rise because the earth is receiving more heat than it is losing. The temperature doesn't start cooling off until the incoming energy ceases to be greater than the outgoing energy (Figure 2.7).

• *The surface properties of the terrain and/or vegetation* - The amount of solar energy reflected or absorbed by any surface varies with the color, texture, transparency, etc. of the surface material (Figure 2.8).

- Color and texture - Both the color and texture of surfaces influence how they absorb or reflect solar radiation heat energy. Black surfaces (like dark soil) absorb more energy than white ones (like dry grass and weeds). Rough surfaces (like tree bark) absorb more energy than smooth ones.

- Transparency - The greater the transparency of a surface, the more solar energy is able to penetrate the medium. This is why water absorbs and holds more heat energy than rock.

- Conductivity - The greater the conductivity of a material (rock vs. wood), the more deeply heat energy is absorbed by it.

- Absorbed heat - Some materials absorb and retain heat more efficiently than others. Water can absorb five times the heat energy of rock, and holds heat longer.
- Surface moisture - Evaporating moisture has a cooling effect. During condensation, heat is released.

When incoming energy exceeds outgoing energy, the temperature will begin to rise.

In Figure 2.8, note how different the surface heating can be, depending on the type of surface. Although water is the coolest (at 65°), its transparency allows it to be heated to a greater depth, efficiently dispersing the solar energy. Further, it is a better "retainer" of heat. The shaded forest is a little hotter (at 75°) followed by the grass/brush surface (at 95°). The plowed field is the hottest (at 115°) because of its dark color, texture, and lack of shade and transparency. All of these areas will impact fire behavior in different ways. A fire will move up the grass/brush slope much faster than it will move across the forested area. The heat of the plowed field may cause upcanyon winds, and if the air is unstable, may even cause whirlwinds. As the time of day changes, or as cloud cover changes, all of these temperature relationships will change.

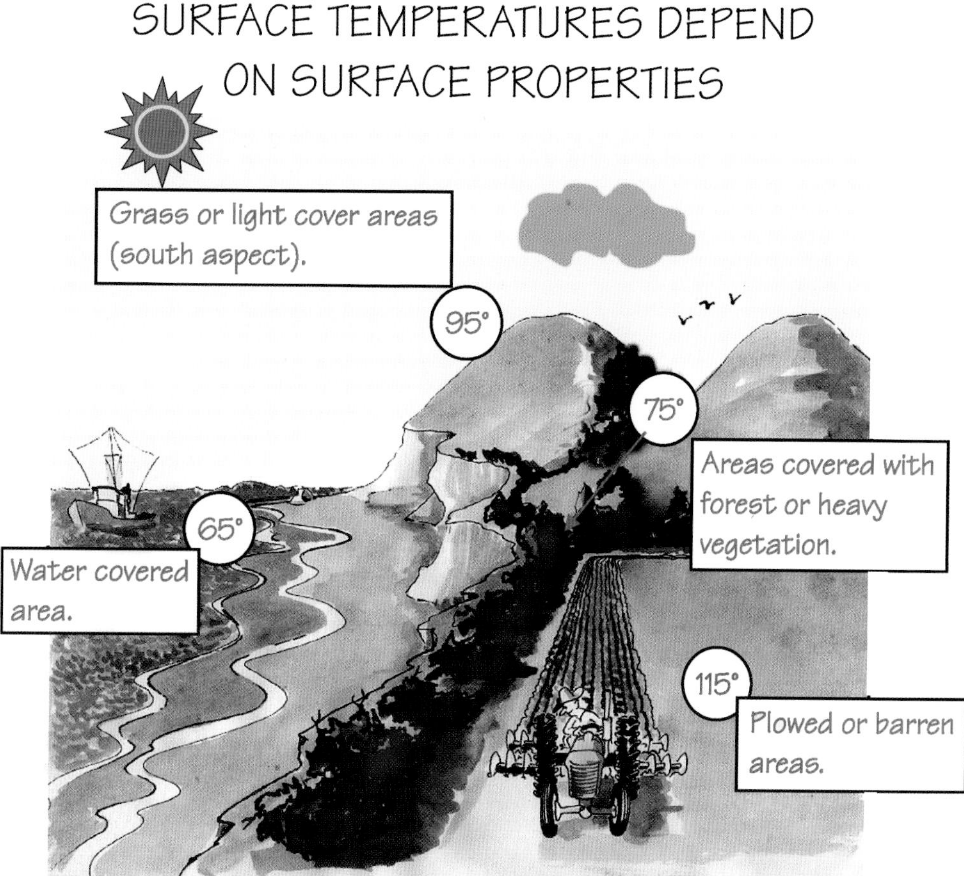

Figure 2.8 The surface temperature is dependent on the type of surface, type of ground cover, aspect and cloud cover.

The color, texture and transparency of the earth's surface absorb/ reflect solar radiation to different degrees.

Approximately thirty percent of the solar radiation absorbed by the earth is radiated back into the atmosphere. The rate at which this energy is lost is controlled by the amount of water vapor in the air.

Variations occur during periods of higher humidity or when clouds are present. During cloudy periods, the amount of heat energy that is radiated back into space is less than on clear nights. The greatest amount of heat is lost at night when there is no cloud cover and the air is relatively dry. This is why the temperature swings in the desert are so wide from day to night. Some heat is lost to the air through convection and conductivity. The condensation of dew also releases heat when it changes from water vapor to liquid.

Winds

Wind has the greatest impact on fire behavior of any of the weather factors. The stronger the wind, the faster the spread of the fire. This is true because wind brings an additional supply of air to the fire. It also flattens (or bends) the flames to preheat the fuel ahead, and causes spot fires by blowing sparks and embers ahead of the main fire into unburned fuels. Wind is the primary factor that influences fire spread (Figure 2.9). Topography affects wind by: causing it to change direction; increase its speed as it funnels through valleys and canyons or any other constriction; form whirlwinds and dust-devils; and form eddies when crossing ridges or canyon intersections (Figure 2.10).

Wind supplies oxygen, bends flames, and causes flying burning material to ignite spot fires.

The passage of a weather front is usually accompanied by a shift in wind direction. The reason for this is that fronts lie in *troughs* of low pressure. The wind's behavior during the frontal passage depends upon: the type of front; its speed; the contrast in temperature of the masses along the frontal line; and local conditions. The passage of a *warm front* will usually bring a wind direction shift of 45 to 90 degrees. The passage of a *cold front* will shift wind direction from less than 45 degrees to as much as 180 degrees.

TO
Jake says,
"Remember, wind is the most important of all of the weather factors."

Fire itself causes local air currents that add to the effect of the prevailing wind on fire spread. The air above the flame becomes heated and rises. This results in fresh air rushing in from the sides, which adds to the intensity of the fire.

Generally, the warmer the wind, the hotter the fire will burn. Hot, dry air draws moisture from fuels and preheats them. Cool

Wind indicators

Indicator	Potential
Strong surface winds	Wind-driven fires and the transport of firebrands.
High, fast-moving clouds	Potential for wind shifts, particularly if the clouds are moving in a direction different from the surface winds.
Approaching cold front	Winds will shift and increase in speed as the cold front approaches. Look for a squall line of thunderstorms as a visual indicator. A dry cold front will not have a squall line; the only indicator will be a shift in the winds.
Cumulonimbus development	Thunderstorms present the potential for strong, erratic downdraft winds. Lightning is associated with these storms. Also, virga may be observed. (Virga is rain that does not reach the ground.) This is a good indicator that downdrafts have begun.
Battling (or opposing) winds or sudden calm	When foehn winds interact with local winds, significant wind reversals are likely. Definite indicator is a wavering smoke column and a sudden calm. This is a dangerous time; reversal of wind direction is likely.

Figure 2.9 Wind indicators and their impact on fire behavior.

air usually has a higher moisture content, so it may dampen fuels to a point at which they will burn slower.

Wind is defined as the horizontal movement of air. More often than not, wind is involved in fatal or near-fatal wildland fires. General winds are named for the direction they are coming from. A north wind is coming from the north, south wind from the south, etc. "You care for whence the wind comes, for once it gets to you, you care not whence it goes!"

Hot, dry winds draw moisture from the fuels.

There are several winds that you should be familiar with. They are:

- ***General winds*** which affect large areas. These are the winds that are reported in daily weather forecasts.

Topography will impact local winds.

- ***Local winds*** which are produced by features of the local terrain.
- ***Surface winds*** are those that are normally measured at 20 feet above the ground.
- ***Mid-flame winds*** which occur at the midpoint of flame height, and ***have the greatest effect on the direction a fire will burn*** (Figure 2.11).

Figure 2.10 Wind direction is influenced by topography.

Warm winds are dry winds.

Wind affects wildland fire spread in several important ways:

- By increasing the supply of oxygen, thus aiding combustion.
- By driving convective heat into adjacent fuels, hastening the drying process.
- By influencing the direction of spread and spotting.
- By bending the flames closer to unburned fuels, preheating, drying and igniting them.
- By carrying away moist air, replacing it with drier air.
- By carrying burning embers to new fuels and increasing fire spread.
- By moving moisture-laden air that may reduce fire behavior.

Winds are identified by the direction the wind is coming from, not for where it is going.

Don't ever forget to factor the predicted winds into your plans. Many firefighters have died because of such an oversight.

High- and Low-Pressure Systems

Winds have their origin on the sun. They're the result of solar radiation heating the earth's surface. Since the earth does not heat in a uniform way, this uneven heating creates atmospheric pressure differences called *pressure gradients.*

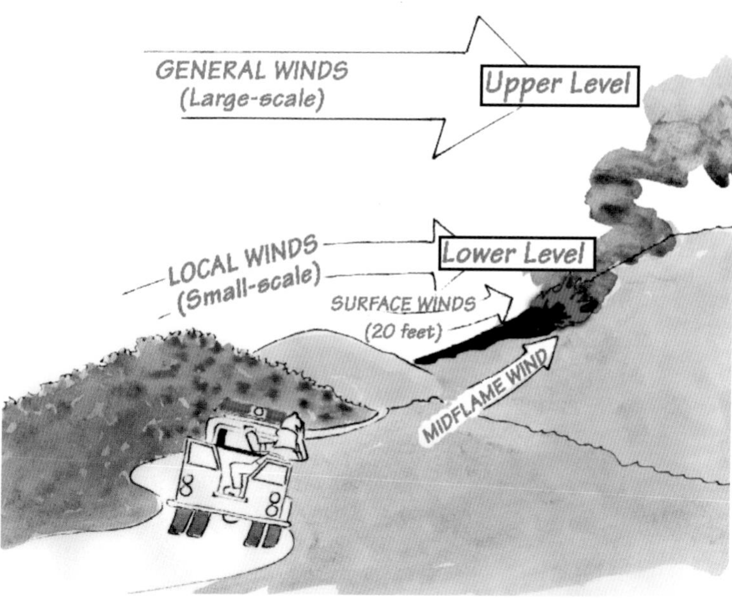

Figure 2.11 The types of winds wildland firefighters must understand.

The atmosphere is constantly trying to equalize and come to a balance. Winds are produced as air from a high-pressure area moves toward a low-pressure area. The greater the difference between the pressure areas, the stronger the winds can be. This process occurs over large areas; sometimes the distance between the highs and lows is hundreds of miles (Figure 2.12).

There are many forces of nature that contribute to the circulation of air masses around the earth. Some of these factors include the rotation of the earth, gravity, and differences in solar heating. As air near the equator rises, air from the cooler polar regions moves into the void. This process develops a natural high-pressure over the poles and low-pressure areas near the equator. Combine this action with the rotation of the earth, and you get an atmosphere that is in a state of constant motion.

LARGE-SCALE PRESSURE SYSTEMS AND FRONTS

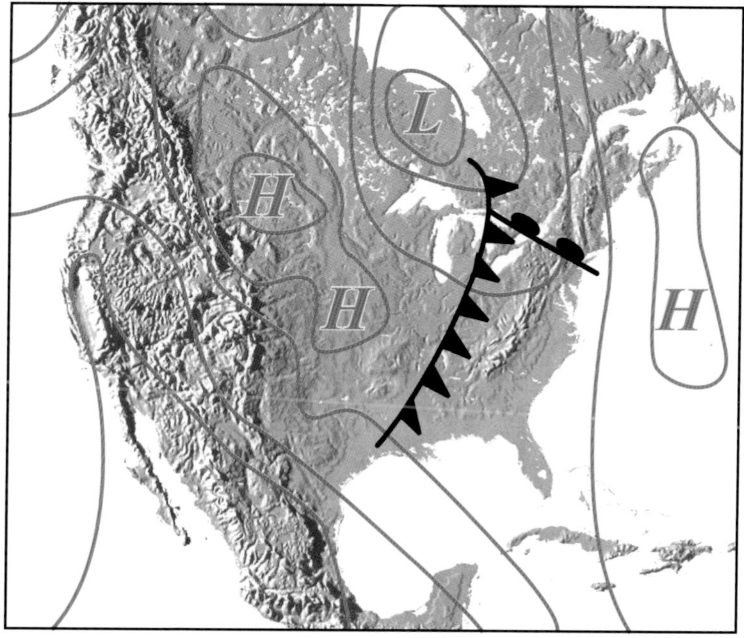

Figure 2.12 Highs and lows are a product of uneven heating of the earth's surface.

The force of the earth's rotation on the moving air causes the air (wind) to rotate in a counter-clockwise direction north of the equator, and clockwise south of the equator. This rotation of the air

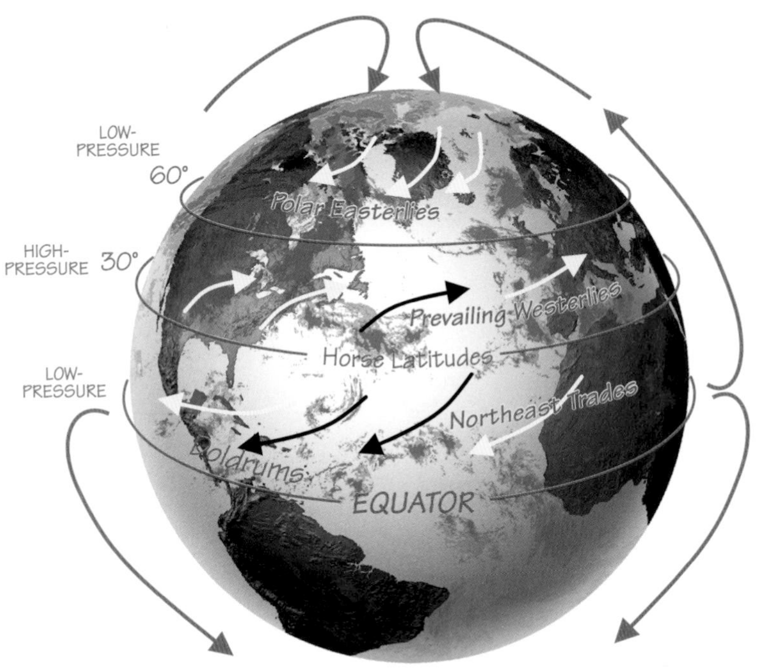

Figure 2.13 General wind circulation created by the rotation of the earth.

(wind) is called the ***Coriolis force*** or effect. This motion also causes the development of regional wind belts. These wind belts in the Northern Hemisphere have several names: the Doldrums, near the equator; the Northeast Trades, between the equator and latitude 30° north; the Prevailing Westerlies from latitude 30° north to near 60° north; and the Polar Easterlies north of 60° north (Figure 2.13).

At some latitudes, the air tends to "pile up" and become belts of high-pressure. These belts or cells are not uniform nor are they stationary. General weather factors are dependent on the location and movement of these high-pressure cells. The maps used to display and track the various pressure cells are similar to a contour map. Points of equal atmospheric pressure are shown with a line, similar to a contour line. These pressure lines are called isobars. The number and closeness of isobars gives an indication of the ***pressure gradient***, or difference in atmospheric pressure between the highs and lows. ***Wind speed is determined by the pressure gradient. Wind direction is determined by the relationship (locations) of the highs to the lows.***

As the warm air near the equator rises, air from the cooler polar regions moves into the void.

The pressure zones (the highs and lows) tend to move from west to east in North America. They are pushed by the upper level westerlies or jet stream winds. Since air moves around a high-pressure cell in a clockwise direction, and counterclockwise around a low-pressure cell, they have a direct effect on the winds that occur at the surface. A good rule of thumb: if you have your back to the general wind, low-pressure is to your left and high pressure is to your right.

Wind speed is determined by the difference in pressure areas.

You can also use weather maps to plot the location of cold and warm air masses. These air masses are delineated by a line called a front. When cold air is displacing warm air, this is a ***cold front***. When warm air is displacing cooler air, this is a ***warm front***. When a warm air mass rides up and over a colder air mass, this area is called an ***occluded front.***

There are also ***troughs*** and ***ridges*** that can be identified on the weather map. An area of low-pressure is called a trough. Ridges are highs, with descending air and a minimum of cloudiness and precipitation. The pressure is higher along the ridge than on either side.

There are semi-permanent "weather centers." These change by the seasons of the year. During the summer, the weather map is dominated by the Pacific and Bermuda Highs and the California Low. During the winter, the map shows the Great Basin and Arctic Highs, with the Pacific and Bermuda Highs shifting their positions. As these "major players" of weather move and shift, they affect all of us (Figure 2.14).

Local Winds

Sea breezes, land breezes, and slope and valley winds are all defined as local winds. These small-scale winds are caused by local temperature differences, and are strongly influenced by terrain. Local winds are just as important for firefighters to understand and consider in their plans as general winds. Local and general winds are related, but the way in which terrain influences local winds can easily obscure the relationship.

SEA-LEVEL PRESSURE PATTERN

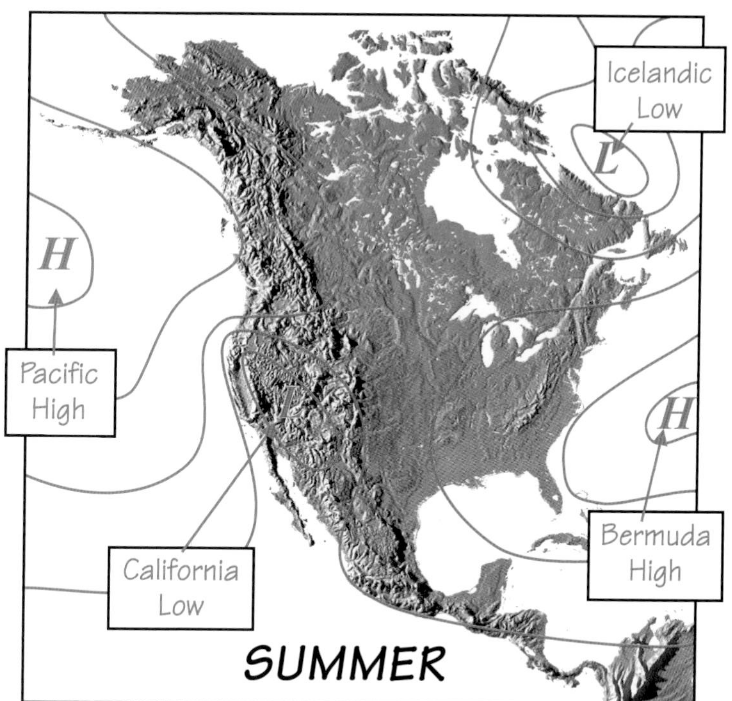

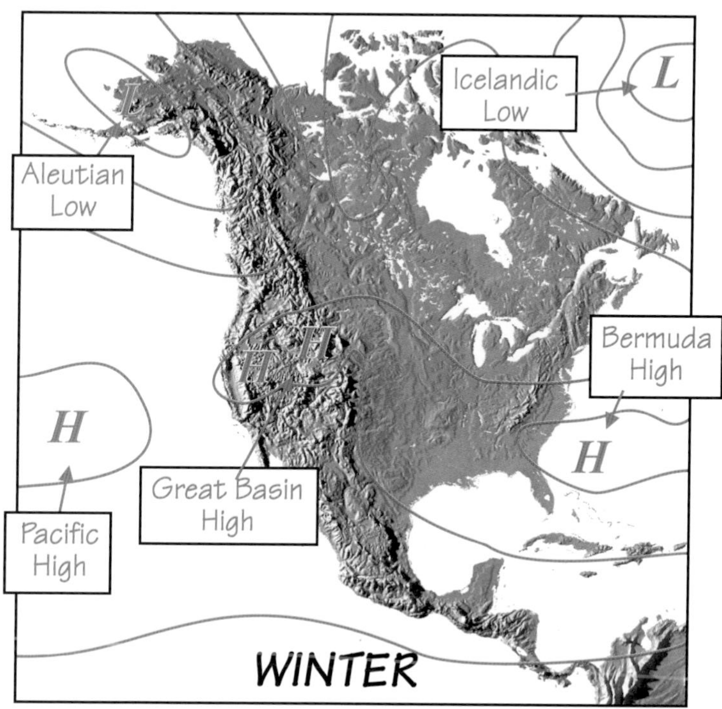

Figure 2.14 The seasonal locations of semi-permanent "weather centers."

Sea breezes and land breezes are essentially the same; they just blow in opposite directions. Sea breezes are light winds that blow from the water onto the land. They usually begin in the mid-morning hours when the land is being heated by the sun and is warmer than the water. As the air over the land also heats up, it rises and draws cooler air from the sea onshore. The sea breeze pushes inland, and becomes strongest in the late afternoon during the period of maximum heating. Sea breezes usually produce winds between 10 to 20 miles per hour. However, near the coast they can be as strong as 20 to 30 miles per hour. Sea breezes are cool, with higher relative humidities. Fog or low clouds may be associated with sea breezes along the coast. Sea breezes can lower the level of fire behavior.

Sea breezes blow from the water onto the land.

The opposite occurs at night when the land is cooler than the nearby water. The wind shifts direction and moves from the land out over the sea. These land breezes usually start 2 to 3 hours after sunset, and usually end shortly after sunrise. They are lighter in speed, with a usual range of 3 to 10 miles per hour (Figure 2.15).

The Daily Cycle of Slope and Valley winds

Slope winds are another local wind phenomenon. They are diurnal (or daily) winds that are produced as the surface of the earth heats or cools. They flow upslope during the day and downslope during the night. This flow of air will continue as long as the air temperature continues to change. Downslope winds are usually gentle (2 to 5 miles per hour), smooth and very shallow (less than 20 feet deep). The principle force behind downslope winds is

DAYTIME
As the sun warms the land, air moves onshore.
Light winds start during mid-morning hours. By late afternoon the winds may be 10 - 20 mph.
NIGHTTIME
At night, the land cools and the local winds reverse direction.
2 - 3 hrs after sunset
3 - 10 mph

Figure 2.15 Sea and land breezes are created when the temperature of the water and nearby land differ.

gravity. Upslope winds are a little stronger, usually between 3 to 8 miles per hour. Relatively calm periods exist during the evening and morning hours when slope winds are changing direction. Slope winds are mainly influenced by aspect, time of year, steepness of slope and current weather conditions.

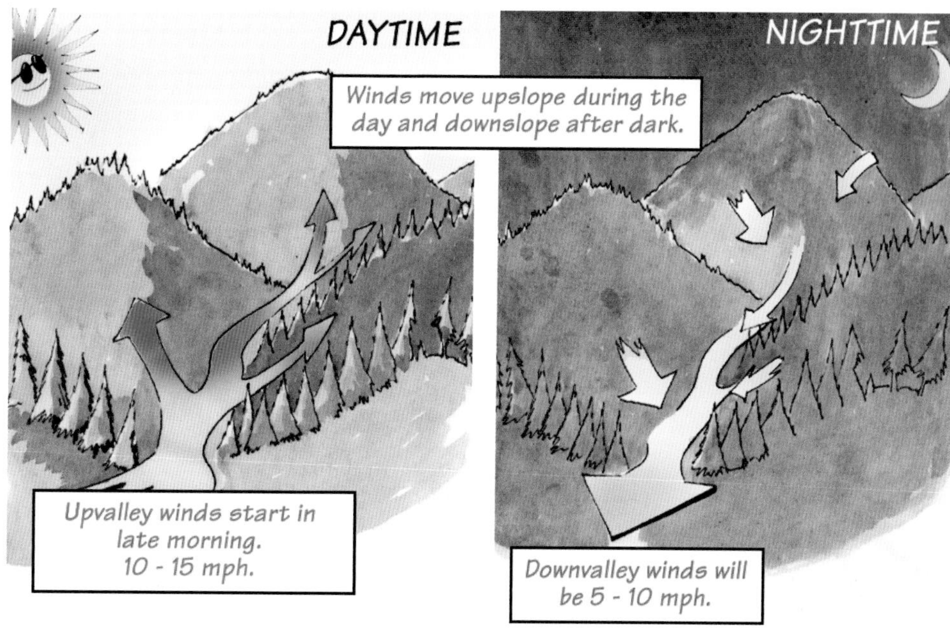

Figure 2.16 Valley heating moves air up valleys and canyons during the day, and cooling air moves down valleys and canyons at night.

Like slope winds, ***valley winds*** are also created by solar heating. The sun warms higher elevations first. As the warmer air at these elevations begins to rise, it draws in cooler air from below to fill the void. These upvalley winds start in the late morning or early afternoon, depending on the size of the valley; winds in large valleys tend to start earlier. Upvalley winds are stronger than upslope winds, ranging from 10 to 15 miles per hour (Figure 2.16).

Normally, afternoon winds are upslope and nighttime winds are downslope.

The winds change direction at night. The earth is still radiating heat back into the atmosphere at lower elevations hours after the higher elevations have begun to cool. Warmer air at the lower elevations continues to rise, drawing cooler air from above. These downvalley winds range from 5 to 10 miles per hour.

Jake says, "Do not underestimate the impact that valley winds can have on fire behavior."

The main difference between upslope winds and upvalley winds is that the upvalley winds do not start until most of the air in the valley is warmed. Slope and valley winds do not happen on all slopes and valleys at the same time and with the same speed. Aspect is the primary factor which drives these differences. Solar heating will generate breezes on east facing slopes long before south or west facing slopes.

Problem Winds

There are four wind conditions that are especially dangerous to firefighters: cold front winds; foehn or gravity winds; winds associated with thunderstorms; and whirlwinds, also referred to as "dust devils" or "firewhirls." Each of these winds can cause dramatic changes in fire behavior, endangering both life and property.

National Center for Atmospheric Research

Figure 2.17 When a front like this moves into your area, the winds will change direction and speed. A dry cold front will not show itself. Be aware of what this may bring to your fire.

Cold Front Winds

Weather fronts are boundaries between two adjacent air masses (Figure 2.17). Weather fronts are centered on low-pressure areas. A cold front is the boundary between two dissimilar air masses when the encroaching air mass is cooler. Wind changes can be dramatic as a cold front passes and may adversely affect fire behavior, especially if the cold air mass is dry. These fronts normally move at about 20 to 30 miles per hour as they are pushed by upper level westerly winds. Figures 2.18 and 2.19 show how the wind directions differ depending on the orientation with the front and low.

Notice how the winds change direction and speed as the front moves across the area. In advance of a warm front, winds are out of the northeast or east. Winds in advance of a cold front shift gradually from southeast to south, and then to southwest. As the cold front passes, winds shift rapidly to the west, then northwest. This change could have a dramatic impact on any fire in the area.

Cold fronts must be monitored closely. The passage of a cold front can create serious danger.

Figure 2.18 Weather fronts moving through the area will affect fire behavior.

Windspeeds increase in strength as a cold front approaches, and usually become quite strong and gusty when the front passes through an area. When the barometric pressure associated with the low-pressure area is very low, there is a greater possibility that strong upper-level winds will "mix" down to the surface. Typically, cold front windspeeds range from 10 to 30 miles per hour, but can be much higher if associated with a strong front.

Winds shift direction and speed when a cold front passes by.

Thunderstorms and precipitation may be associated with cold fronts. Summertime storms are normally dry, and usually will not bring enough rain to provide much relief from the fire season. They also may not "show themselves" with a line of clouds, but they do bring winds that could have a dramatic impact on your safety. Take seriously any prediction of approaching cold fronts in your area. Keep your eye on the sky and track the weather for any hints of change.

> *Jake says, "You can not always see a cold front as it approaches your fire. If one is predicted, be prepared for a shift in wind speed and direction. Cold fronts have killed!"*

Foehn Winds

Foehn (pronounced "*fane*") or gravity wind, is a warm, dry wind that occurs when air spills over high elevations and moves downhill. Gravity winds also occur when a high-pressure system is located in and around mountain ranges. The airflow around the high-pressure system causes some of the air to spill over the higher elevations, resulting in a strong wind moving downhill at high speed. As the foehn wind flows downslope in the atmosphere it is compressed, becoming warmer and drier. This causes the fuels to dry out. As the temperature increases, wind speed may reach 50 to 70 miles per hour. Many of the largest, most damaging and most costly fires in the West have been caused by gravity or foehn winds. These winds have earned reputations and local names: Santa Ana, Chinook and Mono.

WIND SHIFTS WITH THE PASSAGE OF WEATHER FRONTS

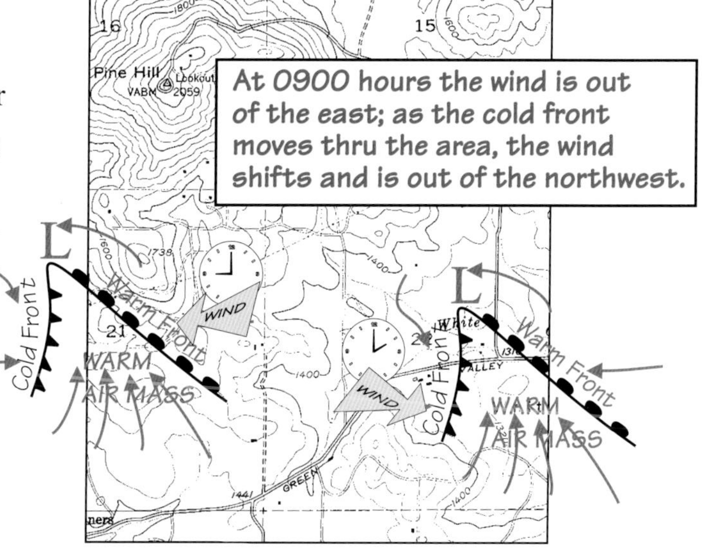

Figure 2.19 Winds will change direction and speed as a front moves through an area. The frontal system moves as it is pushed by strong upper level winds.

Jake says, "Foehn winds are winds with interesting names and well deserved 'bad' reputations."

- ***Santa Ana winds*** occur in the southern portions of California. They are created when there is a strong high-pressure area in the Great Basin and a low-pressure area located above the Pacific Ocean along the Southern California coast (Figure 2.20). This condition would normally produce a strong general wind. However, this wind is heated and dried as it crosses the coastal mountain range and drives toward the ocean. The results are predictable and potentially disastrous. As the Santa Ana Winds approach the ocean, they begin to subside. At this point, the Santa Ana and marine onshore winds begin to "battle" for supremacy. During this battle for supremacy, the winds can abruptly change 180 degrees in direction. Such a change can "ruin your whole day," if you have not planned for it.

Guy M. Delaney

Figure 2.20 Santa Ana winds are pushing this fire toward the ocean. The wind heats and dries as it comes over the mountains.

- ***Chinook winds*** occur in the Rocky Mountains. They blow out of the northwest and up the western slopes, producing clouds over the mountain peaks. As they race down the eastern slope, they become hot, dry and gusting.

- ***North winds*** in northern California are produced in the same way as the other foehn winds, and can sometimes be an indication (or warning) of developing Santa Ana winds.

- ***Eastern winds*** are foehn winds of Washington and Oregon. They are produced as general winds pass over the Cascade Mountain Range.

Thunderstorm Downdrafts

If you are fighting fire in an area where thunderstorms are either present or predicted, be careful! If you are between a fire and an approaching thunderstorm, you are in a very dangerous environment. You should be concerned long before you can see the lightning or hear the thunder. This cannot be stressed enough!

You should be concerned long before you see or hear a thunderstorm.

Like wildland fires, all thunderstorms start small and usually innocently. Thunderstorms (cumulonimbus clouds) develop when there is adequate atmospheric moisture, instability, and a force that will cause the air to rise. The most common lifting force is convection brought on by solar heating. Thunderstorms can rise to 30,000 to 40,000 feet in the west and 40,000 to 60,000 feet in the east (Figure 2.21). When they develop to this size, they can store tremendous energy. This energy is the source for lightning, rain, hail, and strong, gusty winds.

Downdrafts in a thunderstorm develop as air moves down through the center of the cloud. As this air descends in a thunderstorm, it is cooled by evaporation. Cooling makes the air heavier, accelerating its downward movement. This denser air is also pulled by the gravitational force of the earth. Picture the thunderstorm downdraft as a pitcher of cold air being poured onto a table; as the air strikes the table (or terrain) it spreads in all directions (Figure 2.22). These cool, gusty winds (sometimes called a gust front or microburst) can be felt up to 10 miles from the storm center. This effect can reach much farther in the mountains because of the funneling effect created by ridges and canyons. The winds created by the downdrafts contain more moisture than surrounding air, and can range from 25 to 35 miles per hour, with gusts to over 60 miles per hour. The wind speed is the dominating force that you must be concerned with.

A thunderstorm's effect may extend 25 to 30 miles from the actual storm.

Downbursts are caused by thunderstorms collapsing. When this happens, cool air is released in a downward direction. If this occurs over a fire, it can cause real havoc.

Thunderstorms have a "life expectancy" of less than 12 hours. Thunderstorms created by convection generally don't form until afternoon, when there is sufficient solar heating to lift the moisture-laden air to great heights. They begin to break down before midnight. There are several indicators that a thunderstorm is mature and that downdrafts may become a serious concern:

- The base of the cloud begins to "roll" on the downwind side.

National Center for Atmospheric Research

Figure 2.21 If you are fighting fire near a thunderstorm, you must be very careful. They can produce some very dangerous wind conditions,

Figure 2.22 When a thunderstorm breaks down, strong downdrafts develop. It is like pouring cold air from a pitcher.

• Virga begins to "hang" under a darkened, ragged cloud base (Figure 2.23).

• A dust cloud can be seen on the ground as the first gusts of wind reach the surface.

Thunderstorms may either remain stationary or move with the prevailing winds. Indrafts can continue moving into the base of the cloud even as downdrafts are pouring from the center of the storm. The downdraft winds will normally be stronger on the downwind side of the storm as they can combine with the force of the prevailing wind.

Whirlwinds, Dust Devils and Firewhirls

Whirlwinds, dust devils and firewhirls are miniature cyclones that are caused when strong convection currents and updrafts develop. Dust devils are whirlwinds that have "kicked up" dust; firewhirls are whirlwinds that have picked up burning material, or are present in or near the flame front of a fire. They can occur under the following conditions:

Gusty winds, good visibility and clouds "growing" vertically are all signs of unstable air.

• Mostly clear skies - a sign that the air might be unstable.

• Light surface winds - strong surface winds hamper the development of strong convective columns.

• Strong solar heating.

• On the backside of a ridge, or on the lee slope from the prevailing wind.

National Center for Atmospheric Research

Figure 2.23 Virga is an indicator that a thunderstorm has "matured."

Virga under a thunderstorm is an indication the storm is mature...downdraft winds may be in your future.

Whirlwinds, dust devils and firewhirls can range in size from a few feet across and hundreds of feet in height, to over 100 feet in diameter and thousands of feet high. Whirlwinds and dust devils are common in areas that have just burned over; the black ash bed is a better absorber of solar radiation, thus encouraging local heating (Figure 2.24).

Firewhirls contain very hot gases, flames and burning material. They can lift this material to considerable height and deposit it in some very "inconvenient" locations, like across your fireline (Figure 2.25).

Whirlwinds and dust devils are common in areas that have just burned.

The Effect of Topography on Wind

As mentioned earlier, the topography of an area has definite effects on wind speed and direction. These effects can be categorized as mechanical, turbulent, or frictional. There are three mechanical or diverting effects that topography has on wind (directional channeling, venturi, and wave action):

Directional channeling of a general wind occurs when the wind direction changes and flows in the direction of a canyon (Figure 2.26).

Venturi effect occurs as a general wind flows through a pass, saddle or gorge. The wind speed increases as it funnels through the opening.

Kari Greer

Figure 2.24 Dust devils are a definite indication that the atmosphere is very unstable.

Figure 2.25 Firewhirls can carry fire across control lines or cause spotting.

Wave action is caused when a general wind moves across a mountain range at a 90° angle. A wind wave is formed on the lee side of the range, causing very turbulent air. This wind wave can hamper air operations and can be the source of strong up-and-downdrafts. The real impact on firefighting may occur when the wind wave influences the normal daily heating in the basin on the lee side of the mountain range. If the convected air from the basin mixes with the turbulence of the wind wave, the strong upper level general winds may be drawn to the surface and have a spectacular impact on fire behavior. One of the best indicators that a wave action is occurring is the presence of an "indicator cloud" called **altocumulus lenticular**. This unique cloud remains stationary as the air moves through it. It is found over the ridgetop, or on the lee side of the mountain range.

Wind wave action may cause very turbulent air on the lee side of a ridge or mountain range, and create spectacular effects on fire behavior.

There are three types of turbulent effects (lee side turbulence or eddying, canyon or valley winds, and thermal):

Lee side turbulence or ***eddying*** occurs on the lee side of an obstruction like a range of mountains. These swirling winds (eddies) can move either vertically or horizontally (Figure 2.27). They are caused by a combination of wind speed, orientation of the wind to the obstruction, and stability of the atmosphere.

Strong winds in canyons can also cause eddies, which occur to the lee of spur ridges. Eddies are most pronounced during the late afternoon when canyon or valley winds are at their peak (Figure 2.28).

Thermal turbulence is caused by differences in surface heating. This type of turbulence causes low-level winds to become gusty and erratic.

Finally, there is ***frictional drag*** which occurs when wind is slowed by friction as it nears and moves over the surface. The "roughness" of the earth's surface causes this drag. Such things as vegetation, terrain features and man-made structures slow winds. The rougher the terrain or vegetation, the more the wind speed is reduced. Windspeed over open, level ground decreases rapidly as you get close to the surface. On the ground there is virtually no wind.

DIRECTIONAL CHANNELING

Figure 2.26 The general wind may be from one direction, but the winds in the canyons may be from a very different direction.

Wind Speed and Direction

Earlier in the chapter, we discussed four different types of winds that you should be familiar with. Let's review them briefly before we tie in the concepts of speed and direction. ***General winds*** are those winds that are blowing over large areas. These are the winds that are reported in daily weather forecasts. ***Local winds*** are the result of local factors such as temperature

Figure 2.27 The wave action created when a strong upper level wind passes over a mountain range can cause strong updrafts and downdrafts.

Figure 2.28 Note how the smoke and fire are being moved up this draw and that there is some eddying at the ridge line.

differences and terrain (Figure 2.29). ***Surface winds*** are local winds impacted by general winds. Weather forecasters have standardized the definition of surface winds as meaning winds which are 20 feet above the average vegetation cover. ***Mid-flame winds*** are those which occur at the midpoint of the flame height. They have a direct impact on the movement of the fire front, and thus on the speed and direction of fire spread.

General winds impact local winds in several ways. They may combine with locally developed upslope or valley winds. The combined wind speed will be a product of the general wind speed (minus frictional drag) plus the local wind speed. The reverse would occur if the winds were coming from the opposite direction. If an inversion layer is present, the general winds may only impact local winds above the inversion layer.

Winds in canyons can eddy and be from various directions.

How Winds Affect Fire

Specific winds affect a wildland fire in different ways. The winds that most influence the direction in which the ***flame front*** moves are the mid-flame winds. Mid-flame winds are used in calculating the rate of fire spread. Local or general winds, however, may have a greater influence on the overall direction a fire moves, if spotting is a factor. These are the winds that are going to carry burning embers to other locations.

Rough terrain will impact wind speed and direction.

Wind speed and direction can have a dramatic effect on fire behavior. That is why you need to track them. Although wind is also one of the hardest weather elements to predict, you can obtain speed and direction of winds aloft from the weather service. Weather balloons are released and tracked twice a day. As these balloons rise through the atmosphere, they provide very valuable information that can be used to produce a wind profile (Figure 2.30). Note that the wind can be very dynamic; it can blow at different speeds and from different directions at different heights.

Even though you may have a current fire weather forecast, you should develop the skills to accurately predict weather yourself. Make a habit of taking and recording your own weather information. Only in this way can you really get a sense of how the weather is impacting, or can impact, your fire suppression activities. It is also the only way you will be able to detect a potential change in the weather. You cannot rely only on "how the weather feels." At a minimum you must track the temperature, relative humidity, wind speed, and direction to really detect change.

Trevor Wilson

Figure 2.29 Note how the surface wind is pushing this fire to the southwest and the upper level wind is carrying the smoke to the east.

Relative Humidity

Temperature is defined by Webster's Dictionary as "degree of hotness or coldness measured on a definite scale." Temperature is also an indicator of the molecular activity of a substance. Humidity is the moistness or dampness of the air. The temperature of the air is directly related to how much moisture the air can hold. As noted earlier, the surface of the earth is heated by solar radiation each day. However, this heating is not uniform. The amount of heating is governed by the amount of moisture and pollutants in the air; the angle of the sun; and the type, color, texture, etc. of the surface being radiated.

Relative Humidity: the ratio of the amount of water vapor present in the air compared to the greatest amount possible at the same temperature.

Moisture in the atmosphere can appear in three states: solid (snow or ice crystals), liquid (rain) and gaseous vapor (water vapor). We can easily measure the amount of rain or snow that falls to the earth. However, water vapor is measured (or calculated) as the percentage of moisture in the air. The moisture in the air has a direct effect on fuel moisture and fire behavior.

The primary source of moisture in the atmosphere is the ocean. Lakes, rivers, and other smaller bodies of water also contribute to the moisture in the atmosphere, though to a lesser degree. Yet another source of moisture is transpiration of plants—water vapor which is given off by plants during their metabolic processes, similar to the moisture human beings give off when exhaling or sweating.

Air temperature and its corresponding moisture content has a direct effect on how a fire will burn. The hotter the air, the lower the humidity. The hotter the air, the more moisture it can contain. Hot air will draw moisture from dead fuels, lowering their moisture content.

Solar radiation changes the temperature of the fuels, and fuel temperature changes the temperature of the air.

Air temperature also has an effect on firefighters. More safety precautions must be taken when fighting fire on hot days (see the chapter on fireline safety).

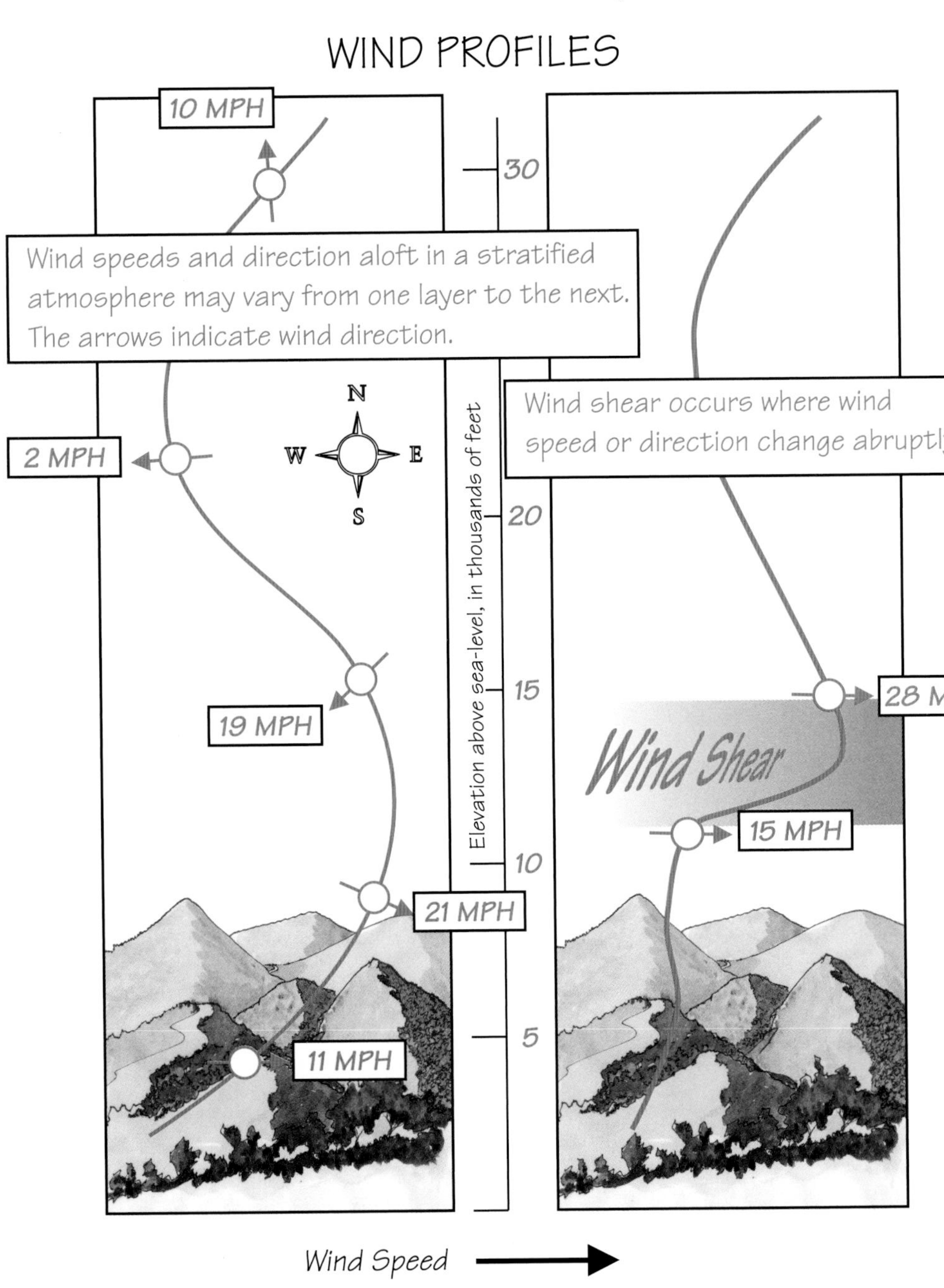

Figure 2.30 Wind profile information can be obtained from the weather forecasters. It tells you what is happening above you.

Relative humidity is most important as a fire-weather factor in the air layer near the ground, where it influences both fuels and fire behavior.

Season, time of day, slope, aspect, elevation, clouds, and vegetation all cause important variations in relative humidity. If the relative humidity is at 30 percent or below, fires will burn freely. If relative humidity is at 10 percent or below, the fire danger is critical and extreme fire behavior is possible.

Relative humidity is defined as "the ratio of the amount of water vapor actually present in the air compared to the greatest amount possible at the same temperature." The importance of relative humidity, which represents how dry or wet the air actually is, is critical and has a direct impact on fire behavior. As temperature goes up, the relative humidity goes down (Figure 2.31). Before you can understand or predict fire behavior, you need to know the actual and predicted relative humidity.

Relative humidity represents how wet or dry the air actually is.

For every 20° increase in temperature, the relative humidity drops by about one half. Conversely, the relative humidity doubles for every 20° decrease in temperature. Knowing this relationship will help you predict how the relative humidity will change during the day or night.

Relative humidity is a key factor in how wildland fuels will burn.

RELATIONSHIP BETWEEN AIR TEMPERATURE, FUEL MOISTURE AND RELATIVE HUMIDITY, AND TIME OF DAY

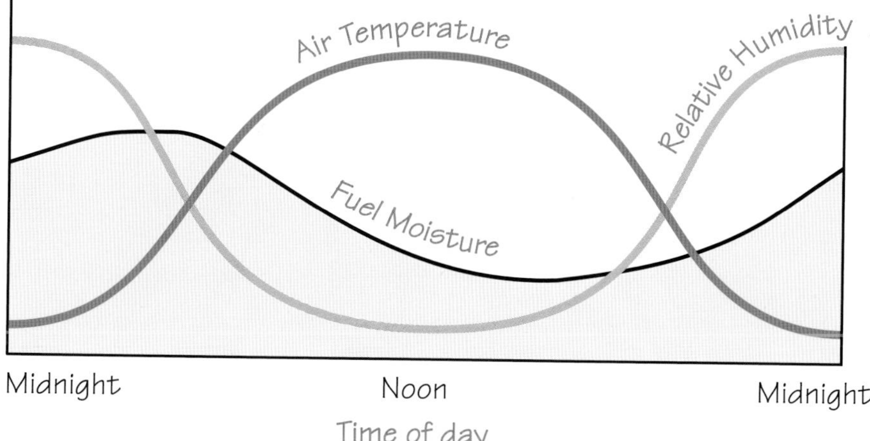

Figure 2.31 As the air temperature increases, the relative humidity drops. The fuel moisture decreases, but at a slower rate.

Fine fuels are especially responsive to changes in relative humidity.

Moisture in the form of water vapor (measured as relative humidity) is always present in the air. The amount of moisture that is in the air affects the amount that is in the fuel. Moisture is absorbed by the air from wetter fuels and from the air by drier fuels. The moisture content of fuels is an important consideration in firefighting. Wet and most green fuels will not burn freely. The drier the fuel, the easier it will ignite and burn. Air is usually drier during the day than it is at night. Fires generally burn slower at night because moisture is absorbed by the lighter fuels from the damp night air.

Jake says, "Track relative humidity...it will give you an indication that fire behavior may be a changin'."

Absorption of moisture by the fuels, downslope winds, lower temperatures, and other day/night weather differences can aid the firefighter at night. This explains why a fire that under normal conditions resists suppression efforts during the day, can often be easily suppressed after dark. Special efforts should be made to contain a fire before more unfavorable burning conditions recur the following day. If a fire is difficult to control during the day, an all-out effort must be made at night.

The amount of moisture in a specific air mass is constant. The relative humidity will vary as the temperature changes.

Dew Point is the temperature at which the air is 100% saturated with water vapor, and either begins to condense into clouds or fog, or begins to form dew. If the air temperature goes up, the dew point remains constant, but the air becomes drier. When the dew point temperature is high, the air has much more moisture in it than if the dew point temperature is low (Figure 2.32).

If you know the dew point, you can determine how much moisture there is in the air at a given temperature. An air mass with a dew point of 75° has considerably more moisture in it than an air mass with a dew point of 45° (Figure 2.33). You should be concerned if the dew point temperature is decreasing over time. This means the air is becoming drier, and this change may impact fire behavior. If you know the dew point and temperature, you can determine the relative humidity of an air mass. If you know the dew point or relative humidity, and the predicted afternoon temperature, you can predict the afternoon relative humidity.

Dew point is the temperature at which the air is saturated or at 100 percent.

Warm air can hold more moisture. So, as the air temperature rises during the day, the relative humidity will decrease and have an impact on fire behavior. Conversely, as the temperature decreases during the evening and night, the relative humidity will increase.

Terrain, fuel type, and aspect vary from fire to fire, or even on various parts of a large fire. Therefore, temperature and relative humidity, which are influenced by these factors, are seldom constant throughout the fire ground. On the other hand, the air mass and its associated water content can normally be considered a constant. So how do all of these various factors impact one another?

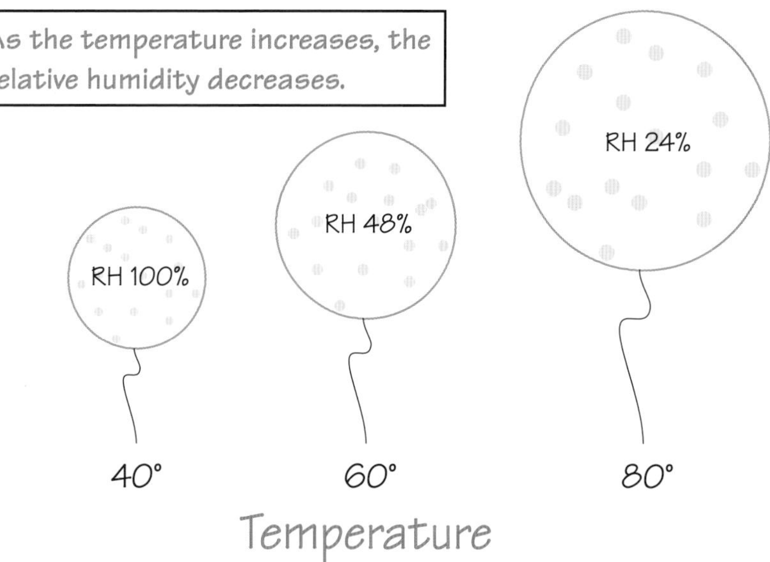

Figure 2.32 As the temperature rises, the dew point (temperature at which the air is saturated) remains constant, but the percentage of water vapor per volume of air decreases.

- ***Terrain*** - Aspect and elevation affect both temperature and relative humidity. Studies indicate that southern aspects are generally five degrees warmer than northern aspects, primarily due to more direct solar radiation. Fuel cover variations may increase this difference. Since southern aspects are warmer, the relative humidity will be less. Generally, daytime temperatures decrease with altitude, with a corresponding rise in relative humidity. This trend usually reverses at night. As cooler air flows downslope and collects in the valleys, an inversion develops. In the inversion, colder air with higher relative humidity is layered below warmer, drier air. (Inversion layers will be discussed in greater detail later in this chapter.)

If you know the dew point and predicted temperature, you can predict the relative humidity.

- ***Vegetation*** - The type, nature and density of vegetation affects temperature and relative humidity by absorbing or reflecting incoming solar radiation. This effect involves both incoming solar radiation during the day and outgoing radiation at night. Low, sparse vegetative cover has much less of an effect than dense, tall vegetation. In a dense forest with a continuous canopy, the crowns of the trees effectively become the contact surface for solar radiation. Air temperatures within a dense forest are likely to be 5° to 8° cooler than temperatures in

The highest daytime temperatures and lower relative humidities are found near the crowns of the trees.

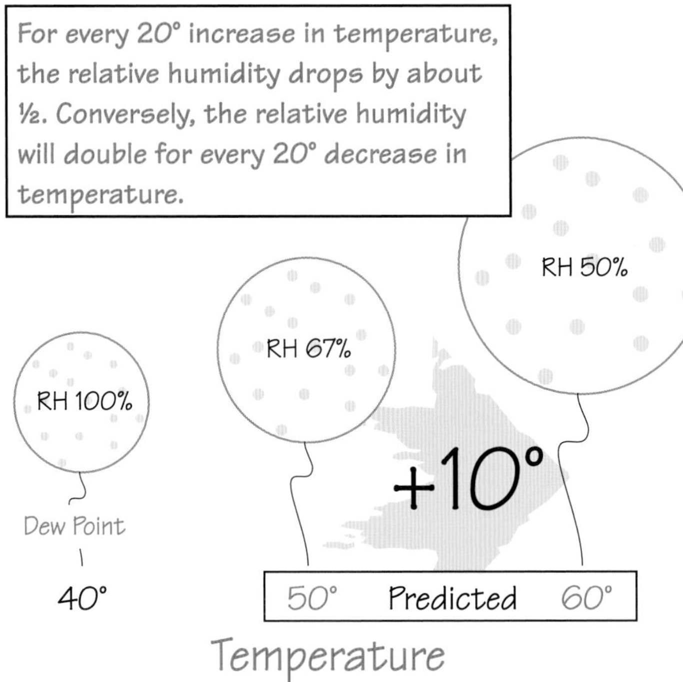
PREDICTING RELATIVE HUMIDITY IF YOU KNOW THE DEW POINT AND PREDICTED TEMPERATURE
For every 20° increase in temperature, the relative humidity drops by about ½. Conversely, the relative humidity will double for every 20° decrease in temperature.
RH 100%
RH 67%
RH 50%
+10°
Dew Point
40°
50°
Predicted
60°
Temperature

Figure 2.33 The dew point of this air mass is 40 degrees. If the temperature is predicted to be 60 degrees later in the day, the percent of moisture per volume of air will be reduced by 50 percent.

nearby cleared areas; the relative humidity may be about 5% to 20% higher. The highest daytime temperatures and lowest relative humidities are found near the crowns of the trees. There will be a gradual decrease in temperature (and rise in relative humidity) as you move closer to the forest floor. The reverse happens at night. A dense forest slows the rate of outgoing radiation, and it will remain warmer longer under the forest canopy than in open areas nearby.

• ***Clouds*** - Clouds affect temperature and relative humidity in much the same way as vegetative cover. The clouds reflect incoming radiation during the day, decreasing the amount of heat that reaches the earth's surface. A thick cloud cover can lower summer daytime temperatures as much as 15° to 20°, with a corresponding 15% to 20% rise in relative humidity. Clouds also reflect outgoing radiation back to earth at night, maintaining higher temperatures in the evening (Figure 2.34).

• ***Wind*** - Wind affects temperature and relative humidity by mixing air near the ground with air at higher altitudes. By mixing air from different elevations it helps reduce extremes in temperature. If the wind is blowing, daytime temperatures will be cooler with higher relative humidities; nights will be warmer with lower relative humidities.

SOLAR RADIATION DISTRIBUTION
Thick cloud cover can lower summer daytime temperatures as much as 15° to 20°, with a corresponding 15% to 20% rise in relative humidity.

Figure 2.34 Cloud cover reflects incoming solar radiation so it does not reach the earth's surface, and holds down the amount of outgoing radiation that can leave the earth at night.

Atmospheric Stability

Up to this point, we have discussed temperature and its relationship to relative humidity in a two-dimensional way; temperature goes up and relative humidity goes down, or vice versa. But actually, it's more complicated than that. Atmospheric pressure is also a player in how things change. The temperature we have been discussing is the temperature at ground level. Now we will discuss temperature and winds at higher elevations and their potential impact on fire behavior.

Atmospheric stability is defined as the resistance of the atmosphere to vertical motion. You've seen satellite images showing cloud movement. This is a two-dimensional view of cloud formations from space. When they show the development of severe thunderstorms during tornado season in the south, you can begin to see and appreciate vertical motion at its best (or worst).

Several factors affect the movement of air masses. They are the earth's rotation, and changes in temperature and pressure. Wind is defined as the horizontal movement of air. It is caused when air from a high-pressure area moves toward an area of low pressure. The vertical movement of air is controlled by the stability of the atmosphere. Depending on the stability of the atmosphere, air can rise, sink, or remain at the same elevation. ***Stable air resists vertical movement; unstable air encourages it.***

Clouds reflect incoming radiation during the day and decrease the amount of heat that reaches the earth's surface.

Lapse Rates

Lapse rate (or adiabatic lapse rate) is defined as the change in temperature caused by a change in altitude. However, lapse rate is affected both by elevation and the moisture in the air. We already know that temperature decreases as elevation increases. Moisture in the air affects how quickly the temperature changes.

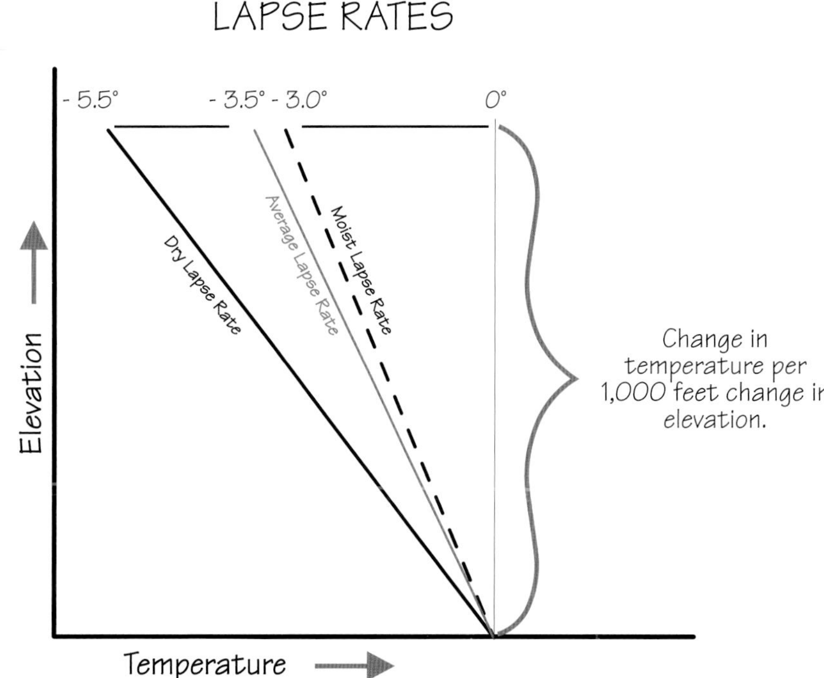

Figure 2.35 Temperature decreases as elevation increases. The moisture in the air has an effect on the rate of temperature change.

There are three types of lapse rates: dry, moist and average. The temperature lapse rate of dry air is ±5.5° per 1,000 feet of elevation change. This is a constant, and is called the ***dry lapse rate*** (Figure 2.35). The ***moist lapse rate*** is about ±3° per

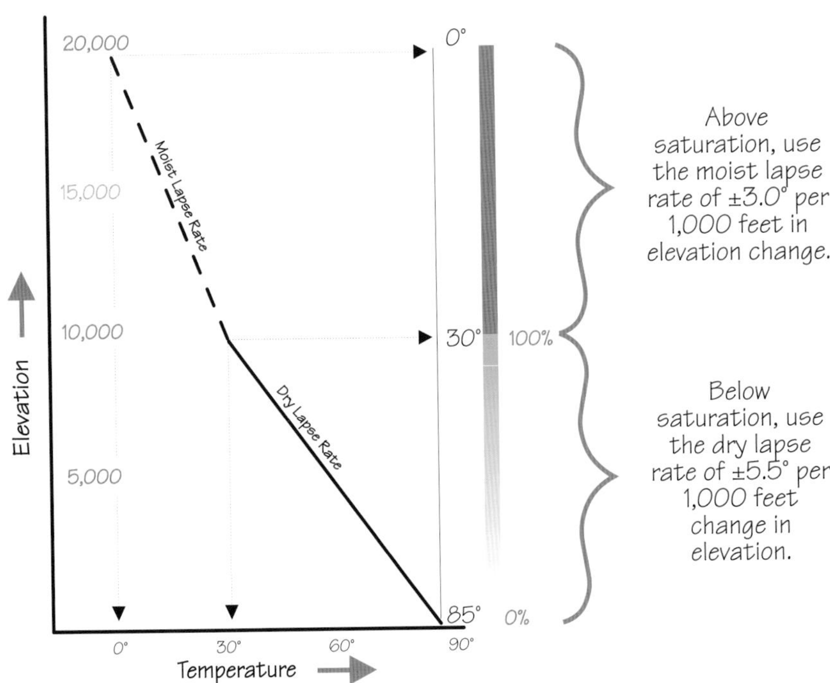

Figure 2.36 Once the atmosphere reaches saturation, the moist lapse rate controls cooling.

1,000 feet of elevation change. Remember that when air cools, its relative humidity increases. Once the relative humidity reaches 100 percent, the rate of cooling decreases. Therefore, when air rises, its temperature changes at the dry lapse rate until it reaches 100 percent relative humidity, or saturation. If it continues to rise, it will cool at the moist lapse rate (Figure 2.36).

The atmosphere may not have a temperature distribution that fits the theoretical dry or moist lapse rate. The ***average lapse rate*** is about ±3.5° per 1,000 feet of elevation change. Lapse rate temperature changes by time of day, depending on the amount of solar heating and the vertical temperature differences of the air mass. Since you may not know the elevation at which the atmosphere is saturated, you can use the average lapse rate when projecting a temperature above or below an observation site.

Wind is defined as the horizontal movement of air.

Stable and Unstable Atmospheric Conditions

In the broad scheme of meteorology, the motions that create weather are an attempt by the atmosphere to gain balance (i.e., "highs" versus "lows," hot air versus cool air, dry air versus moist air). Part of this balancing act involves the vertical movement of air masses. If "things" are in balance, there is little vertical movement of air. This condition is called stable air. If "things" are not in balance (especially the temperature of the air mass), the air is considered unstable. Since hot air rises, if a mass of air is heated, it will rise until it reaches a level where the surrounding air is of the same temperature.

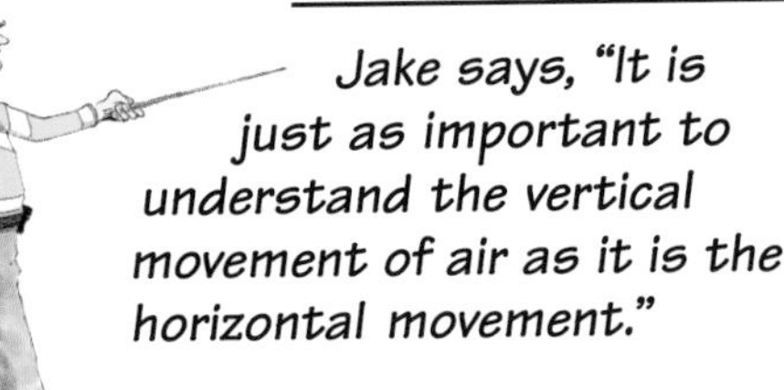

Jake says, "It is just as important to understand the vertical movement of air as it is the horizontal movement."

Of course, it is a little more complicated than that. Rising air is cooled at a lapse rate of between 3.0° to 5.5° per 1,000 feet in elevation. So, if we compare the actual lapse rate of the air with the dry adiabatic lapse rate (±5.5° per 1,000 feet) we can determine whether the air is stable or unstable (Figures 2.37). The lapse rate indicators are as follows:

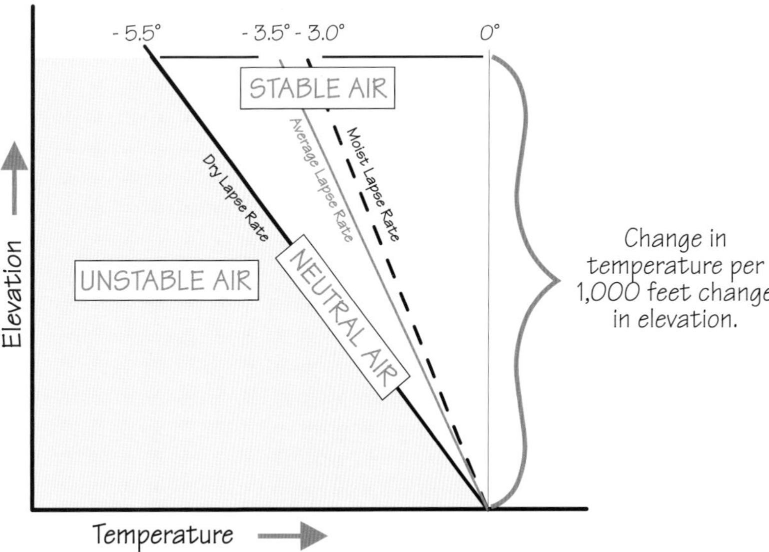

Figure 2.37 The lapse rate is an indicator of whether the air is stable or unstable.

- If the temperature decrease is GREATER than 5.5° per 1,000 feet, the air is UNSTABLE and there will be vertical movement of air. This is important because unstable conditions have a direct effect on fire behavior. Winds may be more pronounced and spotting may occur over greater distances.

- If the temperature decrease is LESS than 5.5° per 1,000 feet, the air is STABLE and there will be little vertical movement of air. Fire behavior will remain about the same or decrease.

- If the temperature decrease is 5.5° per 1,000 feet, the air is NEUTRAL and there will be little vertical movement of air. This is usually a short-term condition as the air moves between stable and unstable states.

Atmospheric stability is defined as the resistance of the atmosphere to vertical movement.

Remember, stable air resists vertical movement. If an air mass is forced to lift as it goes over a mountain range, it will tend to settle back to its original level on the other side. If the atmosphere is unstable, any air mass that is lifted will tend to rise until its temperature decreases (at the rate of 5.5° per 1,000 feet) to equal the air temperature around it (Figure 2.38).

Normally, we attribute the changing stability of an air mass to changes in surface heating and cooling. Cooling of the upper air can also cause the atmosphere to become unstable, while warming aloft will create more stable conditions. This is easily understood if you

Figure 2.38 Vertical movement of air differs depending on whether the air is stable or unstable.

remember that weather changes are driven by the desire of air masses to equalize. If the upper levels of the atmosphere are colder, the difference between the surface air and the air aloft is greater. The warmer air will attempt to rise at a faster rate in order to heat the colder air, thus contributing to unstable conditions.

Unstable air can intensify fire behavior by ***increasing***:

Unstable air intensifies fire behavior.

- The chance of dust devils and firewhirls. Dust devils are common under unstable conditions, and can scatter fire and increase the chance of spot fires. Firewhirls are less common.

- The potential for gusty surface winds. When the air is unstable, strong winds aloft can be brought down toward the surface and may produce strong, gusty surface winds. This can occur very quickly and may catch you by surprise.

- The height and strength of convection columns. When the atmosphere is unstable, convection columns will rise to greater heights and with greater speed. This causes stronger inward drafts (winds) at the surface. Fire behavior will be intensified.

- The chance of spot fires. Tall, well-developed convection columns lift more firebrands, thus increasing the potential for more spotting at longer distances.

Visual indicators are the easiest way to recognize whether the air is stable or unstable (Figure 2.39).

Haines Index

The Haines Index can be used to gauge the relative fire danger. The Haines Index calculates the temperature, dew point and dryness at two levels in the atmosphere, and gives a rating between 2 and 6. The drier and more unstable the atmosphere, the higher the Haines Index. The potential for large fire growth is very low when the index is 2 or 3, low when 4, moderate when 5, and high when 6.

Visible indicators of stable/unstable air

Stable	Unstable
Clouds in layers, no vertical motion; stratus-type clouds.	Clouds grow vertically; cumulus-type clouds.
Smoke column drifts apart after limited rise.	Upward and downward currents; smoke rises to great heights; gusty winds.
Poor visibility in lower levels due to accumulation of haze and smoke.	Good visibility.
Steady winds.	Dust deviles and/or firewhirls.
Cooler at lower elevations: fog layers.	Warmer at lower elevations.

Figure 2.39 Visible indicators of stable and unstable air.

If we look at how the Haines Index compares to actual fire history, statistical analysis shows that only 10 percent of large fires occurred when the index was 2 or 3. The Haines Index was usually in this very low range approximately 62 percent of the fire season. In other words, for more than half of the fire season, the chance of a large fire occurring is low. In contrast, 45 percent of the large fires occurred when the Haines Index was at 6. This high rating existed about 6 percent of the fire season.

Stable air resists vertical movement.

The Haines Index is not something you will be calculating in the field. However, some fire weather forecasters are including it in their forecasts.

Inversion Layers and Thermal Belts

An inversion layer is a layer of very stable air where, contrary to normal behavior, the ***temperature rises as the altitude increases***. Temperatures in an inversion layer may increase as much as 15 degrees per 1,000 feet in altitude. A thermal belt is the "warm area" on a mountain slope associated with an inversion layer.

If an air mass is "lifted" going over a mountain range, it will tend to return to its original level.

An inversion layer acts as a lid or blanket over cooler air. Air rises when warm air is surrounded by cooler air. Cool air will not rise through the warm air. Because of this, inversion layers severely limit vertical motion in the atmosphere. Smoke generated by a fire will rise to the inversion layer (or to a point where temperatures equalize), then flatten out and spread horizontally because it has lost its ability to rise (Figure 2.40). There are three types of inversion layers. They are categorized by how they are formed and whether they are located at the earth's surface or aloft.

The Haines Index is a measure of air stability. Many large fires occur when the air is unstable, and the Haines Index is 6.

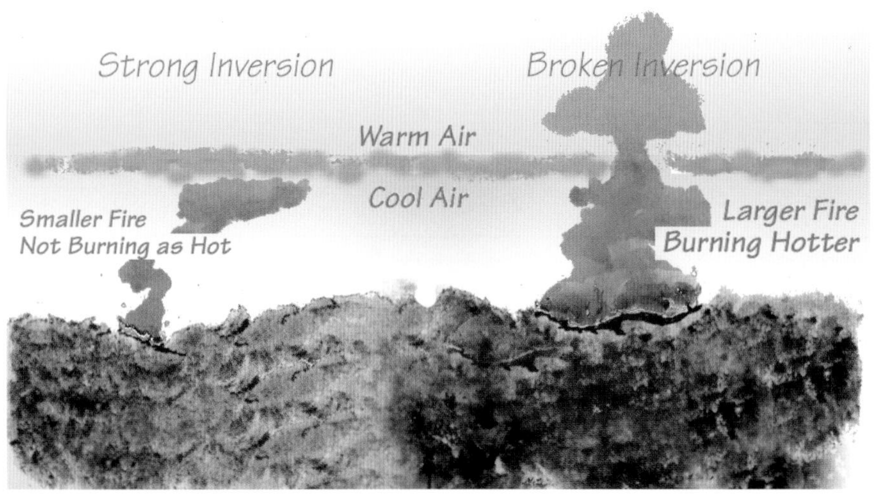

Figure 2.40 A strong inversion is when warmer air "caps" cooler air. Fires under such an inversion will usually burn slower. If the fire "breaks the inversion" a draft is set up and the fire behavior will increase.

Nighttime Inversion Layers

Nighttime (or radiation) inversion layers are the most common type. They are created when heat is radiated away from the earth's surface at night, and the surface air is subsequently cooled by cold land masses. This occurs mainly in the inland valleys and mountainous areas. Conduction is also a factor in the formation of nighttime inversion layers. Air at the earth's surface cools faster than it does at higher levels (air is a poor conductor of heat), thus creating a condition where there is cooler air below warmer air. The layer closest to the earth deepens as the night progresses. This is why the inversion layer is thickest in the early morning hours. During this time, the air is very stable because the different air temperatures of the two layers will not support vertical movement (Figure 2.41).

Jake says, "Listen for the Haines Index number during the forecast. If it is 5 or 6, watch out."

Inversion layers are deeper and more pronounced in mountainous areas and valleys, because as the air cools on the slopes, it "flows" into the valley below. This helps deepen even further the thickness of the "inverted air mass." Since cloud cover

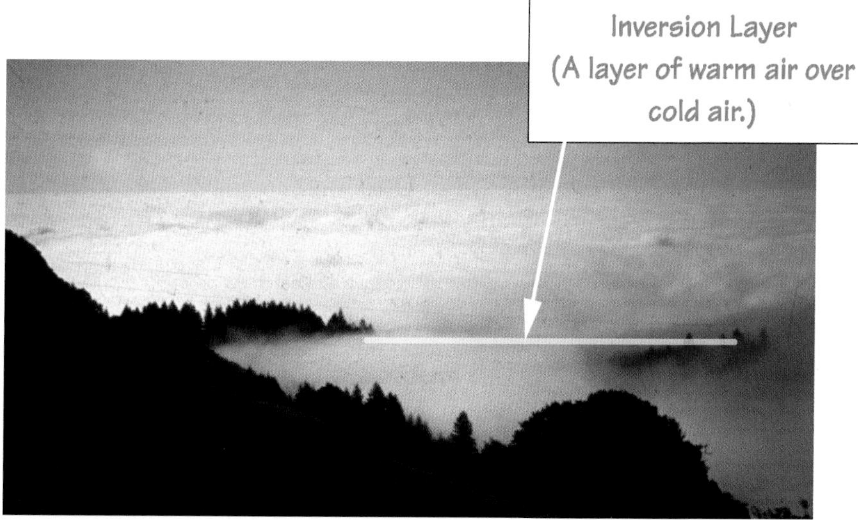

Figure 2.41 Inversion layers form when cooling of the earth at night results in cooler air underneath warmer air. Inversion layers are thickest in the early morning.

In an inversion layer the temperature RISES as the altitude increases.

reduces the amount of radiated heat, and wind mixes and moves air, inversion layers are shallower on cloudy, windy nights. Stable air conditions can develop under cloudy skies, but their degree of development is less than with clear or fair skies. A good guideline to remember is that cooling from below promotes stable air, while heating from below promotes unstable air (Figure 2.42).

Figure 2.42 Cooling from below promotes stable air; heating from below promotes unstable air.

Inversion layers usually begin to weaken after sunrise when the sun begins to warm the earth's surface. As the earth's surface warms, so does the air closest to the surface. Both conduction and convection take place. Once the air on the surface becomes warmer than the air above it, it will begin to rise. Inversions usually break down before noon. When an inversion layer breaks down allowing more vertical air movement, fire behavior increases. This change may be

The presence of an inversion layer indicates that the air is very stable.

either gradual or sudden. Winds may increase while the relative humidity decreases.

If you're fighting fire in an area where inversion layers are present, you should monitor the weather on at least an hourly basis so that you can detect potential changes in the weather factors and be prepared to respond to associated changes in fire behavior. You don't want to be surprised by an event that is very predictable.

Nighttime inversions occur mainly in the mountains.

Marine Inversion Layers

Marine inversion layers have their roots in the ocean, thus they are found in coastal areas. Cool, moist air from the ocean moves over nearby low-lying lands. This layer of air will vary in depth, from several hundred to several thousand feet. The thicker the layer, the farther inland it will invade. Because the marine layer is made up of cool, moist air, it moves in under warmer, drier and relatively unstable air. This intrusion of cool, moist air may persist during the day, but is generally strongest and most noticeable at night. Fog and stratus clouds often form in the marine layer at night, and move inland into coastal basins and valleys. Fog usually forms if the marine layer is shallow; and low stratus clouds if the layer is deep.

The intrusion of a marine layer is usually helpful during firefighting. The cool, moist air reduces fire behavior. However, a marine layer can cause a problem when firefighters are trying to burn-out an area at night, because this same cool, moist air will interfere with (or prevent) a good, clean burn. Another time that marine air can cause real problems, is when a Santa Ana or foehn wind is degrading. This degradation occurs over several hours. During this time, you may have what are called "battling winds," that are pushing back and forth over an area. This can cause real havoc for firefighters because there can be relatively strong wind shifts (up to 180° in direction) every few minutes.

Inversion layers are normally "thickest" during the early morning hours.

Subsidence Inversion Layers

Subsidence layering is a slow sinking motion of a high level air mass over a broad area. It is associated with the presence of a high-pressure system. Once again, this is a slow process, and only occurs when the high-pressure system is stationary for several days. This situation usually occurs during the summer and fall months.

As this air mass sinks, it compresses and becomes warmer, and the relative humidity falls dramatically. The inversion layer will affect peaks and ridges first. If it persists, it could have a significant impact on fire behavior. Subsidence inversion layers can hold smoke in a wide area for days and even weeks. They also prevent nighttime relative humidity recovery. Therefore, fuels in the area will become much drier than normal.

Another effect of subsidence can be foehn winds. Foehn winds occur on the lee slopes of prominent mountain ranges when stable air pours over ridges and through canyons, creating very strong, dry winds.

Cooling from below promotes stable air; heating from below promotes unstable air.

Thermal Belts

A thermal belt is that area of a mountain slope where a nighttime inversion layer "bathes" it with warmer and drier air. This area on the mountain will experience the least amount of temperature swing between nighttime lows and afternoon highs. The temperature above and below the thermal belt will be cooler than the area within the thermal belt. The thickness of a thermal belt is controlled by the depth of the associated inversion layer (Figure 2.43).

The presence of fog indicates stable air.

Figure 2.43 The existence of a thermal belt will affect fire behavior.

Knowing the location of the thermal belt is important to you as a firefighter. The fire behavior in this area will be much more intense than in areas above or below it. This difference in fire behavior is normally caused by warmer temperatures and lower relative humidities. Since inversion layers are associated with stable air, the winds in thermal belts are normally light and steady.

Knowing the location of the thermal belt is important. Fire behavior in this area will be much more intense.

Clouds

Clouds are the result of weather factors that may also have an effect on fire behavior. You should look up and around often to assess the clouds in the area. They may give you a clue as to why something is happening, or what is about to happen. Read the clouds as you would a contour map.

"Battling winds" should be of concern on the fireline. Winds can quickly change directions.

Clouds are made up of water and/or ice particles in the atmosphere, and result from the condensation of water vapor in the air. Most clouds are formed when a rising air mass cools to the point of saturation. Low level clouds are formed when warm, moist air is cooled to the dew point (or saturation) by passing over cooler air. This cooling takes place near the ground. However, if the winds are strong enough, the mixing of air masses may produce clouds hundreds or thousands of feet above the surface. Clouds can also be formed by the addition of water vapor to an air mass. This usually occurs as dry air passes over warm water and water vapor is evaporated into the air. Rain falling through a cool air layer will also add moisture to the air mass.

Clouds are classified by their structure and their height in the atmosphere. Cirro clouds (cirrus, cirrocumulus, and cirrostratus) are high clouds with bases above 20,000 feet. Alto clouds (altostratus, altocumulus, and nimbostratus) are middle level clouds with bases between 6,500 to 20,000 feet. Strato clouds (stratus and stratocumulus) are low level clouds with bases at or below 6,500 feet. Cumulus and cumulonimbus clouds are vertically developed clouds which may have bases between 1,500 to 10,000 feet. Fog, on the other hand, is essentially a cloud that touches the surface of the ground.

Subsidence inversions are slow to develop, but may persist for days and even weeks.

Cloud Identification

There are many types of clouds. However, there are six cloud formations that are especially important for the firefighter to know and understand. These are "indicator" clouds, which will help you predict future fire behavior.

Clouds can predict fire behavior!

Cumulonimbus or thunderstorm clouds - Cumulonimbus clouds begin as small, fair-weather cumulus clouds that grow as the atmosphere becomes more unstable. These cloud formations can occur individually as "air mass thunderstorms" or as a "line or wall" of thunderstorms associated with a cold front. Thunderstorms may produce strong, gusty winds, sometimes gusting to as much as 60 miles per hour. Thunderstorms can have a very devastating impact on a fire. By the time you notice them, they usually have already begun to influence fire behavior. Therefore, you must heed and monitor fire weather forecasts warning of impending thunderstorm activity (Figure 2.44).

National Center for Atmospheric Research

Figure 2.44 Cumulonimbus clouds.

Thunderstorms bring windy conditions.

Cirrostratus clouds - These are very high, wispy clouds that frequently precede a warm front. These clouds are formed when the tops of thunderstorms are blown off by strong winds. They may be carried for 500 miles or more. These cloud formations thicken and lower as they approach. They indicate that rain is possible within the next day or two (Figure 2.45).

Altocumulus castellanus clouds - These are mid-level clouds that form turrets or "little towers," usually arranged in lines. They indicate instability and increased moisture. When seen in the morning, these clouds are a strong indicator that thunderstorms are possible later in the day (Figure 2.46).

Altocumulus floccus clouds - These clouds can be identified by their white or gray colored scattered "tufts," with rounded and slightly bulging upper parts. These

National Center for Atmospheric Research

Figure 2.45 Cirrostratus clouds.

Figure 2.46 Altocumulus castellanus clouds.

National Center for Atmospheric Research

Figure 2.47 Altocumulus floccus clouds.

National Center for Atmospheric Research

Figure 2.48 Altocumulus lenticularus clouds.

clouds resemble very small ragged cumulus, and are often accompanied by trails of virga (rain which doesn't reach the ground) from their bases. Like the altocumulus castellanus clouds, they are an indicator of instability, increased moisture and potential thunderstorm activity (Figure 2.47).

Altocumulus lenticularus clouds - These "lens-shaped" clouds are formed as strong, mid-level winds cross over the top of north/south oriented mountain ranges. They are usually stationary, and positioned over the lee side of a mountain range. Altocumulus lenticularus clouds are indicative of high wind speeds at the altitudes at which they occur. However, these winds can descend or "surface," producing strong ground level winds that can cause long-range spotting. A visual indicator of the possibility that these winds may surface is a sheared smoke column (Figure 2.48).

Stratus clouds - These occur as a single, gray, fairly uniform and featureless layer of low-level clouds. Sometimes they are dark and threatening, although at most, they only produce a drizzle or light rain. Stratus clouds are an indicator of stable air (Figure 2.49).

Thunderstorm Development

A thunderstorm is defined as "a storm accompanied by lightning and thunder." Thunderstorms also bring with them strong, gusty winds that may have a serious impact on fire behavior. You can usually detect a developing thunderstorm by recognizing characteristic cloud formations.

Figure 2.49 Stratus clouds.

There are four ways thunderstorms are formed. These are classified as lifting processes. Each of the processes differ in what causes air to rise and develop into a potential thunderstorm. Most of the time several of these lifting forces are involved with the formation of a thunderstorm.

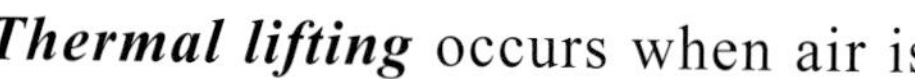

Thermal lifting occurs when air is heated. If there is adequate moisture in this rising air mass, it will form a cloud when it reaches saturation. Thermal lifting is most prevalent during the summer and occurs in areas of flat terrain being heated by solar radiation. It can also occur when the air is heated by a large wildland fire. The formation of cumulus clouds over a fire is an indication of extreme fire behavior and the potential for a thunderstorm (Figure 2.50).

A thunderstorm in your area WILL impact your fire! Be careful.

Orographic lifting occurs when an air mass that is moving horizontally encounters a mountain range and is forced upward. The air mass cools as it moves upward, and forms a cloud when it reaches saturation. Orographic and thermal lifting may work together in the formation of a thunderstorm (Figure 2.51).

THERMAL LIFTING

Figure 2.50 Thermal lifting is the result of heat generated by solar radiation or a large wildland fire.

Frontal lifting occurs when a cold front moves under a warmer air mass and forces it upward. Again, if the air is forced high enough, a cloud will form when the air becomes saturated (Figure 2.52).

Most of the time more than one of the "lifting mechanisms" are involved in the development of a thunderstorm.

Convergence lifting is always associated with a low-pressure system. As the atmosphere attempts to equalize, air moves from high-pressure zones toward low-pressure zones. If more air moves (horizontally) into a low-pressure area than moves out, the "excess air" is forced upward. This lifting cools the rising air, and when it reaches saturation, a cloud is formed. Convergence lifting is the reason that low-pressure systems usually include cloudy conditions (Figure 2.53).

As air moves over a mountain range and is lifted, it drops its moisture.

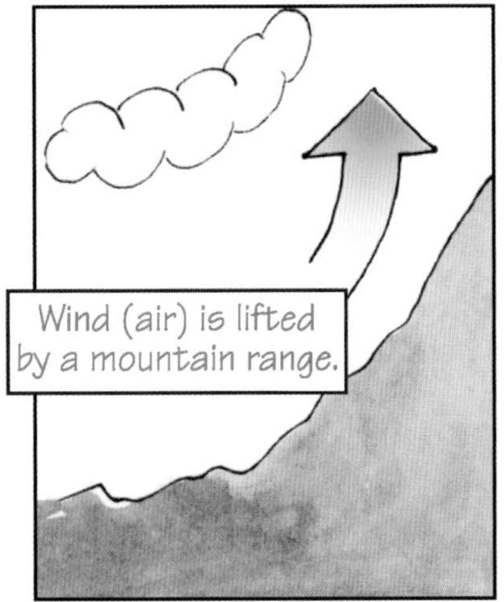

Figure 2.51 Orographic lifting is the result of an air mass being lifted by a mountain range.

Stages of a Thunderstorm

There are three stages in the life of a thunderstorm. They are the cumulus or forming stage, the mature stage, and the decaying or dissipating stage. Each of these stages have different weather phenomena associated with them, and each may affect fire behavior in different ways.

Frontal lifting involves a cold front moving under warmer air.

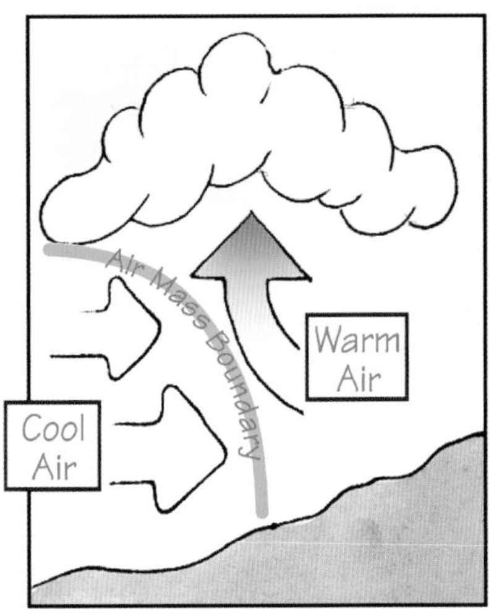

Figure 1.52 Frontal lifting occurs when a cold front moves under a warmer air mass.

Cumulus stage - This stage begins with the formation of a cumulus cloud on the top of a rising air mass. As the cloud grows taller, it begins to look like a head of cauliflower. During this stage of development there are strong indraft winds moving into the base of the cloud. This action may increase surface winds. As updrafts in the cloud increase, wind may begin to change and blow in the direction of the developing thunderstorm (Figure 2.54).

Mature stage - The mature stage of a thunderstorm is the most active and dynamic. As the cloud develops, rain or virga begins to fall from its base. As the rain falls, it causes air to drop with it; this is the beginning of the cloud downdraft. The cloud now has downdrafts and updrafts associated with it. The updraft is warmer than the surrounding air; the downdraft cooler. The cloud develops an anvil shaped top, and lightning begins. At this point, the cumulonimbus cloud has developed and a thunderstorm is produced. The downdrafts that reach the ground are cool and gusty. These winds can extend up to 10 miles from the thunderstorm, and reach 20 to 35 miles per hour, with gusts to 60 miles per

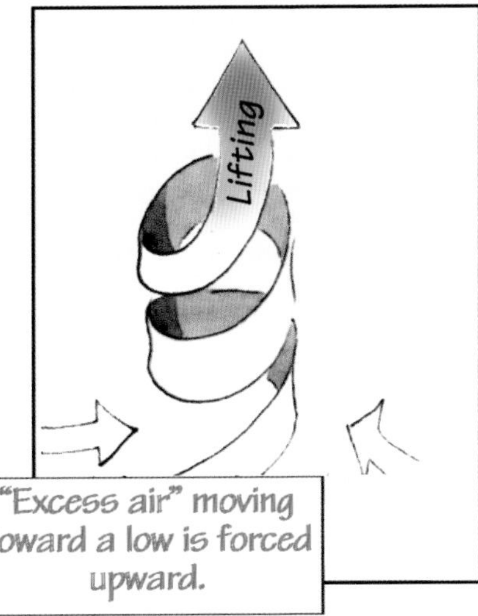

CONVERGENCE
LIFTING
Lifting
"Excess air" moving toward a low is forced upward.

Figure 2.53 Convergence lifting is associated with low pressure fronts.

A thunderstorm in your area WILL impact your fire! Be careful!

When a thunderstorm is "mature" it has an anvil shaped top.

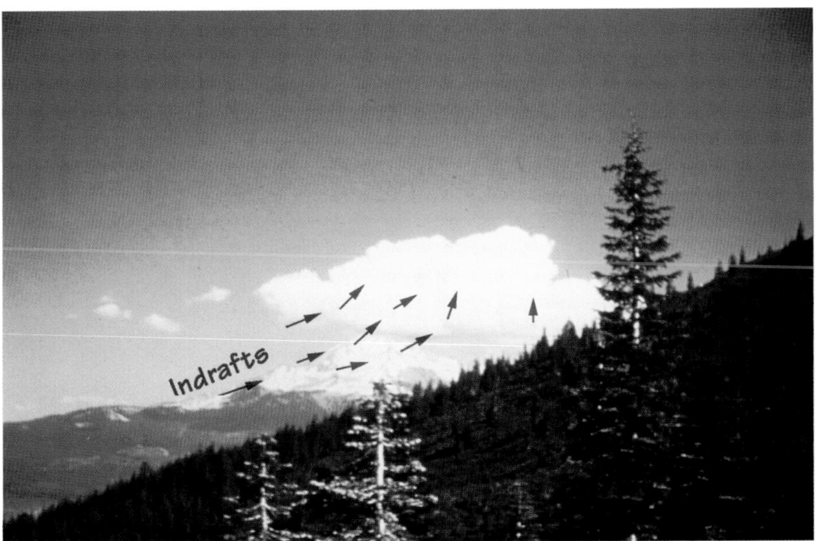

Figure 2.54 The start of a thunderstorm; indrafts may affect fire behavior.

TO
Jake says, "Never underestimate the impact a nearby thunderstorm may have on your fire. Be careful!"

hour. You can appreciate what winds of this strength can do to the fire behavior on a nearby fire!

Dissipating stage - This stage starts when the thermal lifting weakens, the updrafting slows, and the source of moisture and energy ends. The thunderstorm now generates only downdrafts, and the rain slows and eventually stops. The temperature difference between the cloud and the surrounding air equalizes, and the cloud dissipates.

3

Topography and Fuels

Topography is the configuration of the earth's surface, including its relief and the position of its natural and man-made features. Like the weather, topography and fuels have a direct bearing on fire behavior. However, unlike the continual fluctuations in weather, the basic characteristics of topography and fuels usually don't change during the life of a fire. Yet, the effect of topography and fuels will vary in response to other factors. For instance, fire behavior in canyons or on ridges will change depending on wind speed and direction, or on the relationship of these features to one another. The same fuel located in different areas will burn differently depending on the amount and angle of solar radiation reaching it, as well as the relative humidity in the atmosphere.

It is essential that you understand how weather and other factors can cause a fire to behave differently, even within the same topographical area and fuel environment. The potential impact of these changes should be incorporated into your strategy and tactics (Figure 3.1).

Figure 3.1 Both topography and fuels are playing major roles in the behavior of this fire.

The three principal environmental elements affecting wildland fires are ***weather***, ***topography*** and ***fuels***.

Slope has a major impact on fire behavior.

Topography

As noted above, the topography of an area is generally considered a fixed and known factor in firefighting. The variables occur with respect to where the fire is at any particular time, be it in a box canyon or on a south-facing slope. Topography has a direct bearing on how much solar radiation is absorbed by the surface of the earth, because it determines the angle at which solar radiation strikes the surface. This in turn modifies the microclimate which influences the type of vegetation (or fuel) in any given area. These factors should be considered when deciding when and where to fight a fire.

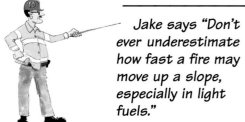

Jake says "Don't ever underestimate how fast a fire may move up a slope, especially in light fuels."

Slope

Slope brings the upslope fuels closer to the heat source.

The slope (degree of incline) on which a fire is burning is a major factor in the rate of spread (Figure 3.2). Slope contributes to preheating and ignition by "presenting the fuels" to a flame front. In the absence of a wind, flames and heat will rise straight up. Slope, in

The rate of spread increases with slope. A fire will double in rate of spread on a 30% slope. On a 55% slope it will double again.

Figure 3.2 On slight slopes (to 5%) rate of spread is not increased; on moderate slopes (30%) rate of spread is increased by a factor of 2; on steep slopes (over 55%) rate of spread will again double.

effect, angles or lifts the fuels to a position that places them closer to the upward moving flames, increasing the level of heating and making them easier to ignite. The steeper the slope, the closer the fuels are to the flame (on the uphill side), and therefore, the faster the rate of spread and the narrower the flame front. Conversely, downhill fire will burn slower because the fuels on the declining slope are positioned below the flame front, and farther away from the heat. Slope has a direct effect on fire spread by preheating fuels above the fire by radiation and convection.

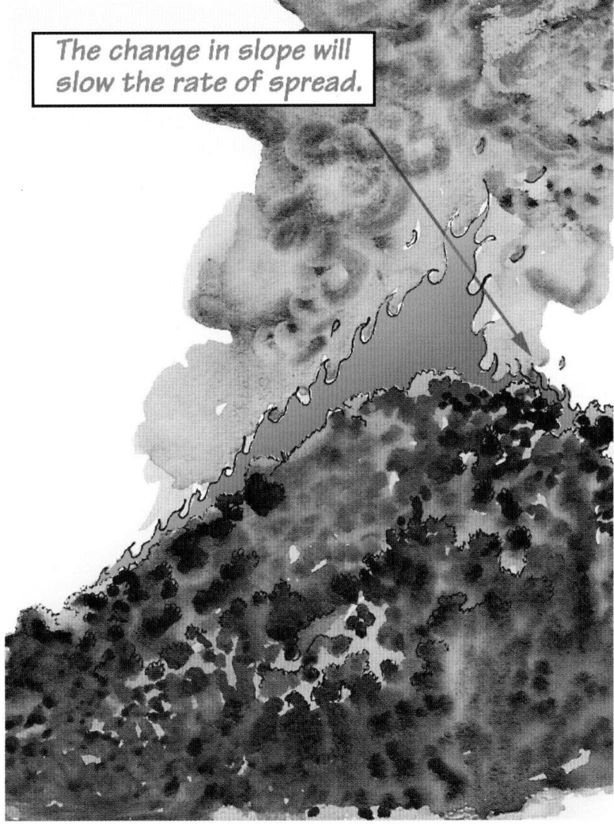

The change in slope will slow the rate of spread.

Figure 3.3 Ridges affect the rate of spread. As the fire crests the ridge, it usually will slow. Ridgetops are usually good places to construct fireline.

Another problem associated with slope is that burning materials can roll downhill, thus contributing to the spread of the fire and endangering firefighters below. This becomes an even greater concern as the fuels get heavier or the slopes become steeper.

Slope reversal is when a fire burns to a point where the slope changes, such as the top of a ridge or the bottom of a canyon (Figure 3.3). At either of these points the rate of spread will usually change. If a fire burns up a slope and hits the top of a ridge, it usually slows as it begins to back down the other side. Conversely, when a backing fire crosses the bottom of a canyon, it usually picks up speed as it runs up the other side of the canyon.

Position on a slope also has an effect on fuel availability. There is usually more fuel at base and mid-slope sites than on the top of a ridge because of greater variations in temperature and relative humidity. This has a direct impact on the potential size of fires. More fires will burn over 10 acres if they start at the base of a slope than if they started mid-slope, because there is normally more fuel available at the base (Figure 3.4).

The position of a fire on a slope has an effect on fuel availability.

Aspect

Aspect is the direction a slope is facing. The orientation of a slope to the sun has a direct bearing on the amount of solar radiation

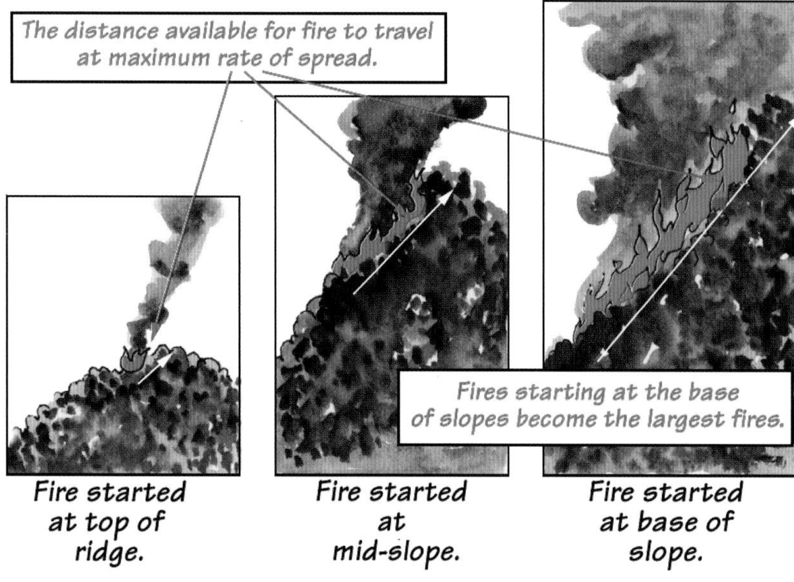

Figure 3.4 Fires starting at the base of slopes become the largest fires.

that reaches the surface. Although the aspect is fixed, the amount of solar radiation received varies with the position of the sun; therefore, the impact of aspect on fire behavior changes throughout the day (Figure 3.5).

The amount and type of fuel available varies greatly depending on aspect.

Northern aspects usually have heavier fuels.

- ***Northern aspect*** - On slopes facing north you normally find heavier fuels with high moisture content. These sites also have the lowest average temperatures, lowest rate of spread, latest curing dates, and last areas of snow melt.

Eastern aspects heat earlier in the day.

- ***Eastern aspect*** - The fuels on slopes facing east are usually transitional, between those found on northern slopes and those found on southern slopes. Surface heating occurs earlier in the day because these slopes are the first to receive solar radiation. They are also the first to cool in the afternoon. Eastern slopes are normally on the lee, or facing away from the general wind direction.

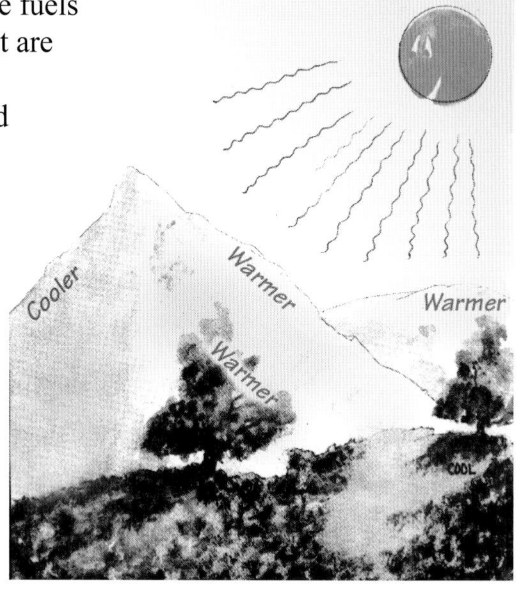

Figure 3.5 Slopes receiving direct sunlight will be hotter, increasing the flammability of the fuels.

South and west-facing slopes will normally burn hotter in the afternoon.

- ***Southern aspect*** - The fuels on southern slopes usually have the lighter and flashier fuels with the lowest moisture content.

This is primarily because they receive the highest levels of solar radiation and, therefore have the highest average temperatures. Fires on these aspects generally have the fastest rate of spread. Snow is gone from south facing slopes first, and fuels cure earlier in the fire season.

- ***Western aspect*** - The fuels on a west facing slope are transitional between the light fuels of southern slopes and the heavier fuels on northern slopes. Surface heating occurs in the afternoon and cooling begins later in the day. Westerly slopes are generally on the windward side of a mountain.

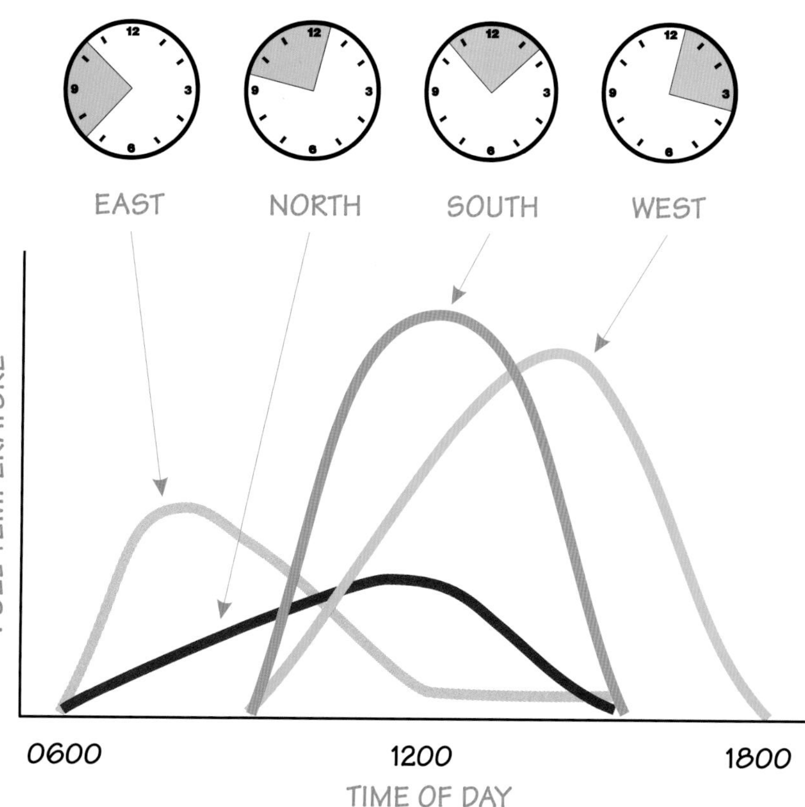

Figure 3.6 Flammability of fuels by time of day and aspect. Take special note that the south and west-facing slopes are the warmest during the primary burning hours. This will impact fire behavior.

South and southwest aspects are the most vulnerable to fire. This is due to the fact that the fuels are usually lighter (easier to ignite), flashy (contributing to a faster burn rate), and warmer (again, easier to ignite). Winds from the southwest are usually hot and dry, with lower relative humidities.

The time of day and aspect influence the flammability of the fuels (Figure 3.6). In general, the period when a south-facing slope will burn with the greatest intensity is very predictable. If you are fighting fire on a southwestern slope in the early afternoon on a hot summer day, you have all the ingredients for a major run or "blowup" of the fire.

Jake says, "Take time to really understand what this chart tells you. If you are fighting fire on a south-facing slope in the afternoon, you better really be careful."

Shape of the Land

Configuration, or the "lay of the land," affects wind patterns and fire spread. Ridges, saddles, canyons and chimneys, the elevation, and barriers all have a direct influence on how a fire burns.

Canyons, saddles and ridges can change fire behavior...rapidly!

Ridges

Ridge lines provide a break between slopes. Considering terrain and weather conditions, ridge lines are often one of the best places to construct fireline in steep country. When a fire burns to the ridge line, it encounters an opposing slope and possibly an opposing wind. As noted earlier, the change in slope slows the spread. If an opposing wind is present, it may reduce the level of spotting. However, if there are converging winds along a ridge line, the winds can be erratic and spotting may become a problem.

Winds on ridge tops can be erratic.

Canyons and Chimneys

Narrow canyons and intersecting drainages are always difficult and dangerous sites to fight fire. Canyons have a tendency to funnel and accelerate winds (Figure 3.7). During the heat of the day, winds are drawn up canyons and draws, much like the action of a chimney over a fireplace. As the heated air moves up and out of the canyon, it "draws" more air in through the mouth of the canyon increasing the intensity of the fire. When these winds cross openings to drainages they develop eddies and other erratic conditions.

Jake says, "Stay away from chimneys; you can be easily fooled. Don't say after a disaster, 'Oh, we were in a chimney?'"

Figure 3.7 Canyons are not the best location to try to fight fire. Radiant heat will preheat the far side of the canyon; wind currents will carry burning material to the other side, etc.

The narrowness of a canyon increases the potential for preheating, spotting and ignition on the opposite slope. Couple this with the potential for erratic wind and you can see that a narrow canyon is usually not the location of choice to attempt to fight fire!

Figure 3.8 Chimneys can be deadly. Fuels are preheated more effectively and the wind is funneled by the shape of the topography.

Chimneys have been involved in the injury or death of many firefighters (Figure 3.8). A chimney is a steep, narrow chute canyon. This is one of the places where the combination of topography, weather and fuel can really create havoc. First, normal upslope air is funneled and accelerated in a chimney. If the air is unstable and surface heating is moving the wind upslope, more air is drawn into the base of the canyon. The greater air flow spreads the fire even faster and increases the likelihood of spotting (Figure 3.9). The narrowness of a chimney allows a fire burning within it to preheat the fuels quicker, again increasing the rate of spread. If the type and quantity of fuels is

Chimneys are sometimes hard to spot when they are full of unburned fuel.

Craig Glazier

Figure 3.9 The topography and local winds are forcing this fire to burn up to the ridge faster than it can spread laterally.

How terrain may impact fire behavior

Type of terrain	Interpretations
Narrow canyons	Surface wind will normally follow the direction of the canyon, which may be different from that of the prevailing wind. Wind eddies and strong upslope air movement may be expected at sharp bends and at intersecting draws. Radiant heat transfer from one slope to another encourages spotting and area ignition.
Wide canyons	Prevailing wind direction will not be altered to any great extent by the direction of the canyon. Significant differences will occur between general fire conditions on north or south aspects.
Box canyons or chutes	Fires starting near the base of box canyons or chutes will react similarly to a fire in a stove. Air will be drawn in from the bottom, creating very strong upslope drafts. These conditions may also occur at the heads of narrow canyons and at the heads of high mountain valleys.
Ridges	Fires burning along lateral ridges may change direction when they reach a point where the ridge drops off into a canyon. This change of direction is caused by the flow of air in the canyon. In some cases, a whirling motion by the fire may result from a strong flow of air around the point of the ridge.

Figure 3.10 How terrain may impact fire behavior.

conducive to a rapidly spreading fire, it will race up a chimney faster than firefighters can escape (Figure 3.10).

Fuels at lower elevations dry out earlier in the year.

The dangerous conditions that can develop in a box canyon or a steep narrow canyon are:

- Steep slopes allow fire to spread faster upslope.
- Narrow canyons increase the possibility for spotting or fire spreading to the other side of the canyon.
- Chutes, saddles and box canyons provide the conditions for the chimney effect to occur.

- Accelerated upslope drafts causing intense burning and/or extreme fire behavior.

Elevation

Elevation plays a large role in determining both the condition and amount of fuel in an area. It affects fuel density, as vegetation types and fuel loading patterns change with elevation. Fuels at lower elevations dry out earlier in the year than those at higher elevations because of prolonged exposure to solar radiation, higher temperature, and lower relative humidity. Elevation affects the length of the fire season; the lower the elevation, the longer the season. Elevation can affect fire behavior in various ways. It influences how air moves from the "warming" valleys to the "cooler" ridges, and can affect the positioning of warm/cool air masses in thermal belts (the layering of air masses). Elevation also affects the amount of precipitation received, the exposure to winds and its relationship to the surrounding terrain. At extremely high elevation such as high mountain peaks and ridges, fuels may be absent.

Elevations affect the length of the fire season.

Barriers

A barrier is a break in the fuel, or change in fuel type or condition (fuel moisture and/or density), that can slow or stop the spread of a wildland fire. Barriers can be natural or man-made. Use barriers in the development of your plan; they may save you time and money. Some of the more common barriers are:

Use existing barriers when you can.

- ***Rocks or bare soil conditions*** - Rock outcroppings, bluffs, and bare patches of land may provide adequate barriers.

- ***Lakes, streams, and moist soil situations*** - Use lakes for barriers when you can. Streams may prove to be adequate barriers on flanks and during periods of moderate fire behavior.

- ***Roads, trails, and other improvements*** - It is a common practice to use man-made barriers for control lines. You can "fire" from roads and trails, or use orchards and vineyards as barriers.

- ***Change in fuel type and fuel moisture conditions*** - Fighting fire in light fuels is usually more effective than fighting fire in

Figure 3.11 Note how they have constructed their fireline in the lighter fuels on the flat.

heavy fuels (Figure 3.11). Some fuel conditions may act as partial barriers. In the morning hours, lighter fuels may have a high moisture content and resist combustion. This condition may be considered a partial barrier if "used" during the morning hours.

• ***Old burns*** - Previously burned areas usually provide a beneficial change in fuel types. Use them if it is to your advantage.

The factors of topography that affect wildland fire behavior are; ***slope***, ***aspect***, the ***shape of the country*** (configuration), ***elevation***, ***barriers***, and ***the position of the fire***.

FUELS

The type, amount and condition of the fuels are important factors to evaluate in the development of your plan to attack and control a wildland fire. Fuel, whether it is in the form of grass, timber or someone's home, is what is burning or what has the potential to burn, and needs your serious consideration. But, before we get into fuels, we need to discuss the chemistry of fire itself: the principles of how heat spreads and the basics of wildland fire suppression strategy and tactics.

Fuels is what burns.

The Chemistry of Fire

When enough heat (just over 500 degrees F) is applied to a flammable fuel in the presence of air (oxygen), fire will be produced. The heat source may be from lightning, a match, a smoldering fire or even spontaneous combustion. The three elements of the fire triangle are: ***heat***, ***fuel*** and ***oxygen***.

TO
Jake says, "Firefighting is all about breaking the fire triangle....it is that simple!"

Fire is a chemical reaction called *rapid oxidation*. Fire is produced only when all three elements are present in the right amounts. Fire cannot exist if any one of the elements (heat, fuel, or air) is absent.

BREAKING THE FIRE TRIANGLE

The basic principle of fire suppression is to remove one or more of these elements in the quickest and most effective manner (Figure 3.12). Heat is the causal agent in the start of any fire. Once a fire has started, it can produce its own *life-giving* heat. Wildland fires start from a variety of heat sources. Some of the more common heat sources are matches, cigarettes, sparks from an engine exhaust, burn piles and lightning (Figure 3.13).

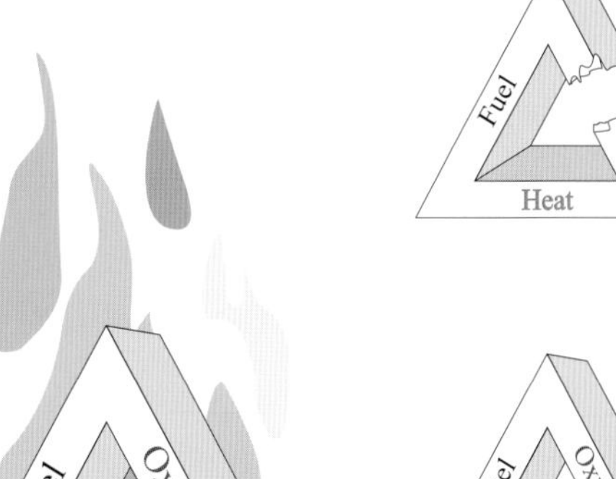

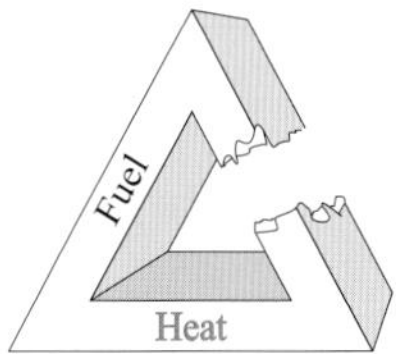

Oxygen is removed by covering with dirt, retardant or foam.

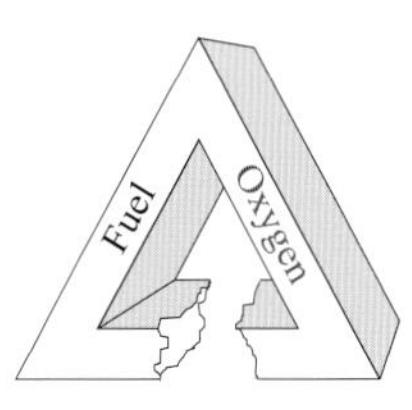

Heat is removed by cooling with water, foam, retardant or dirt.

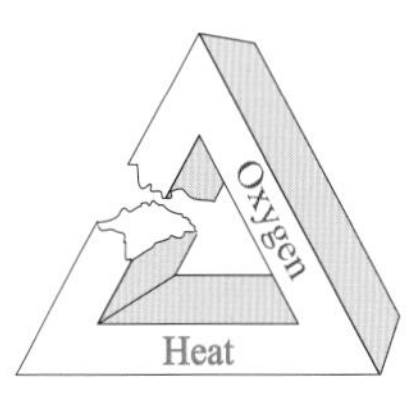

Fuel is removed by cutting a fireline between the fuel that is burning and unburned fuel.

Figure 3.12 Fire can exist only if all elements are present in the right amounts. Remove any one element, and the fire is extinguished.

As a firefighter, it is your mission to break the fire triangle as quickly, efficiently and safely as possible. You can use water or dirt to cool (take away the ***heat***) and smother (eliminate the ***oxygen*** supply). Hand tools or a bulldozer can be used to cut fireline down to mineral soil, thus removing the ***fuel***. You can also use fire, in the form of a backfire or burning operation, to remove fuels. Most of the time, a combination of tools or suppression methods will be used to break the fire triangle.

Fire needs heat, fuel and oxygen to burn.

Figure 3.13 Lightning causes a high percentage of fires in some parts of the country.

Heat Transfer

The movement of heat transferring from burning to unburned fuels is what causes a fire to spread. Heat moves in three very different

Figure 3.14 Heat is transferred in three ways: conduction, convection and radiation.

ways: ***conduction***, ***convection*** and ***radiation*** (Figures 3.14 and 3.15).

The best example of ***conductive*** heat is the heat you feel on your end of a wire when the other end is in a fire. Conductive heat travels through a wire like electricity. Conductive heat travels by contact through one object to another, and the material directly determines how well the heat will be insulated or conducted. Since wood and other fuels typically found in a wildland area are poor conductors of heat, conduction does not play a major role in the spread of wildland fire.

The best example of ***convective*** heat is the heat you can feel above a campfire. Convective heat moves as a heated air mass. Heated air rises, sometimes with an enough force to carry burning material with it. Convective heat plays the biggest

How heat moves

	Conduction	Convection	Radiation
Explanation of how heat moves	Molecules move heat through a solid object.	Heated air rises.	Heat radiates from a heat source. The air is not heated, but solid objects close by will increase in temperature.
Examples	Hold one end of a wire, heating the other end. Heat moves through the wire.	Hold your hand over an open flame.	Hold your hand out to the side of a radiant heater. The air will not be heated, but you will be able to feel the heat.
How does this affect wildland firefighting?	Heat will transfer through wood, although not very quickly. Not a major factor in the spread of a wildland fire.	Ground fire will move into a tree or up a slope by convection. Burning embers are carried by convection, causing spot fires.	Heat can radiate through a closed window of a house, igniting the drapes or other flammable items.

Figure 3.15 Heat travels in three different ways. Know and understand how each will affect your firefighting operations.

role in the spread of wildland fire as superheated smoke and gasses pre-heat fuels, cause spot fires, and move fire into the crowns of trees (Figure 3.16).

The best examples of ***radiation*** heating are the earth receiving heat from the sun; and the warmth you get when standing near a campfire. Radiation moves as "rays or waves." Radiation can pre-heat fuels or ignite unburned fuels across a fireline or though a window. Do not underestimate the importance of radiation in the spread of wildland fire.

Figure 3.16 The majority of the heat energy in this fire is moving in the form of convective heat. The convection column is carrying burning embers and material high into the atmosphere. Radiated heat is also preheating fuels.

Fuel Characteristics

The smaller the fuel, the easier it will ignite.

Fuel characteristics determine the potential fire intensity, and influence the rate of spread. Wildland fuels are commonly divided into two main groups: light or fast-burning fuels (like dry grass), and heavy or slow-burning fuels (like a stand of timber). There is another fuel that must be considered: buildings (homes, barns, sheds, commercial buildings, etc.). Structure firefighting strategy and tactics will be covered in detail in the chapter on wildland/urban interface fires.

Light or fast-burning fuels are dry grass, dead leaves, tree needles, brush, and small trees (Figure 3.17). Light fuels can cause rapid spread of fire and serve as kindling for heavier fuels. Some green fuels, such as tree needles, sage, chamise, ceanothus, and other brush types, have high oil contents. They are extremely fast burning if they are not in the active growing period when the internal fuel moisture (water content) is high. Fires in light fuels are usually easier

Figure 3.17 Never underestimate the danger of fighting a fire in dry grass. It can move with extreme speed.

Figure 3.18 Heavy fuels can burn with a lot of intensity.

to suppress. ***However, don't ever underestimate a fire in light fuels. It can ignite easily, move extremely fast, and burn with great intensity. A very high percentage of fire fatalities are the result of underestimating fire in light fuels.*** Light fuels are also referred to as "fine fuels" or "flashy fuels."

Heavy or slower burning fuels are logs, stumps, branchwood, logging slash or debris, and deep duff. Heavy fuels burn readily and produce large volumes of heat when dry. Fires in heavy fuels are much harder to suppress. Heavy fuels are usually an accumulation of dead and down (lying on the ground) fuels that can burn very intensely (Figure 3.18).

Figure 3.19 A low-intensity surface fire can clean up a forest, preventing catastrophic fires later on.

Fuels can be divided into three groups or levels: ***ground fuels***, ***surface fuels***, and ***aerial fuels***. Each of these fuel levels has an influence on the ease of ignition and the combustibility of the fuels.

- ***Ground fuels*** - Deep duff, roots, and rotten buried logs are considered ground fuels. Ground fuels do not usually play a major role in fire behavior because most burn with low intensity. However, don't be fooled into thinking that they can't be a significant problem. Roots can carry fire under constructed firelines, and ground fuels can hold fire through the winter, even under a blanket of snow. These factors should be considered when constructing fireline.

If all the fuel is not consumed, a reburn is possible.

- ***Surface fuels*** - Grass, forest litter, and brush up to about six feet in height are considered surface fuels. Surface fuels are where fires usually start. They are responsible for most fire spread and for "carrying fire" to the aerial fuels. Surface fuels are often responsible for faster rates of spread (Figure 3.19). Surface fuels that can carry fire up into aerial fuels are called "ladder fuels."

- ***Aerial fuels*** - Aerial fuels are those fuels that are six feet or more in height. This includes limbs, leaves, trunks, and crowns of heavy brush and timber. The "tightness" of the aerial fuel canopy plays a role in how a fire burns. If the canopy is closed, very little solar radiation penetrates to the ground. This condition holds down the fire intensity because the surface fuels are not dried by direct solar radiation, and fuel moisture tends to be higher. But, if weather conditions and the surface fire are generating sufficient heat, a closed canopy increases the likelihood that the fire will move into the canopy and begin to crown.

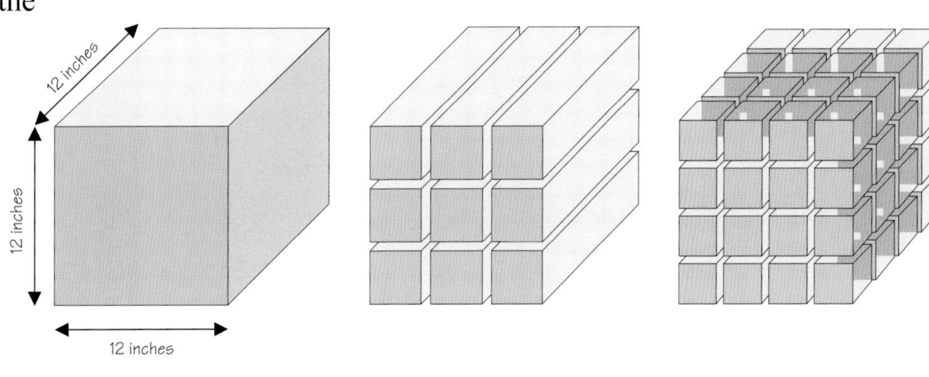

Figure 3.20 The size of a fuel has a direct relationship to the surface area. All three of these blocks have a volume of 1 cubic foot. But, because of their shape and size, the surface area varies greatly.

Size and Shape

The size and shape of the fuels also can affect how they burn. Smaller fuels have greater surface (area to volume) than larger fuels (Figure 3.20). It is easier to start a fire using kindling than to put a match to a log. Less heat is required to ignite small pieces of fuel (Figure 3.21).

Figure 3.21 Which firefighter is going to get the campfire started first? The finer the fuel, the less heat it takes to ignite it.

Some comparisons of the fire behavior of small fuels to larger fuels are: fuel moisture content changes more rapidly in small fuels; the heat required to reach sustained ignition in smaller fuels is less; the burning period of small fuels is shorter; and spot fires can start more easily in small fuels than in large fuels.

The warmer the fuel, the easier it will ignite.

The size and shape of firebrands affects how much spotting will occur and over what distances. Smaller embers normally produce short-range spotting since they burn up more quickly. Cones, bark plates, pine needles, and wood-shakes produce firebrands that can be lifted by convection columns and carried great distances. They also have sufficient size that they will continue to burn for longer periods. The shape of a firebrand is also important. Flat brands stay airborne longer, whereas round firebrands are more likely to roll down hillsides.

Wildland fuel loads range from 1 to 3 tons per acre in grass, to over 200 tons per acre in timber.

Figure 3.22 Ladder fuels allow a fire to move from the ground into the crowns of the trees.

Ladder Fuels

Ladder fuels are those that will allow a surface fire to "climb" or move into the crowns of trees. Limbs, brush, and young trees fall into this category (Figure 3.22). The higher the intensity of the surface fire, the fewer ladder fuels necessary to move the fire into the tree crowns. The spacing of the crowns of the trees can also affect how easily a fire will move through them. The tighter the spacing, the more easily a fire will climb into the crowns.

Smaller fuels accommodate changes in fuel moisture faster.

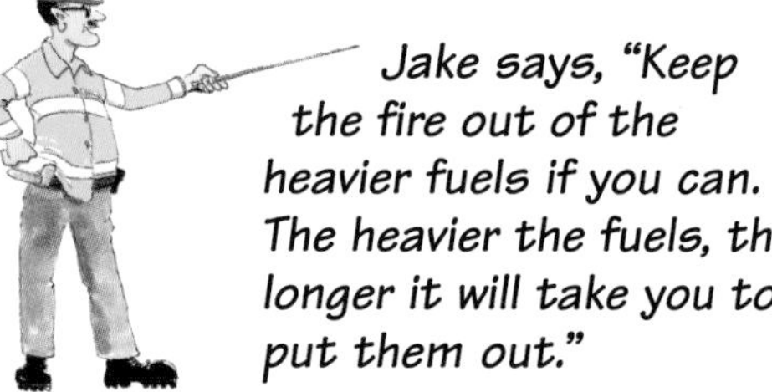

Jake says, "Keep the fire out of the heavier fuels if you can. The heavier the fuels, the longer it will take you to put them out."

Fuel Loading

Fuel loading is the amount (volume) of fuel available for a fire to burn. It is usually measured in tons per acre. The volume or quantity of the fuel in an area is a factor that must be considered. The more fuel, the more heat output. Normally, the greater the volume of fuel, the more intensely the fire will burn. Grassy fuels can range from 1 to 5 tons per acre; brush ranges from 20 to 40 tons per acre; slash can range from 30 to 200 tons per acre; and timber can range from 100 to 600 tons per acre. Heavy fuel loading increases resistance to control.

Your main concern when developing strategy and tactics is the fuel loading that is subject to burning under the current and predicted weather conditions. These are primarily the dead surface fuels 3

Figure 3.23 Grasses and litter range from 0 to ¼ inch in diameter.

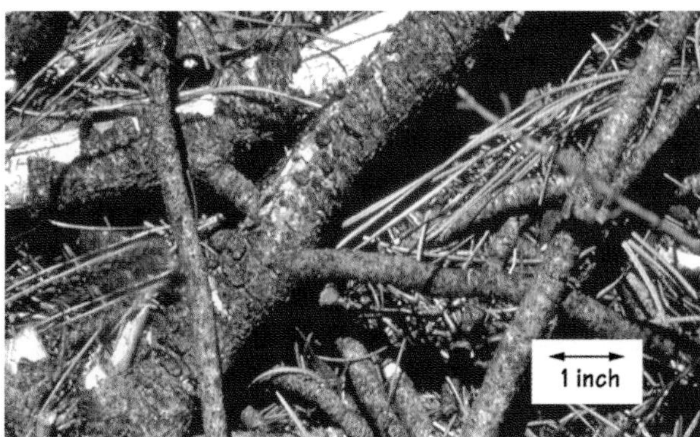

Figure 3.24 Twigs and small stems range from ¼- to 1 inch in diameter.

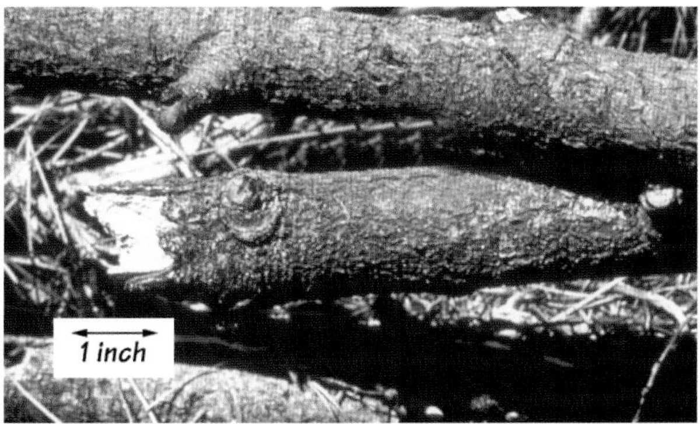

Figure 3.25 Branches range from 1 to 3 inches in diameter.

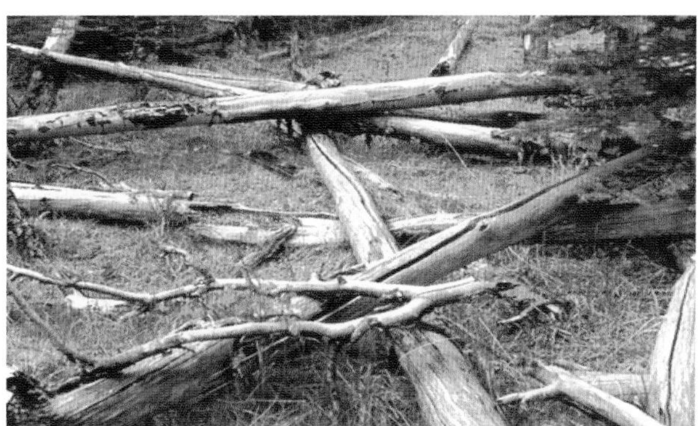

Figure 3.26 Large stems and branches 3 inches or more in diameter.

inches and less in diameter, and live fuels under one fourth inch in diameter.

Fine fuels heat up faster than heavier fuels.

Fuel loading can be separated into four size classifications. The size of the fuel affects how it will respond to various forms of moisture (rain, dew and relative humidity). Size also impacts how quickly a dead fuel will lose moisture. The four size classifications are:

- ***Grasses and litter*** - 0 to ¼-inch in diameter (Figure 3.23).
- ***Twigs and small stems*** - ¼- to 1 inch in diameter (Figure 3.24).
- ***Branches*** - 1 to 3 inches in diameter (Figure 3.25).
- ***Large stems and branches*** - more than 3 inches in diameter (Figure 3.26).

Shade is a site-specific factor.

Fuel Temperature

The warmer the fuels, the less heat it takes to ignite them, thus ignition is easier. This is particularly important when dealing with light fuels. Their small size and mass allow them to heat up quicker and lose their moisture content faster. There can be as much as a 50 to 80 degree difference between the surface temperature of fuel in the sun and in the shade, with a corresponding difference in fuel moisture. Remember, fire is a chemical reaction. For every 18-degree increase in temperature, the chemical reaction, that is fire, doubles in speed. This will influence both intensity and the rate of spread. Fuel temperature is changed by solar radiation. As the sun moves through the sky, the temperature of the fuels will change.

Winds increase drying.

Since it is impractical to measure fuel temperature while fighting fire, here are some general guidelines you can use:

- Fine fuels are more quickly and easily heated by air temperature and direct sunlight.
- During the hottest part of the day, all fuels on the south and west facing slopes will have higher fuel temperatures than those on the north and east facing slopes.
- Heavy fuels react slower than fine fuels to changes in air temperature and direct sunlight.
- Heavy fuels will usually have a lower fuel temperature than the surrounding fine fuels in the daytime, and be warmer at night.
- Of fuels exposed to direct sunlight, ground fuels will usually have a higher temperature than those above the ground (the aerial fuels).

Fuel Moisture

Dead fuel moisture is influenced by relative humidity, precipitation, temperature and wind.

The amount of moisture in the fuels will affect how easily a fuel will ignite and how intensely it will burn. There are two types of fuel moisture: ***dead fuel moisture*** and ***live fuel moisture***.

Dead Fuel Moisture

Dead fuel moisture is the moisture content in fuel material that is dead. The dead fuels and the atmosphere are constantly "exchanging" moisture. If the air is wetter than the fuels, the fuels will take on moisture. If the air is drier than the fuels, the fuels will give up moisture and begin to dry out. There are several factors that regulate the rate at which this exchange occurs:

- The amount of difference between the moisture content in the air and fuel - The greater the difference, the quicker the levels will change.
- The wind - If the wind is blowing, both the movement of air around the fuels and the level of moisture exchange will be greater. Conversely, if the air is calm, the exchange will be much slower.
- The size of the fuels - The smaller the fuel, the quicker the moisture content will change.
- The compactness of the fuels - If the fuels are tightly packed, there will be less air movement around the fuels and thus slower moisture exchange.
- The relationship of the fuels to damp or moist soil - If the fuel is on damp or moist soil, a change in atmospheric moisture will have less impact.

The fuel factors that affect fire behavior are: moisture content, the size and shape, and loading.

The three factors of fuel that affect the start and spread of wildland fire are: ***moisture content***, ***the size and shape of the fuel***, and ***fuel loading***.

Aspect or lay of the land impacts the angle at which solar radiation strikes the earth.

In the chapter on fire weather, you learned that the various air masses were always attempting to equalize temperature, pressure, moisture content, etc. This same concept applies to fuel moisture. If the moisture in the air and fuels are the same, it is said that they have reached equilibrium. Equilibrium can only occur in the finer fuels because it takes so long for the moisture content to change in larger fuels.

There are several environmental factors that directly influence dead fuel moisture: relative humidity, precipitation, temperature and wind. Other factors have an indirect influence on fuel moisture: time of day, cloud cover, amount of shade, fuel canopy, aspect, slope, solar radiation, and elevation.

Aspect influences temperature, relative humidity and fuel moisture.

The factors that influence fuel moisture vary from site-to-site. ***Shade*** is one of those site-specific factors which has a direct influence on temperature. Take, for example, a warm summer day with an air temperature of 85°. The surface temperature can vary as much as 50°, from 110° in the shade to 160° in an unshaded area. Meanwhile, the dead fuel moisture can be 8% in the shade and only 3% in unshaded areas. It may not seem like a lot, but that 5% can mean a significant difference in terms of fire behavior.

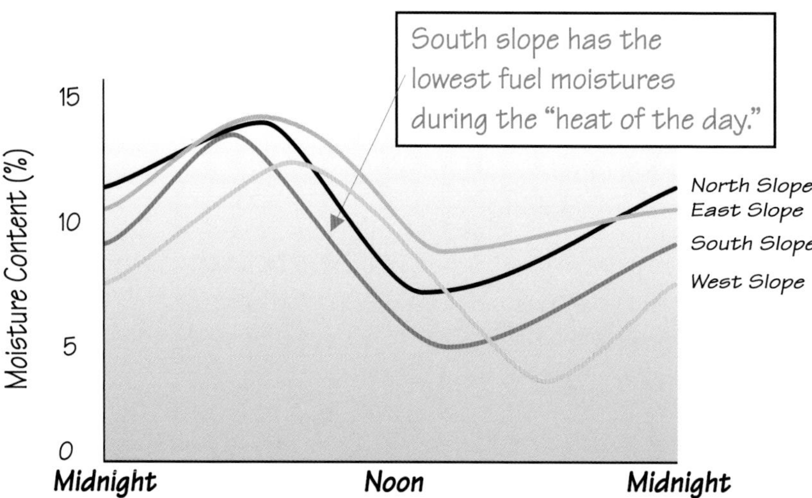

Figure 3.27 Aspect has an influence on fuel moisture. Note how the southern and western aspects are driest during the heat of the day.

Another influencing factor is ***aspect*** or lay of the land. Aspect impacts the angle at which solar radiation strikes the earth. This in turn influences temperature, relative humidity, and fuel moisture (Figure 3.27).

Southern slopes receive more heating during the afternoon than any other aspect. This means temperatures will be higher, while relative humidity and fuel moisture will be lower. Flat areas receive about the same intense heating as a southern slope. As you can see from Figure 3.27, the fuels on east facing slopes reach their lowest fuel moisture contents by early afternoon; south and west slopes have the lowest fuel moistures during the heat of the day. South and southwest aspects have the lowest average fuel moisture contents.

The ***time of year*** indirectly influences fuel moisture in that at different times of the year, the sun is at different angles in relation to the earth's surface. During the winter, the sun is low in the southern skies as it moves from east to west. On June 21, it is at its peak. Figure 3.28 shows how the change in angle changes the heating of the earth. The actual amount of solar radiation that reaches the earth varies with the amount of average cloud cover and the distance you are from the Equator. The farther north you are, the less solar radiation you will receive.

Another influence on fuel moisture is ***elevation***. Remember the lapse rate. Temperature changes about 3½ degrees for each 1,000-foot change in elevation. If temperature changes within an air mass, there is a corresponding change in relative humidity and fuel moisture.

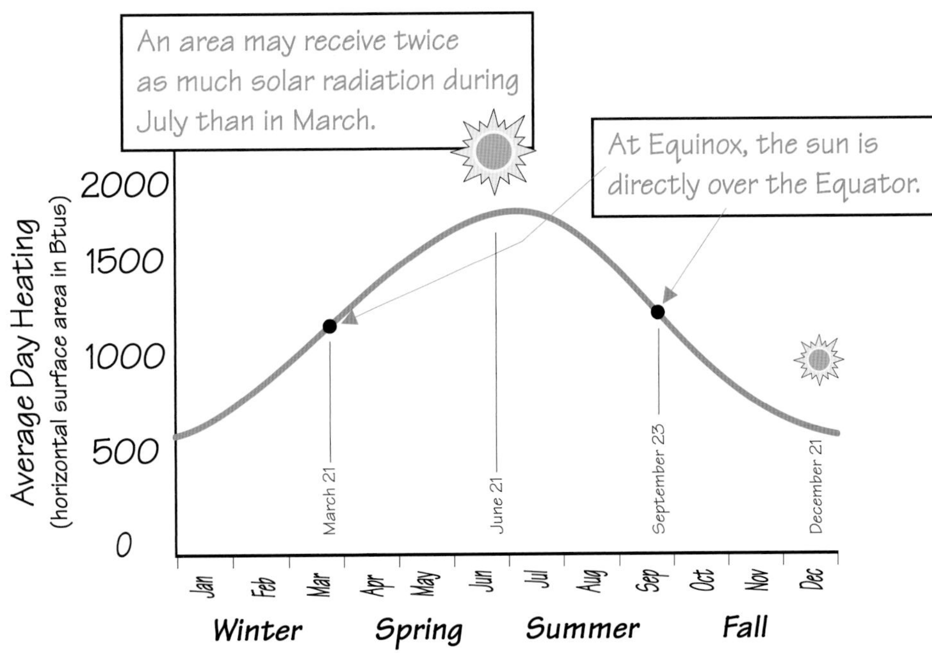

Figure 3.28 The amount of solar heating varies by time of the year. This is the result of the sun being at different angles in relationship to the earth. The sun is at its peak on June 21 (the first day of summer), and at its lowest point in the southern skies on December 21 (the shortest day of the year and the first day of winter).

Slope also has an influence on fuel moisture. The closer the slope is to perpendicular, the greater the amount of solar radiation. The higher the level of radiation, the higher the temperature and the lower the fuel moisture.

Wind influences fuel moisture in several ways. It can increase the rate of evaporation by moving moist, damp air located near the fuels, and replacing it with drier air, which dries the fuels. If the wind is one of the foehn (drying) winds, it can dry the fuels very quickly. However, if the wind is strong enough, it will actually cool the fuels and slow the drying process.

Time of the year also influences fuel moisture.

Precipitation is a factor in fuel moisture. Three main factors influence the level of impact: the amount of rain, the duration of the rain, and the size of the fuels (Figure 3.29). As noted previously, the finer fuels will "take on" moisture much faster than the larger fuels. The opposite occurs when the skies clear.

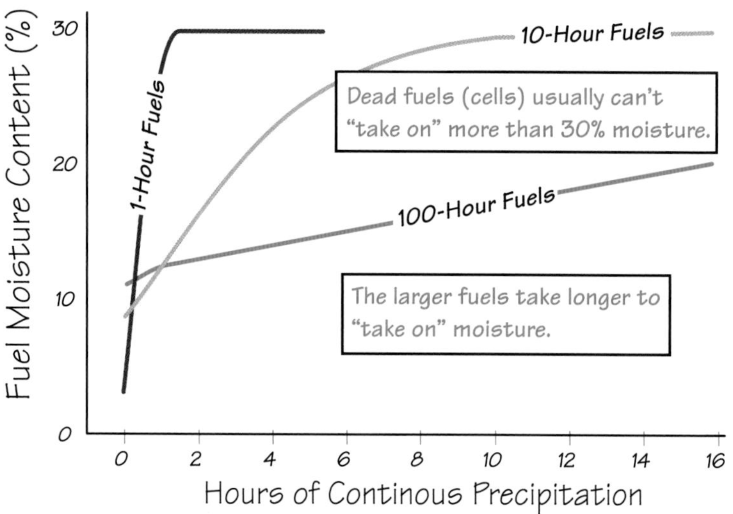

Figure 3.29 Rain will change fuel moisture. Three factors influence how: the amount of rain, the duration of the rain, and the size of the fuels involved.

Figure 3.29 shows the average level of influence on the three fuel size classes. Note how quickly the lighter fuels reach their maximum fuel moisture capacity, as compared to the 100-hour fuels which only reach about 20 percent fuel moisture after 16 hours of continuous rain. Live fuel moisture is the moisture of living, growing fuels. Both are measured in percentage of weight. Dead fuel moisture is changed by the moisture content of the air. Time lag is the time it takes for the moisture content of fuels and the surrounding air to equalize. Time lag is expressed as a rate (usually in hours). Dead grass is considered a 1 hour fuel. It takes about one hour of exposure to change the dead fuel moisture in grass up or down (Figures 3.30 and 3.31).

Dead fuel moisture...time lag relationship to fuel size

Time lag	Diameter of fuel	Examples
1 hour	less than 1/4"	Annual grass.
10 hours	1/4" - 1"	Coastal sage, juniper, and chaparral.
100 hours	1" - 3"	Logging slash.
1,000 hours	3" - 8"	Logs and mature standing timber.

Figure 3.30 The relationship between dead fuel moisture, the size of the fuel, and the associated time lag for the fuel to reach equilibrium with atmospheric moisture.

Logs are considered a 1,000-hour fuel. It takes about 1,000 hours of exposure to a (higher/lower) air moisture content to change the dead fuel moisture in logs. Live fuel moisture is changed by the growing cycle of the vegetation, and varies greatly between species and seasons (Figure 3.32).

Be on the alert for extreme burning conditions when fuel moistures reach these levels:

- 1-hour fuels when the relative humidity is at 25 percent or below.
- 10-hour fuels are at 7 percent or below.
- 1,000-hour fuels are at 13 percent or below.

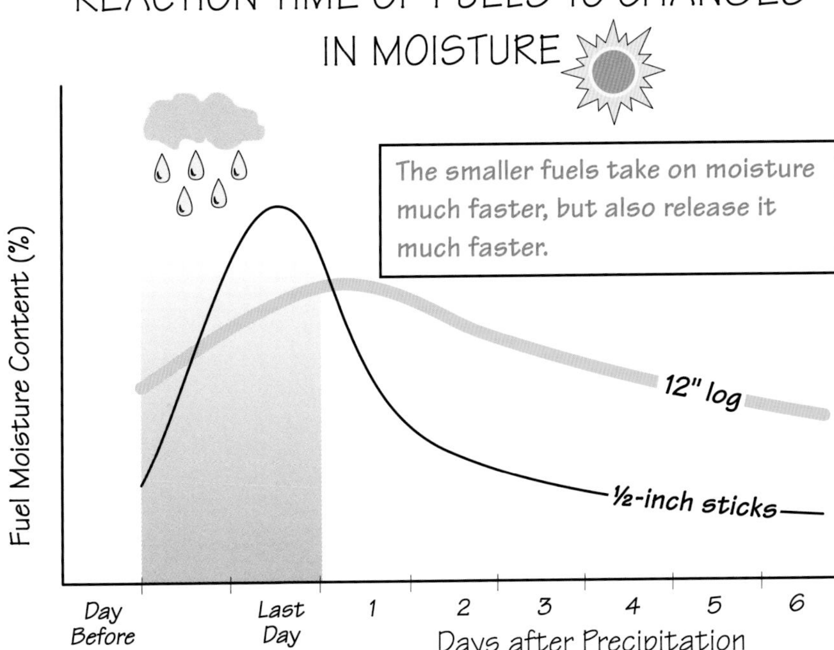

Figure 3.31 Different fuels react to moisture at different rates when they are different sizes. Smaller fuels react quicker than larger fuels. Grass that is too damp to burn in the morning can readily burn after a couple of hours of sun.

Living Fuels

There are two categories of living fuels: herbaceous and woody. Herbaceous plants are those that do not have woody material, like annual grasses or perennial plants. The woody plant material that you should be most concerned with are the twigs, leaves, needles, and small limbs that will be consumed by the flame front of a fire.

Plants that "shed" their leaves each fall are called deciduous. The shed leaves are dead and constitute fine fuels. Evergreen plants are those that do not annually shed all of their leaves or needles, but they do shed some of them. The new growth has high fuel moistures, whereas the older foliage will have a much lower moisture content.

When humidities get into the teens, you must be very careful.

The fuel moisture content of woody vegetation foliage is found in new growth, shoots and stems. The level of fuel moisture content in these fuels usually decreases as the growing season progresses. The lowest amounts occur in late summer or early autumn. Moisture content can range from 2% to 30% in dead fuels, and 30% to over 300% in live fuels.

Why the difference in fuel moistures between new foliage, older foliage, and dead fuels? Living cells, like those found in new and older foliage, have the capability of holding a lot of water, which they use to produce and transport food, and to cool the plant. Dead cells are much less "elastic," and thus become saturated with moisture at about 30 percent.

Fire behavior potential when related to relative humidity; 1-hour fuel moisture and 10-hour fuel moisture

Relative humidity %	1 hour fuel moisture %	10-hour fuel moisture %	Fire behavior potential
over 60	over 20	over 15	Very little chance of ignition; some spotting may occur from burn piles with winds above 9 mph.
46 to 60	15 to 19	13 to 15	Low ignition hazard; campfires become dangerous; glowing brands cause ignition when RH is less than 50%.
41 to 45	11 to 14	10 to 12	Medium ignition hazard; matches will readily start fire; mild burning conditions.
26 to 40	8 to 10	8 to 9	High ignition hazard, matches always dangerous; some crowning, spotting caused by gusty winds; moderate burning conditions.
15 to 25	5 to 7	5 to 7	Quick ignition, rapid buildup, high intensity and spread; an increase in wind causes increased spotting, crowning, loss of control; fire moves up bark of trees igniting aerial fuels; long-range spotting in pine stands; dangerous burning conditions.
less than 15	less than 5	less than 5	All sources of ignition dangerous; aggressive burning, spot fires occur often and spread rapidly, extreme fire behavior; critical burning conditions.

Figure 3.32 Fire behavior potential when related to relative humidity: 1-hour fuel moisture and 10-hour fuel moisture.

Grasses, shrubs and trees have a vertical orientation.

Grasses

Most wildland fires occur in grasses or light fuels. It may be the sole fuel, or it may serve to carry the fire into brush or timber fuels. Three factors control how "well" the grass will burn: ***level of curing*** (grass that is not cured will not burn), ***amount of fuel*** (the more fuel, the hotter and longer the grass will burn), and the ***dead fuel moisture*** (grasses with fuel moistures above 30% will not normally burn). Fuel moisture in grasses in the "living stage" is usually above 100%, but will drop to about 30% as it is curing (in the transition stage), and below 30% in the cured stage. Generally, more than one third of the grass has to be dead and dry before grass will carry fire (Figure 3.33).

Figure 3.33 Grass is the main carrier of wildland fires. Once a field like this cures, it will burn with extreme speed.

In the West, we think in terms of a summer fire season. The grasses begin to mature and die in the spring. As the summer turns warm, the grasses turn brown and can easily burn. Burning conditions can change from day-to-day based on the moisture in the air. In the Midwest, or other areas where the summer is usually wet, the grasses burn during two distinct seasons. The first occurs as the snow retreats and the dead grasses dry out. As soon as the soil warms enough to germinate grass seed, the grass will begin to grow or "green up." Once this starts, the number and intensity of grass fires diminishes until the fall when they cure and can burn again (Figure 3.34).

MOISTURE CONTENT OF GRASSES
for Wet and Dry Summer Areas
Grasses in dry summer area.
Grasses in wet summer area.
Moisture Content (%)
250
200
150
100
50
30
0
Living Stage
Transition Stage
"Green Up"
Dead Stage
Green
Darker
Straw
Coloration
Winter Spring Summer Fall
Fire Season(s)
Dry Summer Areas
Wet Summer Areas

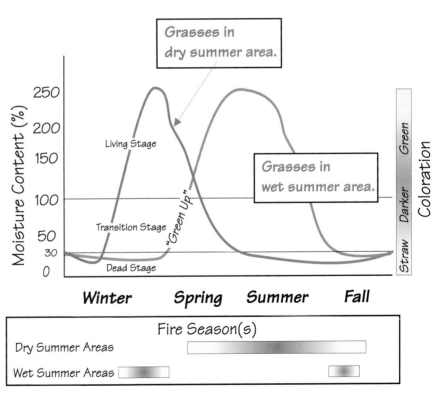

Figure 3.34 Moisture content of grass changes based on the geographic area and the time of year.

Compactness and Arrangement

Compactness refers to the spacing between fuel particles. The closeness and arrangement of the fuel influences how easily it will ignite and how well it will burn. If the fuels are very compact, there is less surface area exposed to the flames and the oxygen necessary to support combustion. You can expect a slower rate of spread. Fuels that are looser in their arrangement are drier, easier to ignite, and burn with a higher rate of spread.

Horizontal fuels include litter and slash.

Fuel bed depth is the average height of surface fuel that is available for combustion. Fuel orientation refers to the horizontal or vertical orientation of the fuel. Fuel bed depth and orientation are significant fuel properties for predicting whether a fire will ignite; as well as its rate of spread and intensity. Grasses and shrubs have a vertical orientation, and the fire intensity in this type of fuel increases with its depth. Timber litter and slash are horizontally oriented fuels and slowly increase in depth as the load increases.

Figure 3.35 Horizontally oriented fuels include timber litter and slash.

The ***horizontal continuity*** or distribution of fuels has an impact on how a fire spreads; the rate of spread; and whether the fire will move along the surface or through the crowns; or both (Figures 3.35 and 3.36). A fuel bed may either be continuous or patchy. If the fuels are widespread or patchy, a fire will have some difficulty moving from one "island of fuel" to another. For a fire to move rapidly through patchy fuels, it will take a strong wind moving up a steep slope. Continuous fuel beds provide available fuels at the surface and in the crowns. You should be most concerned when the surface fuels are continuous.

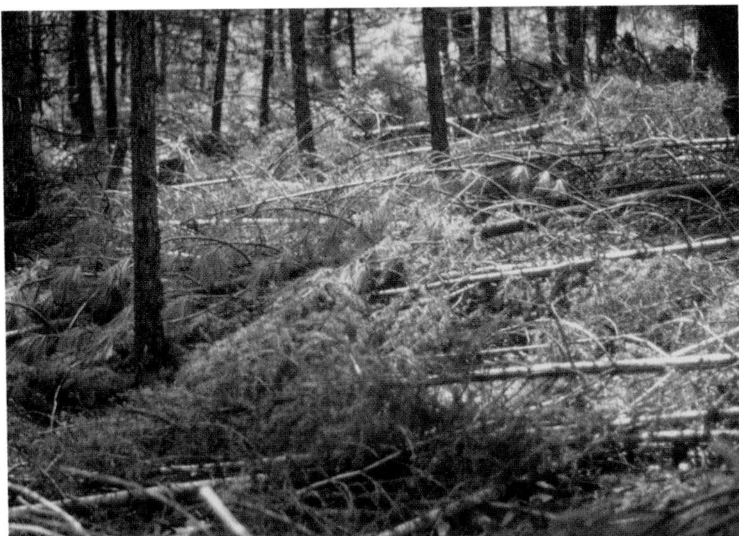

Figure 3.36 This slash is a prime example of horizontal fuels.

As discussed previously, if the aerial fuels are continuous, it will affect how the surface fuels burn. A fully closed canopy may block all solar radiation from reaching the surface, hold moisture longer, and drastically reduce the amount of wind that can influence surface fires. The greater the crown closure, the greater the reduction in wind. However, once a surface fire begins to "torch" individual trees or move into the aerial fuels, you have a completely different situation.

The ***vertical arrangement*** and continuity of the fuels are very important characteristics (Figures 3.37 and 3.38). They determine whether a fire will be able to move into the aerial fuels with ease or not. When there are fuels available for combustion throughout the vertical fuel bed, it is said that there are ***ladder fuels***.

Figure 3.37 Vertically oriented fuels include grass, shrubs and trees.

Vertically oriented fuels affect how a fire will move into the aerial fuels.

There are several strata of fuel in a mature forest stand or even in a brush field. There is decomposing organic matter in the top layer of soil. There are surface litter and grasses. There are the low-level twigs and branches of the brush, understory shrubs, and young trees. Then there is the sub-canopy. The sub-canopy can be the crowns of the brush or young trees, and the lower branches of mature trees. The top stratum is the canopy of the dominant trees. Fire may burn through one or more of these strata. If one or more layers are not burned initially, it leaves fuel available to burn later if conditions change. In most cases when lower fuel levels burn, the upper fuel levels are dried out and perhaps scorched. The upper-level fuels are still able to carry fire if there is sufficient heat to ignite them.

Figure 3.38 This eucalyptus plantation is an example of vertically arranged fuels.

Having an environment where the fire has moved through an area and left one or more of the fuel layers available to burn can be a potentially dangerous situation. For example, if a low intensity fire moved through an area consuming the ground litter, it would dry out the fuels above it. If conditions changed later in

Jake says, "Don't be fooled by an area that has been burned...reburns occur more often than you may think."

the day, and an ignition source became available, a fire of much greater intensity could move through the area again in the higher fuel strata. This is called a ***reburn***.

You must concern yourself with the potential for reburns. When cutting line in an area that has not completely burned, you may think that you are making a direct attack (fighting fire right on the fire's edge), which is one of the safer strategies. But, in fact, you are making an indirect attack (fighting fire some distance from the fire's edge), because you have unburned fuel on both sides of the fireline you are constructing. This is a very dangerous scenario. Maintain situational awareness at all times.

4

Wildland Fire Behavior

To be an effective firefighter, you need to understand and be able to predict what the fire is going to do under various conditions. Fire behavior is really the discussion of those factors you learned about in the first two chapters, but here you will be told about how they all interact with each other.

Reading this chapter will not make you a fire behavior analyst, but it should give you important information that will assist you in developing appropriate suppression strategies and tactics for varying conditions (Figure 4.1). The key to success is to automatically recognize the factors, both present and predicted, that will affect how you should fight a fire. The main goal for this chapter is that you develop an appreciation and understanding for what is happening around you, and what will be happening around you as the fire progresses.

Tom Dean

Figure 4.1 There is some serious fire behavior in this picture of the early stages of Miller's Reach #2 (Alaska 1996). Understanding what your fire will do under existing and predicted weather factors is critical.

Fire Environment

The fire environment controls the level of fire behavior.

The fire environment includes all of the factors that allow a fire to start, burn and spread. There are eighteen primary factors that influence how a wildland fire will burn (Figure 4.2). The fire's rate of spread, intensity, and other characteristics respond to these. Some of these factors can, in turn, be influenced by the fire itself. This process is constantly evolving and changing. Always expect conditions to change on a wildland fire.

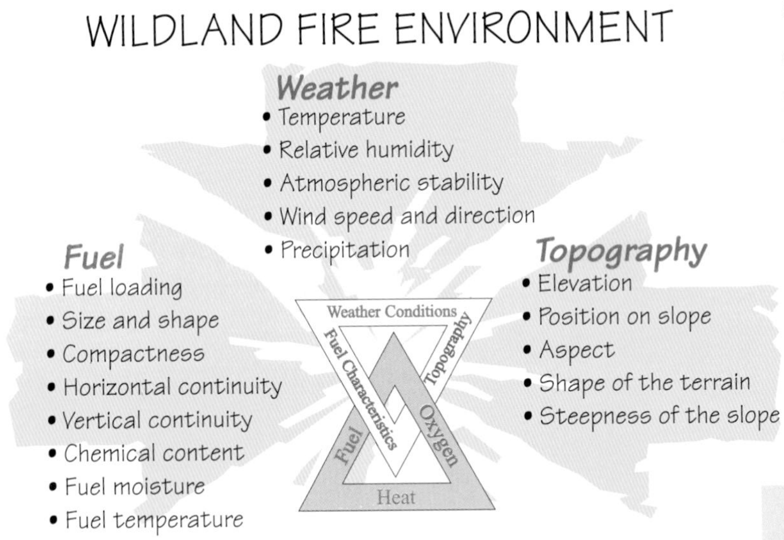

Figure 4.2 The wildland fire environment influences the behavior of a fire. All of the factors interact with each other to set fire behavior at a specific point in time. As the factors change, the fire behavior also changes. For instance, as time passes, so does the position of the sun in relationship to the terrain; and its influence on fuels and air masses.

The ***weather*** related factors that influence fire behavior are:

- Temperature
- Relative humidity
- Atmospheric stability
- Wind speed and direction
- Precipitation

The most important weather factor is wind.

Weather is the most variable component of the fire environment. Weather varies as air masses move through an area. It also varies by the time of day. ***The most important weather components are wind, temperature, and relative humidity.***

The ***topographic*** factors that contribute to fire behavior are:

- Elevation
- Position on the slope
- Aspect
- Shape of the terrain
- Steepness of slope

The most important topographic factors are slope and aspect.

These factors are much more constant, but change as the fire moves through an area (up and over hills, from one aspect to another, etc.). ***The most important topographic components are steepness of slope and aspect.***

The ***fuel*** components that affect fire behavior are:

- Fuel loading
- Size and shape of the fuel
- Compactness
- Horizontal continuity
- Vertical continuity
- Chemical content
- Fuel moisture
- Fuel temperature

The most important fuel components are fuel moisture and temperature.

Fuel components change as the fire moves through the area (changes in fuel types from grass to brush to timber, etc.), and some change when several of the weather factors change. ***The most important fuel components are fuel moisture and temperature.***

EFFECTS OF TIME AND SPACE

	Time	Space	Effect
Weather • Temperature • Relative humidity • Atmospheric stability • Wind speed and direction • Precipitation	The weather factors change constantly. This will affect fuel moisture and curing.	Significant changes occur with topography and weather patterns.	**Causes a change in rate of spread, direction of spread and fire intensity.**
Topography • Elevation • Position on slope • Aspect • Shape of the terrain • Steepness of the slope	Consider these factors constant.	Changes will occur, especially in hilly and mountainous terrain.	**Changes the direction and rate of spread.**
Fuels • Fuel loading • Size and shape • Compactness • Horizontal continuity • Vertical continuity • Chemical content • Fuel moisture • Fuel temperature	Fuel moisture of dead and live fuels will change. Changes can be caused by insects, logging, fires, and severe weather.	Weather and topography will alter the fuel type.	**Fire intensity will increase as more fuel becomes available to burn.**

Figure 4.3 Some fire behavior factors change with time and others are "fixed."

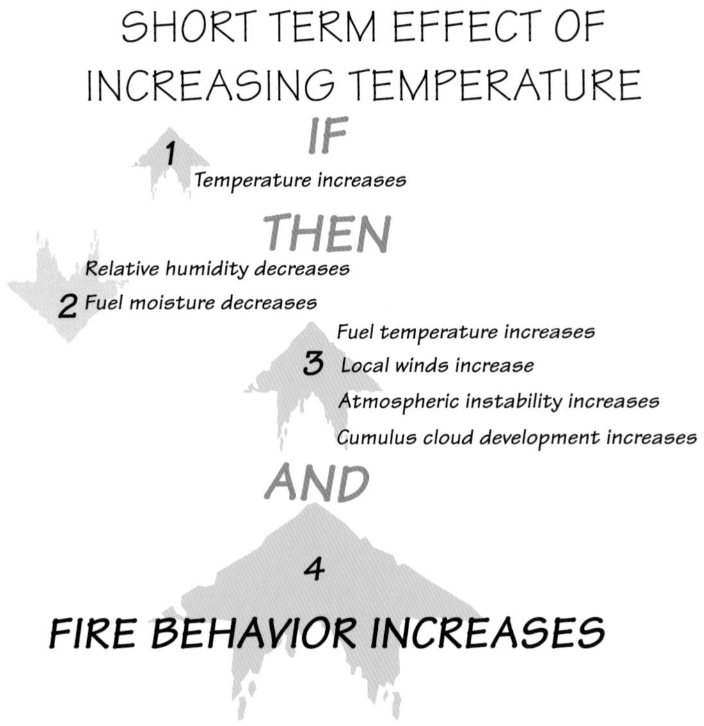

SHORT TERM EFFECT OF INCREASING TEMPERATURE

IF

1 Temperature increases

THEN

Relative humidity decreases
2 Fuel moisture decreases

Fuel temperature increases
3 Local winds increase
Atmospheric instability increases
Cumulus cloud development increases

AND

4

FIRE BEHAVIOR INCREASES

Figure 4.4 If the temperature increases, relative humidity and fuel moisture decrease. When the temperature increases, the fuel temperature, local winds, instability and cloud development will increase. Fire behavior will also increase.

Each of these components will vary with the time of day. It is critical that you recognize that changes will be occurring, and that you realize what the effect of those changes will be. Some of these changes are very predictable; others are very subtle (Figure 4.3).

You have to look at the various factors, not as a listing on a wall, but more as items on a mobile. Picture each of these dynamic elements hanging as parts of a child's mobile; not only do they move, but they also change in "size." If one changes, it causes all of the other factors to change their relationships with the others. Figure 4.4 shows how a simple rise in the temperature affects several other factors associated with fire behavior. Just picture the potential impact on fire behavior when, let's say, the slope and wind all come into alignment on a south-facing slope in the middle of a hot afternoon!

Always expect conditions to change.

Ignition

Ignition of wildland fuels requires an ignition source. There are many such sources: lightning, discarded matches and cigarettes, campfires, debris burn piles, sparks from an exhaust system, etc. The source must be hot enough to heat the fuels to the point of ignition and produce self-sustaining combustion. Combustion usually occurs at about 500 degrees in wildland fuels. The primary factors which affect the ignition of wildland fuels are:

Jake says, "Fire behavior is three-dimensional. A change in one factor causes all of the relationships of the others to shift."

• ***Size and shape of the fuels*** - The smaller/finer they are, the easier they ignite.

- ***Compactness or arrangement of the fuels*** - If they are loose and open, they will ignite easier.

- ***Fuel moisture content*** - The drier they are, the easier they ignite because heat is not expended to drive the moisture out.

- ***Fuel temperature*** - The warmer the fuel, the less heat it takes to bring it to its ignition point.

The fuel size, arrangement, moisture content and temperature all play a part in ignition.

Firebrands and Spotting

Wildland fires are mainly spread by an advancing flame front. Spotting is also often involved. Spotting occurs when burning material is carried ahead of the fire and deposited in unburned material (Figure 4.5). If the firebrand is hot enough when it lands, a small fire will be started. Firebrands are moved by four different forces:

- Popping
- Convection
- Winds
- Gravity

Trevor Wilson

Figure 4.5 The spot fires shown above were the product of a convection column lifting hot firebrands and dropping them outside the fire's perimeter. These spot fires are from a convection column as it is pushed by upper-level winds over the left flank of the fire.

Have you ever been next to a campfire when the fire "popped," and burning embers jumped from the fire? This type of firebrand is usually a short-range problem, but can cause a fire to move across firelines. The popping is a mechanical event. Moisture in the wood is heated and a small steam explosion occurs, throwing a burning piece of wood away from the fire.

A bigger problem is firebrands that are lifted by convective heating and dropped some distance ahead of the main fire. As a fire increases in intensity, so does the "lifting force" of its convection column. If the fuels being consumed are of such a nature that they develop firebrands (wood-shake roofs, oak leaves, etc.), the amount of material lifted into the air can be considerable. There is usually very little spotting during a grass fire,

Spotting can be a real problem.

Figure 4.6 Normally, you will not have too much spotting in a grass fire. The grass is so fine, it is consumed before it has a chance to start a spot fire.

Figure 4.7 This wind-driven fire is moving southeast up the canyon. Note that the topography and wind are in alignment.

because grass is too light a material to develop firebrands (Figure 4.6).

Wind can also pick up firebrands and move them some distance. Winds alone will not move firebrands great distances, because this effect is usually a horizontal action (Figure 4.7). The exception to this is firewhirls. As mentioned earlier, firewhirls are small cyclones sometimes present during unstable conditions that can pick up burning material and transport it quite a distance. Winds in combination with convection can be a real problem; convection lifts the burning material high into the atmosphere, where the winds are able to carry it great distances. In the northern hemisphere, there is normally a counter-clockwise movement of air as it rises. This means that there will be a higher percentage of spot fires on the left side of the fire's head.

Gravity can also spread a fire by moving burning material, especially if the material is round. Burning pine cones or logs are most affected by gravity. It is not uncommon that, as the ground litter is burned away, burning material is dislodged and begins to roll downhill (Figure 4.8).

In steep terrain, rolling pine cones and logs can carry fire across firelines.

Fire Behavior Terminology

Before we discuss fire intensity in depth, we need to outline several terms that describe levels of fire behavior:

- ***Spread*** - The movement of the fire, usually measured by the rate of spread. Rate of spread has a time component (e.g., three miles-per-hour; two chains-per-hour; etc.) (A chain is a surveying term. A chain is 66 feet in length.)

- ***Smoldering*** - Fire burning without flame and spreading very slowly.

- ***Creeping*** - Fire burning with low flame and spreading slowly.

- ***Running*** - Fire spreading rapidly with a well defined head.

- ***Backing*** - Fire moving away from the head, downhill, or against the wind.

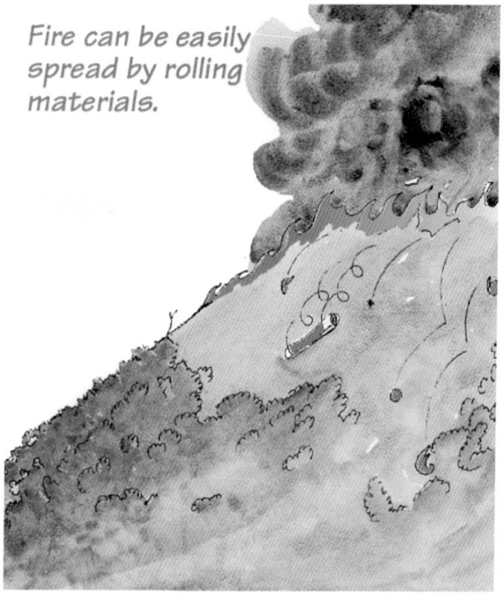

Fire can be easily spread by rolling materials.

Figure 4.8 When a fire is on a steep slope, burning material can very easily roll down the hill and spread the fire.

Undercut fireline can be very difficult to construct and maintain but is necessary on a slope.

- ***Torching*** - Fire burning on the surface, but periodically igniting the crown of a single or small group of trees or shrubs before returning to the surface. Although sometimes confused with crowning, this behavior is not as serious as a crown fire, because it lacks a strong horizontal travel component.

Drought adds to the availability of dry, unburned fuels.

- ***Spotting*** - Sparks or embers produced by the main fire are carried by winds or convection column across the control line.

- ***Crowning*** - Fire that advances across tops of trees or shrubs more or less independent of surface fire. High fire intensity and high forward rate of spread. Use the terms "crown fire" or "crowning" with care, because it describes a very serious fire situation.

- ***Blowup*** - Sudden increase in fire intensity or rate of spread sufficient to preclude direct control or to upset existing suppression plans.

Fire Intensity

Fire intensity is the rate of heat energy released during combustion. Fire intensity is measured in two ways: fireline intensity and flame length (See Figures 4.15 and 4.16). Fireline intensity is the amount of heat released per foot of fire front per second. In Figure 4.9, this fire is just beginning to "crown." The heat that is being generated is considerable. Fire intensity is affected by:

Kari Greer

Figure 4.9 This "crowning" fire is generating considerable energy. If there is sufficient wind or ground fuels, it could remain in the crowns of the trees for some time.

- ***Fuel loading*** - The type and amount of fuel available to burn.

- ***Compactness or arrangement of fuels*** - If the fuels are tightly arranged, with little space for air or convective heat to move, the fire will be less intense.

- ***Fuel moisture content*** - If a considerable amount of heat is needed to decrease the fuel moisture content sufficiently to allow for combustion, the fire will burn with less intensity.

- ***Slope and wind speed*** - Fires burn more intensely moving upslope or downwind, because convective heating is more efficient.

Note how some of the same factors affect both fire intensity and ignition.

Rate of Spread

There are several factors that influence the fire's rate of spread and burn pattern or shape:

Wind is a major factor in how fast and intensely a fire will burn.

- ***Fire intensity*** - If the intensity increases, so does the heat generation and the rate of spread.

- ***Wind speed*** - The stronger the wind, the more lean it puts on the flame and the convection column, and the faster the rate of spread.

- ***Slope*** - The steeper the slope, the closer the flame and the convection column are to the fuel, and the faster the rate of spread.

If the fuel changes, the rate of spread will probably change.

- ***Fuel type changes*** - If the fire burns from one fuel type to another (grass to brush, timber slash to standing timber, etc.), the rate of spread will change, thereby affecting the burn pattern.

- ***Natural or man-made barriers that stop or slow the spread*** - Roads, streams, lakes, rock outcroppings, etc. will all affect the burn pattern.

- ***Spotting*** - If the fire is generating spot fires ahead of or below the fire, the burn pattern will be affected.

A fire burning in uniform fuels on flat ground (with the wind being the only variable) burns with predictable elliptical shaped patterns. There is a ratio between the wind speed and the length and width of the burn pattern. These patterns also hold true when slope is the only variable. Figure 4.10 outlines the shapes of burn patterns for several wind speeds, as well as the ratios between wind speed and the rate of spread. Figure 4.11 shows a spot fire moving up a slope; it is in the shape of a perfect oval.

Spotting usually occurs ahead of the fire.

ELLIPTICAL BURN PATTERNS

Wind speed (mph)

0 X

2 X

6 X

WIND

16 X

Origin of fire

Figure 4.10 Wildland fires generally burn in elliptical patterns. Wind and slope will change the general shape of the ellipse.

Karen Wattenmaker

Figure 4.11 Slope and wind influence a fire in a similar manner. Wind pushes a fire to burn in an oval pattern. Here, slope is doing the same thing.

Vertical Dimension

When trying to predict how far a fire will spread, you usually think in two dimensions: width and length. But, there is a third dimension, the ***vertical*** dimension, that must be considered (Figure 4.12).

When a fire "goes vertical" it has a whole new personality.

As the fire develops in size or intensity, it involves an ever-growing three dimensional area. There are three factors that control the development of this third dimension:

An unstable atmosphere allows a fire to develop "vertically" and grow rapidly.

VERTICAL DIMENSIONS OF A WILDLAND FIRE

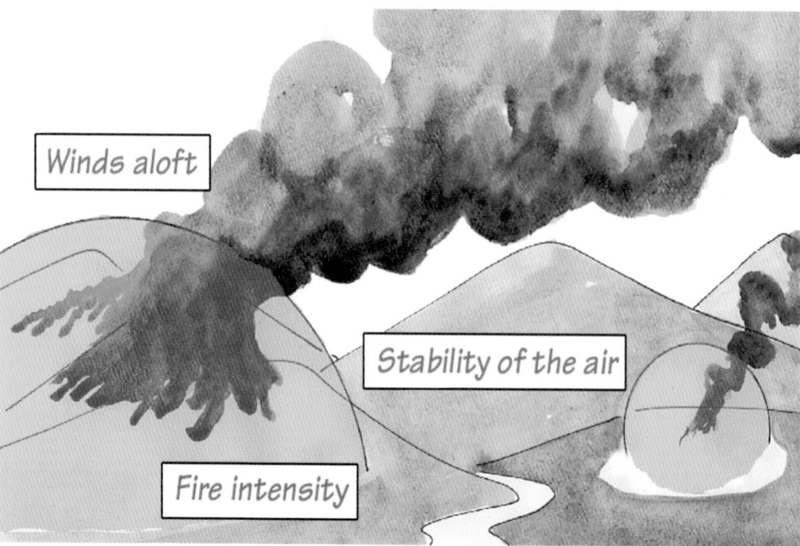

Figure 4.12 As a fire evolves, it can develop a third dimension; the vertical dimension, The extent of this development is controlled by fire intensity, the stability of the air, and the strength of winds aloft.

The faster a fire is moving or the longer the flame length, the higher the level of danger.

- ***Fire intensity*** - The fire heats the air, causing it to rise through the atmosphere. Usually, the hotter the fire, the higher the heat (smoke) column will rise. As the hot air rises, cooler air from the surrounding areas will be drawn inward towards the fire to fill the void. These "indrafts" can actually fan the fire, increasing the intensity even more. When a fire burns with high intensity, it can change the environment around it. It is common for an intense fire to generate thunderclouds with lightning and strong winds. On the other hand, if the fire is of low intensity, both the "indraft" and smoke column will be weak. There will be little or no change in the local environment, because the environment is controlling the development of the fire.

- ***Stability of the air*** - Stability or instability is a major factor in the vertical dimension of the fire. If the air is unstable, with lots of vertical movement, the fire will easily develop a vertical component. However, a strong inversion layer will act as a lid over the fire, restricting its vertical development.

Don't underestimate the influence unstable air can have on a fire.

- ***Winds aloft*** - Strong upper-level winds tend to limit the vertical development of a fire. They cut off the vertical movement of the convective heat (smoke) column.

Direction of Spread

Why does a fire burn in a particular direction? We've been taught that wind pushes a fire, and that a fire burns up a slope faster than on level ground. But, more specifically, the reason a fire burns in a particular direction hinges on the "presentation of the fuel." Did you ever "experiment" with the wick of a candle at Christmas time and cause a perfectly shaped candle to develop a "leak?" In messing with the wick of the candle, you changed the relationship of the heat source to the fuel. The spot where the candle began to "leak" received more heat, thus melting the wax faster (Figure 4.13).

Air Movement
More heat energy strikes the right edge of the candle, melting it.

Figure 4.13 Air movement (wind) is pushing the flame to the right. Thus, the right side of the candle is receiving more heat. In this same way, if wind pushes the flames closer to the fuels on the right, the fire burns to the right.

There are three factors that influence the direction a fire will burn: wind, slope, and condition of the fuel. The three are closely related, though they spread a fire in different ways. Wind "pushes" or bends the fire, bringing the radiated and convective energy closer to the fuel bed. Slope brings the fuel closer to the heat source. Meanwhile, the fuel condition determines how vulnerable it is to fire. The drier the fuel, the more readily it will ignite.

Fire moves in a particular direction, because the heat source is closer to some fuels than others. Once the closer fuels ignite, the fire tends to move in that direction. In the chapter on topography, you learned that changes in topography can often provide a fire break. Fire will move up a slope fairly quickly, because the slope brings the fuels closer to the fire. Yet, once the fire reaches the ridge line, the fuels start falling away from the heat source. The fire may stop moving in that direction. In comparison, picture the fuel presentation in a draw. Both the slope of the draw and the increased angle or degree of steepness of the draw walls present the fuel closer to the fire. Fire moving up through a draw can move extremely fast (Figure 4.14).

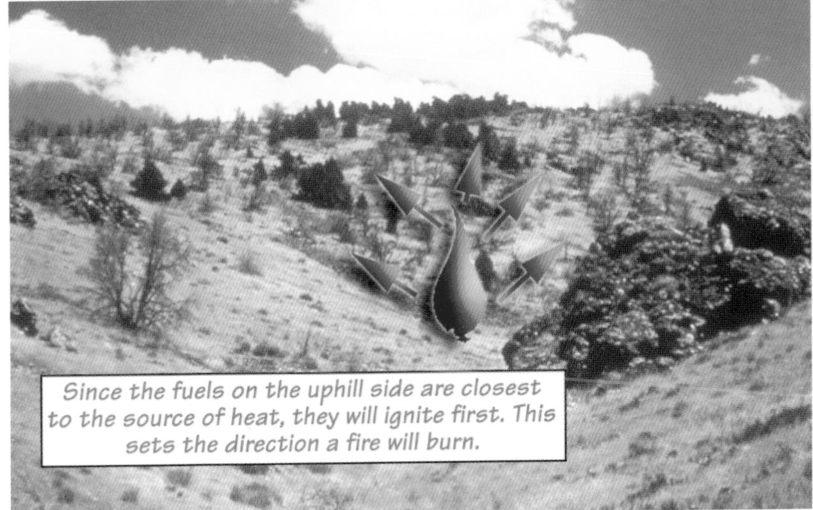

Figure 4.14 The "presentation of fuels" that this slope and draw provides will control the direction the fire will burn. The narrower the draw or steeper the slope, the faster the preheating will be...it is predictable that the fire will move even quicker up the slope.

Fire Behavior Characteristics

Fire behavior characteristics determine how and where you should fight fire, as well as the strategy and tactics you should use (Figure 4.15). The rate of spread and the flame length have a lot to do with the overall fire behavior. These fire behavior characteristics also dictate the level of firefighter safety. The greater the rate of spread and/or flame length, the greater the need to be vigilant.

Fire suppression interpretations from flame length

Flame Length	Interpretation
Less than 4 feet	Fires can generally be attacked at the head or flanks by firefighters using hand tools. Handline should hold fire.
4 to 8 feet	Fires are too intense for direct attack on the head with hand tools. Handline cannot be relied upon to hold the fire. Bulldozers, engines, and retardant drops can be effective.
8 to 11 feet	Fires may present serious control problems: torching, crowning,and spotting. Control efforts at the head will probably be ineffective.
Over 11 feet	Crowning, spotting and major fire runs are probable. Control efforts at the head of the fire are ineffective.

Figure 4.15 The length of the flame will give you a good indication of the problems you will have in suppressing the fire.

Figure 4.16 Flame length and flame height are different measurements. Flame length is measured along the long-axis of the flame, where flame height is measured as the distance above the ground.

Rate of spread and flame length are interrelated. If the rate of spread increases, so does the flame length and fire intensity. Flame length does not necessarily dictate rate of spread, but is a visual indicator of the surface fire intensity (Figure 4.16).

Once again, fire behavior characteristics impact your strategy and tactics. For example, if a high rate of spread will carry a fire to a ridge faster than you can construct a fireline, you will have to either pick another location for the fireline or start a flanking attack. If the flame length is over four feet, it will pretty much eliminate hand tools as a method of control. Rate of spread, fire intensity, and flame length all respond to changes in fuel type, moisture content, slope, and wind speed. You will see more about how these factors are interrelated as we continue.

When a fire develops a vertical dimension, its behavior can be difficult to predict.

Fire Characteristics Chart

The fire characteristics chart was developed to show the relationship between rate of spread, flame length, heat per unit area, and fire intensity. Heat release per unit area remains constant for a given fuel type and moisture content. Therefore, once you determine the type and moisture content of the fuel you're dealing with, you can quickly determine changes in rate of spread and flame length.

There are two types of fire characteristics charts: one for light fuels (Figure 4.17), and one for heavy fuels. The relationships between the various factors are the same, but the scales are different. The little figures on the chart for light fuels provide a quick reference to fireline limitations. These charts are sometimes referred to as the

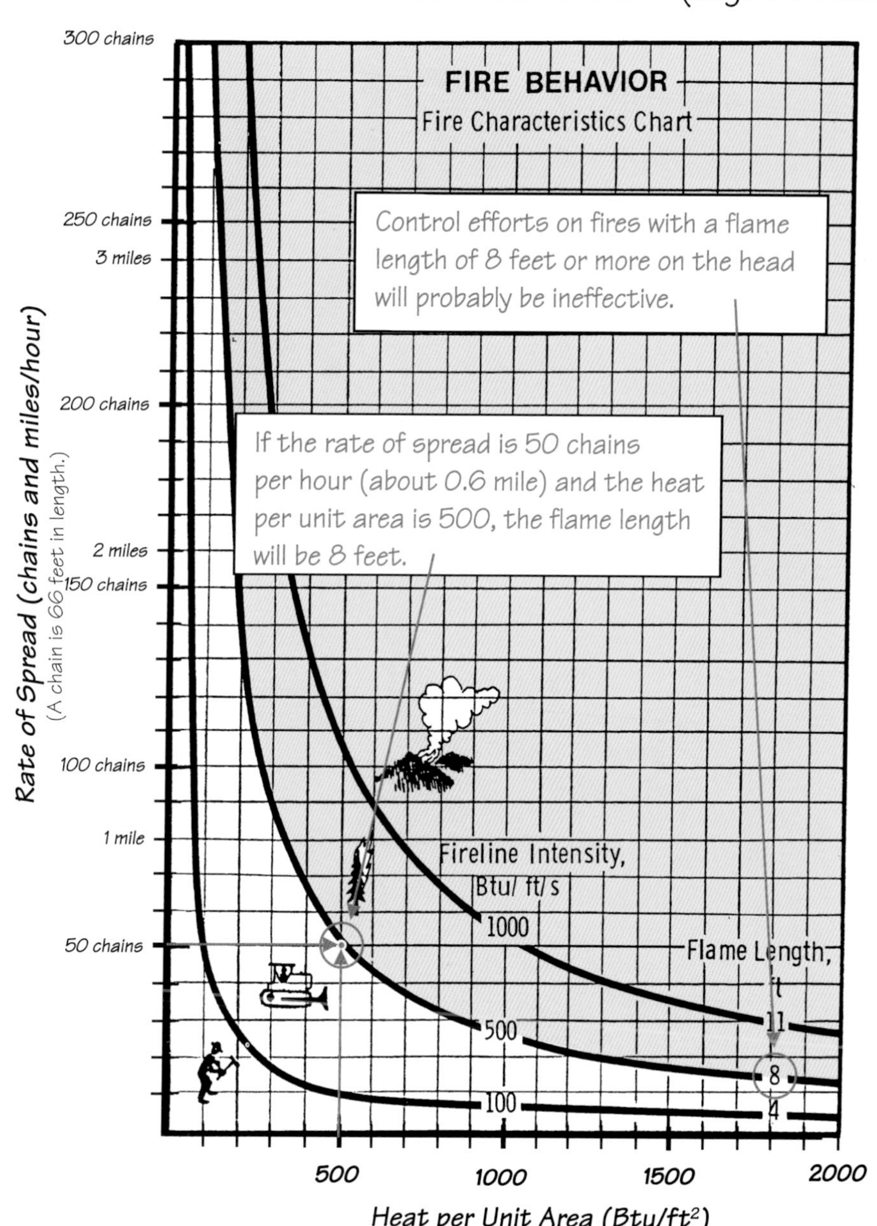

Figure 4.17 This chart for light fuels, although complicated, shows you the relationship between rate of spread, heat generated and flame length.

"haul charts," meaning you can haul hand tools, or you can haul bulldozers, or you can haul butt.

Surface Fire Intensity

Typical energy release from wildland fuels is about 8,000 BTUs per pound. If the fuel contains resins, oils or waxes, this can add significantly to the energy output. This energy is released during combustion, with most of it going into the rising convection column or radiating in all directions. A small portion (usually less than 20 percent) is effective in heating unburned fuels. The unburned fuels are "heated" by radiated heat generated from the flaming zone of the fire (Figure 4.18). If there is a wind, or the fire is burning on a slope, convective heat and direct flame contact play a part in the heating of unburned fuels. The relationship between the flame zone and unburned fuel is changed in the same way by a 50 percent slope or a 5 mile per hour wind (Figure 4.19).

ENERGY GENERATED
80%
Only about 20% of the heat energy in the form radiation is heating unburned fuels. Most is dissipated into the atmosphere.
20%
Kari Greer

Figure 4.18 About 80% of the energy released from a wildland fire goes into the atmosphere in the convection column.

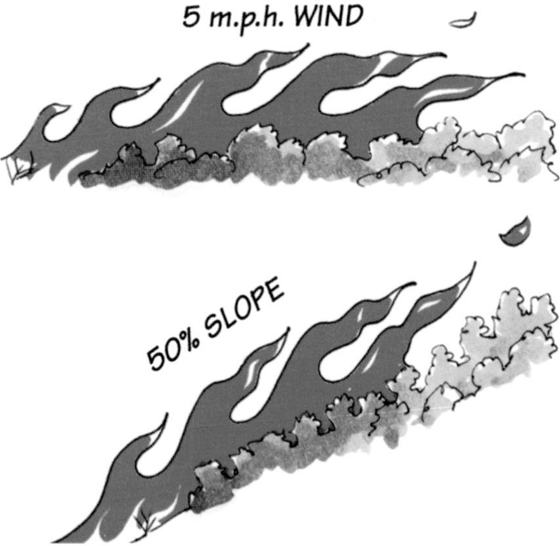

Figure 4.19 A 5 mile-per-hour wind will impact rate of spread in the same way as a 50 percent slope.

The rate at which unburned fuels are preheated changes based on fuel type, its moisture content, fuel temperature, etc. If the intensity of the fire increases, the rate at which a fuel is preheated increases. Increased heat production by the fire, more efficient heat transfer, and a reduction in the amount of heat required for ignition will increase the rate of spread.

Fire Environment Indicators

There are seven general indicators which can help you predict the possibility of extreme fire behavior. To ensure that you are tracking these indicators, you must take into consideration the seven factors that constitute the fire environment (Figure 4.20). These factors are:

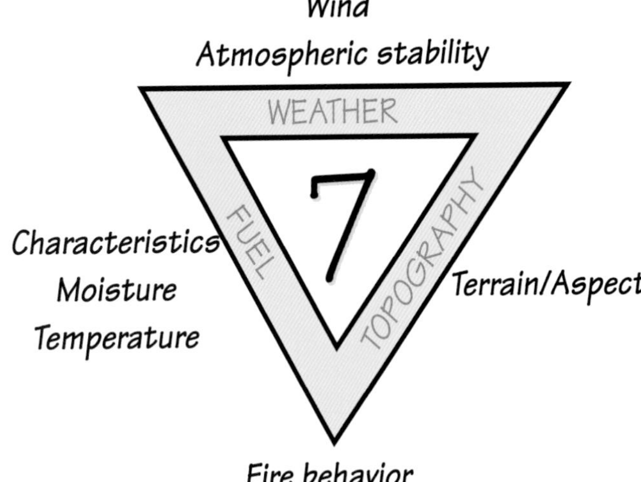

Figure 4.20 There are seven general areas within the fire environment that can indicate the potential for extreme fire behavior.

- ***Fuel characteristics*** - What is the fuel? How is it arranged? Is it compact? Is it loosely arranged?
- ***Fuel moisture*** - This includes both dead and live fuel moistures.
- ***Fuel temperature*** - Are the fuels shaded? Can you expect the fuels to warm in the next few hours, and thus change the fire behavior?
- ***Topography/Aspect*** - Is it steep? Will the fire be moving into better or worse terrain? Is the fire burning on a south-facing slope in the afternoon?
- ***Wind*** - What are the present and predicted mid-flame winds, local winds, and general winds?
- ***Atmospheric stability*** - Is there an inversion layer, and if so, when will it lift? Are there clouds that indicate unstable air which may influence the fire?
- ***Fire behavior*** - How is the fire burning? Is it making runs, or just "skunking around?" Do you expect the fire behavior to change? Is the level of fire behavior modifying the direction and speed of local winds?

There are seven factors that may predict the possibility of extreme fire behavior.

Jake says, "Fire environment indicators are like road signs; they can warn you of pending danger."

Fuel Characteristic Indicators

None of these extreme events will occur if there is not enough fuel to sustain it. Fuel characteristics change very slowly, but the type of fuel can change as the fire moves through an area. Look and see what is burning currently, and what type of fuel is close by and may burn soon.

Fine fuels are involved with the spread of most wildland fires. ***Continuous fine fuels provide the potential for rapid rates of spread.*** Remember, fine flashy fuels are one of the common denominators of fatal fires.

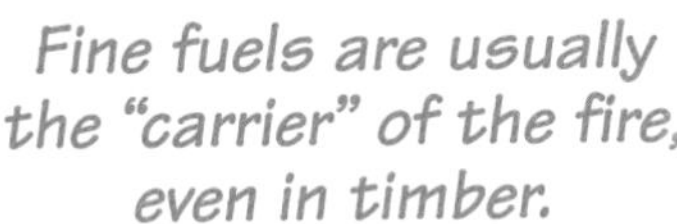

Fine fuels are usually the "carrier" of the fire, even in timber.

Do not underestimate the potential for a reburn and the danger it presents.

Heavy fuels can also spell disaster. An accumulation of dead-and-down fuels can set the stage for an intense fire that may burn for a long period of time (Figure 4.21). The ratio of dead-to-live fuels, and areas that have been logged, should also be of concern.

Another important factor is the arrangement of the fuels. Are there ***ladder fuels that will allow the fire to move into the tops of the fuels*** (trees, brush, etc.) and start a crown fire? The higher the intensity of the fire on the ground, the fewer ladder fuels necessary for moving the fire into the tops of the fuel. Brush and young tree reproduction are some of the more common ladder fuels.

The finer fuels will dry first.

Spacing of the fuels is another factor controlling crowning. If the spacing of the crowns is less than 20 feet, you have the potential for a crown fire. However, you can still have a crown fire with wider spacing, especially on a steep slope or when there is a strong wind present.

There are also some special fuel conditions that you should consider. Are there snags in the area? They are good sources of firebrands, and are also places where firebrands will ignite spot fires. Has the area been hit by bugs or frost, killing a lot of trees? Is there snow damage that has increased the ratio of dead-to-live fuels? Is the area suffering from drought? All of these are special factors that play a role in the fire environment (Figure 4.22).

Figure 4.21 There are miles and miles of dead and dying heavy fuel available for a fire. A fire in this setting has the potential to run for hours, consuming thousands of acres of quality timber and valuable homes.

Figure 4.22 There are thousands of bug-killed trees in this picture. It will take years for this forest to return to a healthy condition. A similar blight condition led to the fire disaster in Southern California in the fall of 2003.

One of the more deadly situations is when a low intensity fire has moved through an area burning the understory. The upper-level fuels have been dried out and can burn with extreme intensity. Reburns have been the cause of many serious accidents.

Fuel Moisture Indicators

Fuel moisture determines if fuels will be "available" to burn. Fine fuel moisture changes constantly and quickly, and is a primary reason why fine fuels are involved in many fatal fires. A good way to track the fuel moisture of fine fuels is to measure the relative humidity. In most areas, a relative humidity of 25 percent or less will allow fine fuels to burn with considerable intensity. The larger or 10-hour fuels are less reactive to relative humidity, but they can be "tested" by bending them. If they break easily, they are dry and will burn. If they bend, their moisture content is higher. They may still burn, but with less intensity. 10-hour fuels with a fuel moisture content below 7 percent are an indicator of dry conditions.

In most areas, when the relative humidity gets below 25 percent, things could get interesting.

The large, dead fuels (usually called the 1000-hour fuels) also play a part in the equation. In the high country, where there is considerable snowfall, these fuels don't begin to influence fire behavior until summer, after they have had time to dry out. If there is a drought, or there was an early spring, these fuels may come into play earlier. ***If you are told the 1000-hour fuels are 20 percent, you know that they will burn.*** If they get into the low teens, they are very dry and should be of real concern.

Aspect can be very influential on how a fire will burn.

Wildland fuels normally dry in a sequence. The finer fuels, such as pine needles and grasses dry out first. The next fuels to dry are the 10-hour and larger fuels. Live fuels are the last fuel class to be effected in the drying sequence. You can see that, as the season progresses and conditions become drier, more fuels will become "available" to be burned.

Fuel Temperature Indicators

The temperature of fuels is directly related to how much heat it will take for ignition to occur. If the fuels are cold, it will take more heat energy to bring the fuels up to the point of ignition.

If a fuel is shaded, it will be considerably cooler than adjacent fuel that is exposed to direct sunlight and solar radiation. This condition changes as the sun moves through the sky. The topographic aspect, as it relates to the position of the sun, is an indicator of fuel temperature. You should pay attention to: aspect, the position of the sun, and fuel temperature, as each of these factors will definitely affect fire behavior.

Terrain Indicators

The terrain or topography of an area directly affects both the rate and direction of fire spread. Although the topography is fixed, where the fire burns in the terrain is constantly changing. You must have a feel for the terrain in which the fire is burning, as well as what is ahead. Study the terrain for its relationship to the wind and sun (Figure 4.23). Watch for terrain indicators that warn of changes in fire behavior:

Stay away from chutes, saddles and narrow canyons.

Figure 4.23 There are several terrain indicators to watch for: steep slopes, box canyons, chutes, saddles and narrow canyons. Some of them may be hidden by fuels.

• ***Steep slopes*** have a dramatic effect on fire behavior. Not only do steep slopes allow a fire to move quickly uphill, they also allow burning material to roll downhill. Steep terrain is one of the common denominators in fatal fires.

• ***Box canyons or chutes*** can act like a chimney, channeling both wind and fire. This channeling can cause very rapid rates-of-spread.

• ***Saddles*** are another terrain feature that you should be concerned with. They also cause a channeling effect that can contribute to rapid rates-of-spread. Fireline constructed on ridge lines most likely will cut across saddles. These low spots on the ridge line are not places to park vehicles or spend a lot of time if fire is below you.

Jake says,
"Sometimes you will not be able to see that you are in a chute because it is hidden by the fuel cover. Look carefully; don't be surprised!"

- ***Narrow canyons*** pose another problem. As a fire backs down one side of the canyon, its narrowness allows for preheating of the other slope. Narrow canyons increase the chance of spot fires, which can jeopardize the safety of anyone in the canyon floor. When the fire does reach the bottom of the canyon, slope reversal can occur. Rapid rates of spread can be expected. Intersecting drainages should be of concern, because they can cause erratic winds and increased fire behavior.

Wind Indicators

Wind is the most influential factor when it comes to the fire's rate and direction of spread. Moderate to strong winds provide a means for wind-driven fire runs, and for transporting firebrands. Generally, a wind of just 10 miles per hour can cause rapid rates of spread.

Wind-driven fires move in a very predictable direction.

Wind can be measured in several ways. It can be directly measured using the wind gauge in your weather kit, or you can use the Beauford Scale to estimate the wind speed (Figure 4.24). In general terms, a wind of 5 miles per hour will cause the leaves of a tree to flutter and move. A wind of 10 miles per hour will cause standing single trees to sway from side to side. At 15 miles per hour, there will be considerable movement in the brush and timber, and dust will be stirred up.

Knowing what the wind is doing and what wind is predicted is important.

The clouds can also be an indicator of wind conditions. Lenticular clouds indicate strong winds at high elevations. These clouds form when strong winds pass over a mountain range. There is the potential that the wind may surface on the lee side (the side under the lenticular cloud), causing strong downslope winds.

Fast moving clouds indicate high winds, and may be signaling a wind shift. Be especially alert if the clouds are moving in a direction different from the surface winds.

A fire can develop its own wind.

Approaching cold fronts are another concern. As they move through an area, they bring shifting winds and increased wind speeds.

MODIFIED BEAUFORD SCALE FOR ESTIMATING 20-FOOT WIND SPEED

Wind class	Range of speeds (mph)	Nomenclature
1	3 or less	VERY LIGHT - Smoke rises nearly vertically. Leaves of quaking aspen in constant motion; small branches of bushes sway; slender branchlets and twigs of trees move gently; tall grasses and weeds sway and bend with wind; wind vane barely moves.
2	4 - 7	LIGHT - Trees of pole-size in the open sway gently; wind felt distinctly on face; loose scraps of paper move; wind flutters small flags.
3	8 - 12	GENTLE BREEZE - Trees of pole-size in the open sway very noticeably; large branches of pole-size trees in the open toss; tops of trees in dense stands sway; wind extends small flag; a few crested waves form on lakes.
4	13 - 18	MODERATE BREEZE - Trees of pole-size in the open sway violently; whole trees in dense stands sway noticeably; dust is raised in the road.
5	19 - 24	FRESH - Branchlets are broken from trees; inconvenience is felt in walking against wind.
6	25 - 31	STRONG - Tree damage increases with occasional breaking of exposed tops and branches; progress impeded when walking against wind; light structural damage to buildings.
7	32 - 38	MODERATE GALE - Severe damage to tree tops; very difficult to walk into wind; significant structural damage occurs.
8	39 or over	FRESH GALE - Surfaced strong foehn winds; intense stress on all exposed objects, vegetation, buildings; canopy offers virtually no protection; wind flow is systematic in disturbing everything in its path.

Figure 4.24 The Beauford Wind Scale is an internationally recognized method of describing a wind condition and estimating the wind speed using indicators.

Some cold fronts have squall lines of thunderstorms that you can see as they approach. But, the first indication that a dry, cold front is passing through is a change in wind direction. The effects of a passing cold front should not be underestimated. If you have any active fire near control lines–especially on the south and east flanks–be very careful.

Thunderstorm buildups in an area can cause strong, erratic winds. If you see mid-level, puffy clouds in the morning, they may be altocumulus castellanus clouds. These clouds signify instability, and are a strong indicator that thunderstorms are possible later in the day.

A thunderstorm can develop very quickly; sometimes in less than 30 minutes. As a thunderstorm builds, air is moved toward the base. Once the cloud is mature, violent downdrafts may occur. Downdrafts are strong winds that move downward inside the storm. If you see virga (rain that isn't reaching the ground), the downdrafts more than likely have begun. When they hit the earth's surface, they can travel in all directions. If this occurs close to or over a fire, it can be "spectacular"; fire will go in all directions. If you have a thunderstorm within 10 miles of your fire, be prepared for strong, erratic winds.

> *Jake says, "When a foehn wind is blowing, you know where the fire is going. Do not forget that you may have a wind direction reversal when the foehn winds begin to die. If you forget this, you may be caught with your pants down!"*

Foehn or gravity winds can also cause problems. These regional winds are so notorious, they have been given special names such as, Santa Ana, Mono and Chinook. Foehn winds are the result of high-pressure systems moving air over mountain ranges. As this occurs, the air is heated and dried by compression. These winds can last for hours or days, yet their patterns are very predictable. When they blow, the wind is normally from one direction. As they begin to degrade or weaken, local winds may begin to move in. This can cause wind reversals in just minutes (Figure 4.25). This "battling of winds" is especially prevalent along the Southern California coast, where the Santa Anas do battle with the onshore winds from the Pacific Ocean. A sudden calming of winds, or a shifting or wavering smoke column, are both indications that the wind direction will be changing. This type of event doesn't have to be right on the coast. It can occur some miles inland.

USDA Forest Service

Figure 4.25 One of the more famous of the foehn winds is the Santa Ana. When it blows, you know where the fire is going. When they are weakening, the normal onshore winds come ashore and can reverse the fire's direction by 180 degrees. This is a satellite picture of the series of fires in Southern California in the fall of 2003. Note how the Santa Ana winds are blowing from the north-northeast down over the mountains and out to sea.

Atmospheric Stability Indicators

Stability refers to the resistance of the atmosphere to vertical (up and down) movement. The chance of extreme fire behavior increases as the atmosphere becomes unstable. An unstable atmosphere allows a fire to develop "vertically" and grow rapidly. Some of the best indicators of stable and unstable conditions are listed in Figure 2.39 on page 91.

Another important indicator of changing conditions is the breaking down of an inversion layer. An inversion layer exists when the atmosphere is stable (no vertical movement of air) and a warmer layer of air tops a colder air mass. When this warmer air mass is relatively thin, it is called a thermal belt. In this belt the air will be warmer and drier. Fires burning in an area that falls within this belt will burn with greater intensity than fires above or below it.

The presence of a thunderstorm in your vicinity WILL impact you in one way or another.

When an inversion is present, fire behavior below it will normally be subdued. However, conditions will be very smoky and very hard on firefighters who have to work at a level within the inversion layer. As the day begins to warm up and the air becomes unstable, the inversion layer will begin to break down. You may recall from the chapter on weather that the atmosphere becomes unstable when winds begin to mix the air layers, and when solar radiation begins to heat the earth's surface, causing the air to lift. If you know the country, you will be able to predict when an inversion layer will begin to break up. This is the time to be alert and cautious.

Fire Behavior Indicators

Fire develops in stages. There are several fire behavior indicators which can help you predict what a fire will do. These indicators are:

Keep a lookout far enough away from the fire to observe the convection column.

- ***Smoke column*** - One of the best indicators of fire behavior is the smoke (convective) column (Figure 4.26). If it is ***leaning***, this is an indication of a wind-driven fire, one with a rapid rate of spread and short-range spotting. If the smoke column rises to some height and then is flattened off, this is a ***sheared*** smoke column. Strong upper-level winds are cutting off the top of the smoke column. With this situation, you have the potential for long-range spotting and strong surface winds (if the upper-level winds come down to the surface). If the smoke column is ***well developed***, rises to considerable height, and is capped with a very white cloud, you can expect intense burning conditions and unpredictable fire spread in all directions. If there is a light rain or the wind becomes calm, "watch out!" There is the possibility of downbursts. If the smoke column is ***changing color*** or ***beginning to rotate*** or ***roll***, this is an indicator that conditions are changing and the fire behavior is going to pick up.

- ***Trees torching*** - If trees are beginning to torch either one at a time or in groups, the situation is getting worse. This change means the relative humidity is dropping (Figure 4.27).

- ***A smoldering fire that is beginning to pick up*** - If the fire you are on has been just "skunking around" and not doing much, then suddenly begins to burn with some vigor, conditions have changed and it is time to "watch out!" This increase in fire behavior could be the result of high temperatures, lower relative humidity, or an increase in wind. More than likely, it is a combination of all three factors.

READING THE COLUMN OF SMOKE

Leaning smoke column

- A wind-driven fire.
- Rapid rates-of-spread.
- Short-range spotting.

Sheared smoke column

- Smoke rises straight up.
- Column is sheared off by strong winds aloft.
- Potential for long-range spotting.
- Strong winds could surface.

Santila Studios

Well-developed smoke column

- Intense burning conditions.
- Unpredictable fire spread in all directions.
- Look for capped top.
- Strong downburst potential.

Changing smoke column

- Beware when the column begins to change color or rotate.

Kurt Taylor

Collapsing smoke column

- Beware when the column begins to collapse, The downdraft winds can cause the fire to run in all directions.

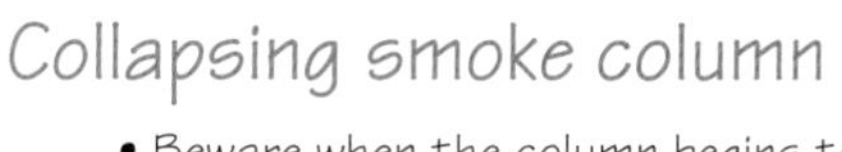

Figure 4.26 The smoke column is a good indicator of wind conditions and fire behavior.

- ***Firewhirls*** - The presence of firewhirls is not just an indicator of unstable air, but an indicator that the fire is going vertical; it is developing a strong convection column. The fire may be on the verge of crowning. Firewhirls may cause an increase in the level of spotting (Figure 4.28).

Flat firebrands can travel some distance. Wood-shakes produce a flat firebrand.

Figure 4.27 When you start noticing individual trees "torching," it could be an indication that the humidity may be dropping.

• ***Spot fires*** - If you are beginning to have more spot fires than you can handle, this is a sign that conditions are deteriorating rapidly.

It is your responsibility to track the weather and the associated fire behavior. You may have to use lookouts to aid you. As you track the fire's behavior, several of these indicators may catch your attention at the same time. However, even one indicator may be a "heads up" to changing conditions. If you sense a change, you must review your strategies and tactics to see if they, too, must change.

Wes Schultz

Figure 4.28 A firewhirl doesn't have to be large to be an indicator to start reassessing your condition. The fire behavior may be on the verge of changing in a dramatic way.

Extreme Fire Behavior Contributing Factors

The previous section discussed the characteristics or warning signs of extreme fire behavior. This section will attempt to show how ***various contributing factors work together to actually cause a fire to move into the third dimension***.

Under certain circumstances, fires can exhibit special fire behavior conditions. These include area ignition, crowning, extreme spread and blowups. Figure 4.29 is a listing of fire behavior indicators. Figure 4.30 outlines the burning characteristics of a wildland fire based on time of day.

When a fire begins to develop vertically into a third dimension, it begins to respond to processes that the fire is creating. For example, if a fire generates enough heat to break through an inversion layer, the vertical rush of hot gases will be replaced with cooler air. This rush of air toward the fire is in the form of a wind, generated by the fire itself.

There are several environmental factors that contribute to extreme fire behavior. They are: the availability of fuel; a wind event; and a dry, unstable air mass.

- ***Available fuels*** - The fire needs fuel to generate the heat intensity needed to produce a rapid rate of spread, lift firebrands to considerable heights, or develop strong convective wind patterns. If fuels have been stressed or killed by drought, bugs, disease, a freeze, or wind, the available fuels will be increased.

If individual tree torching is starting, things are changing...watch out!

FIRE BEHAVIOR INDICATORS

Indicator	Interpretation
Leaning smoke column	Rapid rates of spread and short-range spotting.
Sheared smoke column	Winds aloft may cause long-range spotting.
Well-developed smoke column	Intense burning and unpredictable fire spread in any direction can occur. When the power of the fire becomes stronger than the power of the local winds it is called a plume-dominated fire. The danger of a plume-dominated fire is the potential for downbursts.
Changing smoke column	A smoke column that is changing to a darker color, beginning to rotate faster or split, can indicate the fire is increasing in intensity.
Trees are beginning to torch	Torching trees are an indicator that a fire is starting to transition from a surface fire to a crown fire.
Smoldering fire is picking up	Weather factors are changing. An increase in fire intensity can be anticipated.
Firewhirls are being observed	Firewhirls are another indicator that a fire has the potential to move from a surface fire to a crown fire.
Frequent spot fires	Spot fires increase fire spread and complexity.

Figure 4.29 There are numerous fire behavior indicators. Learn to recognize them and what they may mean and how they will impact the strategies and tactics you are using.

BURNING CHARACTERISTICS BASED ON TIME OF DAY

Time of day	General prediction
1000 to 1800 hours	All factors of fire intensity are at their highest: air is dry; fuels are dry; temperature is high; and winds may be strong.
1800 to 0400 hours	Factors favorable for fire control: winds usually moderate; air is cooler; relative humidity usually increases; and fuel moisture may increase.
0400 to 0600 hours	Time of the day when a fire can most easily be controlled. Burning usually remains slow until dawn.
0600 to 1000 hours	Fire intensity begins to increase, making fire control more difficult.

Figure 4.30 The burning characteristics of a wildland fire normally change based on the time of day. Use these general predictions in planning your attack.

The stronger the wind, the farther a firebrand may be carried.

- ***Wind*** - Winds capable of producing extreme fire behavior can be the result of several weather conditions. The winds could be associated with the passage of a front (e.g., South Canyon and Point fires), or as a result of a foehn or gravity wind (Santa Ana or Chinook winds). The wind provides oxygen, but more importantly, it facilitates the transfer of heat from the burning fuels to the unburned fuels. This transfer can be in the form of a direct transfer, (e.g., moving the flames closer to the fuels, or by carrying firebrands out in front of the fire).

- ***Low atmospheric moisture content*** - Dry air means drier fuels. The drier the fuels, the easier it is for them to ignite.

- ***Unstable air mass*** - If there is instability in the atmosphere, there is a greater chance that the fire will develop a strong convective action. That, in turn, creates a wind that will further intensify the fire.

Crown Fires

Crown fires are one of nature's most dramatic and spectacular events. They are also the fastest spreading of all fires. It is not uncommon for a crown fire to move at a rate of 5 miles per hour or more. Crown fires not only occur in heavy timber fuels, but also in the canopies of tall shrubs, brush, chaparral, plantations, and such.

Crown fires develop in three progressive stages: passive, active and independent. Crown fires begin as an individual torching tree or bush. If conditions are right, the torching fuel will cause other trees to torch, and a crown fire is born. Sometimes, the crown fire will move back down to the surface, and then move up into the crowns again later. Each of the stages of a crown fire present a different challenge.

Winds are associated with most deadly fires.

- ***Passive crown fire*** - A passive crown fire is usually confined to one tree or a small grouping of trees (Figure 4.31). This stage of crown fire is commonly called "torching-out." A passive crown fire is entirely at the mercy of a surface fire and the availability of ladder fuels. This torching-out is usually short-lived and confined to trees which are closely spaced. This type of crown fire may cause localized spotting.

Crown fires can move from the tops of the fuel to the ground and back to the crowns.

Northtree Fire International

Figure 4.31 Crown fires all start with a single tree "torching out."

Crown fires start with torching of individual trees.

Karen Wattenmaker

Figure 4.32 An active crown fire is supported by a ground fire. This extreme fire behavior is spectacular and very dangerous.

Kari Greer

Figure 4.33 This fire has sufficient energy to move independently from the ground fire.

• ***Active crown fire*** - An active crown fire is also dependent on the surface fire, but there is much more transfer of heat from one burning crown to the unburned crowns of other trees (Figure 4.32). The crown fire may run ahead of the surface fire, or the surface fire may precede the crown fire. Active crown fires often pulsate, moving from the crowns down to the surface and back into the crowns again. Be very careful when a fire is in this stage of activity.

• ***Independent crown fire*** - An independent crown fire gets its name from the fact that it is not dependent on the surface fire for its energy (Figure 4.33). The energy that supports this kind of crown fire is developed in the crowns as the fire moves from one crown to another. The surface fire spread is the result of the crown fire. Independent crown fires are not very common, but when they do occur, it is time you drop back and become an observer. There is nothing you can do to stop this type of event. Just get out of its way!

There are several environmental conditions which can contribute to the possibility of a crown fire. One of the conditions is the ***crown flammability*** of the fuel. A higher ***ratio of dead-to-live fuels*** will influence the ease with which a fire will move into the crowns. Low, fine fuel moistures promote crown flammability. The moisture content of the live fuels is also a factor. ***Live fuel moisture*** is controlled by the stage of growth, which is related to the time of year. In the spring when the fuels are actively growing, the fuel moisture content is usually too high to support a crown fire. Some fuels, like chamise, jack pine,

Jake says, "When you are up against a crown fire, you become an observer. Get out of the way of this dangerous situation."

and palmetto/gallberry, include ***volatile oils*** that support combustion. This makes them more susceptible to crown fires. ***Crown closure*** of 75 percent or more will contribute to the chance of a crown fire.

All crown fires start as a surface fire (Figure 4.34). Several factors determine the ease with which crowning occurs. The greater the ***surface fire intensity***, the more likely it is that the fire will move into the crowns. Heavy fuel loading, loosely compacted fuels, low fuel moisture contents, and unstable atmospheric conditions all impact the intensity of the surface fire. The ***vertical arrangement of the fuels*** is also a factor. If there are ladder fuels or low crown foliage, it will take less surface fire intensity to move the fire into the crowns. The ***steepness of the slope*** will affect crowning. The steeper the slope, the shorter the distance between the surface fire and the crowns.

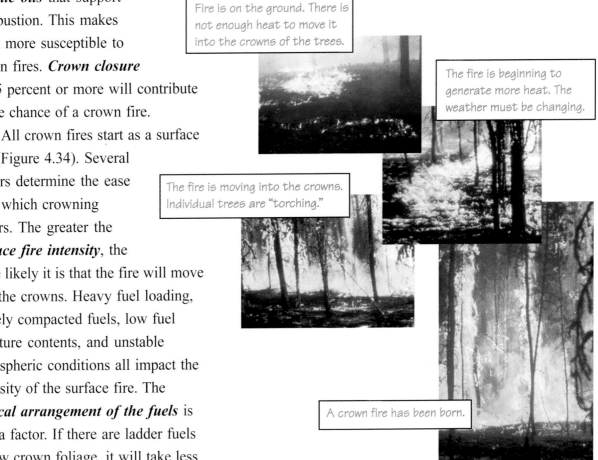

DEVELOPMENT OF A CROWN FIRE
Fire is on the ground. There is not enough heat to move it into the crowns of the trees.
The fire is beginning to generate more heat. The weather must be changing.
The fire is moving into the crowns. Individual trees are "torching."
A crown fire has been born.

Figure 4.34 If conditions are right, a fire may move into the crowns and run for some distance. This series of pictures shows the stages of a ground fire developing into a crown fire.

Even crown fires start as a surface fire.

Another contributing factor in crown fires is ***crown-to-crown heat transfer***. Both active and independent crown fires spread from crown-to-crown. Heat transfer between aerial fuels occurs more readily under certain conditions. ***Crown spacing*** of 20 feet or less (roughly 100 trees per acre) will permit convective and radiant heat transfer to occur at a level of intensity that can maintain a fire spreading through the crowns. ***Crown level winds*** of 20 miles per hour or more will permit and sustain convective and radiant heat transfer. The ease of crown-to-crown heat transfer is also affected by ***steepness of slope***.

Crown fires can occur in most fuel types. They are not limited to timber.

Figure 4.35 This fire has begun to crown and "move out." You will be hard pressed to stop it. Rethink your tactics if they involve being in close to this fire.

For an active or independent crown fire to sustain its run, several conditions are needed:

- Low fuel moisture.
- Relatively close crown spacing.
- Intense surface fire.
- Strong winds and/or steep slopes.
- The ability to spot ahead in discontinuous fuels.

Under the right conditions, an active or independent crown fire may sustain a run for miles (Figure 4.35). In 1988, at the same time Yellowstone National Park was on fire, the Canyon Fire in central Montana had a run of over 20 miles in one evening. Normally, however, changes in environmental conditions or fuels will end crown fires long before they travel miles.

The higher the dead-to-live ratio, the higher the chance for a crown fire.

Spot Fires

When a fire moves into the third dimension, spotting may be the result (Figure 4.36). There are several factors to consider when analyzing spotting potential. They are:

Trevor Wilson

Figure 4.36 This spotting is along a flank. It is a strong indication that there is considerable spotting out in front of the head.

- ***Probability of spotting*** - The probability of spotting is directly related to the intensity of the surface fire, burning conditions, overstory species, crown spacing, etc. On the other hand, cover canopy may intercept the upward travel of firebrands and limit the spotting problem.

- ***Number of firebrands*** - Most firebrands do not start fires. They either burn out before landing, land in an area where there is no fuel, or land in fuel that will not ignite. However, the more firebrands launched into the air, the greater the probability that some of them will ignite spot fires.

- ***Type of firebrands*** - For a firebrand to cause a spot fire, it first must travel beyond the flame front or across fireline. It must also still be burning with enough heat energy to ignite a receptive fuel bed upon landing. Another factor associated with the various types of firebrands is their aerodynamic capability. Dense, "chunky" firebrands don't fly very well, and thus will not travel great distances. On the other hand, flatter firebrands will travel some distance when lifted by convective heat and carried by the wind. One of the worst types of firebrand is produced by burning wood-shingle roofs. These firebrands can be large and flat, and their aerodynamic properties allow them to be easily lifted and carried great distances.

The hotter the fire, the higher firebrands will be lifted.

- ***Convective lifting*** - The hotter a fire, the more intense the vertical convective action, resulting in a stronger convection column and a higher lift to the firebrands.

The firebrands may not necessarily fall down-wind.

- ***Wind*** - Once a firebrand is lifted into the atmosphere, its travel is controlled by the wind, gravity, and the aerodynamics of the firebrand. Generally, the stronger the wind, the farther the firebrand will travel.

Wood-shingles make the worst kind of firebrands.

- ***Fuel bed*** - Firebrands must land on receptive fuels in order for them to ignite. Some examples are fine, dry fuels, such as litter, duff, rotten wood and grass.

- ***Environmental conditions*** - Before a firebrand can start a spot fire, it must not only land in a bed of receptive fuels, but the fuel moisture, wind and slope must favor fire spread in order for the spot fire to become a problem.

Jake says, "If upper-level winds are blowing in a different direction than the surface winds, you may be surprised where some long-range spotting occurs."

There are two types of spotting: short-range and long-range. They are characterized not by actual distance, but rather by how they relate to the original fire. Short-range spotting is defined as spotting that is quickly overrun by the main front of a fire. Long-range spotting may contribute to the spread of the fire, but in the same way a new fire would (Figure 4.37). There is no question that a spot fire 15 miles away from the main fire is the result of long-range spotting. However, going strictly

Jake Oosthuizen

Figure 4.37 Here is an example of long-range spotting. The main fire is some distance to the right.

by our definition, a spot fire at only 50 feet from the main fire that creates a new fire is also defined as long-range spotting. Conversely, a spot fire that is ¼-mile forward of the fire's head, but which is quickly overrun by the fire, is defined as short-range spotting. Long-range spotting can cause you serious problems. Short-range spotting usually does not contribute to the spread of the fire.

Spotting usually occurs downwind. This may not be the same direction that the surface winds are blowing. If the upper-level winds are moving in another direction, the spotting may be downwind in that direction. Also, spot fires do not necessarily "show themselves" right away. They may smolder for hours or even days. Spot fires don't usually travel alone. If there is one spot fire, his buddies may be close by!

Spot fires may hide for days; patrol your fire after it is contained.

There are seven different large fire types which demonstrate different spotting characteristics. Figure 4.38 outlines these characteristics.

Probability of Spotting

The probability of a firebrand starting a fire can be predicted, as you will see shortly. Probability of ignition is the percentage of firebrands "launched" that actually start fires. When the probability of ignition is 70 percent, it means that of ten glowing firebrands that land on receptive fuels, seven of them will, in fact, start a new fire.

The probability of spotting does not reflect the size or shape of the firebrand, nor the type of available fuels. It only deals with the probability that if firebrands are produced, what percentage will start a new fire (Figure 4.39).

Area ignition occurs when hundreds of spot fires start influencing each other, and an area "explodes" in fire.

Area Ignition

Area ignition occurs when an area is "peppered" with spot fires. The spot fires will begin to interact and affect each other. This can

SPOTTING CHARACTERISTICS OF LARGE FIRES

	Large fire types wind velocity	Dominant feature
I	Towering convection column with light surface winds.	Moderate to rapid fire spread persistent until fire environment changes.
II	Towering convection column over slope.	Rapid, short-term spread with convection cut off at the ridge crest.
III	Strong convection column with strong surface winds.	Rapid rate of spread with short-range spotting.
IV	Short, vertical convection column cut off by wind shear.	Steady or rapid fire spread with occasional long-range spotting.
V	Leaning convection column with moderate surface winds.	Rapid spread with both short- and long-range spotting.
VI	No rising convection column under strong surface winds.	Very rapid spread driven by combined fire and wind energy; frequent close spotting.
VII	Mountain topography with strong surface winds.	Rapid spread both up and downslope with frequent spotting and area ignition.

Figure 4.38 Spotting takes several forms, primarily based on what the wind is doing.

The probability of spotting is directly related to the type of fuel being burned.

occur in a canyon, valley, or on flat land. As the fires begin to draw each other closer, and their convective actions connect, the area may explode in fire. When this occurs, the fire may not be moving in any one particular direction, but may just take out a large chunk of real estate. In an area ignition, most of the energy release is vertical, not horizontal.

Crown fires can move very quickly.

PROBABILITY OF IGNITION TABLE

Shading (Percent)	Dry-Bulb Temp. (°F)	FINE DEAD FUEL MOISTURE (in percent) 2	3	4	5	6	7	8	9	10	11	12	13	14	15	16	17
Unshaded <50%	110+	100	100	80	70	60	60	50	40	40	30	30	20	20	20	20	10
	100-109	100	90	80	70	60	60	50	40	40	30	30	20	20	20	10	10
	90-99	100	90	80	70	60	50	40	40	30	30	30	20	20	20	10	10
	80-89	100	90	80	70	60	50	40	40	30	30	20	20	20	10	10	10
	70-79	100	80	70	60	60	50	40	40	30	30	20	20	20	10	10	10
	60-69	90	80	70	60	50	50	40	30	30	20	20	20	20	10	10	10
	50-59	90	80	70	60	50	40	40	30	30	20	20	20	10	10	10	10
	40-49	90	80	70	60	50	40	40	30	30	20	20	20	10	10	10	10
	30-39	80	70	60	50	50	40	30	30	20	20	20	10	10	10	10	10
Shaded >50%	110+	100	90	80	70	60	50	50	40	40	30	30	20	20	20	10	10
	100-109	100	90	80	70	60	50	50	40	30	30	30	20	20	20	10	10
	90-99	100	90	80	70	60	50	40	40	30	30	20	20	20	10	10	10
	80-89	100	80	70	60	60	50	40	40	30	30	20	20	20	10	10	10
	70-79	90	80	70	60	50	50	40	30	30	30	20	20	20	10	10	10
	60-69	90	80	70	60	50	40	40	30	30	20	20	20	10	10	10	10
	50-59	90	80	70	60	50	40	40	30	30	20	20	20	10	10	10	10
	40-49	90	80	60	50	50	40	30	30	30	20	20	20	10	10	10	10
	30-39	80	80	60	50	50	40	30	30	20	20	20	10	10	10	10	10

Figure 4.39 The Probability of Ignition table provides a clue as to the potential for spot fires based on the shading, temperature and fuel moisture. (Also known as Ignition Index.)

Large, unbroken fields of fuels can contribute to blowup conditions.

Extreme Fire Behavior Patterns

Wildland fires that are demonstrating extreme fire behavior can be classified as wind-driven or plume-driven, depending on which force is strongest. In a wind-driven fire, the power of the wind is dominant; in a plume-driven fire, the forces that are developing the plume or convection column are dominant.

Wind-Driven Fires

With a wind-driven fire, you usually know in which direction it will be burning.

Most fires that become large are driven by a wind, and are classified as wind-driven. With a wind-driven fire, you usually have an idea what direction it will burn in. The smoke (convection) column is pushed over by the wind, driving the heat into the unburned fuels (Figure 4.40). Spotting is normally downwind, and is a major contributor to the spread of the fire. This spotting can occur miles ahead of the main fire. A crown fire is very likely. A wind-driven fire may become a plume-driven fire, when the energy released by the fire takes the convective column to such heights that its force becomes greater than the wind.

Whenever you have a wind-driven fire:

- The rate-of-spread is rapid.
- There is long-range spotting.
- The direction of spread is predictable.
- The flanks of the fire may be safe to attack.
- Wind shifts should be a major concern.

Kurt Taylor

Figure 4.40 Wind-driven fires can move very quickly. Note how the smoke column is pushed over and held close to the surface by the wind. Long-range spotting can be a problem. The direction of spread is very predictable.

Plume-Driven Fires

The convection column and the forces that develop it are the dominating forces on this type of event (Figure 4.41). The direction of travel and the rate of spread are harder to predict. Spotting usually does not occur at great distances, but can occur in any direction. Whenever you have a plume-driven fire:

- There can be a sudden increase in fire intensity.
- Spotting occurs in all directions.
- The direction of spread is difficult to predict.
- Downdrafts should be a major concern.

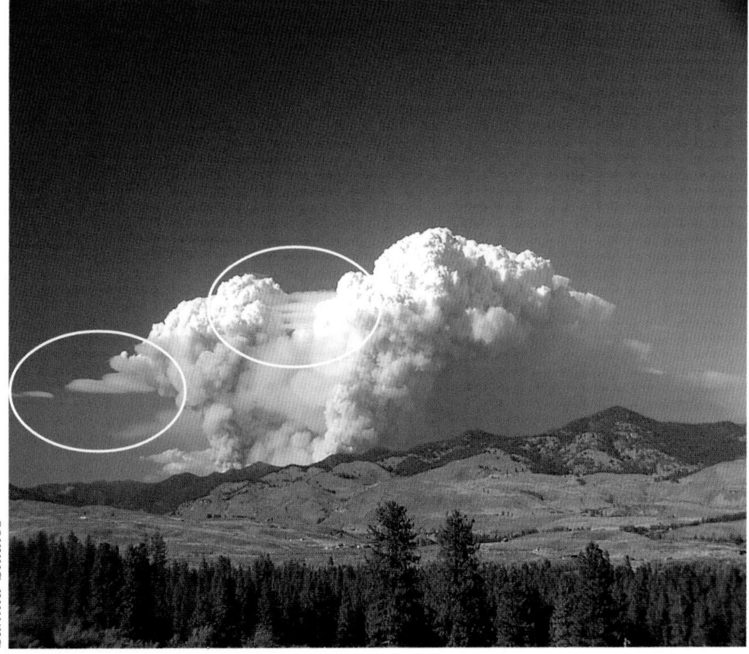

Santila Studios

Figure 4.41 Plume-driven fires occur when the fuels and/ or topography allow a fire to develop great energy in a vertical sense. The smoke column will reach to thousands of feet and usually be capped with a white cloud. The direction of spread for a plume-driven fire is unpredictable. It will go wherever the fuels and topography will take it. Note how severely the twin-columns are sheared off, and the lenticular cloud formations to the left and between the two columns.

There are two types of wind events that can accompany a plume-driven fire: indrafts and downdrafts. ***Indraft*** winds move in to replace heated air that is lifted by the convective action of the fire.

Jake says, "It is much harder to be an accurate fire behavior analyst on a plume-driven fire. It is very difficult to predict what they will do. Be prepared for changes in fire behavior."

Indrafts provide oxygen to the flame front, increasing convective heating and preheating the fuels. This creates a "cycle of reinforcement," where the increases in the wind increases the intensity of the fire, which in turn further increases the wind.

Downdrafts occur below the convective column. When the heated air is lifted by the convection column, and subsequently cooled in the upper atmosphere (by the effect of lapse rate or evaporation of rain), it rushes back toward the surface of the earth. When it hits the ground, the wind spreads in all directions, causing real havoc for firefighters.

If you are working on a plume-driven fire, and you begin to see virga and/or rain, downdrafts may be developing. The indraft winds usually drop off just before the downdraft hits. Survivors of fatal fires have reported feeling light rain and cool air and a short period of calm just before the fire exploded.

Figure 4.42 is the fire progress map for the Harlow Fire that occurred in July 1961 in the Central Sierras of California. The green area represents the two hours from 1600 to 1800 hours of July 11. During this two hour period, the fire consumed 20,000 acres and traveled over 10 miles...all as the result of massive spotting and area ignition.

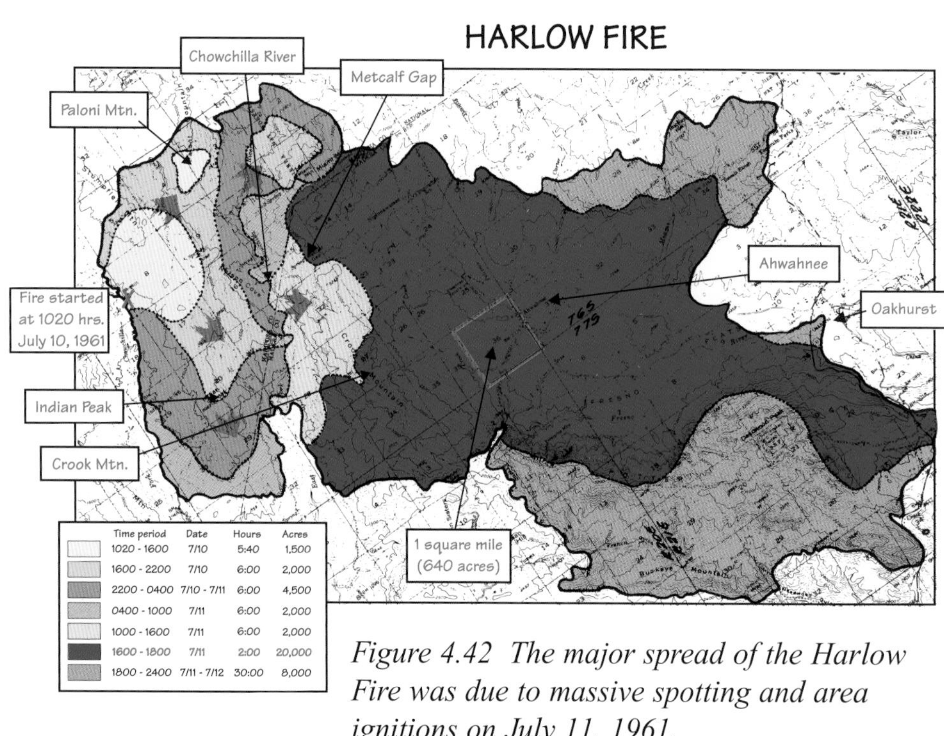

Figure 4.42 The major spread of the Harlow Fire was due to massive spotting and area ignitions on July 11, 1961.

5

Fire Extinguishment Methods

There are several ways to "break the fire triangle" and suppress a fire. You can use water, retardant, Class A foam or dirt to cool or smother the fire, eliminating heat and oxygen. You can cut a fireline with hand tools or bulldozers to remove fuels, or apply a gel barrier to protect some high-value fuels; and you can use fire, itself.

The method or methods you choose will depend on the type of fuel, the fire behavior, the terrain, the management objectives and the firefighting resources that are available to you. Pick and choose, using the right tool for the right job at the right time.

Choose the right tools for the right job at the right time.

Use of Water

Water is the most effective fire-suppressing agent known. Water cools the fire, breaking the "heat" side of the fire triangle. It can also dilute the "oxygen" side of the fire triangle when water vapor is created. Water, however, will be most effective if applied as part of an overall strategy that includes a fireline cut to mineral soil. Never consider a "wet line" as the final control line.

Jake says, "Many a fire has been lost because the wet line was not followed up with a line cut to mineral soil."

Water Management

To get the most out of the water you have available, apply just enough to do the job. Remember, it doesn't take a gallon of water to put out a match. It takes very little water to cover large volumes of fuel, especially if it is broken up into a spray, because each small droplet can cool many "units of fuel" at the same time. ***One volume of water will cool 300 volumes of burning fuel, IF applied properly.*** Apply the water to cool the fuel, but realize that it will continue to burn unless the water is spread over all the fuel, cooling it below the ignition point.

Jake says, "Water is still the best tool in the firefighter's tool chest."

A single drop of water will extinguish a match, if it is applied to the base of the flame (Figure 5.1). There is always a base of a flame; sometimes it will be difficult to reach. If grass or light brush is burning, apply a spray to the base of the flame, where it meets the fuel. If your target is a tree, start at the base, working up the trunk (Figure 5.2). Work as close to the burning fuel and with the finest spray possible. Apply just enough water to extinguish the fire. Shut off the nozzle when you don't have a specific target. ***Don't waste water! Make every drop count; most supplies are limited.***

Figure 5.1 Make every drop of water count.

Don't waste water. If you don't have a "target," shut the nozzle off.

Karen Wattenmaker

Figure 5.2 The key to water conservation is making every drop of water count. Match the water to the fire.

The nozzle should be used properly. Most nozzles provide a "combination" of streams (straight-stream, a spray or a fog). Straight streams are used for reach, but are usually not very efficient (Figure 5.3). They tend to apply more water than is needed. Sprays should be used to attack most wildland fires. The spray pattern uses water more efficiently by cooling larger areas with less water. Fogs can provide some protection from intense heat, and are

Apply water to the base of the flame.

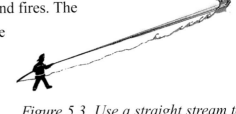

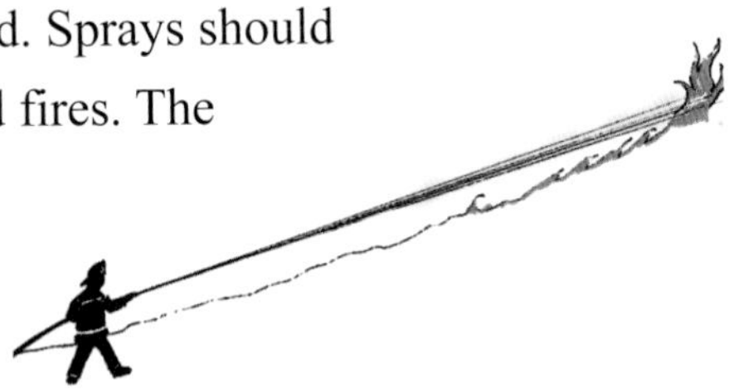

Figure 5.3 Use a straight stream to reach a hot spot or flare-up.

mostly used in structural firefighting (Figure 5.4).

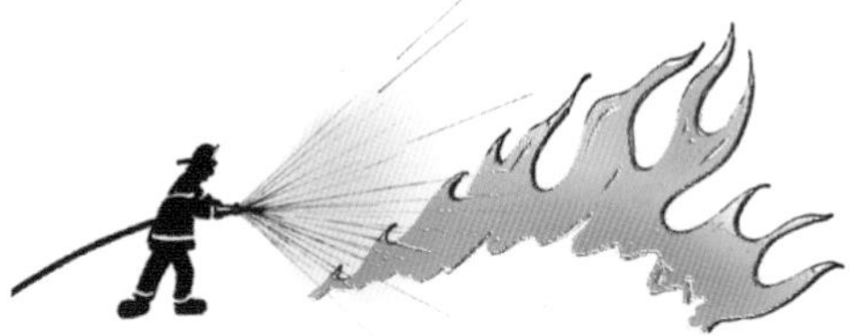

Figure 5.4 Use a wide spray or fog pattern when working in close to the fire.

Wide spray patterns can be used to shield you from heat.

Select the right stream for the situation or fire conditions. Use the straight-stream to reach in and cool an area of fire, so that you can move in and start using a spray. Also, use a straight-stream if there is fire high up in a tree, if there is a hotspot up the line that you need to knock down, or to penetrate deep into soil or ash pile during mop-up. Vary the stream as you encounter different situations (Figure 5.5).

Jake says, "A little spray goes a long ways."

There will be times when you will not have enough water volume to cool down the fire. Unless you have a specific reason to keep applying water to an area when it isn't having any effect, you may be wasting it. If you can't provide enough water to cool down the fuel and gain the upper hand, you will have to change your attack.

Figure 5.5 Vary the stream pattern to meet the need. The key is to apply the proper amount of water to do the job.

Use a spray-stream for most of your work. Use a straight-stream for reach.

You want to apply the right amount of water, but move as fast as you can. Move along the fire's perimeter, remembering that you are not there just to knock the fire down, but to put it out. Constantly check the fire edge you have just put out for small "smokes" that will indicate there is still enough heat in the area to reignite the fire. Apply water and shut it off. This will allow you to see if you are applying just enough water, or overkilling and drowning it. As you move from a smoke or hotspot, shut off the flow of water until you have a specific target.

If you are working in the "black," be careful of hot material.

In lighter fuels, you can sometimes work from inside the burn or "black." Move along the fire's perimeter, parallel to the fireline (Figures 5.6, 5.7 and 5.8). If the fuels are

Figure 5.6 Sometimes you can work from the "black."

Figure 5.7 In this case, the light fuels and fire behavior allow a safe attack from inside the black.

Figure 5.8 There are times when working from the "green" is the best positioning for an attack.

heavier, you will have to work from the unburned area or "green." Move in and again start moving parallel to the fireline. Use the spray to attack the base of the fire. A spray pattern can also be used to shield you from heat. The water droplets will absorb radiated heat.

Jake says, "If the firefighter on the nozzle holds the hose over his or her shoulder, they will have a greater percentage of re-ignitions behind them. Hold the nozzle close to the ground...you get greater penetration of the water into the fuels."

Use the water stream to put in a water "scratch line." This is where you try to ensure that the very edge of the fire perimeter is dead out. Don't drown the area. Experience will give you a sense of when you have applied enough water to do the job.

Wetting Agents, Foams and Gels

Water is a very effective fire-suppressing agent. Water is one of the best firefighting tools we have. There are several chemicals that can be added to water to make it more effective. Historically, wetting agents have been added to allow water to penetrate the fuels or ash piles. Then came Class A foam concentrates that produced a wet, foamy bubble solution that could "hold" moisture on unburned fuels, allowing it to cool and penetrate longer. The latest additives are the gels. These products involve the use of a chemical that holds a large number of water molecules together creating a thick, wet gel that can stick to fuels like houses, protecting them for days.

Constantly check the line behind you for "smokes" or other sources of heat.

Wetting Agents

Wetting agents are chemicals that, when added to water, will reduce the surface tension of water and increase its penetration and spreading capabilities (Figure 5.9). These types of products are especially effective during the mop-up stage of a fire (Figure 5.10). They are more efficient (and less expensive) for this use than Class A foam. Wetting agents are added directly to the water tank.

Plain water tends to stay on the surface of the fuels or ashes.

Water with a wetting agent added will penetrate the fuels.

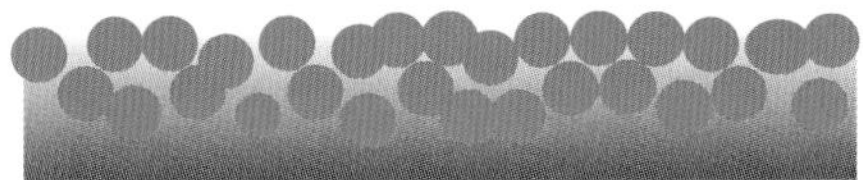

Figure 5.9 The surface tension of water tends to keep it on the surface of the fuels. When a wetting agent is added, the "strength" of the surface tension is reduced, and the water will penetrate deeper into the fuels.

Class A Foams

Water with a Class A foam added to it is even better. ***Don't confuse Class A foams with other classes of foams. They are very different and cannot be interchanged with foams for other types of fires*** (Figure 5.11).

Don't confuse Class A foam with other classes of foam.

Class A foam is a chemical which, when added to water in the right amounts, creates bubbles having a lower density than water. The foam is made by introducing air into the water/foam mixture.

Bubbles shield the fuel and gradually release the moisture they contain.

The bubbles stick to the fuels and gradually release the moisture they contain. The bubbly water absorbs heat more efficiently than plain water. The layer of bubbles may provide a barrier to oxygen needed for combustion and can insulate the fuels from heat (Figure 5.12). The reduced rate of water release results in more efficient conversion of water to steam, enhancing the cooling effect. In addition, the wetting agent contained in the foam allows the water to penetrate the fuels and reach deep-seated heat (Figure 5.13). The bubbles can also provide a protective barrier for some unburned fuels. The ratio of Class A foam concentrate to water is normally 0.1 to 1.0 percent.

Karen Wattenmaker

Figure 5.10 A wetting agent allows the water to penetrate deep into the ash pile, cooling it.

The classifications of fire types

Class A Fires	Wood or other combustible materials.
Class B Fires	Petroleum fires - those involving gas, oil or other petroleum products.
Class C Fires	Electrical fires - those that involve charged electrical lines.
Class D Fires	Metal fires - those that involve burning metals like magnesium.

Figure 5.11 There are four different types of fire. Wildland fires are Class A fires. Class A foam is for use on Class A type fires.

Safety when Handling Foam Concentrates

Even though foam concentrates are similar to common household detergents and shampoos, be very careful when handling them. Always use goggles and waterproof gloves. If possible, use disposable coveralls. Do not wear your leather boots. If some of the concentrate is spilled on them, your feet will become soapy and slippery. You should be trained in the proper handling of these products, and follow the manufacturer's instructions.

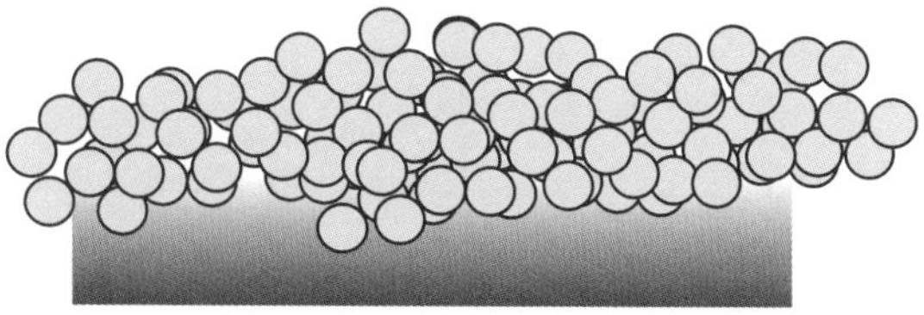

Figure 5.12 Class A foam can provide a layer of air bubbles. This layer can insulate the fuels from heat.

Environmental Concerns

When appropriately applied as part of a firefighting operation, Class A foams are biodegradable, and have a minimal effect on the forest vegetation and soils. However, due to the sensitivity of some aquatic plants, animals and fish, the direct application of foam into bodies of water should be avoided. Leave at least a 100-foot buffer zone from high-water marks.

Plain Water
Surface Tension
Unable to Penetrate
Water with
Wetting Agent
or Class A Foam
Less Surface Tension
Able to Penetrate

Figure 5.13 Wetting agents and Class A foam allow water to penetrate more effectively.

Various Foam Mixing Systems

Class A foams work best if they are mixed in the appropriate amounts on the discharge side of the pump. There are several proportioning systems on the market which can be classed as either ***manual regulation*** or ***automatic regulation***.

The manual regulation systems incorporate batch mixing, a suction-side proportioner, in-line proportioners or eductors, and around-the-pump proportioners. The automatic regulating systems are designed to accurately proportion foam concentrate for a wide range of water flows and pressures. This minimizes the necessity for the firefighter to constantly adjust the proportioner. All automatic systems add the concentrate to the discharge side of the pump, thus eliminating the exposure of the tank and pump to foam solutions. There are three automatic regulating systems on the market: ***balanced-pressure bladder tank*** proportioners; ***balanced-pressure pump*** proportioners; and ***electronically controlled direct injection*** proportioners. Become familiar with the foam induction system you will be using.

Know the foam induction system that you will be using.

Vary the foam solution to match the fire problem.

Compressed Air Foam Systems (CAFS)

Compressed air foam systems inject air into the foam solution with an air compressor. These systems are usually mounted on engines or water tenders because of the complexity of the system and the addition of the air compressor. They produce high-quality foam starting at a 0.3 percent solution (Figure 5.14).

Compressed air foam systems inject air into the foam solution.

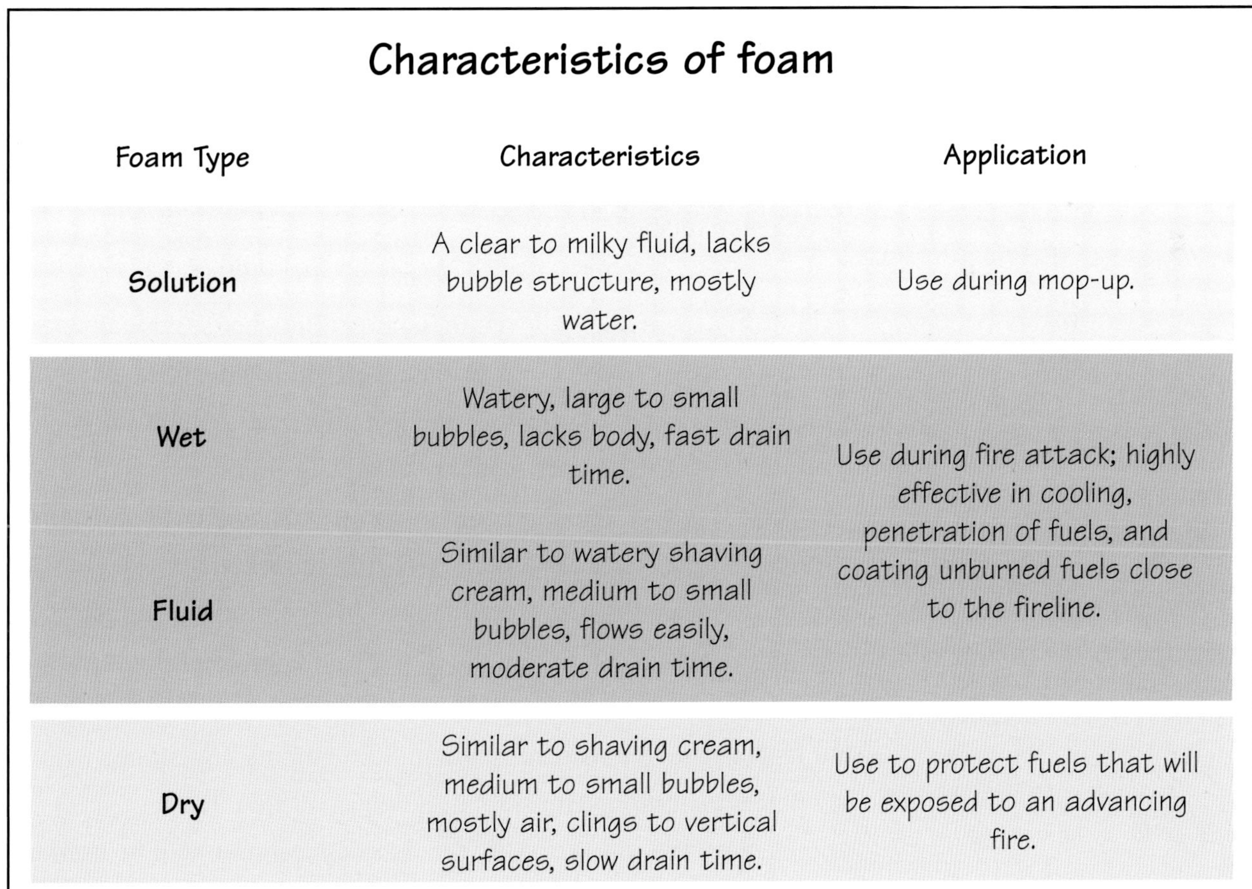

Characteristics of foam

Foam Type	Characteristics	Application
Solution	A clear to milky fluid, lacks bubble structure, mostly water.	Use during mop-up.
Wet	Watery, large to small bubbles, lacks body, fast drain time.	Use during fire attack; highly effective in cooling, penetration of fuels, and coating unburned fuels close to the fireline.
Fluid	Similar to watery shaving cream, medium to small bubbles, flows easily, moderate drain time.	
Dry	Similar to shaving cream, medium to small bubbles, mostly air, clings to vertical surfaces, slow drain time.	Use to protect fuels that will be exposed to an advancing fire.

Figure 5.14 Various characteristics of Class A foam solutions.

Nozzles

Conventional nozzles can be used to apply Class A foams. You can use straight-stream, spray or fog settings on nozzles. The foam applied with these nozzles will give you a "wet" water that will enhance the wetting and penetration of fuels, and the spread of the water. It will not, however, give enough foam structure (bubbles) to provide insulation or heat reflection. Conventional nozzles can produce foam with a mixing ratio from 0.1 to 0.3 percent foam solution.

Aspirating nozzles inject air into a fine stream of foam solution at the nozzle tip. This produces light, fluffy bubbles. Bubble size may vary based on the foam solution mix ratio. Light foams have larger bubbles that travel only a short distance from the nozzle. Wetter foams will be needed if they have to be propelled some distance. Aspirating nozzles can produce foam from 0.2 to 1.0 percent solution. The ratio is varied, depending on the type of application, fuels and fire behavior.

PROS of foam: more effective use of water; extends the "life" of water; reduces mop-up time; and easy to use.

Use of Class A Foams in Wildland Firefighting

Class A foams increase the effectiveness of water in ALL firefighting applications (Figure 5.15). Water alone "breaks" one side of the fire triangle. By adding foam to the equation, all three sides of the fire triangle can be attacked. Foam will cool, smother, and insulate the fuels.

Experience will be your best teacher. Fire could creep under a light, fluffy foam laid out over an unburned fuel bed. You will have to learn whether to use Class A foam as a wetting agent to allow the water to soak deep into the fuel and ash pile, or as a light, fluffy foam to protect exposures from an advancing fire.

CONS of foam: can irritate eyes; corrosive to some metals; and may be harmful to fish and animals when in high concentrations.

The ability of foam to continue to wet and cool fuels long after application is especially effective during initial attack. The best application is one that applies the foam in a continuous and rapid sweeping motion, knocking down the flame, and blanketing the smoldering fuels. This allows the foam solution to work, penetrating the fuels to prevent their re-ignition. Start gaining experience with a 0.5 percent solution. Adjust up or down as needed. Remember, if you increase the solution ratio, the foam concentrate will be used up quicker. Don't adjust your foam ratio above 1.0 percent. Experience shows that this actually reduces the effectiveness of the foam.

Advantages and disadvantages of Class A foam

Advantages	Disadvantages
Increases the effectiveness of water.	Can be irritating to the skin and eyes.
Extends the useful "life" of water.	Corrosive to some metals and may speed deterioration of some types of seal materials.
Provides a short-term heat barrier.	May have harmful environmental effect in high concentrations.
Effective on all types of Class A fires.	Reduced life expectancy of leather goods such as footwear, and can be very slippery.
Reduces suppression and mop-up time.	
Relatively easy to use (mixing and handling).	
Visible from ground and air.	

Figure 5.15 The advantages and disadvantages of Class A foam.

Foam attacks all three sides of the fire triangle.

Steam is a sign you have reached the source of heat.

Use wet foam during mop-up.

As you advance along the fire front, apply some of the foam to the unburned fuels adjacent to the burned area. This will cool and coat these fuels, to help prevent them from igniting. When penetration of a fuel bed is needed, move the nozzle closer to the fuels, applying more foam...this will aid in extinguishment (Figure 5.16). Move quickly. As soon as steam is visible, you know you have reached the seat of the heat.

Figure 5.16 Class A foam is being used to suppress this grass fire. The foam allows the water to penetrate the matted grass. This firefighter should be holding the nozzle closer to the fuels, if possible.

Wet line constructed using foam can be used to "fire" from during an indirect attack. Apply foam at least 2½ times as wide as the expected flame length from the burning operation. Apply foam to trees and brush that are adjacent to the wet line. Don't use a dry foam during mop-up. This will slow penetration into the hot fuels, not reaching deep-seated heat.

Dry foam is not for mop-up.

Foam is also very effective when used to fight a structure fire. Apply a dry foam to the exterior of a building. Loft the foam on the walls, roof, under the eaves, and any other exposed areas. If the fire is in the interior, use a wetter foam. This will increase the cooling and penetration actions.

Firefighting Gels

Firefighting gel is just "thick" water.

Firefighting gels are now being approved for use on the fireline. Since water has such a great ability to absorb heat, if you can build up a layer of water, it can't be penetrated with a blowtorch. Think of gel as very "thickened" water (Figure 5.17).

Gels are normally too thick to pump through hose lines.

Gels involve the use of a long-stringed "super absorbent" polymer that can absorb water at a ratio of at least 50 to 1 (water to polymer by weight). Water is mixed with a polymer that attracts the water molecules and holds them together. The best mix is about 97 percent hydration; just enough below saturation that the polymer still wants to absorb more water, and it attempts to steal water molecules from adjoining polymer strings. This is what holds the gel together. If you under-hydrate, you get a mess that can only be applied with a putty knife or shovel. If you over-hydrate, you get a very watery gel that will not stick properly to fuels or building surfaces. The recommended gel ratio for structure protection is about 3 percent gel concentrate. So, for every 100 gallons of water, you should use 3 gallons of concentrate (Figure 5.18). If you are going to apply gel to vegetation, you can use a 1 percent mixing ratio.

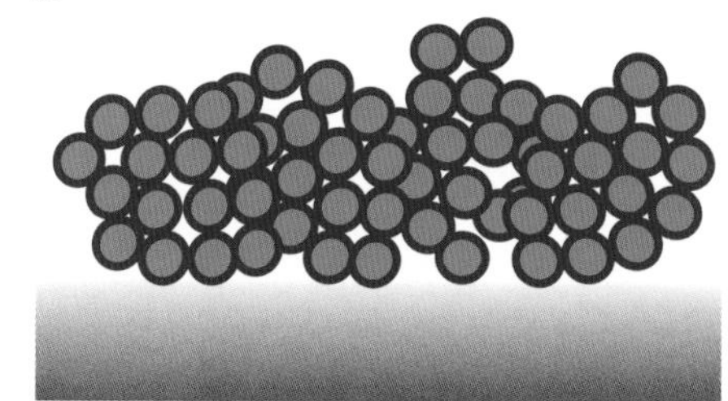

Figure 5.17 Water has a tremendous ability to absorb heat. A gel, properly mixed and applied, can provide a very effective barrier.

The recommended mix for gels is 3 gallons of gel concentrate for every 100 gallons of water.

Figure 5.18 Firefighting gels are becoming more commonplace on the fireline. They are very effective in protecting exposed structures for up to 72 hours, depending on the weather and the percentage of the mix.

The polymers come in two forms, powder or liquid. The powdered form is much more difficult to use. It is used when you can batch mix in quantity, such as an airtanker application, for example. The liquid form usually involves mixing the powder with mineral oil. Since the polymer is attracted to water, not oil, the oil holds the polymer in suspension. This mixture is then mixed with water at the nozzle…and you have gel.

Since most gels used on the fireline involve a mixture of polymer and mineral oil, you have to keep this mixture from separating out. Like paint, the polymer will tend, over time, to settle to the bottom of the container. It is recommended that you agitate the container every six months or less. Without periodic agitation, the polymer particles can settle to the bottom of the storage container, and you run the risk of mixing water with just mineral oil, which, of course, will not produce gel. Also, once the polymer has settled to the bottom of the container, it is very difficult to get it suspended evenly back into the solution.

Properly mixed and applied, gel can be a very effective tool to protect structures and other valuable flammable fuels. But, gels are not "firefighter fool-proof." In other words, if you don't know what you are doing, you'd better not attempt to mix and apply gels. You must be properly trained in their mixing and use. You can't use the old adage, "if a little is good, a lot is better."

Don't attempt to use gels unless you are properly trained.

It is nearly impossible to pump gels through fire hose, so they have to be inducted at the nozzle. Although it can be a real "pain" having to carry the gel concentrate around with you, gels are proving to be an effective asset to firefighters. Gels cover and protect, whereas wet water and Class A foams allow water to penetrate the fuels better.

Gels can remain effective for many hours.

When protecting a structure, it is sometimes not necessary to treat the whole house. If the house has a wood roof and siding, single-pane windows, and very little clearance, then full protection is warranted. But if the roof is fire resistant and the siding is stucco, you may only have to treat the windows, attic vents, etc. You may also be better off treating the fuels adjacent to the house. This creates a deeper defensible space.

You have a greater "window" of application with gels. If you wet a house or fuels down with water, you have only a short impact period. If you use a Class A foam at about a 5 percent mixture, you have a much longer window. If you use gels, you may have up to 24 hours before you have to rehydrate or reapply.

Gels can damage paint.

The gels will lose their effectiveness over time. Wind and air temperature will eventually evaporate the water. But, the polymer is still there, and if you lightly mist the area, they will rehydrated. If you use too strong of a spray, you will rehydrate, but also wash the gel off of the surface.

Gels and fire retardant mixtures should be removed from the home once the fire threat has passed. Gels and fire retardants are chemicals that can discolor paints. Again, you have to know what you are doing. You don't want to just come in and wash the stuff into the creek.

Gels are expensive.

Not all of the gels are the same. So read the labels and follow the instructions for safe mixing, application and removal. Of all the water additives available to firefighters, gels are the most unforgiving. You have to know what you are doing before you attempt to use gels. You must become very familiar with the products and system your agency uses before you begin fighting fire.

One last point on gels; they are very expensive. You might not be that concerned with the cost, but you should be. As an employee of some organization, you must consider the costs of what you are doing. Any tool that you are given has a proper and most effective and efficient use. Don't use gels if water will work, and only apply the gels to surfaces that need protecting.

Fireline is used to isolate burned and unburned fuels.

Use of Fireline

You want to put a fireline in only once.

Control line is a comprehensive term used for all the constructed or naturally existing fire barriers and treated fire edges used to control the fire. Some examples of existing control lines include streams, lakes, ponds, rock slides, areas of sparse fuels, roads, canals or previously burned (cold) fireline.

Fireline refers to any cleared strips or portion of a control line from which flammable material has been removed by scraping or digging down to mineral soil.

A fireline is constructed for two purposes: to create a "safe strip" from which to start burning out to remove fuels between the fireline and advancing fire; and to isolate the burned area from the unburned area. The goal is to create a gap in the flammable materials which prevents the fire from continuing to spread. Fireline can be constructed using hand tools or mechanized equipment. Some things you should think about before you begin: You want to put this line in only once! Where should it be placed? How wide does it have to be? What method will you use? Do you have the firefighting resources to construct AND hold the line? ***The success of your attack is often dependent on where the fireline is placed and how it is constructed.***

Final control lines should be "to mineral soil."

Fireline Descriptions

Different types of fireline are constructed to control the spread of fire using a variety of methods and equipment. They are constructed around flanks, heads, slopovers, islands, pockets, fingers and spots and are described as follows:

- ***Wet line*** is a line that has been constructed using water or foam (Figure 5.19). A wet line is constructed to extinguish the flame front or to be used as a safe strip to burn from. Except in VERY light fuel, a wet line should not be considered the final control line. The final control line should be cut through the fuel to mineral soil.

- ***Retardant line*** is usually constructed by an airtanker or helicopter (Figure 5.20). Treat a retardant line just like a wet line. Follow up with some ground action.

- ***Scratch line*** is a hasty, narrow line cut in the fuels to temporarily stop the spread of the fire. It can be widened later and become the final control line.

- ***Hand line*** is constructed using hand tools (Figure 5.21).

Figure 5.19 This engine company is constructing a wet line. It should be followed up with a line cut to mineral soil.

Karen Wattenmaker

Figure 5.20 Aerial delivery of fire retardants is very effective. Large helicopters and airtankers drop millions of gallons of retardant each year. In most cases, these drops should be followed up with a line to mineral soil.

Figure 5.21 Hand line and cat line are both lines to mineral soil. One is constructed by hand and the other by bulldozer.

Figure 5.22 This is an undercut line. Its position on a slope makes it difficult to construct and hold.

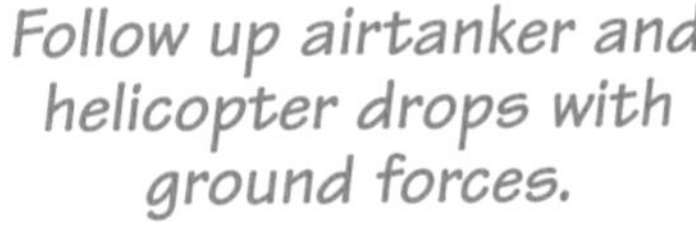

Follow up airtanker and helicopter drops with ground forces.

- ***Undercut line*** is a line that is constructed on a hillside when there is the possibility of burning materials rolling down and crossing the fireline. Undercut line incorporates a trench into its construction. It can also be called a trenched line (Figure 5.22).

- ***Cat line*** is constructed using bulldozers or tractor-plows. In lighter fuels, other types of mechanized equipment, such as graders and scrapers, can be used (Figure 5.23).

- ***Cold line*** is a fireline that has been controlled. The fire has been mopped-up for a safe distance inside the line and can be considered safe to leave.

- ***Hot line*** is line that still has active fire along it (Figure 5.24).

- ***Open line*** refers to an open fire front, where no line has been constructed.

- ***Detonation line*** is made by specially trained detonation crews that use explosives to construct a fireline.

Figure 5.23 Fireline constructed with a bulldozer is called a "cat line."

- ***Blackline Concept*** - Fuels that remain between the main fire and the control line are burned-out, or allowed to burn to the control line. This method ensures that fuels and heat remain inside the control line and prevents the fire from making a run at the control line. This action provides for safety. Line is not completed until fuels are burned-out to prevent flare-ups.

- ***Anchor Point*** - An anchor point is an advantageous location, usually a barrier to the fire spread, from which to start building a fireline. An anchor point is used to reduce the chance of firefighters being flanked by the fire.

Parts of the Fire

Various parts of a fire are identified by specific names (Figure 5.25). Common terminology is important for accurate communication on the fireline. The following are commonly used terms:

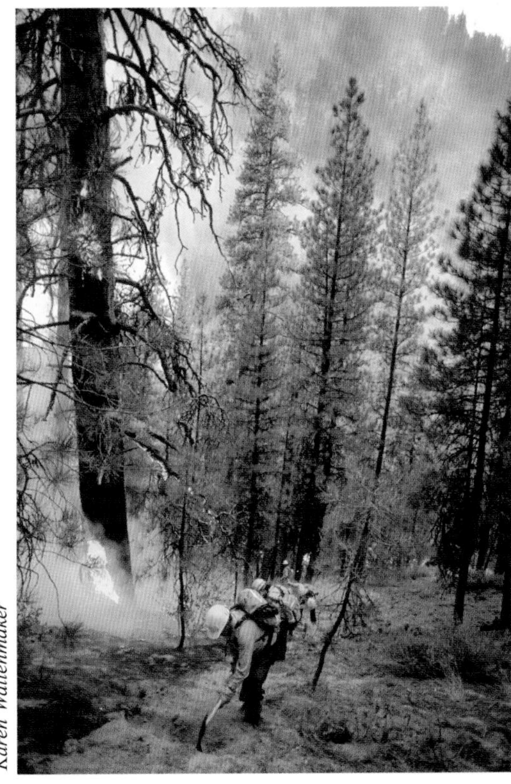

Karen Wattenmaker

Figure 5.24 This is considered a hot fireline, because ground crew members are working close to active burning.

- ***Origin*** is where the fire started. Protect this area for the fire cause determination and investigation.

- ***Head of the fire*** is the "running edge" of the fire, usually spreading with the greatest speed. It is driven by the wind or topography. It is not uncommon to have two or more heads on a fire.

- ***Rear of the fire*** is usually opposite from the head, closest to the origin and nearest the source of the wind. The rear edge of the fire is usually burning slower than other sectors of the fire. Sometimes this is called the ***heel*** or ***base*** of the fire.

Jake says, "Until you have a control line to mineral soil around the whole fire, you are not done!"

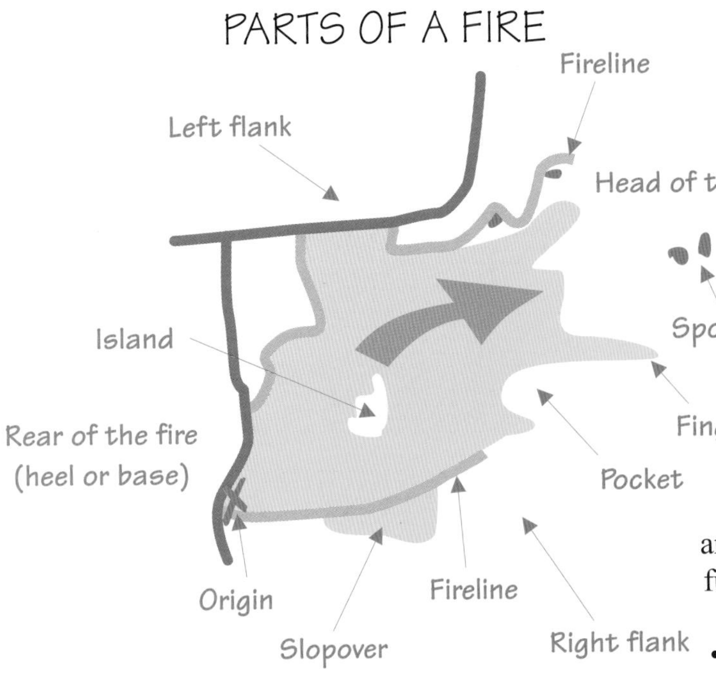

Figure 5.25 These are the names of the various parts of a fire. It is important that everyone uses the same terminology.

- ***Flanks*** are the sides of the fire. Usually they are not burning as hot as the head. The left flank is the left side looking toward the head from the origin or base of the fire. The right flank is on the right side of the fire.

- ***Perimeter*** is the total length of the outside edge of the burning or burned area.

- ***Slopovers*** occur when the fire crosses the fireline. The fire and slopover are adjacent to each other with no unburned fuel between them.

- ***Fingers*** are caused by a shift of wind or change in topography. They develop behind the head, extending from the flanks. They may become second heads.

- ***Islands*** are patches of unburned fuels inside the fire's perimeter. Some form of suppression action must be taken on islands that are close to the line.

Slopovers occur when the fire crosses fireline.

- ***Pockets*** are deep indentations of unburned fuel along the fire's perimeter. Normally, fireline will be constructed across pockets and they are then burned out.

- ***Spot fires*** are burning areas outside the main fire perimeter, usually caused by wind blown embers or rolling debris.

Principles of Fireline Placement

Line placement is critical to the containment and control of the fire. There are several simple principles of fireline placement. Follow them as best you can; they will make your task safer, easier and more successful.

Always anchor fireline.

- ***Anchor firelines*** - Unless you are assigned to "hot spot" (to slow the spread of fast-burning fingers), anchor your line to a barrier or other control line. If you don't, the fire will probably move around your line and outflank you. Anchor your line!

- ***Utilize natural or existing barriers*** - These are roads, trails, streams, lakes, rock outcroppings, or any other break in fuel that can be used as a fireline. Use reason here - don't just use "natural or existing barriers"; use them if they will help you. The key is not to construct line if a barrier already exists.

Jake says, "Do not ignore the rear or heel of the fire. If the wind changes, it may become the head."

- ***Go around heavy fuel concentrations*** - If possible, keep your fireline in the lightest fuels you can. Don't make the task of line construction and mop-up any harder than you have to. Avoid snag patches. Burning snags will throw firebrands into the area and across firelines. Snags are also "magnets" for burning embers.

Some form of suppression action must be taken on islands close to the line.

- ***Include spot fires in your control lines*** - If an area is saturated with spot fires, construct your line around them and burn-out. Make your job easier; there is no reason to construct fireline around every small spot fire.

- ***Keep the line as straight as possible*** - The key here is to avoid sharp turns in your line. Bends, sharp turns and corners put more fuel closer to the burning fuel. Fire is more likely to cross firelines at turns and bends. Some turns and bends in fireline are inevitable, however. Construct sweeping turns, or widen the line. Keep a special watch on these areas, especially when the fire is still active.

Keep lines straight; sharp turns and corners will be problems.

- ***Locate the fireline close to the flame front*** - Attack the fire as close to the flame front as possible. This action will tend to keep the fire smaller. It is also much safer to work close to the "black"; you can use it as a safety island.

Use existing barriers when possible and practical.

- ***Construct line at the base of a hill*** - By constructing the fireline at the base of a hill, you avoid the need for an undercut line to catch burning material that may roll down the hill.

- ***Construct line on the backside of a ridge*** - By placing the fireline on the backside of a hill or ridge, you can construct a narrower line, because the heat from the advancing fire will be less likely to cross the fireline.

Construct fireline at the base of hills, or on backs of ridge tops. Avoid side-hill fireline if possible.

- ***Plan for fire spread*** - Allow enough time and distance to construct, burn out, and hold the line. If you don't plan ahead, all of your effort will be wasted. By allowing time for the necessary tasks to be accomplished, you will save acres in the long run. Remember, you want to construct the line only once!

Other Considerations of Fireline Construction

In wildland firefighting, the construction of fireline is often the primary goal. You may encounter a critical situation such as saving lives and/or property, but ultimately, constructing and holding line is your primary mission. Selecting line placement may be as simple as using roads, trails, or other existing barriers. Or, it may be quite complicated, such as when you have to deal with heavy fuel concentrations, steep topography, or certain land use restrictions. Some of the factors that will influence the type of fireline you construct are:

Why are you constructing the fireline here?

- ***Safety of personnel*** - Safety is not limited to such concerns as downhill fireline construction or working below bulldozers; it also involves the qualifications of the personnel. Safety is # 1 in everything you do. Never move on to the next step of an action plan until your blueprint for safety is established.

Safety is first!

- ***Values at risk*** - The value of what is being threatened by fire has a significant influence on where and when you will construct fireline. If a fire is moving toward homes, you may place a priority on constructing line to protect them first. The key is to know the values at risk and to prioritize them accordingly.

Values at risk is a distant second.

Jake says, "If the cost of suppression is going to exceed the value of what is being protected, rethink your plan."

- ***Land use restrictions*** - If an area has been designated wilderness, involves archeologically sensitive materials, or if the landowner has special restrictions, you may be limited to construction methods that have little or no impact on the land. If there are such restrictions in your area, become familiar with them. Use the most effective line construction technique that is allowed under the restrictions. It will be imperative to ask permission before using techniques that may be prohibited.

- ***Terrain features and access*** - You may be restricted from an area by the terrain; it may be too steep for bulldozers or unsafe for hand crews to work. The terrain may be too rocky for a mobile attack with engines. Or, the fire may be deep in a narrow canyon where airtankers cannot safely operate. Unfavorable terrain features may require your retreat to areas that are more favorable.

Make the task of line construction easy on yourself.

- ***Availability of resources*** - You can't use certain types of firefighting resources if they aren't available. Design your plan around the tools and apparatus that you will have at hand.

- ***Fire behavior*** - The level of fire activity will have a significant influence on where and what type of fireline is constructed. If the fire is running hard, lines should be placed far enough ahead of the fire so you have adequate time to complete construction. The fire's flame length, intensity and resistance to control will influence the width of the fireline.

Jake says, "If you don't have a plan to hold the line you are constructing, don't waste your time and energy. Do something productive!"

- ***Speed*** - The old saying, "the quickest route between two points is a straight line," may not always apply to firefighting. If this straight line takes you through a snag patch or heavy brush, you will want to go around it. You do want to construct line as fast as possible, but never construct line that you won't be able to hold. It is better to put in one mile of line and hold it than to put in three miles of line and lose it.

Fireline Width

The width of the fireline is dictated by the fuel, topography and fire behavior. ***As a general rule, the fireline should be at least 1½ times as wide as the predominant fuel is tall*** (Figure 5.26)**.** In areas where you normally expect extreme fire behavior, the width of the fireline should be 2 or more times the fuel height. Many times firelines will have to be much wider, compensating for expected flame length and wind direction. Line width is not controlled by the fact that you may expect spotting.

Fireline should be at least 1½-times the height of the ground fuels.

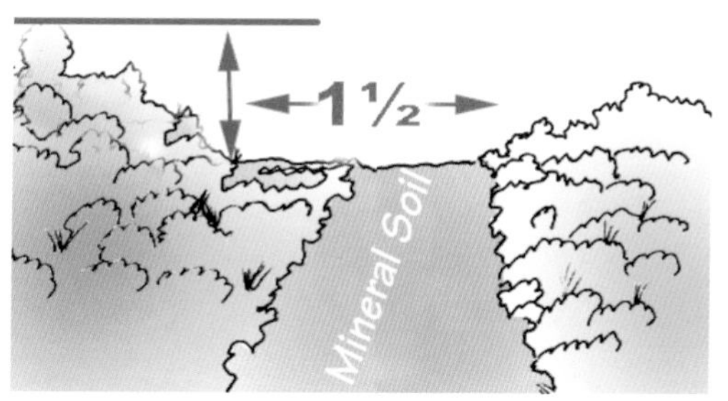

*Figure 5.26 The fireline should generally be **1½** times the height of the ground fuels. When working in timber, this rule is not practical.*

The steeper the slope, the wider the fireline.

Fire can cross a fireline in several ways (Figure 5.27). Radiant and convective heat may ignite fuels outside the line if it is too narrow or does not have adequate overhead clearances. Convective heat may carry burning brands across the line, starting spot fires. The fire may cross a fireline which has not been cut to mineral soil. Burning material may roll down a slope, crossing an improperly constructed undercut line. Burning snags may fall and cross the line. Gusts of wind, whirlwinds and firewhirls may carry brands across the

Larger fires are harder to control.

line. Firefighters sometimes accidentally spread burning embers across firelines. And, fire may pass under a fireline by burning along a buried tree root. The potential for one or more of these events will dictate line width. See Figure 5.28 for general guidelines for the width of firelines.

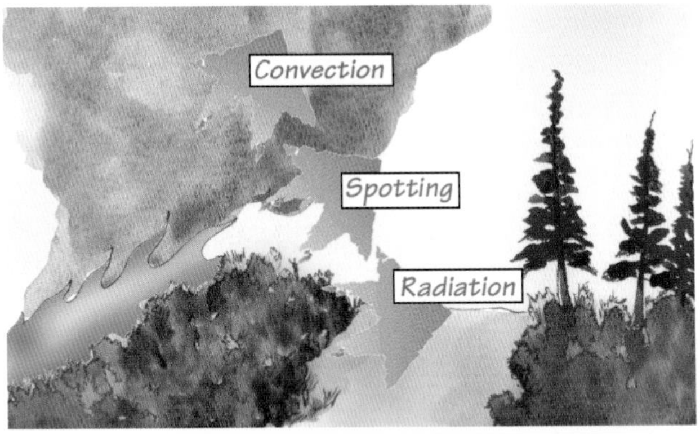

Figure 5.27 Fire can cross a fireline in three ways...by radiation, by spotting and by convective heat.

Convective heat may "launch" burning material across your fireline.

Do not underestimate the impact of radiated heat.

Guidelines for width of fireline

Fuel type	Width of cleared area	Width in mineral soil
Grass	2 to 3 feet	2 to 3 feet
Medium brush	4 to 6 feet	6 to 8 inches
Heavy brush	9 feet	1 to 2 feet
Very heavy brush or logging slash	12 feet	3 feet
Timber	20 feet	3 feet

Figure 5.28 Guidelines for the width of fireline and the width to mineral soil of various fuel types.

There are two fireline widths: the width of the cleared line and the width to mineral soil.

Anything that affects how a fire burns must be considered in determining the width of line needed to hold or control the fire. The hotter or faster the fire burns, the wider the control line must be. There are six factors that determine the width of the fireline:

- ***Fuel*** - The type of fuel, its height, density, size and condition will dictate fireline width. See Figure 5.29 for the effects of fuels on fireline width.

The effect of fuels on fireline width

Consideration	Concern
Kind of fuel	Some fuels burn hotter than others because of their oil content. The hotter the fuel burns, the wider the control line needs to be.
Height and density of fuels	The higher and denser the fuel, the higher and hotter the flame will burn, and the wider the control line needs to be.
Size of the fuels	Heavier fuels, such as logs, heavy limbs, and thick-stemmed brush do not ignite easily. However, once ignited, they burn very hot for a long time and may require wide control lines.
Condition of the fuels	The condition of fuel (whether it is dead or alive or dry) affects fire intensity. The drier the fuel, the hotter it will burn, thus the wider the fireline will have to be.

Figure 5.29 How various fuels will affect the width of fireline.

- ***Slope or topography*** - When a fireline is to be built above a fire burning on a slope, the steeper the slope the wider the line must be. This is because the fire usually burns faster and more intensely on steeper slopes. When a fireline is to be built below a fire burning on a slope, the width of the line is not dictated by the slope, but rather by the need for trenching. The steeper the slope, the deeper and wider the trench must be. Trenching is necessary to prevent rolling burning material from crossing the fireline.

Slope impacts the width of the fireline.

- ***Weather conditions*** - Weather conditions affect the intensity of the fire. The hotter the fire is burning, the wider the line should be.

- ***Part of the fire to be controlled*** - A fire burns hottest, with a longer flame length, on the head of the fire. The flanks generally burn with less intensity. This dictates wider firelines on the head.

The hotter or faster the fire burns, the wider the fireline must be.

- ***Size of fire being controlled*** - The amount of heat generated by a large fire has a bearing on the width of the line necessary to control the fire. The larger the fire, the wider the line.

- ***Possibility of cooling*** - The width of the fireline can be reduced if water is available for cooling the fuels.

When constructing fireline, consider the fuel, its type, height, density, size and condition.

Fireline Construction

Fireline can be constructed with hand tools, mechanized equipment, water, or retardant. The only reliable line is one which has been cut to mineral soil; that will catch rolling material; and that is on the fire's edge. See Figure 5.30 for a listing of fireline construction techniques.

As a general rule, construct line moving uphill! If there is no practical alternative to constructing line downhill, do so with extreme caution. Many firefighters have lost their lives attacking wildland fires from above. The following are guidelines for downhill line construction. They also apply to fireline that is being constructed some distance from the fire's edge, where fire behavior cannot be observed and responded to:

Jake says, "Attacking a fire from above is very dangerous. Do it only as a last resort, and with several ways to escape, if that becomes necessary."

- The decision to make a downhill attack should be made by an experienced and knowledgeable firefighter, after thorough scouting. Supervisor authorization and reliable communications are essential.

- Don't attempt to construct downhill line if the fire is directly below the starting point. This is a very unsafe position. Relocate the line or pick another strategy.

Take special precautions when constructing downhill fireline.

- The fireline should not lie in or adjacent to a chimney or chute that could burn-out while the crew is nearby.

- Reliable communications must be established between the crew working downhill and the crews working toward them from below. When neither crew can see the fire, communications must be established between the crews, the line supervisor AND a competent lookout posted where the fire's behavior can be observed.

- The firefighters should be able to quickly reach a safety zone from any point along the line, if the fire crosses below them.

When possible, construct fireline moving uphill.

- A downhill line should be anchored at the top. Avoid undercut line if at all possible.

Fireline construction techniques

Technique	Why?
Cast charred fuels inside the fireline.	You can't tell when charred fuels may still be hot. Don't take the chance of allowing the fire to cross the fireline in this manner.
Usually preferable to cast unburned fuels outside the fireline. However, cast inside if this will reduce the hazard or assist in burning-out operations.	The key is to reduce the fuels available for the fire to consume and generate heat. If you were to pile all the castings from your line construction on the inside of the fireline, the heat generated might be enough for the fire to cross the line.
Fall or isolate snags and reduce other aerial fuels before burn-out.	Snags throw fire and are magnets for burning embers. Reducing the aerial fuels will reduce the possibility of a fire crossing your line by the convective action of the fire and keep the fire on the ground if it does escape.
Trench firelines to catch rolling material.	If you have to construct line on a hillside and there is the possibility of rolling material, trenches are a must. If you don't have a way to catch burning material, your line is incomplete and unsecured.
Cool burning fuels close to the fireline with dirt or water.	This reduces the potential for the fire to cross your line.
Do not bury burning fuels.	Soil can insulate and hold the heat in a fuel for days. Covering burning fuels also hides the fact that a heat source is present. This could lead to a flareup or escape.
Match width and depth of line to fuels, fire behavior, and your control objectives.	If you are just constructing a scratch line, it can be narrow and shallow. If this is to be the final control line, construct it wide enough to do the job. However, don't overbuild.

Figure 5.30 Techniques for the construction of fireline.

- Line firing should be done as the line is constructed, beginning at the top (anchor point). The burned-out area provides a safety zone for the firefighters. It also reduces the potential of the fire crossing the line at a later time.

Trench undercut line to catch rolling material.

Figure 5.31 Explosives can be used, under special conditions, to construct fireline.

• Constantly review the 18 situations that shout "watch out," and ensure full compliance with the standard Fire Orders.

Fireline Explosives

In some cases, special explosives can be used to construct fireline (Figure 5.31). As you can imagine, this method can only be used in special conditions and only by someone who is certified to conduct such an operation.

Use of Fire

The use of fire to fight fire is very common in wildland firefighting. It is not normally used to fight structure fires. There are two types of uses: burning-out (or firing-out) and backfiring. Burning-out involves the use of fire to remove the unburned fuels between the fire's edge and the control line (Figure 5.32). You can use fire to great advantage in cleaning up, straightening line, and widening natural or existing barriers. Backfiring is a special technique requiring extensive planning. Backfiring is used to control or turn a high-intensity fire front that will overrun firelines if it can't be slowed or stopped. ***The key to a successful backfire is that the main fire DRAWS the backfire to it.*** Except in extreme emergencies, the decision and timing to backfire lies with the Incident Commander or Operations Section Chief.

Explosives can be used to construct fireline.

Use firing operations to your advantage.

Fire can be used in two ways:

- ***Burning-out*** - is done to remove fuels between the fire and the fireline. It can also reduce mop-up time, incorporate spot fires into the fire perimeter, and widen fireline. Burning-out can sometimes be used to create a safety zone (Figures 5.33 and 5.34).

- ***Backfiring*** - is an indirect attack tactic used to slow a fast burning fire. A backfire is set ahead of the main fire in such a way that indrafts to the main fire push the backfire toward the main fire. This can reduce the fuel available to the main fire under controlled circumstances. This tactic has to be well timed and coordinated with other operations on the fire. Backfires are used when it is too dangerous or otherwise impossible to directly attack the fire (e.g., heavy fuels, extreme fire behavior), or to buy time until sufficient forces are

Figure 5.32 This firing operation is attempting to keep the fire above this road.

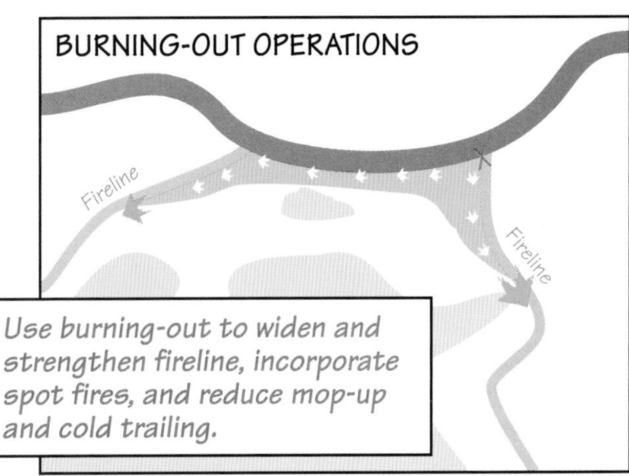

BURNING-OUT OPERATIONS
Fireline
Fireline
Use burning-out to widen and strengthen fireline, incorporate spot fires, and reduce mop-up and cold trailing.

Figure 5.33 Burning-out is used to clean up or widen fireline.

available (Figure 5.35). A properly executed backfire can slow or even turn a fast running fire. The decision to backfire is one that is normally made by the Operations Section Chief or Incident Commander.

Figure 5.34 A classic burning-out operation. The firing crew will slowly take the fire to the bottom of the hill, then carry the fire to the left.

Burning- or Firing-out Techniques

Burning- or firing-out entails the starting and spreading of fire from a control line toward the fire's edge. You can use a wet line, road, stream, or constructed fireline as the base of this operation. You can use fusees, drip torches, matches, helitorches, or most any ignition device to start your burning operation. To ensure a successful firing operation, you must regulate the heat. If you generate too much heat, you may lose control. If you don't generate enough heat, your burn may not be clean and may require extensive mop-up.

Burning-out is the most commonly used firing operation.

The objectives of burning-out are: use as a direct method of attack; to strengthen, widen and secure control lines; to reduce the required holding force; to reduce mop-up and the need to cold trail; to cut across fingers, incorporate spot fires, move a line to light fuels, or utilize natural or existing barriers; and, to provide safety zones and

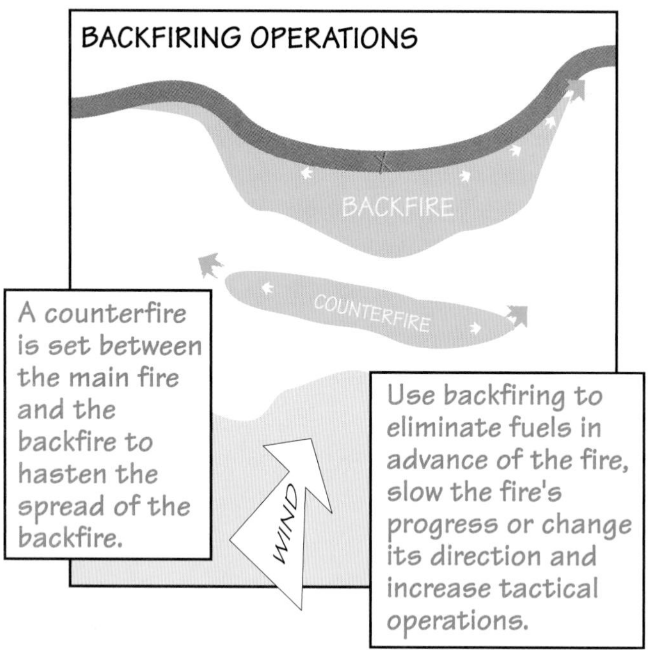

BACKFIRING OPERATIONS
BACKFIRE
COUNTERFIRE
A counterfire is set between the main fire and the backfire to hasten the spread of the backfire.
WIND
Use backfiring to eliminate fuels in advance of the fire, slow the fire's progress or change its direction and increase tactical operations.

Figure 5.35 Backfiring is a special type of firing operation.

escape routes. Burning operations must not adversely affect the actions of other firefighting forces. Keep those around you informed if you have been authorized to do some burning. If not, other firefighters may see the firing operation and think it is a flare-up or slopover. Many burning operations have been stopped by an unwelcome retardant drop.

Some general rules for burning operations:

- Firing must not jeopardize safety or adversely affect other divisions on the fire.

- Do not start the burning until the line has been prepared and there are adequate firefighting forces available to hold the line. This means falling snags, removing ladder fuels if necessary, and briefing the participants on the plan.

- Always have an anchor point for your operations. However, you may not actually start the burning at the anchor point.

Remember the "blackline" concept.

- Whenever possible, fire from the top down in steep terrain; fire into the wind; from the lee side or tops of ridges; from the bottom of wide canyons; and from roads or benches.

- Adjust to fit changes in conditions and situation.

- There are potential problem areas or situations that you must consider when firing a section of line (Figure 5.36).

Always be careful when you have fire below you.

There are several firing patterns that can be used when backfiring. Each one of these firing patterns has been developed to meet a specific need. They are as follows:

- ***Backing fire*** - A low intensity fire is allowed to back into the wind from an established fireline or barrier. This is a slow, time consuming technique, but it is one of the safest ways to use fire (Figure 5.37). Some of the hazards and drawbacks of using backing fire are the possibility of wind shifts, spotting and rolling materials, and the amount of time this method takes.

Backfiring relies on "draft winds" to draw the two fires together.

"Watch-outs" on firing operations

Situation	Why?
Adverse line location - line is located midslope, is undercut, or has hooks and sharp bends.	Burning from midslope or undercut line gives you very little control over the burning operation. Topography, fuels and weather will control the heat generation. Firing through hooks and around bends is difficult to hold within control lines.
Firing in a narrow canyon bottom, through saddles, or across long slopes.	When firing in narrow canyons, you run the risk of extensive spotting across the canyon. When burning through a saddle, you should burn BOTH lines from the tops to the middle. You have little control over a firing operation on long slopes.
The fuel varies, with heavy buildup near the line, is dense and difficult to work in, little or no ladder fuels to generate heat or there are lots of snags close to the line.	If there are heavy fuels close to the line, you may not be able to control heat generation. You may have to slow the rate of your burning to compensate. If spots develop in heavy, inaccessible fuels, controlling them will be difficult. You need a clean burn; that means burning all of the fuel. If ladder fuels are not available, you run the risk of creating weak line. Snags spread fire; snags are magnets for spot fires. Remove them if possible.
The weather is changing; wind is shifting or increasing; thunderstorms are in the area; the humidity is changing.	ADJUSTMENT is necessary. If the change in wind is going to jeopardize the firing operation, you may have to stop. When thunderstorms are close by, winds could shift 180 degrees in minutes. THINK SAFETY! If the humidity is going down, spotting may increase. If the humidity is going up, you may not be able to get the fuels to burn.
There are structures, improvements, or high-value areas within or close to the line.	It is very difficult to explain to a homeowner why you intentionally burned his house down. Ensure that lines are constructed around any area you don't want burned and forces are placed to control any spots. You may have to increase the level of fuel preparation in these areas. Keep a close eye on the area long after the firing operation has moved through.
Loss of contact with your supervisor and the adjacent forces; a vital piece of equipment breaks down; or you don't have the right type or enough firefighting forces to hold the line.	You will have to adjust your plan and reestablish communications. Secure additional firefighting resources or revise the plan, utilizing the resources at hand.

Figure 5.36 Watch-outs on firing operations.

- ***Flanking fire*** - A flanking fire is set along a control line parallel to the wind and allowed to spread at right angles to, and towards, the main fire (Figure 5.38). This type of burning is a little more intense because the control fire is generally burning with, rather than against, the wind. The hazards of a flanking fire include spotting potential and wind shifts.

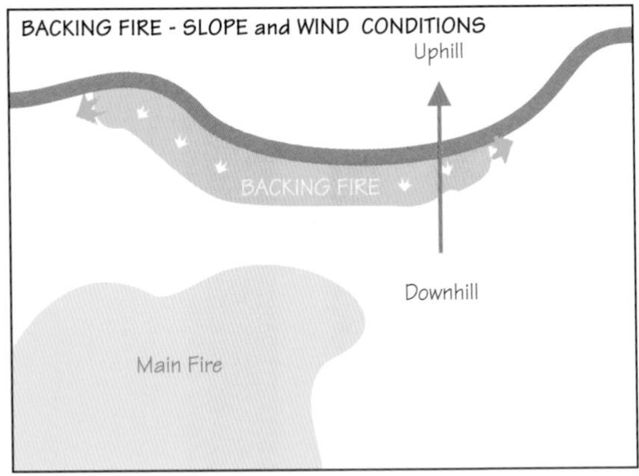

Figure 5.37 Backing fire is usually slow-burning and easier to control.

When working a flanking fire, watch for spot fires and the hazards of a shifting wind.

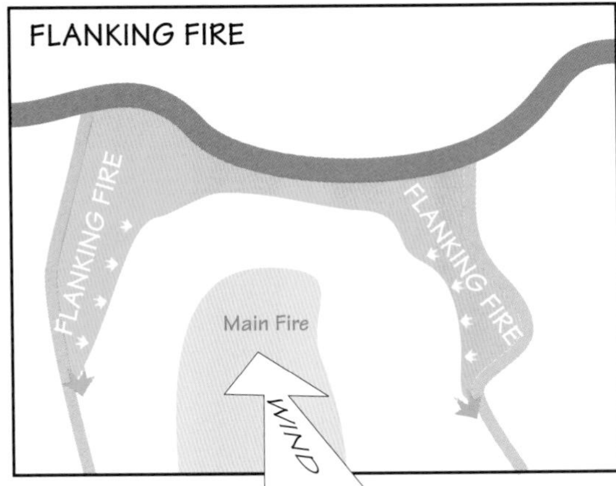

Figure 5.38 Flanking fire is used along the flanks of the fire. The wind is more of a factor in this method of burning. Flanking fires usually burn with a little more intensity than a backing fire.

- ***Head firing*** - When employing a head firing pattern, a fire is ignited at a control line and allowed to burn with the wind (Figure 5.39). This can result in a very intense fire. This technique may be useful if a wind shift is expected and you want to strengthen your control lines quickly.

TO
Jake says, "Don't start more fire than you can handle."

- ***Spot firing*** - When using a spot firing technique, spots of fire, rather than strips, are ignited. This method is sometimes used when ignition is being done from the air.

- ***Strip firing*** - The strip firing technique is used when you want to widen a fireline quickly. In this firing pattern, fire is set in strips parallel to the fireline (Figure 5.40). With

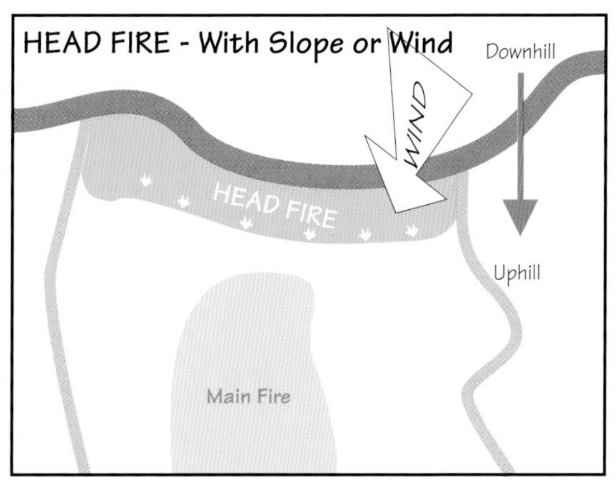

Figure 5.39 *Head firing allows the fire to burn with the wind. They can sometimes burn with great intensity.*

close coordination of the firefighters doing the firing, the intensity of this type of firing can be controlled. If several firefighters are used to light several strips of fire at the same time, the "1-2-3" firing scheme should be used. The direction of slope and wind dictate the sequence of burning (Figure 5.41).

Strip firing is the most common firing technique.

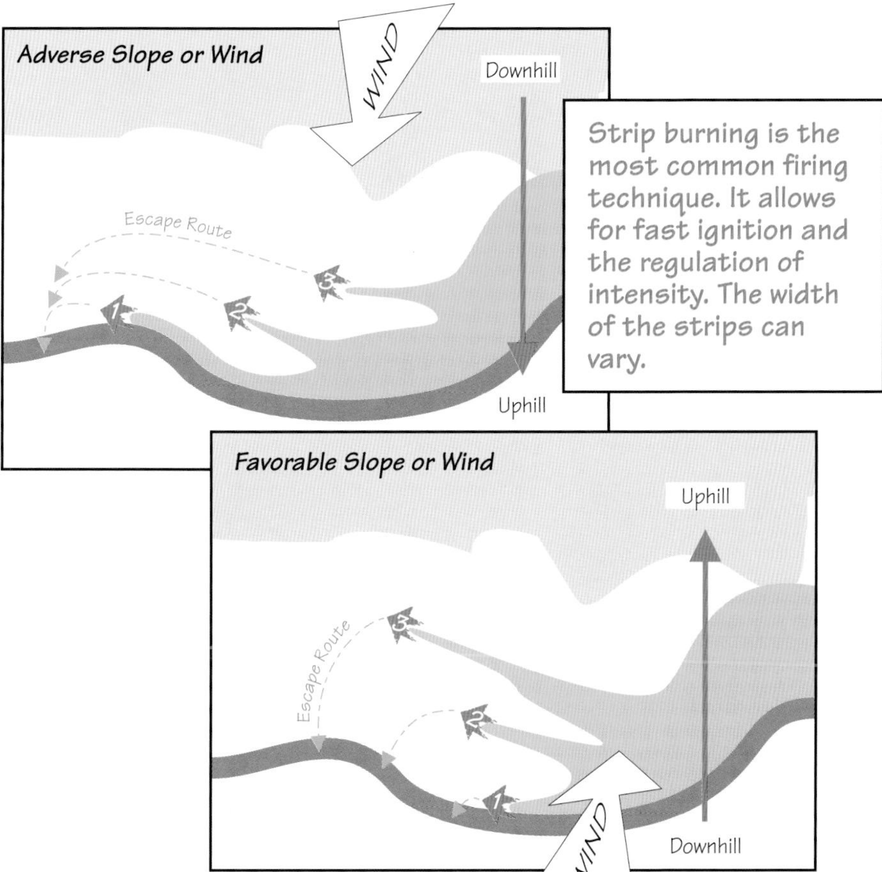

Figure 5.40 *Strip firing is used when you need to widen fireline quickly.*

Use strip firing to widen fireline.

The firing order is dictated by the direction of slope and wind.

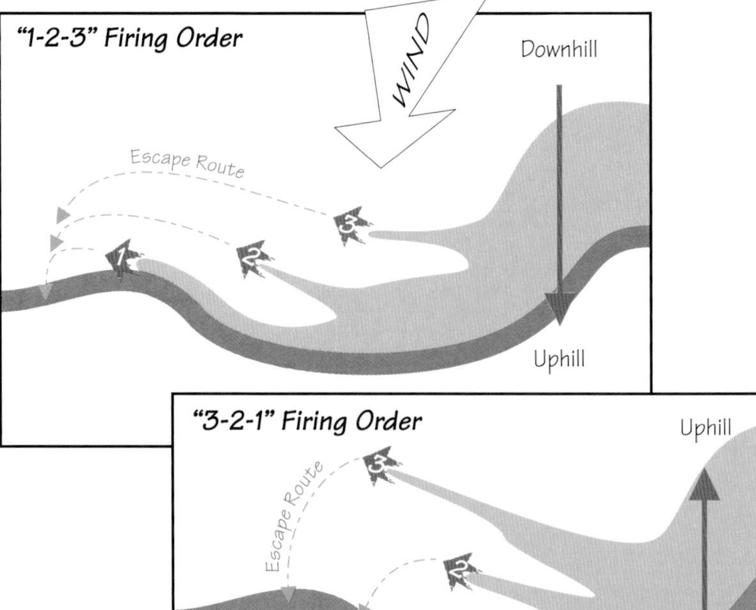
THE "1-2-3" FIRING CONCEPTS
"1-2-3" Firing Order
WIND
Downhill
Escape Route
Uphill
"3-2-1" Firing Order
Uphill
Escape Route
Downhill
WIND
Strip burning involves the use of the "1-2-3" firing concept, where several firefighters are used to ignite strips of fire. The direction of slope and wind dictate the sequence of burning. Lighter #1 is always closest to the line.

Figure 5.41 The direction of the wind and slope dictate the sequence of burning. Ensure that all of those involved understand the escape routes.

Use ring firing to move the main fire around the structure(s).

- ***Ring firing*** - This type of firing is used when you are trying to save a valuable resource like a structure, or a historic or archeological site. This method of firing isn't anchored by the fireline. It is designed to create an unburned island (Figure 5.42).

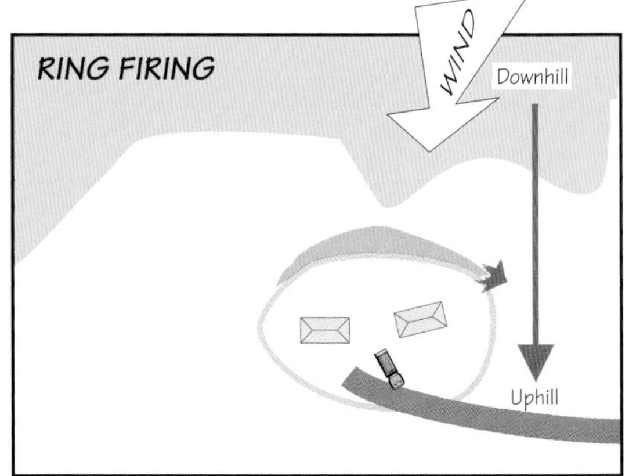
RING FIRING
WIND
Downhill
Uphill

Figure 5.42 Ring firing is used when you need to protect a structure(s) or other significant site(s).

There are several other firing techniques, but they are not commonly used and require very experienced people in order to be used safely and properly.

As noted earlier, the key to a successful burning operation is how well you regulate the heat. If your firing operation burns with low intensity, you may not be getting a clean enough burn and the main fire could break through your line (Figure 5.43). Conversely, you could misjudge the conditions and start a real "barn burner," adding to your problems by adding to the intensity of the main fire.

Figure 5.43 These firefighters are burning to stop the advancing wildfire.

Some general rules for firing operations are as follows:

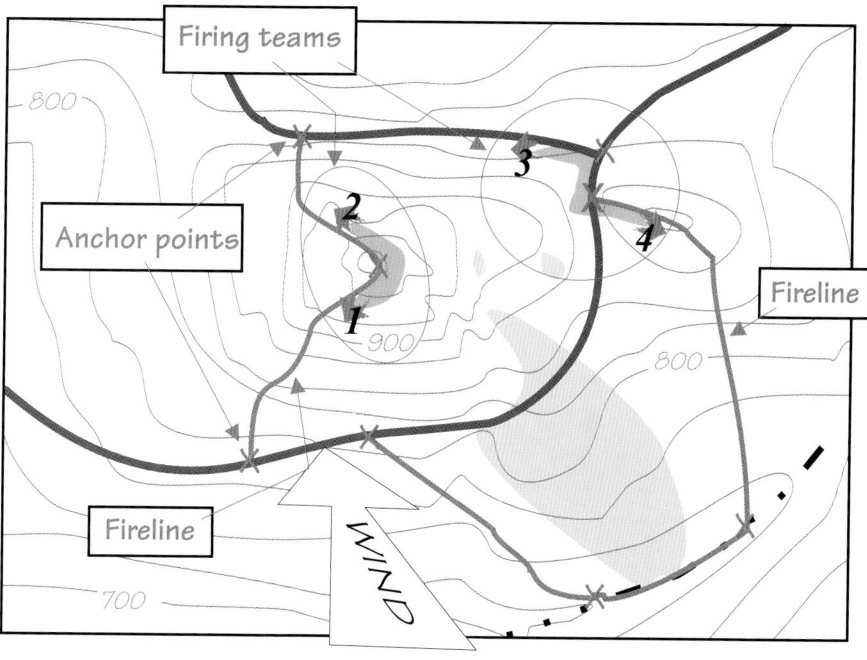

Figure 5.44 Note the number of anchor points. They are necessary to tie all of the various firing operations together. The firing sequence and coordination are critical.

- Always begin firing from an anchor point; except ring firing (Figure 5.44).
- Begin at the head, working down the flanks to the heel if you can.
- Burn downhill when you can.
- Burn from the back side of ridges (not the top) when you can.
- Burn into saddles simultaneously from both directions.

Don't start more fire than you can handle.

Karen Wattenmaker

Figure 5.45 They are firing this road to hold this very hot fire inside the fireline.

- Adjust firing to fit the situation.
- If conditions are favorable, fire without delay; later may be too late.
- Fire short sections of line so that if you have a problem, you have the best chance of controlling it (Figure 5.45).
- The rate of firing should be consistent with your ability to hold it. Don't introduce more fire than your forces can handle.
- Fire around structures first before firing-out control lines.

There are some things that will affect any firing operation. Consider the following problems as warning signs:

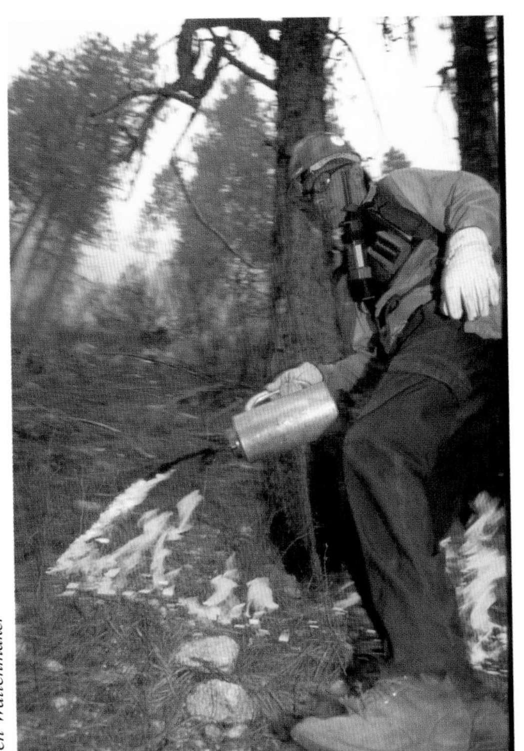

Karen Wattenmaker

Figure 5.46 This firefighter is firing mid-slope, probably in an effort to slow an advancing fire, just over the ridge.

- There are sharp doglegs or hooks in the control line to be fired. Burn them first to reduce the spotting potential.
- The fireline is in a poor location (e.g. it is mid-slope and hasn't been trenched), or there are heavy fuels close to the line (Figure 5.46).
- The weather is likely to change. In this case, you may have to adjust or delay the firing schedule.
- There are structures or other valuable improvements in the area. In this case, you may have to change the plan or take actions to make the structures safe.
- The fuel type varies.
- You will be firing in box canyons or through draws.
- The firing pattern is wrong for the conditions; there may be erratic fire

behavior, or firefighting personnel may be potentially endangered.

If the conditions are unfavorable, the winds are against you, or if you have to fire in a canyon, get an expert who has had special training in dealing with these conditions. There are techniques that can be used in these situations, but you need an expert to advise and assist you.

Time your firing operations to take into consideration favorable weather conditions.

The decision to use fire in your plan should not be made casually. Several of these techniques take special timing and coordination. Unless these techniques are used cautiously and skillfully, you could turn a good tool into a real disaster. Once you introduce your own fire, you change the dynamics of the whole event. Think out your plan and what it will do to the main fire before you light the match or fusee.

Special note: There are laws in some areas prohibiting the use of firing operations. In some areas, there are laws which require firefighters to pay damages if a firing operation escapes; in other areas firefighters are protected from suits. In some states, the permission of the private land owner is needed. Know the local laws (Figure 5.47).

Figure 5.47 In some areas you are prohibited by law from using fire as a means of control. Know if the law allows firing before you light the torch.

Use the firing pattern that matches the conditions. The wrong method may spell disaster.

Notes:

6

Initial Attack Strategy and Tactics

Almost all fires start small. The objective is to keep them small, controlling them before they cause injury to the public or serious damage to property and resources. Being ready to respond is the firefighter's first responsibility. Once "at scene," the firefighter must select the correct strategy and tactics to control the fire in the safest, most efficient and cost-effective manner possible (Figure 6.1).

Strategies are overall plans and actions to control the fire. The strategy may be to protect threatened structures and use a direct attack when possible. ***Tactics*** are techniques to accomplish the strategy; the art of deployment or method of direct action on the fireline.

Tom McIsaac

Figure 6.1 Here you have an excellent view of the fire. Most of the time you will not know what you have until you actually go "at scene."

There should be no doubt in your mind that you will be fighting not just a fire, but a fire driven and influenced by the weather. When you consider the influences of weather, you need to be thinking in terms of past, present, and more importantly, the future. How will the weather conditions of the last few hours or days contribute to the spread of the fire? What kind of impact is weather having now? What impact will it have over the duration of the fire? In addition, a post-fire analysis of what occurred and why, and how it was affected by weather, will enhance your strategic thinking skills for the next fire you respond to.

You must select the correct strategy and tactics to control the fire in the safest, most efficient and cost-effective manner.

At the start of each day, you should routinely assess the weather conditions and compare them to what has been predicted. Discuss the potential fire weather for the day with your crew, supervisor, and dispatch office. With experience, you will develop an instinct for what to expect, including a localized sense for the prediction and how it relates to your "first-in area." Weather forecasters use specific sites to develop their general predictions. Your local area may be influenced by a lake, the shadow of a mountain range, or have other variables that the weather forecasts do not take into consideration. Take the forecast and compare it to the actual weather in your area. Develop a set of corrections that fine-tune the forecast for you. For instance, you may find that a forecast is usually 3° higher in temperature and is 5 percent lower in relative humidity than is generally true for your area. Use this information to adjust your forecast. If you are in an area you are unfamiliar with, you should discuss the forecast with a knowledgeable local to get a clearer idea of what the forecast may mean in that area.

Jake says, "Base your plan on what is going to be! Give serious thought on how the weather will impact your fire."

Before the fire response, the firefighter must ensure that personal safety gear, the engine, and all of its equipment are ready for response. Always be aware of the status of other firefighting resources and any fires in your area. Are they at normal levels, or are you going to be responding with fewer resources or with personnel who are strangers to you and don't know your area?

Make sure your safety gear is ready for use.

Know the weather forecast.

Taking the Dispatch Information

The command and control center (Figure 6.2) is about to give a dispatch; listen to and write down the information important to you:

- Type of fire.

- Location of the fire. This could be in the form of a street address, local landmark, Public Lands Survey section, or be very vague, depending on how the call was received.

Figure 6.2 The dispatchers will give you all of the information they have...it may not be complete...it may even be wrong.

Be ready to "copy" the dispatch information. Ask for clarification if you do not think you have it right.

- Access, or route of travel. This is especially important if you will be going into an area that you are not familiar with. Try to get a map system reference and use your maps.

- What other forces have been dispatched? Is it the normal first alarm, or will you be fighting fire with "new" people? Have aircraft been dispatched?

Start sizing-up the fire as soon as you leave the "barn."

- Are there any special hazards, like chemicals, downed power lines, etc.?

- If this is an "out-of-county" dispatch, be sure to get order and request numbers. This number is your authority to respond. It will follow you throughout your response. This will help you get paid.

If the dispatchers have any other information, they will usually provide it. ***Write the information down. Don't rely on your memory.*** If you did not get all of the information, have the dispatcher repeat it. It is important that you know where you are going and how to get there.

Do the weather predictions match what you see?

En route to the Fire

The goal is to get to the emergency as quickly as possible. This doesn't mean breaking the sound barrier. It means you take the shortest, safest route of travel. If you get in an accident, you will not get there as fast as you should have.

Your size-up of the fire begins when you leave the station. Sense the wind direction and speed, and the dryness of the air (relative humidity).

Jake says, "Your first responsibility is to get to the fire safely. If you get in an accident...you failed."

Compare what you observe with what was forecast; a significant difference means "watch out!"

You've heard the dispatch, so you know what has been sent. Listen for the reports from lookouts or other responding units. Understanding what you've heard is all part of size-up.

As you get close to the fire, you can tell a great deal from the smoke column. Is it small and wispy, or is it "building and boiling?" Figure 6.3 outlines what the color and shape of the smoke column MAY tell you. "Where there's smoke, there is (usually) fire," but you will not know exactly what kind of fire you have until you get there.

Size-up and Report on Conditions

Size-up is not a one-time event. You should constantly review conditions.

Size-up is the development of a mental picture of what is happening; what the fire is doing or will do. Size-up is not a one-time activity, it is a continuous process that is only concluded once the fire is controlled. Wildland firefighting is largely a process of size-up, decision making, and action taking. Analyze what you see, draw conclusions and develop a plan. Constantly evaluate your situation and react appropriately.

Fire behavior will dictate your tactics.

- ***What is the fire behavior and rate of spread?*** Draw some conclusions about how the fire is burning. Is the fire burning with such an intensity that direct attack will not be effective? Is the fire spotting, crowning, or just smoldering? Determining the intensity is critical; it will dictate the strategy and tactics. Consider the time of day in the determination of present and anticipated fire behavior. Figure 6.4 describes rate of spread.

If the fire is spotting, you may be in for a long day.

- ***Is the fire spotting?*** Spot fires are small fires that are started by flying brands from the main fire. Low humidities, low fuel moisture and wind contribute to the potential for spotting. Rotten wood can be ignited from sparks, while larger brands are necessary to start fires in slash, duff and grass. These fuels

What the color and column of smoke may mean

What you see	What it may mean
The smoke column is thin, rising lazily, and the color is light blue to gray.	Probably a campfire.
The smoke column is narrow, thin, and dark gray to black.	Could be diesel-powered heavy logging or construction equipment.
The smoke column is small, thick, and white in color.	This may mean a small grass fire. If the smoke puffs up every so often, it may mean someone is burning leaves or grass and "feeding" it.
The smoke is widening at the base; it is predominantly white, but starting to turn brown or black on its downwind side.	This may indicate the fire is spreading in grass and moving into heavier fuels. Dead brush will burn with a dark brown color; brush with a higher oil content will burn black.
The column of smoke is thick and black, with no spread to the base.	This could be a structure or vehicle fire. It may also be tires.
The smoke is black, but some white or light brown is showing away from the main column.	This may mean your vehicle or structure fire has moved into the grass.
The column is going straight up.	There is little or no wind on the fire.
The column is going up, but the top of the smoke is bent over.	There is little surface wind, but there is wind where the smoke bends. Beware; that wind may surface at any time.
The smoke is bent over at the ground and building in volume and intensity.	The fire is wind-driven with a good fuel supply.
The smoke has built to several thousand feet and a small white cloud has formed on the top.	Don't plan on days off. You are going to be quite busy.

Figure 6.3 The smoke column can give you some idea what you will be confronting.

are readily susceptible to spot fires: rotten wood, either on the ground, in logs or in snags; moss and lichens on trees; slash, particularly when it is compacted; duff or peat; and grass or grain.

Determining rate of spread

Rate of spread	Description
Low	The head of the fire moves less than 100 feet per hour.
Moderate	The head of the fire moves from 100 to 400 feet per hour.
High	The head of the fire moves from 400 to 1,800 feet per hour.
Extreme	The head of the fire moves over 1,800 feet per hour.

Figure 6.4 Use these definitions to describe the rate of spread.

• ***What is the size of the fire?*** Estimating the size of the fire provides an indication of the complexity of the fire. It gives the command and control center an idea of the situation. It also gives the public information people something they can give the press. The size of wildland fires is measured in acres (or hectares in countries that use the metric system). An acre is 43,560 square feet or a square 209 feet on a side (a hectare is 2.47 acres). Estimate the size, being careful not to overstate the true size. (Note: most estimates are overestimations, especially at night.) Try to be accurate in your estimates. It's very awkward to try to explain to someone how a fire "got smaller overnight." Water has never shrunk a wildland fire, but inaccuracy has! Be conservative in your estimates. Figure 6.5 outlines two ways to measure acreage.

By watching the smoke column, you can get a good idea of what is happening above the fire.

• ***Where did the fire start?*** Is the cause obvious? Be sure to make a sketch or mental note of the fire when you first see it. This will help the fire investigators to locate the origin and determine the cause. Protect the origin, restricting access to it.

• ***What are the weather conditions?*** Is there a wind? What is its direction and speed? Is it gusting or constantly changing direction? Is the wind upslope, across slope or downslope? Is this a local wind, or a general airflow across the area? Is the wind what was predicted? What is the relative humidity? This factor affects spotting potential. Is the air stable? Check the smoke column.

Rate of spread and flame length are good indicators for fire behavior.

• ***What is the topography?*** Is the lay of the land flat, steep, rocky or broken up with canyons? This will influence how the fire will spread and how easily the firefighters will be able to traverse it. The steeper the terrain, the harder it will be to fight fire in.

ACREAGE CALCULATIONS

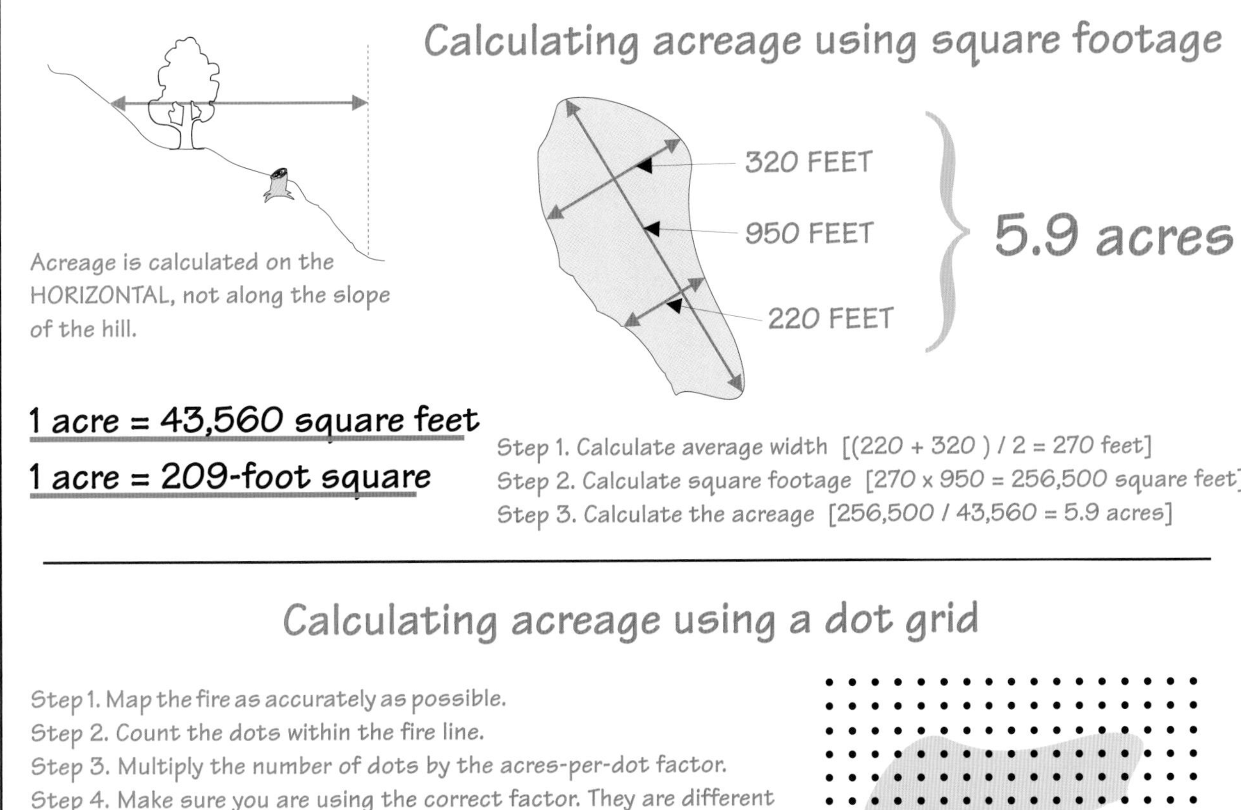

Calculating acreage using square footage
320 FEET
950 FEET
220 FEET
5.9 acres
Acreage is calculated on the HORIZONTAL, not along the slope of the hill.
1 acre = 43,560 square feet
1 acre = 209-foot square
Step 1. Calculate average width [(220 + 320) / 2 = 270 feet]
Step 2. Calculate square footage [270 x 950 = 256,500 square feet]
Step 3. Calculate the acreage [256,500 / 43,560 = 5.9 acres]
Calculating acreage using a dot grid
Step 1. Map the fire as accurately as possible.
Step 2. Count the dots within the fire line.
Step 3. Multiply the number of dots by the acres-per-dot factor.
Step 4. Make sure you are using the correct factor. They are different for each scale of map.
If the acreage factor is 2 acres per dot, and there are 55 dots within the fireline, the fire is 110 acres in size.

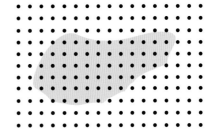

Figure 6.5 Accurately estimating acreage is difficult. The best way to get an accurate acreage figure is to plot the fire on a map.

- ***What are the fuel types?*** Is the fire burning in grass, brush, woodland, timber, logging slash or tundra? What fuels are ahead of the fire? These are major considerations in developing your strategies and tactics.

- ***Are structures or high-value areas being (or going to be) threatened?*** What is the potential for the loss of homes or high-value timber or plantations? Do the structures being threatened have "defensible space?" Do they have flammable roofs? Is there a nearby water source? The presence of high-value areas will change your strategy and tactics. It may also change the level of response - more firefighting resources may be required.

Don't overestimate the fire size. Estimating size is especially tough at night.

- ***Are there natural or existing barriers?*** Are there roads, streams, lakes or other barriers that can be used as a control line?

Can the landowner help in your efforts?

- ***Are there any special safety hazards?*** Are there downed power lines? Other common hazards are rolling rocks, automobile traffic, fleeing evacuees, narrow roads and other access problems. If there are powerlines in the area, they may be a safety issue if aircraft are going to be used.

- ***What is the fire's potential?*** Make an effort to predict where the fire will be in 30 minutes, 1 hour, 2 hours, etc. This will help you in developing your strategy. If you predict the fire to be at Road "X" in two hours and that is where you want to stop it, be sure you will have the appropriate firefighting resources to meet that strategy. If you don't, change your strategy! Estimate in your mind when you think the fire will be contained and controlled.

Jake says, "Ensure that you can manage the resources you have on the fire. We can easily mobilze more than you can manage."

- ***Whose land is threatened?*** If the fire is burning on lands owned by a large landowner, notification may be important. If the fire is burning on lands of an agency with a modified suppression policy, this will impact how you fight the fire. Modified suppression policies are rules developed by agencies to guide the actions firefighters may take in controlling a fire. The rules are intended to reduce the impact of firefighting on the land. This is of concern in wilderness areas, national parks, or on lands of historic significance. Large landowners may be able to provide equipment to help fight the fire.

- ***Have enough firefighting resources been dispatched?*** Are the firefighting resources dispatched to the fire appropriate and adequate to control the fire? What is the timing of their arrival? Are they rested and fed? Are they trained and equipped to accomplish the needed work? Do they know the area? All of these factors will dictate the effectiveness of the first alarm.

What is the fire's potential? Project spread in hourly intervals.

- ***What should you tell the others?*** The size-up should be passed on to your dispatch so that they can use the information to plan their next actions. Give them some information as soon as you arrive. Make it brief, but meaningful. As soon as you have developed more information, give it to them.

To develop an accurate size-up, you may have to scout the fire. Walk the perimeter, or move to a vantage point to get an overall picture.

On a small fire you may be able to see the whole fire as you go "at scene." If you conclude you can handle the situation by yourself and it will take about 30 minutes to control the fire, tell your command and control center:

"At scene; small spot of grass; can handle; out 30 minutes."

Short and to the point. If the fire is larger, with a moderate rate of spread, your report of conditions may be:

Give clear and concise reports. Be sure to communicate your plan to other forces.

"At scene; several acres of brush; moderate rate of spread; investigating; continue the response."

When you have finished your initial size-up you should pass it on to the command and control center. It may sound like this:

"Fire is burning in light grass and brush; moderate rate of spread; approximately 4 acres; homes will be threatened in next 30 minutes; staging area established at the intersection of Highway 1 and Forest Road 7; can handle with first alarm assignment."

As you prepare your plan of attack, you should write down several simple objectives for what you will be attempting to do. Of course, if the fire is a spot of grass and you can put it out with 50 gallons of water or less, this step isn't necessary. But let's say the incident is a little more complicated. You can't see all of the fire, aircraft are en route, and structures may be threatened. This is when you should outline your plan, not just for you, but for the other firefighters on the incident (Figure 6.6).

Figure 6.6 This firefighter may have a plan, but it may not be successful because it failed in its execution. Lay out what you want and tell all those involved what it is.

Operations may vary in size, but the overall priorities stay the same. The four general priorities are:

- ***Life safety*** - The protection of human life is always the first priority. This includes the lives of firefighters.

- ***Property protection*** - The protection of structures is usually the next priority. In some jurisdictions, the protection of property and resources are grouped.

- ***Resource protection*** - The protection of natural resources, timber, plantations, etc., is generally the next priority.

- ***Incident stabilization*** - This priority deals with the organization that is developed to ultimately control the incident.

The safety of the firefighters and the people you are assigned to protect is NUMBER 1.

After developing your plan, develop the strategic objectives for the incident. These are broad statements that describe major actions. ***They don't necessarily state how you plan on attacking the fire, but what you plan to do, or what you want to accomplish. Keep the objectives or the strategic goals simple:***

"Keep the fire from crossing Duncan Hill Road" or *"Protect the structures along Penny Lane."*

Develop a list of objectives that encompass all of the fire, placing them in order of their priority. Even though the development of an incident action plan is usually left to the incident commander or a chief officer, you should develop the initial plan to guide your actions and those of the first alarm response. Use the information from the size-up to set priorities and develop the strategy. Priorities generally follow in this order:

Develop a plan, a strategy and objectives.

- Keeping the fire from moving into heavier or more dangerous fuels.

- Keeping the fire on one side of a canyon or keeping it in one canyon.

- Keeping the fire as small as possible. Consider costs and values at risk.

- Make sure all suppression actions contribute to final control of the fire.

What are the values at risk? Can you justify your plan and its potential costs?

Outline in your mind, or preferably on a note pad, your action plan. Set out your goals:

"Protect the structures along Highway 1; keep the fire from jumping Forest Road 7; anchor the fireline on Forest Road 7 and use a flanking action toward Highway 1."

Assign your resources to accomplish these objectives:

> *"Engines 2 and 3 to protect the structures along Highway 1, Engine 2 Captain is Division A; Engine 45 will patrol Forest Road 7, keeping the fire from crossing the road; Engines 6 and 10, Bulldozer 3 and Handcrew 6 to construct line from Forest Road 7 to Highway 1, Engine 6 Captain will be Division B; Airtanker 1 will provide drops as needed."*

Match the strategy and tactics to the present and predicted fire behavior and conditions.

Make sure all concerned know your plan and the tasks assigned to them.

Putting in fireline is the only way to put out a wildland fire.

Need for a Map

Begin to plot the fire on a map. You would be unwise to begin a long trip in unfamiliar country without a road map. The same thing is true when attacking a wildland fire. Maps provide a way for you to visualize where you are going and what is the best and most effective way to get there. It is your two-dimensional view of reality. Your map provides a way to plot the progress of the fire, your fire behavior predictions, access, the location of any special hazards, and the placement of your resources.

The best type of map is a 7.5-minute topographic map. It has a useful scale and provides a good picture of the lay of the land. If you don't have this kind of map available to you, do the best you can, even if you just sketch some general features on a piece of paper (Figure 6.7).

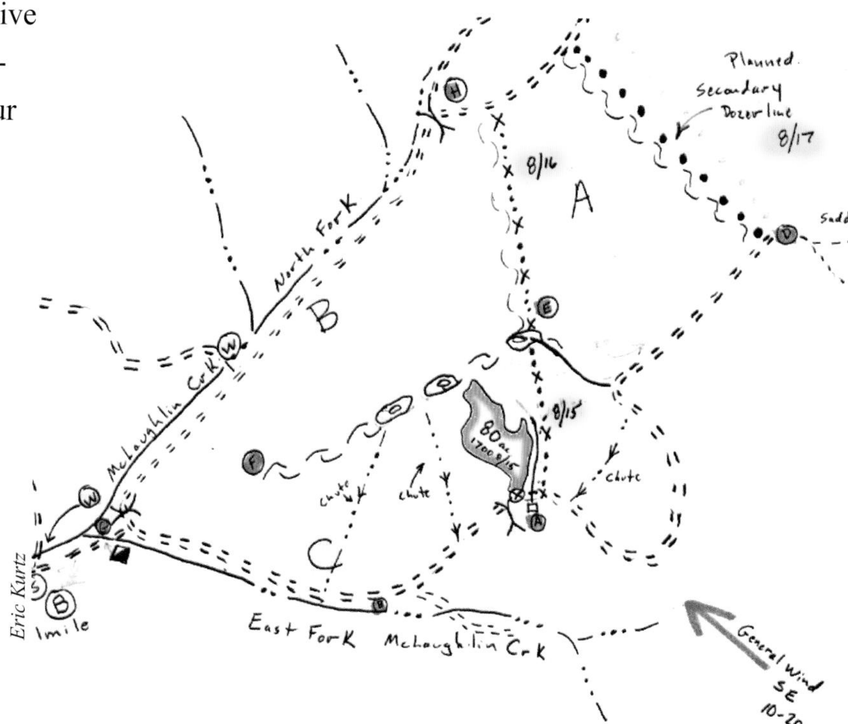

Figure 6.7 If all you have is a blank piece of paper, plot your fire on it. At least it provides a visual reference from which to work. Note the estimated spread is highlighted in yellow.

Initial Attack Tactics

Now you have developed a feel for the situation through your size-up and fire behavior prediction. Your first decision is whether the situation dictates you take offensive or defensive actions.

Even if you have to be defensive, try to construct some fireline.

- ***Offensive Action*** - An offensive attack is one that confines and controls the fire by constructing fireline. Protection of structures may be required, but the majority of the fire suppression effort is focused on line construction. Before deciding on an offensive action, you must be certain that your people can implement it safely, and you must have ample resources available. An offensive action can utilize direct or indirect attack strategies. In structural firefighting, an offensive action is usually an interior attack.

Jake says, "If you need to use all of your resources to protect structures, you aren't going to put in a lot of fireline."

- ***Defensive Action*** - A defensive action is taken when the fire behavior is such that an offensive action would not be productive, loss of structures would occur, or is too dangerous. You might also have to take defensive positions if there are limited firefighting resources, or if the resources are focused on protecting life, property or high-value natural resources. In most cases, a defensive action is taken to minimize the damage, not contain the fire. ***In a defensive action to protect structures, the moving fire controls the action***. In structural firefighting, a defensive action is usually an exterior attack.

When you are in defensive mode, the fire controls your actions.

There will be cases where the attack mode on one division may be defensive, but the rest of the fire is being attacked using various offensive actions. As noted, it is very important that you know which mode you are in—you don't want to mix modes without giving it some real thought (Figure 6.8).

You can be defensive on one part of the fire and offensive on another; don't mix actions on the same part of the line.

You must now determine the proper tactics for attacking the fire. There is the *direct attack method,* where you work directly on the fire's edge; the *parallel attack method,* where you construct the fireline just away (6 to 50 feet) from the fire's edge; and there is the *indirect attack method,* where the fireline is constructed some distance from the fire. You can choose one method, or a combination of methods that best fit the situation.

Figure 6.8 There are considerable differences between offensive and defensive actions. Don't mix them on the same part of the line. Know exactly which mode you are in. The actions being taken by Engine 1 or Engine 2 may be successful, but when you mix them, someone may get hurt.

If the flame length is over four feet, the fire is probably burning too intensely for a direct attack.

You usually will have to deal with heat, smoke and flames.

Direct Attack

This method of attack is conducted in lighter fuels, directly on the flaming edge of the fire (Figure 6.9). You can scrape a fireline to mineral soil to cut the linkage of burning and unburned fuels, or you can cool the fire with water and then construct the final fireline. This method of attack brings with it several advantages and disadvantages (Figure 6.10). The type of fuel and the flame length will dictate your tactic. If the flame length is over 4 feet, the fire is probably burning too intensely for a direct attack. There is a variation of the direct attack that is referred to as the "two-foot method." In this case you construct the fireline a couple of feet from the burning edge. This allows the firefighter to be a little farther away from the heat of the fire, and is normally used when the attack is done without the aid of water.

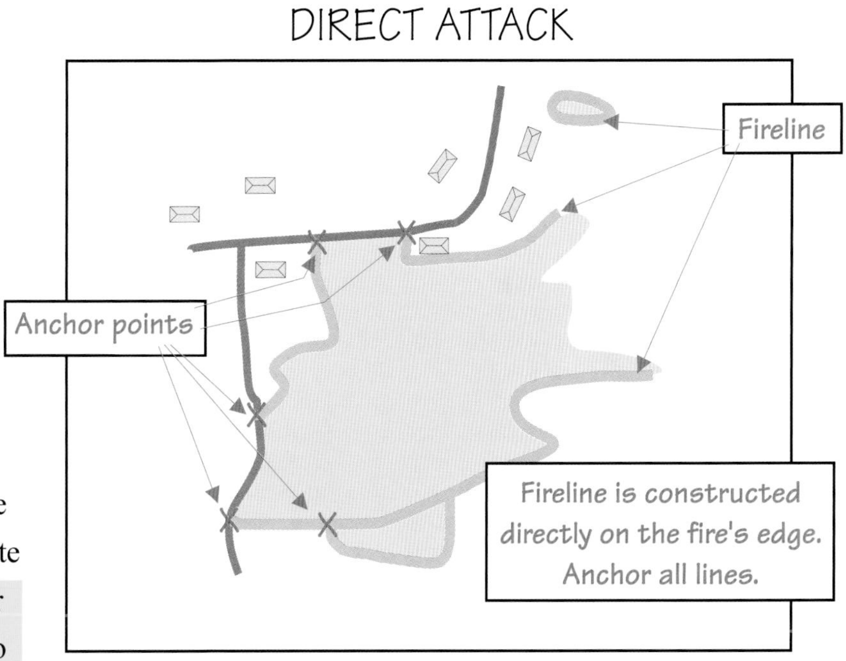

DIRECT ATTACK

Fireline

Anchor points

Fireline is constructed directly on the fire's edge. Anchor all lines.

Figure 6.9 A direct attack focuses all of the actions being taken on the burning edge of the fire.

A direct attack is usually the safest attack mode.

Advantages and disadvantages of direct attack

Advantages	Disadvantages
There is a minimal area burned. No additional area is intentionally burned.	Firefighters can be hampered by heat, smoke, and flames.
Safest place to work. Firefighters can usually escape into the burned area.	Control lines can be very long and irregular, because the line follows edge of the fire.
Full advantage is taken of burned out areas.	Firefighters may accidentally spread burning material across the line.
May reduce the possibility of the fire moving into the crowns of the trees or brush.	Doesn't take advantage of natural or existing barriers.
Eliminates the uncertain elements of burning-out or backfiring.	Usually more mop-up and patrol is required.

Figure 6.10 There are times when the advantages of a direct attack outweigh the disadvantages. Use the method best suited to your situation, and don't be afraid to change methods as the situation changes.

Don't be afraid to back off and go indirect.

A direct attack may mean a smaller fire, but it may mean more mop-up.

The direct attack method of line construction is most commonly used on smaller fires, and on the flanks and rear of larger fires. The key to using this strategy is whether the fire's intensity allows the firefighters to work directly on the fire's edge. Unless special situations dictate otherwise, anchor your line and start construction. There are four deployment strategies for a direct attack: flanking action; tandem action; pincer action; and envelopment action:

When a fire is burning at a fast rate of spread, anchor your line and start flanking the fire.

- ***Flanking Attack*** - The first action is to pick an anchor point, usually the road, and start extinguishing the fire front (Figure 6.11). The firefighter with the nozzle is usually in front of the engine, and to one side. The firefighter must always be in view of the operator. The firefighter and engine may work from the green (unburned area) or "the black" (the area that has already burned). Choose the attack position that best fits the situation. In the flanking action, the engine moves along the line as fast as the fire is extinguished. Be sure all the fire is extinguished. All too often, the attack moves too fast and the fire flares up behind you.

Jake says,
"Anchoring and flanking is one of the most basic of attack strategies. You don't have to get fancy to be effective."

- ***Tandem Attack*** - This attack method is a flanking attack that involves two or more engines, or other firefighting apparatus (airtanker, bulldozer, hand crew, etc.). The lead engine takes the heat out of the fire and the second engine is used to follow behind, picking up hotspots and securing the line (Figure 6.12). The lead engine can move faster knowing the tandem engine will pick up any hotspots. These engines can also "leapfrog" each other, thus allowing the firefighters on the lead engine to "get a break" from the heat and smoke.

Tandem attack forces all work on the same flank.

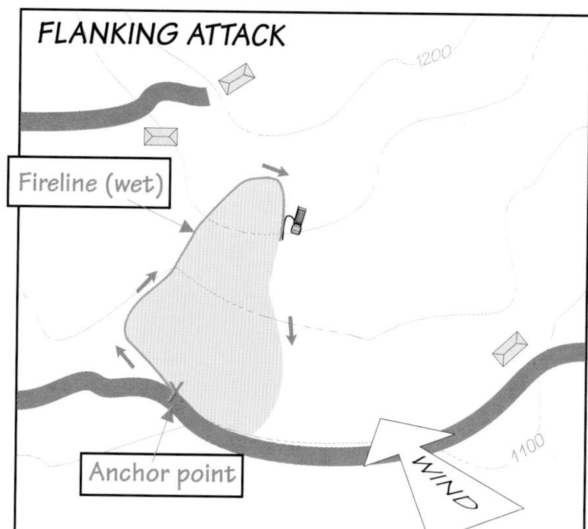

Figure 6.11 A flanking attack starts at the base or origin and moves along one flank putting out the fire as it goes.

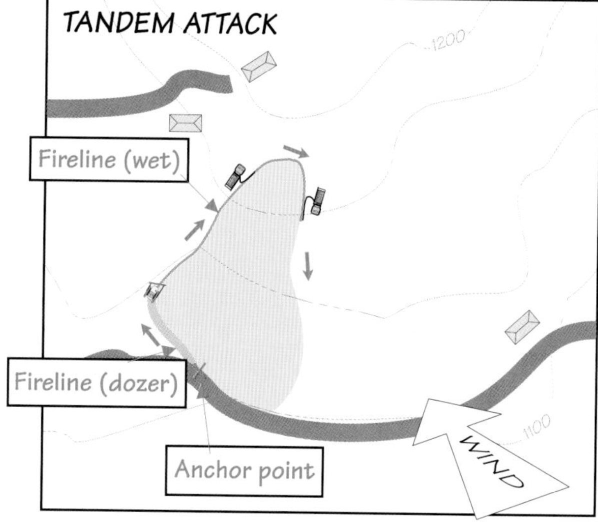

Figure 6.12 A tandem attack is similar to the flanking attack, but there are several resources moving along the same line. The lead engine is knocking the fire down, and the following equipment is backing them up.

- ***Pincer Attack*** - In the pincer attack, both flanks are attacked at the same time (Figure 6.13). This can be from the head or heel of the fire. The point from which the engines start the attack is the anchor point. The attack moves up or down the flank in a coordinated effort, stopping when the fire is contained.

An envelopment attack takes action on several sections of the fire at once.

- ***Envelopment Attack*** - In the envelopment attack, the fire's perimeter is attacked at several places at one time (Figure 6.14). There are numerous anchor points. Critical areas are attacked first using the hotspotting technique, then the engine moves toward another engine, tying the lines together. If this method is used, the attack must be well-coordinated. If not, a section of line may be overlooked and the fire may escape.

Once the fire has been contained, the fireline must be strengthened and secured. Burning material that is close to the line must be extinguished. See the section on mop-up for more information on this very important element of fire control.

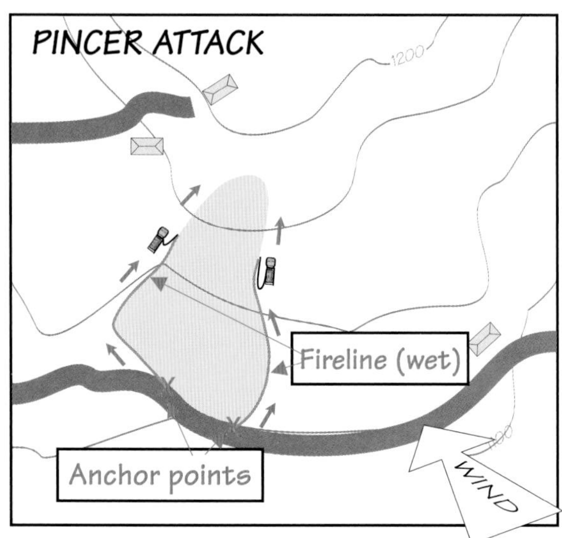

Figure 6.13 The pincer attack involves forces working both flanks simultaneously, trying to "pinch off" the head of the fire.

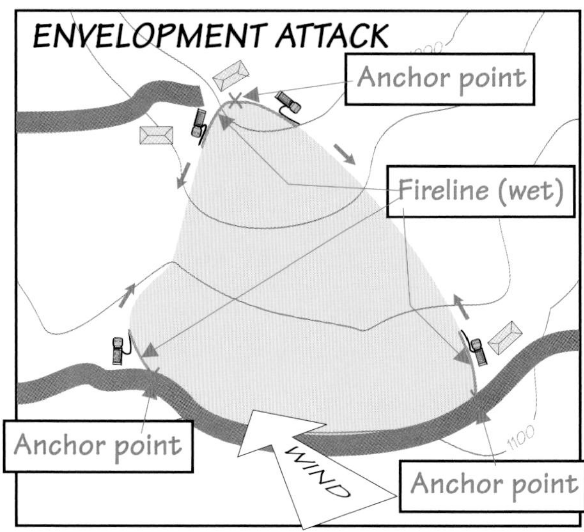

Figure 6.14 The envelopment attack involves forces establishing several anchor points and attacking from several locations at the same time in a coordinated effort.

A parallel attack constructs line within 50 feet of the fire.

Parallel Attack

The parallel attack method of line construction allows you to work close to the fire's edge, but to drop back from the fire's edge when the intensity increases (Figure 6.15). See Figure 6.16 for the advantages and disadvantages of this tactic. Line is usually constructed from 6 to 50 feet from the fire's edge. Materials are cast to the unburned side of the line. Continuous line is constructed to mineral soil. In very light grass, a wet line can be used to fire from. The line is immediately burned-out after construction. You "carry" the fire with you as you go. This is key to this tactic.

PARALLEL ATTACK

Use a parallel attack in light fuels where you can maneuver to your advantage, incorporating spots or pockets; at the base of slopes, when the fire is above you; and only when you have adequate forces. Never use this tactic if there is any danger of not being able to burn-out the fuels between the fire and the fireline. This tactic is especially effective when you can combine engine companies, bulldozers and hand crews together in a coordinated attack.

The parallel attack is usually used to get out of the smoke and heat.

PARALLEL ATTACK

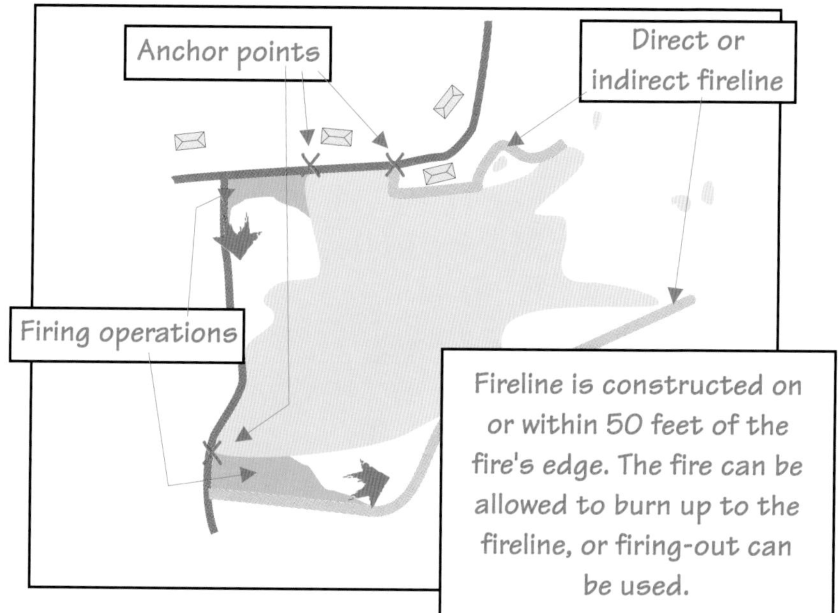
Anchor points
Direct or indirect fireline
Firing operations
Fireline is constructed on or within 50 feet of the fire's edge. The fire can be allowed to burn up to the fireline, or firing-out can be used.

Figure 6.15 The parallel attack method is a form of indirect attack, in that, because of fuels, terrain or fire behavior, you work just off the fire's edge.

Indirect Attack

The indirect attack method is used when a direct attack is not possible or practical. In this method, you select the "ground" on which to meet the fire to gain the greatest advantage in suppression and control (Figure 6.17). This tactic locates the fireline construction some distance from the fire's edge. How far is critical. Topography,

Advantages and disadvantages of the parallel attack method

Advantages	Disadvantages
Firefighters can drop back from the fire's edge, getting away from the smoke and heat.	Fire may cross fireline before it is burned-out.
Can cut fireline across pockets and fingers.	The burned area is not readily available as a safety zone.
May be able to place line in lighter fuels.	Fails to take advantage of fireline that has burned-out on its own.
Usually shorter and straighter line.	Will increase the area burned.

Figure 6.16 One of the biggest advantages of the parallel attack method is that you don't necessarily have to work in the smoke or heat.

More acreage will be burned.

fuel type, fire behavior, and available firefighting resources will dictate fireline placement. See Figure 6.18 for the advantages and disadvantages of an indirect attack.

You find more indirect attack on larger fires.

Since the indirect method is used when the fire is burning too hot for a direct attack, it is the method of choice on most large fires or anytime you confront a very fast and hot-burning fire. Under normal weather conditions, you may find an indirect attack best during the heat of the day, and a direct attack at night.

The indirect attack differs from the parallel attack in that the line is NOT burned-out as you proceed. The burning-out of

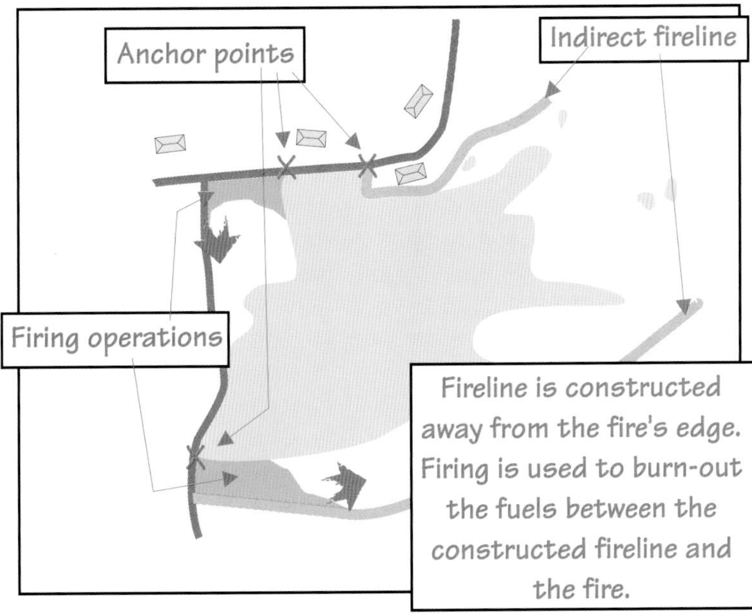

INDIRECT ATTACK
Anchor points
Indirect fireline
Firing operations
Fireline is constructed away from the fire's edge. Firing is used to burn-out the fuels between the constructed fireline and the fire.

Figure 6.17 If the fire is burning extremely hot, or is burning in heavy fuels or terrain that will not allow the use of a direct attack, you may have to go indirect.

When using an indirect attack, you can take advantage of terrain and natural barriers.

the indirect line is handled as a second phase of construction. This is usually timed with weather conditions, the deployment of adequate firefighting forces and the advancement of the fire. Use the procedures outlined in Chapter 5, when burning-out an indirect line.

It is a common practice for an Incident Commander, after sizing-up the fire situation, to decide to use the direct method on one portion of line and use the indirect method on another. Match the method to the problem. Such strategy becomes more common as the fire grows in size and there is a greater variation in weather, fuel types and topography.

An example of using various methods of attack: A grass fire is being pushed by a strong wind and develops a fast-moving head. The plan is to construct an indirect line across the head of the fire, using a backfire from an

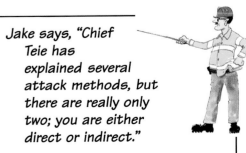

Jake says, "Chief Teie has explained several attack methods, but there are really only two; you are either direct or indirect."

Locate the fireline where you want it to be.

Move to lighter fuels if you can.

Advantages and disadvantages of indirect attack method

Advantages	Disadvantages
Can locate line along favorable topography.	More acreage will be burned.
Takes advantage of natural or existing barriers.	May be dangerous to firefighters, because they are working some distance from the fire and can't observe it.
Firefighters work out of smoke and heat.	Fire may cross line before it is burned-out.
More time to construct line.	Burning-out may leave unburned islands.
Allows line to be constructed in lighter fuels.	Brings into play the dangers of burning-out or backfiring.
May be less danger of slopovers.	Fails to take advantage of line that has already burned-out.

Figure 6.18 There are a lot of advantages to working indirect, but you have to realize you will be adding acres to the fire.

existing road to stop and control it. In the meantime, the flanks are to be controlled by use of direct attack, as in-drafts (air moving toward the fire) along the flanks make for favorable firefighting conditions.

Hotspotting

Hotspotting buys time.

Hotspotting is the stopping of the spread of hot-burning points along the fire's edge (Figure 6.19). These spots are usually in the form of fingers, racing ahead of the main fire. The purpose of this dangerous tactic (working without an anchor point at the head of the fire), is to check these rapidly advancing fingers until firelines are constructed. Hotspotting is also used to protect property and resources of high value.

This tactic utilizes small crews that move out in front of the fire. Their sole mission is to slow the progress of the fire. Helitack crews and retardant drops are often used for this purpose. The firefighters "scratch" a line around hotspots and fingers of fire. These lines most likely will not be part of the final control line.

If firefighters are used to hotspot, extreme caution must be used. Only the most experienced personnel should be utilized in this type of work. Crew protection with water or retardant drops may be critical.

Hotspotting is not for the first year firefighter. You have to be well-trained and experienced.

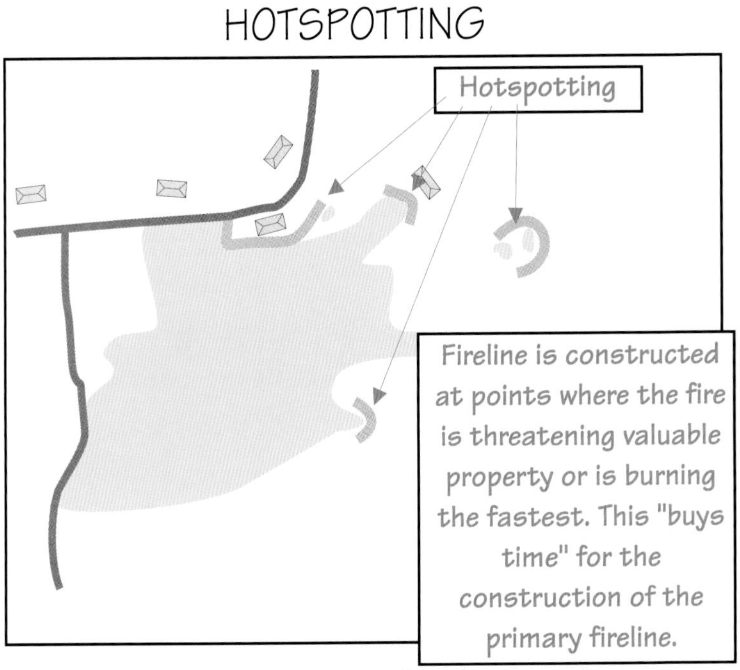

HOTSPOTTING

Figure 6.19 Use hotspotting to slow fingers to buy time until the primary fireline is constructed.

Cold Trailing

Cold trailing means the firefighters are working along a partially dead line. They are inspecting the line for hotspots, improving existing or cutting new line when necessary. Cold trailing is used to ensure islands do not flare up and throw burning brands across firelines. As in hotspotting, line constructed during cold trailing may not be part of the final control lines.

You cold trail to keep an island from burning and throwing burning material across the fireline.

Mop-up

Mop-up is dirty, but it must be done.

When the fire has been contained, the real work begins (Figure 6.20). If all the material near the fireline is not extinguished, you run the risk of the fire rekindling and escaping (Figure 6.21). This is something you don't want to experience or contribute to. Figure 6.22 outlines the rules of mop-up. Remember, it is not uncommon that hot material could still be found on large fires months after the fire was controlled.

Kari Greer

Figure 6.20 Mop-up is where the real work begins. This is the dirty part of the firefight. Do it right.

Mop-up is one of the most important phases of fire suppression because any remaining burning debris may rekindle the fire making all previous efforts worthless. Many fires have been lost because of sloppy mop-up.

Don't think that because you are mopping-up that the fireline is no longer dangerous. That thinking is wrong! In some ways, it may be more dangerous because your guard and your adrenaline may be down. Lives have been lost during mop-up. Every firefighter must remain diligent to the principles of safety and awareness.

Mop-up is scraping, digging, stirring, mixing, separating and turning.

There are two types of mop-up, dry and wet. Both involve separating burning and unburned material, and then extinguishing the material that is still hot. Some mop-up actions can be described in these terms: scraping; digging; stirring; mixing; separating; and turning logs and other heavy material.

Figure 6.21 All too many fires have been lost because the mop-up was not completed. Don't make that mistake. Mop-up isn't that exciting, and it is downright dirty, but it is an important part of the job.

Rules of mopping-up a fire

Rule	Action Plan
Start work on each portion of line as soon as possible.	Start with the most dangerous line first. Work from the fireline toward the center of the fire. Small fires are totally extinguished. On larger fires, mop-up a minimum of 100 feet, or to such a distance that nothing will blow, roll or spot across the line.
Secure and extinguish burning materials.	Arrange burning fuels so they cannot roll across the line. Spread smoldering fuels and apply water so they will cool. Scatter fuels away from the line.
Deal with special hazards INSIDE the line.	Fell snags, extinguish logs and stumps. If you can't fell the snag, carefully clear around the base, so that burning material will not fall into flammable fuels.
Deal with special hazards OUTSIDE the line.	Move slash back, away from the fireline. Fell snags and cover with dirt to prevent ignition. If stumps are close to the line, cover them with dirt.
Reinforce the fireline.	Widen and clean the fireline. Reinforce any undercut line. Burn-out or cold trail islands. Dig out roots that cross under the fireline. Feel for hot material along the fireline.
Check for spot fires.	Constantly check for spot fires, especially downwind from the fireline. Check heavier fuels (logs, snags, slash, etc.) for smoldering material.

Figure 6.22 Here are some pointers on how to conduct a proper and complete mop-up operation.

Dry mop-up is done using hand tools with no water, retardant or other wetting agent. This method involves using tools to separate the burning material, sometimes piling and allowing it to burn up. Or, you can separate it, using the hand tool(s) to scrape the hot material from it. You can also use dirt to cool and smother the burning

Dry mop-up is performed using hand tools and dirt.

material. You may need to develop a "boneyard" or "bone pile" in which you place material that you think is not hot, but needs to be checked more closely. Remember you are looking for hot material, so be careful, you may find a hotspot or two. Three points of caution about boneyards. The first, which has already been discussed, be careful of hot material. The other two cautions are: make sure you don't miss a hotspot, and accidently discard the material into the unburned area; and, the potential exists that there may be enough heat in a boneyard to start the pile on fire.

Set up a "boneyard."

Another form of dry mop-up is called "chunking and piling." This is where you pile the burning material to completely burn itself out. Do this only when you are in an area that is safe, well away from unburned fuels. This method does allow the fire to be "hot" for a longer period of time, and spotting may be a concern.

Separate and spread.

Speading out burning material reduces the heat mass, and aids in extinguishment. This is especially true when working with heavy logs, etc. Separate them, placing them so that they will not roll, and just let them cool. You will see them cooling down very quickly. Do this with great care...one slip and you may be on the injury list.

Wet mop-up is performed using hand tools with water.

You can also bank the fuels, temporarily covering with dirt in an effort to cool them. Banking unburned material protects them from ignition. If you bank the fuels, be sure to mark their location, because you or the next shift will eventually have to locate and uncover them to check for hot material.

Wet mop-up is about the same as dry mop-up, but you use water for cooling and extinguishment (Figure 6.23). Some key points to remember when using this method of mop-up are:

Bryan Day

Figure 6.23 Wet mop-up is when you use hand tools and water to remove the heat.

- Apply water in a fine spray. This conserves water and is actually more effective.

- Apply from control line inwards, so that if something is kicked up, it doesn't cross the fireline into the unburned area.

- Apply from the outside of a hot area inward to the center.

Don't just apply water...use hand tools to mix and stir.

- Use a straight stream to penetrate or reach a target. If the target is a tree or snag, start at the base and work up the tree.

- Use a system that includes spray, stir and spray again. Do this for as long as it takes to cool the ash pile and extinguishment is assured.

- If you can, use a wetting agent to aid in penetration. Be sure that there are no environmental issues before you apply chemicals.

Jake says, "Mop-up is not the time to let down your guard. It can be a very dangerous time."

- Do not totally rely on water to do the mop-up. Use hand tools to move, mix, stir, scrape, etc. In this way, hot material that may lay out of the reach of the water stream will be uncovered and cooled.

- If the fire is in baled hay, sawdust or other concentrations of fuel, they need to be cooled and then separated and spread. This is the only way you can be sure to get to the seat of the heat.

Some of the hazards you may encounter during mop-up are:

- Overhanging or leaning trees.

- Dead trees or dying trees; snags or trees with widowmakers, broken branches or tops, or large pieces of loose bark.

- Trees that may have had their roots burned away, or trees that are caught in other trees.

Wear all of your PPE, including your fire shelter.

- Rolling material (logs, rocks, etc.) on steep slopes. Fire sometimes burns-out material that was holding the logs or rocks on the slope. There will be no noise, just all of a sudden something may be coming your way.

- Bees, hornets or other things that bite and pinch may be lurking. The fire may have disturbed them, and your stirring around really upsets them.

- If you are using water, be careful for dirt and rocks that may be kicked up by the stream. Also, if you direct a stream of water into a hot ash pile, steam or hot debris may flash back toward you. Wear eye protection to prevent embers, ashes, mud, steam and debris from injuring your eyes. Choose a nozzle setting appropriate to the situation, and apply water from the outside to the center of a hot area.

Start at the hottest area and then move to cooler areas.

- Watch your footing on steep slopes, especially if they are wet or if there has been a retardant drop.

Protect your eyes from "blow back."

Work into the fire from the control line.

When mopping-up, always wear your PPE, with emphasis on eye protection. Don't think you can leave your fire shelter with your gear or on the engine. It must be with you at all times when you are on the fireline (Figure 6.24).

Figure 6.24 Even during mop-up you must wear all of your PPE.

So far, we have discussed how you mop-up and some of the safety concerns. Now we will discuss what you are really looking for, the signs of hot material. Mopping-up a small fire is relatively easy, but if the fire is large, you need to attack the problem in a systematic way:

- Start with the hottest area and move toward the cool areas. Plan a beginning and an ending point...work methodically.

- Work inward from the control line.

White ash is a serious indicator of high heat.

- Make sure you examine the entire area assigned to you. If the area is large, you may have to establish a grid or block system; number each block and assign a priority, checking off the blocks as you go.

Here is what you should be looking for. Use your senses of sight, touch, smell and hearing to assist you:

- ***Smoke*** - Look up as well as down. What you may need to mop-up could be the top of a burning tree.

- ***Heat waves*** - You can sometimes see the heat waves distort your view, like a mirage in the desert. This is a sure sign the area needs some work.

Beware of stump holes.

- ***White ash*** - This is a sure sign of high heat. The white ash may be covering some very hot material, so be careful when you begin to stir it up.

- ***Stump holes*** - Old dead stumps and root balls may be completely consumed. You will be like a dentist giving a root canal...dig deep to ensure there are no hidden pockets of hot material. It is not uncommon for a fire to burn under a fireline along a buried root, only to come up in the unburned area and allow the fire to rekindle (Figures 6.25 and 6.26).

Bryan Day

Figure 6.25 This old stump is burning deep down. It will take some work to ensure it is dead out. (Safety note: The sleeves should be rolled down.)

- ***Steam*** - Again, a sure sign for you to work the area.

- ***Gnats*** - These little creatures love to hover over hotspots. If you see a swarm of them, check it out.

- ***Smell*** - There is a difference between smoke from burning fuel and smoke associated with smoldering fuels. Learn the difference and use your nose to find the source.

- ***Sound*** - Sometimes you will only hear the cracking and popping sounds of the fire. There may also be a "hiss" as pitch in a log "boils off."

Bryan Day

Figure 6.26 The steam is a good sign that they are getting to the seat of the fire.

How to Feel Out Fireline

You can use your sense of touch to find heat. A section of line may appear to be dead, but that may not be the case. Experienced wildland firefighters feel out fireline to ensure that

Feeling out is used to find hot pockets of ash.

Northtree Fire International

Figure 6.28 Snag removal can be very dangerous. It is best left to experts.

Patrolling the Fire

Fires should be patrolled, long after the fire has been controlled. The primary purposes for patrolling are:

- Look for weak sections, hotspots close to the line and slopovers.
- To reinforce the line when necessary.
- To look for spot fires outside the fireline.

Patrolling actually begins just after the line has been constructed. The intensity of patrol decreases as the danger decreases. One of the primary purposes of patrolling is to locate spot fires. Spot fires are caused by wind blown embers and ash, radiant heat, rolling materials and by wild animals. There are some situations that contribute to spot fires: extreme dry weather; steep topography; heavy fuels; crown fires; whirlwinds or dust devils; torching-out of lone trees; winds that blow across firelines; punky logs; tree roots; snags and flashy or fine fuels. Some points to remember when establishing patrols: assign specific "beats"; post lookouts to watch for flare ups or spots; constantly move around; mark spot fires so others can find them; pay particular attention to areas where only water was used in suppression; and check for total extinguishment. Consider working in pairs or in a systematic approach.

You patrol to ensure the fire is dead out.

Be prepared to share information with supervisors and/or personnel, and always know where your escape routes and safety zones are.

Environmental Protection and Historical Preservation

Even the little creatures have rights.

For years, wildland firefighters didn't give too much consideration to the environment or to preserving our historical heritage. The damage the fire was doing to the land was considered much worse than any damage done by firefighters. Now there is clear evidence that the environment is truly being damaged by our efforts.

As a firefirefighter, you have to concern yourself with how to reduce the environmental impact of both fire and fire suppression efforts. This section will provide an overview of some of the laws and regulations that impact firefighters during a wildland fire.

Don't do more damage to the environment than absolutely necessary.

Federal Legislation

There are several pieces of federal legislation that regulate how you may fight fire within your own jurisdiction and on federal lands. Some of them have been in place for years; others are relatively new and still evolving. It is your responsibility to know these laws and understand the restrictions they place on you.

Antiquities Act of 1906 - This law is designed to protect our nation's history. It provides that "any person who shall appropriate, excavate, injure, or destroy any historic or prehistoric ruin or monument, or any object of antiquity, situated on lands owned or controlled by the Government of the United States…shall upon being convicted, be fined …or imprisoned…" Even though this law covers federally owned land, most states have enacted similar laws.

Know the law. You must comply with any restrictions in place to reduce environmental impact.

What this means to you as a firefighter is that you must know where historic sites are located and construct your firelines well away from them.

National Historic Preservation Act of 1966 - This Act directs federal agencies to inventory, evaluate, and protect cultural resources under their jurisdiction.

Archaeological Resources Protection Act of 1979 - This act is similar to the Antiquities Act of 1906, but it is specific to "archaeological resources" such as pottery, basketry, bottles, weapons, weapon projectiles, tools, structures or portions of structures, pit houses, rock paintings, rock carvings, intaglios, graves, human skeletal materials, etc. The act covers federal or tribal lands, but most states have similar laws protecting other public and private lands.

Jake says,
"Fighting fire does NOT give you the right to disregard the law."

Once again, this means that you must stay clear of sites with historic value. Under this law, you can not even pick up or move an arrowhead. It is against the law.

As a firefirefighter, you will have to determine where such sites exist. If you happen to stumble across one during your

Figure 6.29 You can't destroy or damage historical sites when fighting fire. Be very cautious and ask for help when working in an area that may have some historical value.

firefight, move out of the area and notify the appropriate authorities (Figure 6.29).

Endangered Species Act **-** The Endangered Species Act was signed into law in 1973. It is designed to regulate a wide range of activities affecting plants and animals designated as endangered or threatened. By definition, an "endangered species" is an animal or plant listed by regulations as being in danger of extinction. A "threatened species" is any animal or plant that is likely to become endangered within the foreseeable future. The law specifically prohibits the following activities involving endangered species:

- Importing into or exporting from the United States.
- Taking (includes harassing, harming, pursuing, hunting, shooting, wounding, trapping, killing, capturing, or collecting).
- Possessing, carrying, transporting, or shipping.
- Delivering or receiving.
- Selling or offering for sale.

If in doubt, ask for advice.

Once an animal or plant has been declared endangered by the US Fish and Wildlife Service, you can't even look at it funny. The penalties are very severe. You can't harm directly or change the habitat of an endangered species. You can't fell a snag that may be the future home of an endangered owl, muddy a stream where an endangered salamander lives, or cut fireline through an area where an endangered mouse may live (Figure 6.30).

This cartoon may be funny, but this piece of legislation is very serious. The federal government is serious about the protection of what it feels are endangered species. You may disagree with this, but it is the law and you would be foolish to ignore it. You should know what is endangered in your area and what to watch for.

Minimum Impact Suppression Tactics

Figure 6.30 When a little critter is endangered, they are protected under federal law. You must make every effort not to disturb them or their homes.

You have a responsibility to protect endangered species.

The goal of minimum impact suppression tactics is to use the minimum amount of force necessary to contain and control the fire, consistent with the law and the agency's policies and procedures. At one time, the primary objective of a firefighter, after life preservation and safety, was putting the fire out, even if the suppression actions did more long-term damage to the land than the fire itself. This was justified by the desire to keep all fires small. But times have changed, and it is no longer correct to fight fire this way.

Jake says, "Do not underestimate the power of little critters and their friends."

As a firefighter, you are responsible for the suppression of fires in the most expedient and prudent manner possible. You have to consider the long-term effects of your actions and modify your plan accordingly. To some, this appears to be counterproductive. You may hear comments like, "What do you mean I can't use my bulldozer to cut that line!" or "Why can't I cut down that snag; it's a safety hazard—let the '*blank*' bird nest somewhere else!" Firefighters are sometimes known for their "bullheadedness," and unwillingness to take "orders" from an "ologist." But these are the new rules, and they must be complied with (Figure 6.31). It goes without saying, however, that the safety of your people takes priority over everything else.

The goal is to minimize the long-term impact of fire suppression activities.

The single most destructive tool in the wildland firefighting arsenal is the bulldozer. It is also one of the most productive line building tools. But using a bulldozer isn't the only activity that can cause damage to the land. You can damage an area by just walking along the same path over and over. Small "scratch lines" used as control lines can become gullies if not properly cared for. Your goal, although it may be difficult to attain, is to leave an area better and cleaner than it was when you arrived.

The single most destructive tool in wildland firefighting is the bulldozer.

Figure 6.31 This "bird" knows her rights. In some areas, the snag will have to be left for nesting.

The intent of the following guidelines is not to present a whole new set of firefighting tactics. They are meant to be used as you would use traditional suppression methods. The goal is to minimize the long-term impact on the environment and historic relics, and still put out the fire as safely and efficiently as possible.

Fire suppression activities can impact the environment in a number of ways:

You can damage fragile soils simply by walking on them.

- Bulldozer blades and tracks cause damage in and around historic sites by cutting deep into the soil, destroying or displacing artifacts.
- When operated off roads, vehicles can crush cultural material.
- Water pressure can cause erosion to sites, chip away at the integrity of structures, and displace artifacts.
- Handline and helispot construction can damage sites.
- Retardants can cause structural damage and stain artifacts, rocks, and foliage.

- Firefighters can damage the terrain by using the same area over and over for foot traffic or by cutting a line.

- Sites that had been hidden for years may become exposed.

- Felling snags may destroy the nesting sites of endangered species.

Use existing barriers as control lines when you can.

The best way to avoid damaging the environment, historically important areas, or artifacts is to stay out of sensitive areas. Use natural or other existing barriers as control lines when possible. If possible, construct firelines so that you do not allow the fire to move through sensitive areas.

If you have to construct fireline in sensitive areas, there are things that you should do to reduce the level of damage. Your plan should address watershed rehabilitation efforts, like stream bank stabilization, and the construction of water bars and other erosion and water control measures. Other things which can minimize environmental damage are slash reduction and contour-log structures for erosion control when necessary.

Notes:

7

Use of Firefighting Resources

Today's wildland firefighters have all kinds of tools that can aid them in fighting fire...everything from wet sacks to fire engines to multi-million-dollar aircraft. But one thing is certain, all of these tools are only as good as the person using them. None of these tools fight fire by themselves; they need a real human to swing, drive or fly them.

Jake says, "Any tool is only as good as its operator."

This chapter will attempt to show you how these tools—some of them very expensive to purchase and use—can be used in an effective, yet efficient manner. You are only as good as your training and experience, and these tools are only as effective as the person using them.

But before we talk about engine companies, airtankers, bulldozers, etc., we must discuss the basic firefighting resource; the firefighter. Tools, like axes, shovels, bulldozers, engines, and crews don't fight fire; the individuals using or operating them fight fire. While firefighters come in all shapes, sizes and levels of training and experience, they have something in common; the same set of basic responsibilities:

- Perform manual and skilled labor to the level of their training.
- Ensure that the objectives and instructions are understood.
- Work in a safe manner.
- Maintain self in the physical condition required.

- Keep personal clothing and equipment in serviceable condition.
- Report close calls, accidents or injuries to supervisor.
- Report hazardous conditions to supervisor.

Use of Engine Companies

The engine company is the most versatile firefighting resource you have.

The engine company is the most versatile element in the fire service. It carries personnel, water, hose, and other firefighting tools. It usually transports the first Incident Commander to the scene of an emergency, and is a communications link to the dispatch office. All other elements, from the Battalion Chief to the airtanker, are there to support the engine company.

Jake says, "The training, experience, motivation and supervision of the firefighters is the key to their success."

Engine companies can be used for: direct attack on the fire; hotspotting; cutting fireline with hand tools; mopping-up and patrolling the fireline; supplying water (or foam) to hose lays and backpumps and resupplying other engines; protecting structures or other valuable resources; or providing medical aid to injured firefighting personnel or members of the public.

Fire Engines

Fire engines come in many sizes, types and colors. There are big ones, small ones, old ones, red ones, white ones, green ones, etc. (Figures 7.1 through 7.6). So, you must know something about this tool before you can properly use it in your firefighting operations. The most important information you need to know is the level of training and experience of the firefighters on the engine. You would be better off with an old engine with experienced firefighters on it than with a shiny, new engine with untrained and/or inexperienced firefighters that have not worked together.

Angelica Silveira

Figure 7.1 This triple-combination engine is the mainstay of the fire service. But they can't, and should not, venture too far off of the pavement.

Most fire engines are large pieces of equipment costing hundreds of thousands of dollars. They are great for what they do, but getting off the pavement and getting down and dirty isn't one of them. These big beasts can be useful in protecting structures, providing crash/fire/rescue for an air operation, and any other work close to a road. But, in most cases, the firefighters assigned to these types of engines are not trained—and more importantly experienced—in wildland firefighting operations. There are exceptions, but even if trained and experienced, this beast of an engine does not readily lend itself to the potential mission.

To meet the requirements of wildland firefighting, agencies have developed some really effective engines. Some of the key elements of a wildland engine are: flexible frame and body, higher clearances for plumbing and running gear, better angles of approach and departure, and in some cases, four-wheel drive.

Not all engines have to be big, fancy, and pack 500 gallons of water. Some of the more effective engines are constructed on light truck chassis.

Other parts of the world meet the need for mobile firefighting equipment in different ways. These two pictures are from South Africa (Figures 7.7 and 7.8). These units are used in forestry work and "veld" fire protection. They are owned and operated by private companies, some of which provide full-service fire protection with

Wes Schultz

Figure 7.2 This Type 3 wildland engine is specifically designed for a quick response to any type of emergency, and for the rugged requirements of off-road use.

Figure 7.3 This watertender is used to carry water to the fireline, and refill fire engines or pump a hose lay.

Wes Schultz

Figure 7.4 There are three Type 3 engine strike teams in camp, ready for assignment to the fireline.

Figure 7.5 This CDF Type 3 engine is very versatile. It can operate off-road and provide quick initial attack on any type fire.

Figure 7.6 This Type 6 engine is very simply designed and meets the requirements for the area it serves.

computer-driven cameras for fire detection and airtankers that drop Class A foam or gels.

ICS Typing of Engines

The Incident Command System (ICS) has a system of "typing" various pieces of firefighting equipment; you will learn more about ICS in Chapter 9. There are seven types of engines (Figure 7.9). They differ in levels of staffing, pump and tank capacities, etc. Type 1 and 2 engines are primarily for protecting structures. Type 3 engines can do either, and Types 4 to 7 engines are primarily for wildland fire protection.

Mobile Attack

Mobile attack with engine companies is a fast and efficient method of controlling wildland fires. In a mobile attack, the engine drives along the edge of the fire and a firefighter walks just ahead of the engine, extinguishing the fire as they

Figure 7.7 This is a medium sized 4x4 unit used to burn blocks and in initial attack.

Figure 7.8 This unit is used by planting crews and initial attack on fires.

move. Other firefighters follow-up to make sure there are no flare-ups. Mobile attack is usually a direct attack on the fire's edge, and water or foam is the extinguishing agent. If an indirect attack is selected, the engine is used to support a firing or holding operation.

Mobile attack can be used only when the fuels are light (grass or scattered brush and sage), and when the topography allows the maneuvering of an engine across it. The engine must have "pump-and-roll" capabilities (be able to pump while moving), and be constructed for the task (have adequate clearance and flexibility). Don't expect the normal triple-combination Class A structural engine to operate "over hill and dale" chasing a fast-running grass fire (Figure 7.10). They are built to stay on the pavement and pump while stationary.

ICS engine types and minimum standards

Type	Pump (gpm)	Water Tank	2 1/2" Hose	1 1/2" Hose	1" Hose	Ladders	Staffing
1	1000	400 gals.	1200 feet	400 feet		48 feet	4
2	250	400 gals.	1000 feet	500 feet		48 feet	3
3	150	500 gals.	-	500 feet	800 feet	-	3
4	50	750 gals.	-	300 feet	300 feet	-	2
5	50	400 gals.	-	300 feet	300 feet	-	2
6	30	150 gals.	-	300 feet	300 feet	-	2
7	10	50 gals.	-		200 feet	-	2

Figure 7.9 These are the minimum specification for ICS engine strike teams. It must be stressed, these are minimum requirements.

A mobile attack is a fast attack method, if the fuel type and topography are right. Don't try a mobile attack if there is a possibility that the engine will become disabled. A mobile attack allows the firefighters to stay close to their engine, with its hose stream protection and equipment. It also allows the firefighters to work close to the burned area, which can be used as a safety zone (in light fuels) in the event of extreme fire behavior. There are several mobile attack methods—flanking, pincer, and envelopment:

If the fuel and terrain allow it, and the equipment is built for it, mobile attack when you can.

Figure 7.10 Match the equipment to the task. Large "road queens" like this triple-combination engine should have stayed on the pavement.

• ***Flanking Attack with an Engine*** - The first action is to pick an anchor point, usually the road, and start extinguishing the fire front. The firefighter with the nozzle is usually in front of the engine, and to one side. The firefighter must always be in view of the operator. The firefighter and engine may work from the "green" (unburned area) or the black (the area that has already burned). Choose the attack position that best fits the situation (Figure 7.11).

In the flanking action, the engine moves along the line as fast as the fire is extinguished. Be sure all the fire is extinguished. All too often, a mobile attack moves too fast and the fire flares up behind the engine.

• ***Tandem Attack with an Engine*** - This attack method is a flanking attack that involves two or more engines or other firefighting apparatus (airtanker, bulldozer, hand crew, etc.). The lead engine takes the heat out of the fire and the second engine is used to follow the first engine, picking up hotspots and securing the line. The lead engine can move faster knowing the tandem engine will pick up any hotspots. These engines can also "leapfrog" each other, thus allowing the firefighters on the lead engine to "get a break" from the heat and smoke.

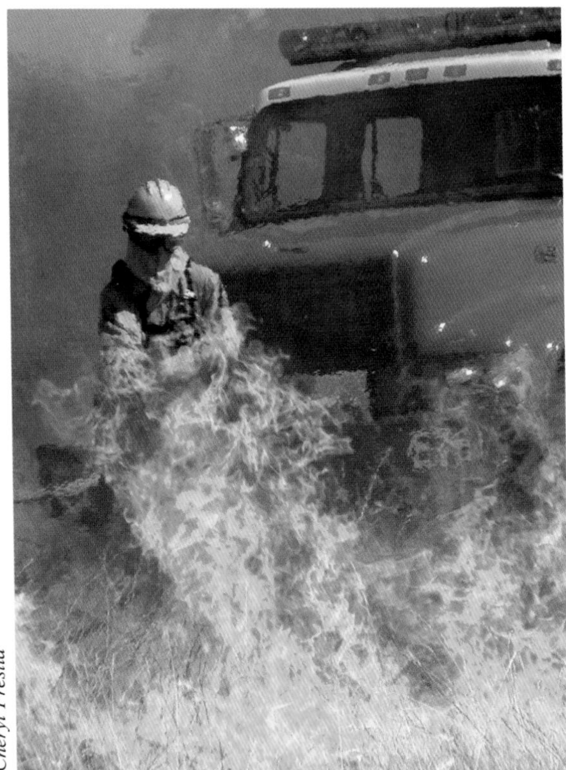

Cheryl Prestia

Figure 7.11 In light fuels and smooth terrain, a mobile attack with engines and bulldozers can be a very effective tactic.

• ***Pincer Attack with Engines*** - In the pincer attack, both flanks are attacked at the same time. This can be from the head or heel of the fire. The point from which the engines start the attack is the anchor point. The attack moves up or down the flank in a coordinated effort, stopping when the fire is contained.

• ***Envelopment Attack with Engines*** - In the envelopment attack, the fire's perimeter is attacked at several places at one time. There are numerous anchor points. Critical areas are attacked first using the hotspotting technique, then the engine moves toward another engine, tying the lines together. If this method is used, the attack must be well-coordinated. If not, a section of line may be overlooked, and the fire may escape.

Fire Hose, Brass Fittings, Nozzles and Accessories

It is necessary that you have the right tools in order to complete a hose lay to control the fire. You have to have the right type of hose, enough hose lengths to accomplish the task, the right type nozzles and brass fittings, and enough firefighters (Figure 7.12).

Figure 7.12 Lay only the amount of hose you need. Remember, you will probably have to pick it all up later.

Fire Hose

Most wildland fire hose lays are made with 1- or 1½-inch single-jacket fire hose. This hose can be made of cotton or one of many synthetic fabrics. The synthetic hose is usually lighter and does not normally have to be dried before storing. Hose comes in 50- or 100-foot lengths. It is rolled into "donuts," packed in various accordion lays, or layed in the hose bed of the engine (Figures 7.13 through 7.16).

Match the hose size and type to the job at hand.

Brass Fittings

There are many types of fittings used in a hose lay operation. There are: in-line tees with or without shut-offs; wyes and siamese valves; adapters; check and bleeder valves; increasers and reducers; connectors; caps and plugs; washers and gaskets; and pressure relief

Figure 7.13 Fire hose comes in all sizes. These rolls of 1" and 1½" single-jacket hose are the most commonly used hose on a wildland fire.

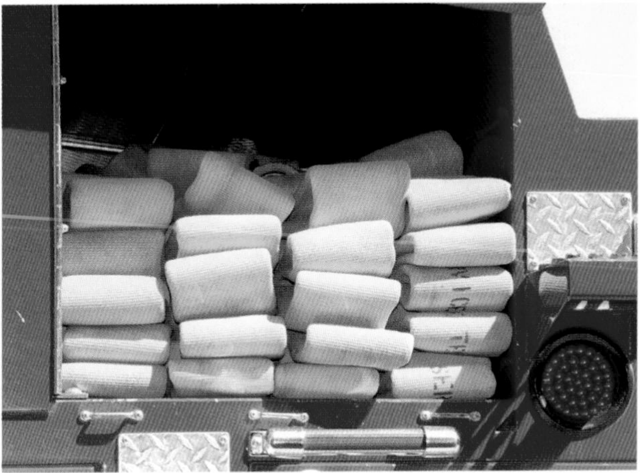

Figure 7.14 This 2½" hose is used as a supply line or on a structure fire, when larger volumes of water are needed.

Figure 7.15 The hose lay on the left is called a "pre-connect." There is usually at least 100 feet of 1½" hose ready for quick action. The hose pack on the right contains several hundred feed of rolled hose.

Figure 7.16 Here is another pre-connect next to a hose reel. Hose reels are convenient, but they don't deliver large volumes of water, which could be a safety concern if a flare up occurs.

Figure 7.17 "Brass" is what is used to connect hose lines together.

valves. Although today they are mostly made of lightweight alloys, these metal accessories are commonly called "brass." Some or all of these fittings may have to be used in the hose lay (Figure 7.17).

The "brass" fittings are either stored on the engine or in the hose packs. When planning your hose lay, consider the brass that will be needed. It is very awkward and time consuming to have to go back to the engine to get it, and also very difficult to put it into a working hose line.

To protect hose and fittings when in use or being transported, roll the hose to protect the exposed threads. The male end should be on the inside of the roll. Drain any water from the hose and roll using accepted methods.

Nozzles

Nozzles are on the "action" part of the hose lay. They control the shape of the stream of water (spray or straight-stream), and usually contain the shut off valve. Some nozzles can also control the volume of water flowing through the nozzle and hose. Combination nozzles are the most versatile, and can provide a protective water cone; straight-stream nozzles have greater reach. Be sure to select the right nozzle for the task at hand. Also, be sure there is a nozzle for each lateral line.

"Brass" is easily damaged, especially the threads.

Accessories

The two main hose lay accessories are the spanner wrench and the hose clamp. You might also find a gravity sock or ejector on your engine. The spanner wrench is used to tighten or loosen fire hose connections. The hose clamp is used to stop the flow of water in a pressurized hose line. There are many shapes, types and designs of these tools. The key is to have both as you start out from the engine.

Hydraulics

Before you can be effective in hose lay operations, you need a basic knowledge of the principles of hydraulics; the science of the motion of fluids. There are several basic rules of hydraulics.

Suction or Lift

When drafting water from a pond or stream, it is important to know the difference in elevation between the pump and the water source. Drafting water is not really an act of sucking water up the hose into the pump. What actually happens is the air is removed from the hose line (a vacuum is formed), and atmospheric pressure (weight of air) pushes water through the line and into the pump. At sea level, there is enough atmospheric pressure to push water up 15 feet in elevation. As you increase in elevation above sea level, atmospheric pressure decreases, thus reducing the vertical distance you can be from the water source. So, if you were at 5,000 feet elevation, the maximum you could lift water is 10 feet. The rule of suction or lift is:

If your hose lay goes up or down hill, the change in elevation will change your nozzle pressure.

> *For every 1,000 feet change in elevation, there is a loss of one foot in suction or lift.*

Elevation Change and Head Pressure

If you are laying a hose lay up (or down) a hill, the change in elevation will also affect the pressure in the hose (Figure 7.18). As you rise in elevation above the pump, the pressure in the hose stream is reduced as a result of head pressure. As an example, if the pump pressure was 43 psi, and the hose lay was 100 feet straight up, the nozzle pressure would be zero. Head pressure would "absorb" all of the pump pressure. As you move down a hill, the pressure will increase as a result of head pressure. In this case, if the pump pressure was 43 psi and the hose lay was 100 feet straight down, the nozzle pressure would be 86 psi.

Rule of Thumb: ± 43 psi for every 100 foot change in elevation.

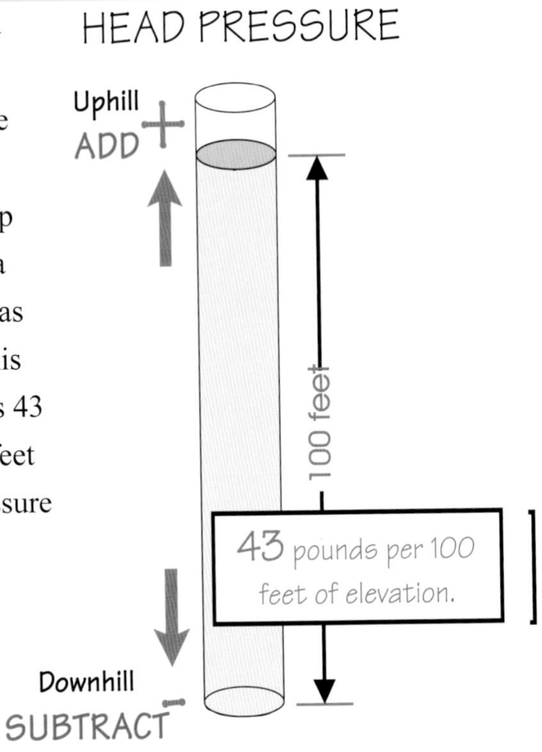

Figure 7.18 Head Pressure is the weight of water in a vertical column, in pounds per square inch (psi). In downhill hose lays, head pressure will add pressure at the nozzle tip; in uphill hose lays, it will reduce pressure at the nozzle tip.

It is important that you calculate the effect elevation change will have on your hose lay. What causes this effect is head pressure; the weight of the water in the hose. It will affect the pump that is trying to raise the water, and the nozzle that is holding it back. If a hose lay rises 100 feet in elevation, there will be a 43-pound decrease in pressure at the nozzle. The rule of elevation change is:

It is impossible to lift water more than 15 feet.

> *For every one foot in elevation change, there is a change of 0.43 pounds per square inch. If the hose lay is downhill, there is a gain in pressure; if the hose lay is uphill, there is a loss of pressure. Some easy-to-remember rules: 43 psi per 100 feet; 22 psi for every 50 feet; 11 psi for every 25 feet and 5 psi for every floor in a structure.*

Nozzle Pressure

Figure 7.19 Watch the nozzle pressure. You not only can burst hose lines, you can give a firefigher on the nozzle an interesting thrill.

Nozzle pressure is the pressure of the water at the nozzle tip (Figure 7.19). Most straight-stream nozzles are most effective at 50 pounds pressure. Most combination-stream nozzles are most effective at 100 pounds pressure. Nozzle pressure is affected by the size of the nozzle opening. If you need to "reach out with your water stream" and can't increase the nozzle pressure, decrease the nozzle opening. By making this change, you will increase the pressure at the nozzle tip, and increase the length of the water stream. The general rule for nozzle pressure is: The smaller the nozzle size, the higher the nozzle pressure.

Friction Loss

Friction loss is the reduction of pressure caused by the friction of water moving through a hose. Friction loss is variable. It is influenced by the diameter of the hose, the smoothness of the hose lining, the length of the hose lay, and the volume of water moving in the hose. Friction causes a 12 psi loss in a 100 foot length of 1 inch hose and a 2 psi loss in a 100-foot length of 1½-inch hose (when flowing 20 gallons per minute, at 50 psi, and the nozzle tip $^{5}/_{16}$-inch). The number and type of brass fittings, and whether you are using foam or wet water will also impact friction loss. A gated wye and other commonly used brass fittings can add up to a 5 psi loss. See Figures 7.20 through 7.22 for a list of friction loss data and two simple hydraulics problems. General rules on friction loss (if all other factors remain constant) are:

Rule of Thumb: -2 psi for every 100 feet of 1½-inch hose.

The smaller the diameter of hose, the greater the friction loss.

The longer the hose lay, the greater the friction loss.

The more fittings in the hose lay, the greater the friction loss.

The smaller the hose line, the more friction loss.

The longer the hose lay, the greater need to calculate pump pressure.

Friction loss table

1 inch hose at 50 psi nozzle pressure						at 100 psi nozzle pressure
Tip size (in inches)	1/8	3/16	1/4	5/16	3/8	1 inch combination
Friction loss per 100 feet	1	2	5	12	25	40
GPM	3	7	12	19	28	42

1½-inch hose at 50 psi nozzle pressure						at 100 psi nozzle pressure
Tip size (in inches)	1/4	5/16	3/8	1/2	5/8	1½-inch combination
Friction loss per 100 feet	1	2	3	10	25	44
GPM	12	29	38	50	81	116

Figure 7.20 Friction loss varies by tip size, nozzle pressure and hose size.

Hose Lays

Hose is used to get the water from a source to the fire. In a wildland fire situation, the water source is usually the engine. There are several ways to apply water from an engine: use of reel lines; use of simple hose lays; use of progressive hose lays; and use of "master stream" appliances.

Reel Lines

Use the reel line when you don't need a lot of water.

Reel lines involve small-diameter hard rubber hose, mounted on a reel. They are relatively easy to deploy and retrieve, but not very practical in a "real" firefight. They are more durable, but their diameter does not allow the water volume needed to extinguish much fire and protect the firefighter. Use reel lines during mop-up, and in situations where water volume is not critical. ***Never use a reel line to start an extended hose lay due to high friction loss.***

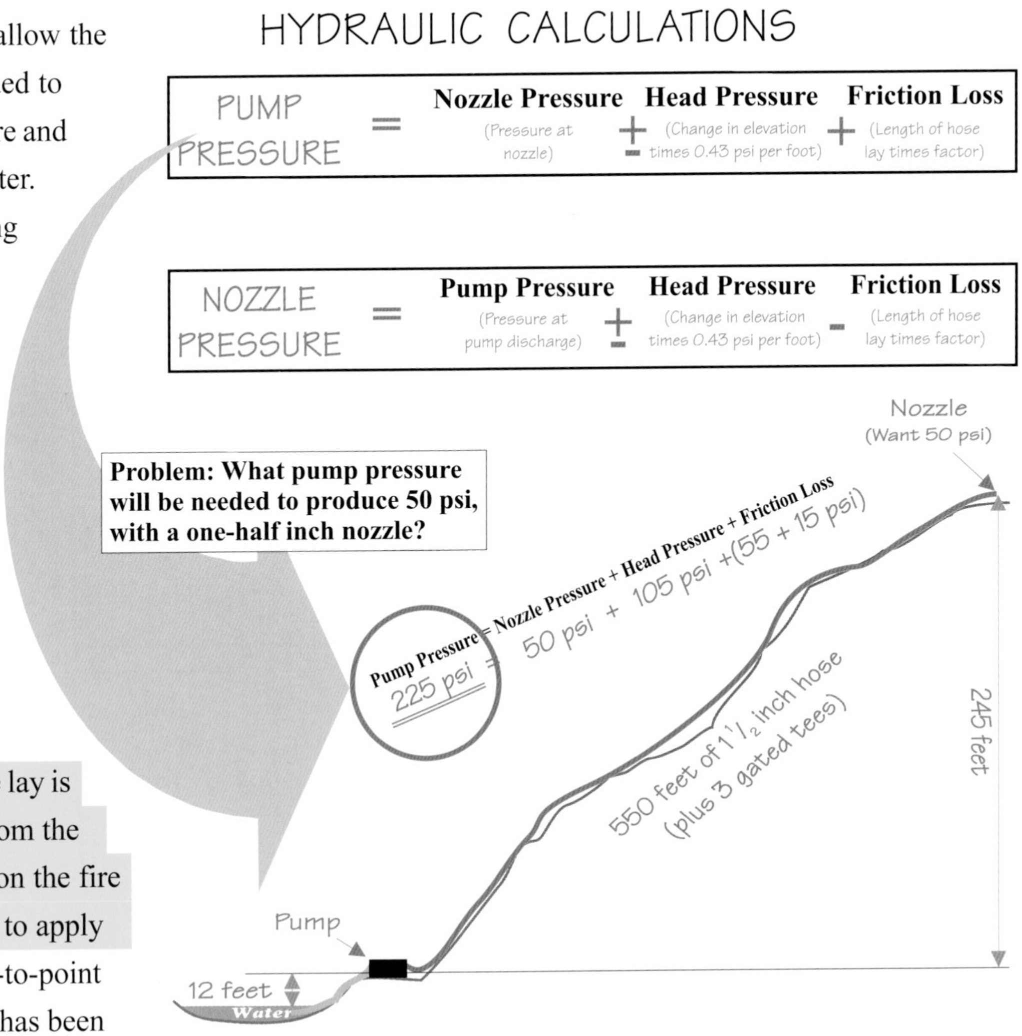

HYDRAULIC CALCULATIONS

PUMP PRESSURE = **Nozzle Pressure** (Pressure at nozzle) ± **Head Pressure** (Change in elevation times 0.43 psi per foot) + **Friction Loss** (Length of hose lay times factor)

NOZZLE PRESSURE = **Pump Pressure** (Pressure at pump discharge) ± **Head Pressure** (Change in elevation times 0.43 psi per foot) - **Friction Loss** (Length of hose lay times factor)

Problem: What pump pressure will be needed to produce 50 psi, with a one-half inch nozzle?

Nozzle (Want 50 psi)

Pump Pressure = Nozzle Pressure + Head Pressure + Friction Loss

225 psi = 50 psi + 105 psi +(55 + 15 psi)

550 feet of 1½ inch hose (plus 3 gated tees)

245 feet

Pump

12 feet

Water

Figure 7.21 This is a simple hydraulic calculation. You can see that it can involve several factors, which will impact the nozzle pressure at the end of the hose.

Simple Hose Lays

A simple hose lay is one that is layed from the pump to the point on the fire to which you want to apply water. It is a point-to-point lay. Once the hose has been put in place, water is flowed through the line. The selection of hose diameter is dictated by the length of the hose lay, the change in elevation, and the water volume desired. Be sure to lay the hose so that it cannot be damaged by hot embers, sharp rocks or cutting tools, or be driven over by heavy equipment.

Progressive Hose Lays

Progressive hose lays are used when mobile attack or other attack methods are not possible. They involve the laying of hose "progressively" while applying water along the fire's perimeter in an effort to contain and control the fire. It is impractical to use less than 1½-inch hose in a progressive hose lay. Wyes and tees should be placed in the line at regular intervals (usually every 300 feet), so that branch lines can be attached and used. These lateral or branch lines

Progressive hose lays are layed "wet."

PUMP PRESSURE CALCULATIONS

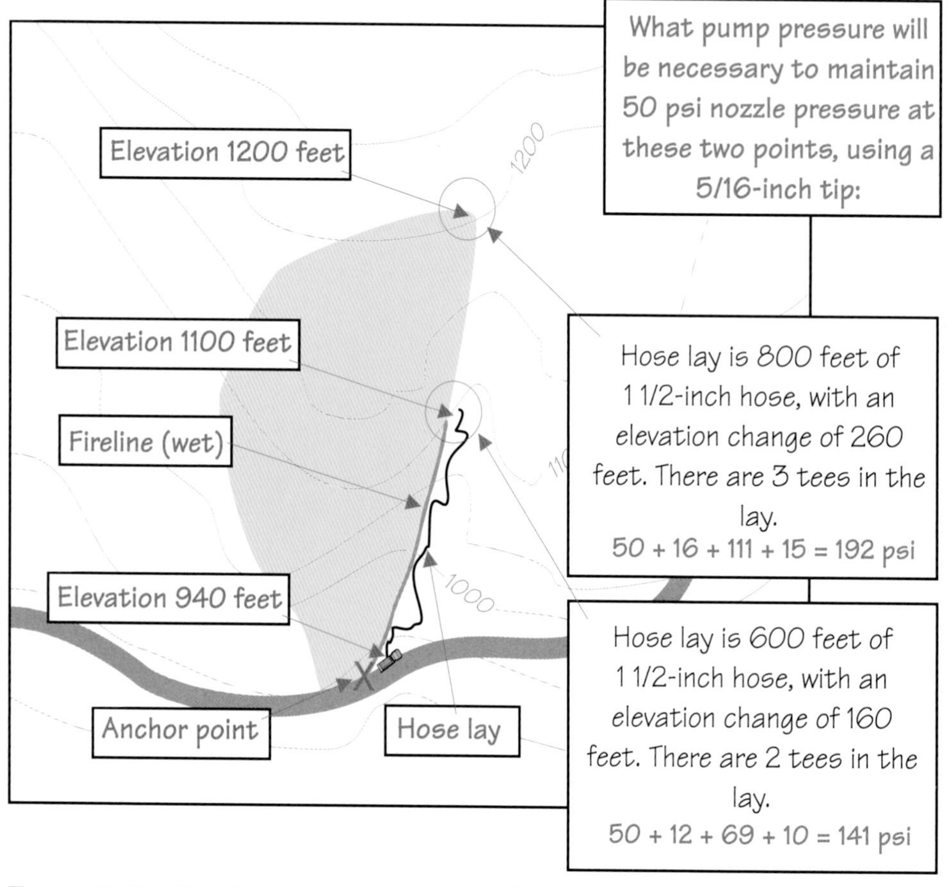
What pump pressure will be necessary to maintain 50 psi nozzle pressure at these two points, using a 5/16-inch tip:
Elevation 1200 feet
1200
Elevation 1100 feet
Fireline (wet)
Elevation 940 feet
1000
Anchor point
Hose lay
Hose lay is 800 feet of 1 1/2-inch hose, with an elevation change of 260 feet. There are 3 tees in the lay.
50 + 16 + 111 + 15 = 192 psi
Hose lay is 600 feet of 1 1/2-inch hose, with an elevation change of 160 feet. There are 2 tees in the lay.
50 + 12 + 69 + 10 = 141 psi

Figure 7.22 Here is a simple hose lay calculation that calculates the changes that are needed as a progressive hose lay is moved up a hill.

normally utilize 1 inch hose. The laterals or branches are used to control spot fires and hotspots, or to begin mop-up.

Progressive hose lays are layed "wet." Water is in the line as it is being "built." As the fire is being extinguished and the hose lay is being extended, the water is shut off at the nozzle end of the line with a hose clamp. As soon as the new length of hose is secured to the line and the nozzle is in place, the hose clamp is removed, providing instant water (Figure 7.23). Some key points to remember when constructing a progressive hose lay:

- Anchor the wet line.
- Keep the hose line free of kinks and entanglements.
- Lay the hose so it will not be burned and rupture.
- Position the hose so that it will not interfere with later line construction.
- Place tees or wyes every 300 feet.
- Know how much water you have to work with.
- Use water sparingly, just enough to do the job. Don't waste it.

Progressive hose lays need several firefighters to accomplish.

There is a firefighter on the nozzle, one firefighter to help lay and maneuver the hose, and one or more to pack hose from the engine to nozzle. There should also be one person stationed at the engine to control the pump. For an extensive hose lay, several engine companies will be necessary.

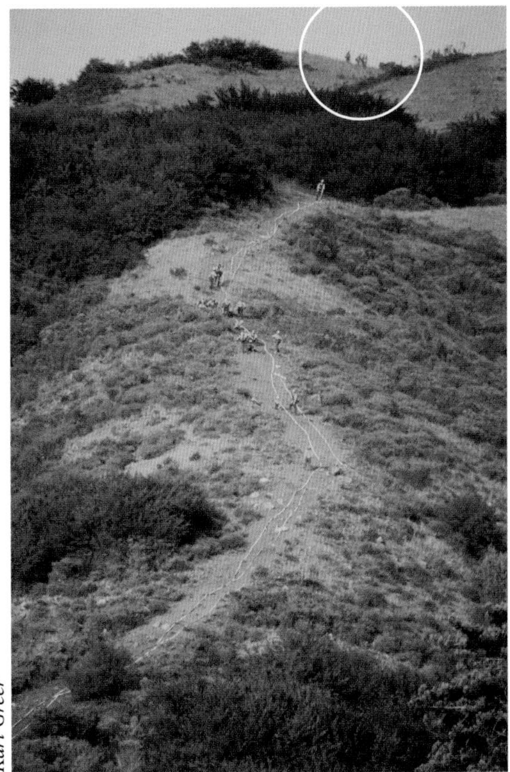
Kari Greer

Figure 7.23 This progressive hose lay is being placed in advance of this line being fired-out. Lateral hose lines are placed every several hundred feet. Their goal is to lay hose to the ridgeline (circle).

Rule of Thumb: place laterals every 300 feet.

Gravity Sock

A gravity sock is a canvas device that is attached to a length of hose and placed in a stream. The sock is filled by the flowing stream filling the hose. The hose stream must be positioned downhill from the sock. If there is at least a 100-foot drop in elevation between the gravity sock and the nozzle, there will be adequate nozzle pressure (about 50 psi). Be sure to secure the sock to stakes, rocks or trees so that it is not carried downstream with the flow.

Ejector

A hydraulic ejector is a device that uses a venturi to draw water into it. The venturi effect is when a high pressure, low volume water stream is passed through a cavity creating a vacuum. If this device is placed in a water source, it pulls water into the vacuum-filled cavity, adding water to the flow. A 1-inch line can be used to provide water to an ejector, and a 1½-inch return line can be filled by the action of the venturi in the ejector. Thus, the engine gets back more water than it pumps out.

An ejector can be used to fill an engine when the water source is too far way to reach with suction hose. If there is enough water in the engine to fill the length of hose needed (both lines, the one to and the one from the ejector), then an ejector can be used.

If you are going to use an ejector, be sure you keep enough water in the tank to fill the hose line.

Hand Tools

There are many types of hand tools. They are primarily used to construct fireline (Figure 7.24). There are cutting and scraping tools, pumping, climbing and firing tools. Each has a specific use.

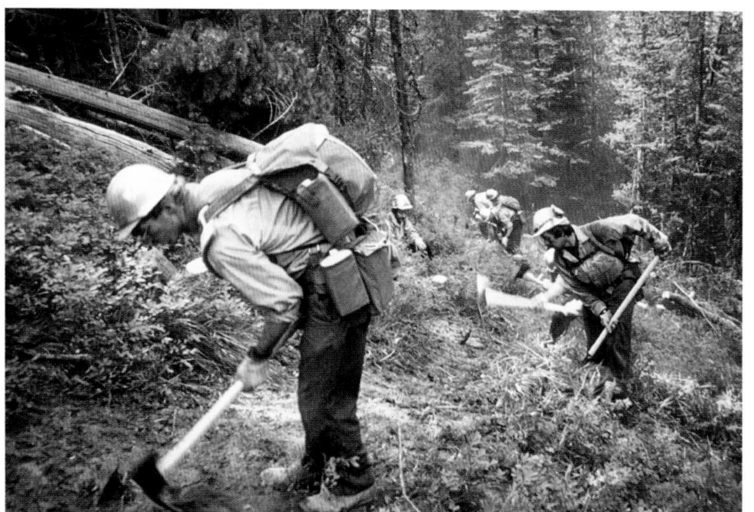

Figure 7.24 Using hand tools is hard work. These firefighters are using Pulsakis, shovels and McLeods to cut this fireline.

Cutting and Scraping Tools

The most common cutting and scraping tools are the axe, Pulaski, brush hook, shovel, McLeod, various rakes, and wire broom. Learn how to use each tool and ensure that they are ready for use. The tool must be sharp, free of rust, and with a smooth handle that is tight in the head.

There is an art to sharpening hand tools. Be sure to review local policy and wear the proper safety gear when working around sharp objects. Don't use power tools to sharpen hand tools unless you are specifically trained and authorized to do so.

Some general points about using cutting and scraping tools:

Keep your hand tools sharp.

- When carrying the tool, carry it at the balance point and on the downhill side.
- Walk and work at least 10 feet from others.

- Carry tools with the cutting edge away from the body, except for the McLeod. Carry it rake side down (puncture wounds are worse than cuts).

- Always have a firm grip on the tool with your feet firmly planted.

Backpumps

Backpumps can be extremely useful in controlling slow-to-moderately spreading fire in light fuels. They can be used to follow-up the engine on a mobile attack, in hotspotting, and in mop-up. Backpumps are especially useful when worked in conjunction with a scraping tool.

Points to remember when using a back pump:

- Do not climb over obstacles or fences while wearing a backpump.

- Adding a wetting agent or foam makes the water go farther.

- Lift carefully: a full, 5-gallon backpump weighs 56 pounds.

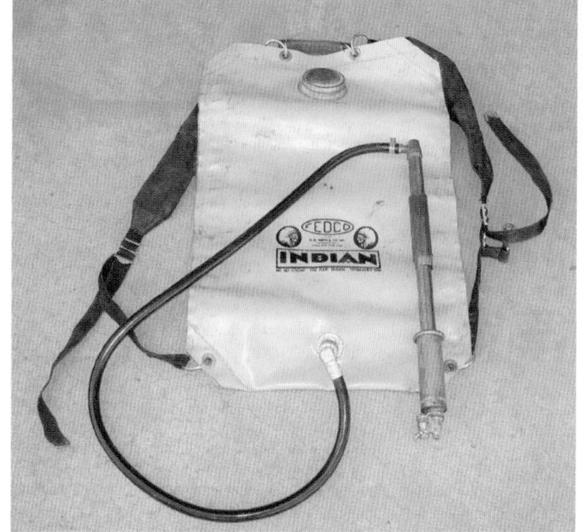

Backpumps are very useful during hotspotting and mop-up.

- When extinguishing a fire in light fuels, use your finger to spread the water stream into a spray, and a swinging motion to fan the water across the burning fuel.

Ladders

Ladders of one size or another are carried on most structural fire engines. Some wildland engines will also be equipped with them. You can use ladders to gain access to the roofs of structures, to climb steep banks and road cuts, and in an emergency, as a litter.

Power Tools

The two most common power tools carried on an engine are a chain saw and portable pump.

Chain Saws

Chain saws are one of the most dangerous tools on the fireline. They are to be used only by those who have had specific training in their use. They are valuable for cutting line in heavy fuels, dropping snags, and bucking logs into small enough sections to roll over during mop-up.

Chain saws are very dangerous. If you haven't been trained in their use, don't use one.

Portable Pumps

There are three types of portable pumps: those that are placed on land near a water source; in the water (submersible); and those that float on the water source (Figure 7.25). Each can be extremely useful. They can be used to "tap" a water source anywhere near the fireline. They can be placed near or in a swimming pool for the purpose of filling the tank on an engine, or protecting a structure. One nice thing about a floating pump; it doesn't need a suction hose.

Kari Greer

Figure 7.25 This portable pump is taking water from a stream and pumping a hose lay. Note how the pump is placed on a piece of plastic to protect the stream from contamination.

There are many sizes, types and styles of portable pumps. Get trained in the use of the type you have on your engine. You have to know what you are doing, or you may be wasting time and damaging equipment.

Protect the environment when using a pump close to a stream.

Firing Tools

There are usually two types of firing tools carried on a wildland engine—the drip torch and fusee—but the common match or butane lighter can come in handy, also. Firefighters should keep matches or a lighter in their web gear for use in an emergency. In all cases when using fire to fight fire, you must have a plan (Figure 7.26).

Drip Torch

The drip torch is very useful when you need to start a backfire or burn-out a fire line. The drip torch (sometimes called a backfire torch) is a device that uses a liquid fuel to ignite dried grass or forest litter (Figure 7.27). It is normally used in a swinging motion, spreading the burning fuel into the area that you want to burn. Some key points to remember when using a drip torch:

- Ensure that the fuel mixture is not any "hotter" than four parts diesel fuel to one part gasoline (other variations: two parts diesel fuel to one part solvent; or two parts crankcase oil and one part gasoline).
- Wear gloves, and hold the torch only by its handle.
- Carry in the upright position until you are ready to use.
- Extinguish the torch when not in use, and don't touch until it cools down.
- Keep the torch over fuel you REALLY WANT TO BURN and away from your clothing.
- Always have an escape route planned.

Figure 7.26 When firing operations are prescribed, be sure you have a plan and everyone knows it. There is nothing worse than firing the wrong side of the fireline. It has happened!

Kari Greer

Figure 7.27 The drip torch is very useful in burning-out. Be sure not to drip any of the fuel on your protective gear.

Fusee

The fusee or road flare can be used to ignite grass and forest litter (Figure 7.28). It can be thrown into the wildland fuels, to start burning farther from the fireline. Each firefighter should carry several fusees for potential use and for safety (to be used to fire-out a safety island). Key points to remember when using a fusee:

If the drip torch fuel is not mixed properly, it will either be too "hot" and thus dangerous, or will not burn properly.

"Strike" the fusee away from your face.

- When igniting, turn your face away and strike AWAY from your body.
- Keep the fusee away from your body and clothing.
- The phosphorus material is extremely hot (1,400 degrees) and readily drips when burning. Be careful!
- Do not stare at the burning tip of the fusee; it can damage your eyes.
- If you are going to do a lot of burning, find a suitable handle to attach to the fusee. This will limit the need to constantly stoop over.

Kari Greer

Figure 7.28 This firefighter is using a fusee to burn-out an area between a line constructed by a tractor-plow and the fire's edge.

- Keep the fusee over fuel you REALLY WANT TO BURN and away from your clothing.
- Avoid breathing the smoke and vapors emitted by the fusee.

Matches

Matches can be used to ignite a small patch of grass or forest litter. A McLeod or other scraping tool can be used to spread the fire into new areas, gathering new fuel as you move. Lighters can be used in the same way, or to light the tip of a twig or stick to use as a torch.

(***Note***: There is a much more detailed discussion of the ***Use and Care of Hand Tools*** in the Appendix.)

Use of Firefighting Aircraft

Airtankers and helicopters are tools to be considered in the wildland firefight. The objective of tactical air operations is to aid in safe, effective and efficient fire suppression activities. Aircraft are most effective when used during initial attack (when the fire is small) AND when they work in conjunction with ground forces. They are also very expensive, so they should be released when they are not effective. Safety is the most important consideration when using aircraft. ***Never compromise safety for the sake of fireline production.***

Airtankers are most effective during initial attack.

Tactical Support with Aircraft

Airtankers and helicopters can provide excellent tactical support for ground forces. Airtankers can deliver retardant or suppressant drops to knock down the fire so that ground forces can safely advance (Figure 7.29). ***They can also hotspot, slowing the fire and trying to buy time, or place a drop to protect a structure or other valuable resource.***

Karen Wattenmaker

Figure 7.29 Airtankers are most effective on small fires. They can also serve a real purpose on larger fires. This P-3 Orion can carry 3,000 gallons of retardant.

Airtankers are most effective during the initial attack phase when the fire is small. This is why aircraft may be *diverted* to new fires, from fires in the extended attack phase. Keep this in mind when establishing your plan of attack.

Helicopters can provide close-in tactical support with water, foam or retardant drops (Figure 7.30). They can move personnel to remote sections of line, transport supplies, be used to start backfires, and evacuate injured personnel. They also provide an

Kurk Taylor

Figure 7.30 This Siskorski Sky Crane is providing close-in support to engine companies protecting structures. It can carry 2,000 gallons of water.

Figure 7.31 Airtankers are limited by steep terrain, power lines, wind and smoke. Alert the pilot to any hazards in the area.

excellent vantage point for reconnaissance and mapping.

Air attack aircraft provide a platform for the air attack officer to control and coordinate the air operations on an incident. The primary function of an air attack officer is air operations safety. Coordination between air and ground forces is essential for maximum effectiveness.

Limitations of Air Operations

There are several factors that will limit the effective use of aircraft: steep terrain; winds over 20 miles per hour; shadows (especially during early morning and late evening); tall trees or snags; power or telephone lines; elevation; smog and dense smoke (Figure 7.31).

The fuels involved in the fire, or those ahead of the fire, can limit the effectiveness of airtanker and helicopter drops, since a heavy canopy limits penetration of the retardant. Airtanker drops are usually ineffective during periods of extreme fire behavior.

Turnaround time (the time it takes an airtanker or helicopter to secure a new load of retardant or suppressant and return to the fire) and the distance from the drop area to the refill point (refill base for airtankers, and water source or heliport for helicopters) could be limitations. If the turnaround time for the airtankers is considerable (more than 30 minutes), more airtankers can be added to shorten the time. A 10-minute turnaround time for a helicopter is considered the limit. For maximum airtanker effort, consider ordering enough airtankers for a drop every five minutes. The air attack supervisor can provide the incident commander with the type and number of airtankers needed.

> Jake says, "Airtankers do not put out fires...they help firefighters put out fires."

Airtankers

Airtankers have only one purpose: to deliver fire retardant or suppressant to the fireline. They can do this very quickly over wide areas and deliver large volumes of retardant/suppressants. Airtankers are very effective on small, initial attack operations, and when there is quick follow-up by ground forces. Except for water-scooping types, airtankers must return to an air attack base for refilling.

Airtankers vary about as much as fire engines. There are small ones, big ones, and mostly old ones. Airtankers really became a part of the firefighter's tool kit in the mid-1950s. First on-scene were the cropdusters. These were 200- to 300-gallon N-3-N biplanes that sprayed crops in the morning and flew fires in the afternoon. Then came a whole series of retired military fighters and bombers used during World War II. There were TBMs, F7Fs, B-17s and PBYs. They have all been retired, and we now use P-3 Orions, C130s, S-2T, CL-215s, AT802, etc. (Figures 7.32 through 7.36). Because each of these aircraft have different capacities, airtankers are rated by their retardant capacity. Figure 7.37 outlines the various ICS type classifications for airtankers.

Airtankers are equipped with *variable flow* tank doors that allow the pilot to release the exact amount of retardant needed for a given width and length of line. Retardant should fall to the ground as light rain or mist.

Trevor Wilson

Figure 7.32 This is the flight line at CDF's maintenance facility in Sacramento, CA. Airtanker 90 is one of the newly converted S-2s with turbine engines.

Brian Weatherford

Figure 7.33 These are Canadian CL-215 water-scooping airtankers owned by the Minnesota Division of Forestry. They were being used by the Montana DNRC on a series of fires in 2003.

Figure 7.34 This is a 600-gallon cropdusting aircraft called a Dromedary.

Kari Greer

Figure 7.35 The National Guard(s) have eight C-130 aircraft outfitted with a 3,000-gallon retardant dropping system called MAFFS. This is the business-end of the system.

Years of experience show that the best drop height is 150 feet above the fuel. Drops are most effective when the retardant has stopped all forward motion before it reaches the fuel. If the drop is higher than 150 feet, the coverage of the fuel will be "thinner." If the drop is too low, the forward motion will limit the coverage, and the retardant drop could uproot trees, dislodge limbs or tops, and injure firefighting personnel.

Never consider a retardant drop the final suppression action on a section of line. Ground action must then follow. Many a fire has been lost because there was no follow-up, and the fire burned through the retardant line.

Jake Oosthuizen

Figure 7.36 This Airtractor AT-802 carries 600 to 800 gallons of retardant.

Retardants and Suppressants

Airtankers are single-purpose...they only deliver retardants or suppressants. For the retardants and suppressants to be effective, the various drops should overlap. If gaps are present, the fire will simply burn through the retardant line (Figure 7.38).

Never consider a retardant line the final suppression effort. Follow up with ground action.

A fire retardant is a substance that, by chemical or physical action, reduces or slows combustion, thus slowing or retarding the rate of spread of the flame front. Most retardants are produced by combining water, several chemicals, and a coloring agent. The main chemical ingredient is a fertilizer. Retardant is not dangerous as a chemical, but it is very slippery, especially on a hard surface like a log or the tailboard of an engine. Retardant weighs just over 9 pounds per gallon. Long-term retardants contain chemicals which continue to retard fire even after the water has evaporated. Request long-term retardant when it may be some time before follow-up by ground forces will be possible. Retardants are most effective when applied ahead of the flame front, not directly on it.

A fire suppressant is a chemical mixture or substance that is applied directly to a fire, usually at the base of the flames. It suppresses the fire by cooling or smothering. Once dry, it is ineffective. Plain water, water mixed with a gel, and water mixed with Class A foam are classified as suppressants.

When you have a fire with high-heat output and flame lengths, long-term retardant will be needed to slow the spread. It is a matter of fuel loading; fire behavior and the associated BTUs of heat vs. water delivery and application.

ICS type classification for airtankers

ICS type	Specifications	Aircraft
Type I airtankers	3,000 gallons or more	DC-7, C-130, P-3
Type II airtankers	1,800 to 2,999 gallons	DC-7, SP2H, P-2
Type III airtankers	800 to 1,799 gallons	S-2, S-2T, CL-215/415
Type IV airtankers	100 to 799 gallons	Air Tractor 802 Thrush, Dromader

Figure 7.37 Airtankers come in many size-classes. These ICS typing classes are used internationally.

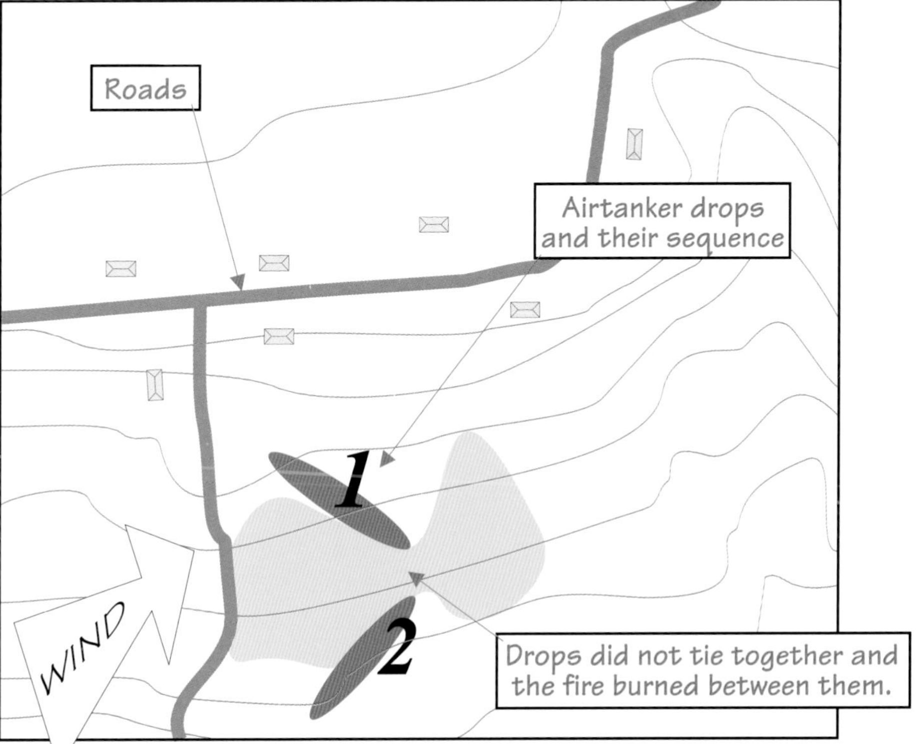

Figure 7.38 Make sure that the airtanker drops tie together. If not, the fire will burn between them and escape containment.

If You Are About to Be Hit by a Retardant Drop

If you are working close-in with airtankers, know when and where they are going to drop (Figure 7.39). Listen for the airtanker. The older aircraft with reciprocating engines are quite loud. If the roar is getting louder, you may be in a drop zone. The newer airtankers powered with turbine engines are very quiet and can sneak up on you. Airtankers can drop several tons of retardant, which can result in critical injury to a firefighter or severe damage to apparatus if caught off-guard by a low drop. If the drop is at the proper height, those caught in the drop will just become "very pink." If you think you are about to be hit with a drop:

Airtankers are single purpose: they just deliver retardant.

Figure 7.39 If you are about to be hit by a retardant drop, lie down behind a rock or other stout object; hold your hand tool to your side, and hold onto you helmet.

- Move out of the target area, if there is time.
- Avoid larger, old trees (limbs or tops may break off with the force of the drop).
- Never stand up in the path of an airtanker drop. This greatly increases the chance of injury.
- Get behind something solid, like rocks or your engine. Lie down on your stomach, facing the oncoming drop; helmet and goggles on; feet spread apart for better body stability; cover face; and hold any tools you have to your side, away from your body.

Drops can be dangerous if they are too low.

After the drop, continue your firefighting actions. If working in the retardant drop area, you will find it very slippery. Be careful and watch your footing. Wipe off your hand tools, especially the handles. If time and water supplies allow, wash the retardant off your engine and any road surfaces or homes hit by the retardant drop.

The vortex from the wingtips of low-flying aircraft (both fixed and rotary wing) can cause violent air turbulence (Figure 7.40). This may cause erratic fire behavior. Report low drops or vortex problems to the air attack officer.

Cammeron Eck

Figure 7.40 This P2V is pretty low as it crosses over the ridge. There could be some turbulence on the ground that could impact fire behavior.

Anchor the retardant line.

Helicopters

Helicopters can provide close-in tactical support. Helicopters are very versatile. They can move personnel to remote sections of line, transport supplies, and be used to start backfires. They can also be used to evacuate injured personnel and provide an excellent vantage point for reconnaissance and mapping. Helicopters are classified in the same manner as the airtankers. Figure 7.41 outlines the specifications for the various sizes of helicopters.

Helicopters in the initial attack role can often be problem solvers, dealing with spot fires, gaps in retardant line, and personnel and

ICS type classification for helicopters

ICS type	Specifications	Aircraft
Type I helicopters	16+ seats; 5,000 lbs. carded weight and 700 gallons of retardant.	Boeing Vertol 234/CH-47, Siskorski S-61, Bell 214B
Type II helicopters	9 to 15 seats; 2,500 to 4,999 lbs. carded weight and 300 to 699 gallons of retardant.	Bell 204/205/212/412, Sikorski S-58T
Type III helicopters	5 to 8 seats; 1,200 to 2,499 lbs. carded weight and 100 to 299 gallons of retardant.	Bell 206, Lama SA-315B, A-Star AS-350D, Hughes 500D
Type IV helicopters	3 to 4 seats; 600 to 1,199 lbs. carded weight and 75 to 99 gallons of retardant.	Bell 47G-3B2, Hiller 12E

Figure 7.41 Helicopters are classified by their carrying capacity.

Jake says, "Use helicopters in your plan, but don't base your success on them...they may not be there when it counts."

structure protection. The effectiveness of a helicopter can easily be overwhelmed by high-heat or extreme fire behavior. They are also a high-maintenance machine, so they may not be available to you when you want them.

Helicopters can be fitted with two types of drop tanks: buckets and fixed-tanks. Buckets range from 100 to 2,000 gallons (Figures 7.42 through 7.47). They are suspended by cables from the helicopter's cargo hook. Fixed-tank systems involve the installation of a tank directly to the frame of the helicopter. Older tanks usually must be filled by hoses on the ground. Newer tank designs include a snorkel system that allows for unaided filling from ponds, portable tanks, streams or lakes. Most helicopters that are used to drop water are also fitted with a Class A foam injection system.

Fixed tanks on helicopters are more accurate.

Figure 7.42 *This is a CDF Super-Huey. It carries a crew of eight and 300 gallons of Class A foam.*

Northtree Fire International

Figure 7.43 *This is a US Army Black Hawk being used on a fire in Oregon.*

Wes Schultz

Figure 7.44 *This is a US Marine Corps Super Stallion that can carry 2,000 gallons of water. It was on the Cedar Fire in 2003.*

Jake Oosthuizen

Figure 7.45 *This is a Russian MI-8 used by the South African National Park Service. It carries a crew of 14 firefighters and can drop 2,000 gallons of water.*

Karen Wattenmaker

Figure 7.46 Whenever you are working around a helicopter that is moving cargo, be especially alert.

Kari Greer

Figure 7.47 This is a CH 47 that can carry 16 firefighters or 1,000 gallons of water.

Helicopters can be used to drop either suppressants or retardant. For tactical missions when direct suppression is intended, water or suppressants can be used. Indirect attack generally requires a retardant that inhibits burning for a longer period. If a reliable water source is close by the drop area, helicopters can deliver more gallons of material per hour than an airtanker.

Helitack crews

Helicopters are most effective if they have an organized helitack crew assigned. This is a highly trained and experienced firefighting group. They are experienced at working in remote locations, using hand tools supported by "drops" from the helicopter. Helitack crews work well taking action to contain spot fires or burn-through in retardant lines. It is a waste of resources to tie a helitack crew to a task that can better be accomplished by an engine company or hand crew. Ensure that you have communications with the helitack

Keep the helicopter and helitack crew together; they are more effective when working as a team.

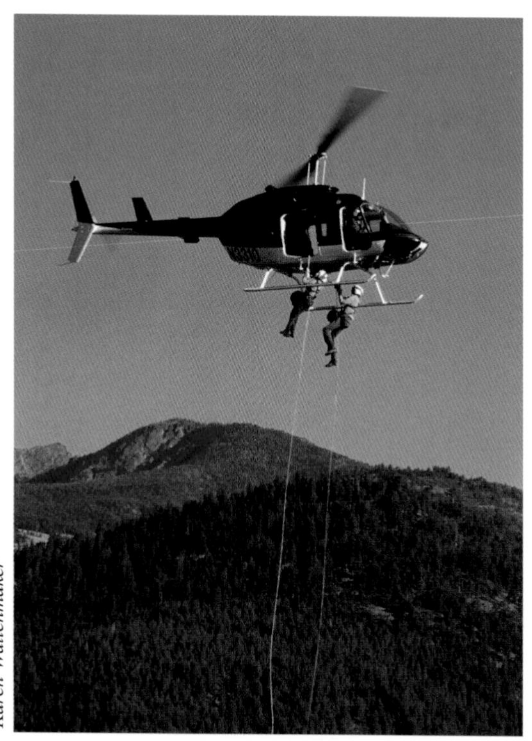

Karen Wattenmaker

Figure 7.48 This is a Bell Jet Ranger that is delivering two firefighters to a remote section of fireline. This repelling technique takes special training.

crew and helicopter. The crews can be flown to a landing site near the fire, or specially trained crews can rappel to the fire (Figure 7.48).

Working around Helicopters

Helicopters are very dangerous to be around, especially if "things are turning." There are several safety rules to keep in mind when in or around helicopters:

• Approach and depart the helicopter from the front or side, in a crouching position, ***in view of the pilot***. A gust of wind can cause the rotor blades to dip lower.

• Approach and depart on the downslope side (to avoid main rotor).

• Approach and depart in the pilot's field of vision (never toward the tail rotor).

• Use a chin strap or secure helmet when working around main rotor.

• Carry tools horizontally, below waist level (never upright or over your shoulder).

Never approach a helicopter without having eye-to-eye contact with the pilot.

• Fasten seat belt upon entering helicopter and leave buckled until pilot signals to exit. Fasten the seat belt behind you before leaving.

• Use the door latches as instructed; caution should be exercised around moving parts and plexiglass (it is easily cracked).

• Keep landing areas clear of loose articles that may "fly" in the rotor downwash.

• Do not throw items from the helicopter.

Never approach a helicopter from the rear.

• Provide wind indicators for takeoff and landings. Flagging tape works well as a wind indicator.

• Wear eye and hearing protection when working close to helicopters.

The downdrafts from low-flying helicopters can cause erratic fire behavior. Be aware that your fire may flare up and jump lines if a helicopter is working low in your area.

Engine Company Support for Fixed-Tank Water-Dropping Operations

You may be required to support a water-dropping operation. Helicopters equipped with a fixed-tank usually have to rely on a fixed water source (fire hydrant, lake, stream, etc.) or water tenders for resupply. These helicopters may have to be refilled using an engine company for support.

The engine company has two functions: providing a water source and crash/fire/rescue services.

When establishing a helispot for refilling helicopters (Figure 7.49), keep these points in mind:

- Have a water source that can provide adequate water for the operation. A Type II helicopter can use 7,200 gallons of water per hour.

- Pick a location that is free of power lines, light poles, trees and other obstructions. The approach clearance must be at least 100 feet, with a departure clearance of at least 300 feet. Ridges work quite well. The site must be secure. All vehicular traffic and sightseers must be controlled. By federal regulations, you cannot operate within 1,000 feet of a school. If a windsock is not available, use flagging tape to indicate wind direction.

- The helipad area should be 20 by 20 feet. It should be cleared of all debris that could fly into personnel working in the area, or into the

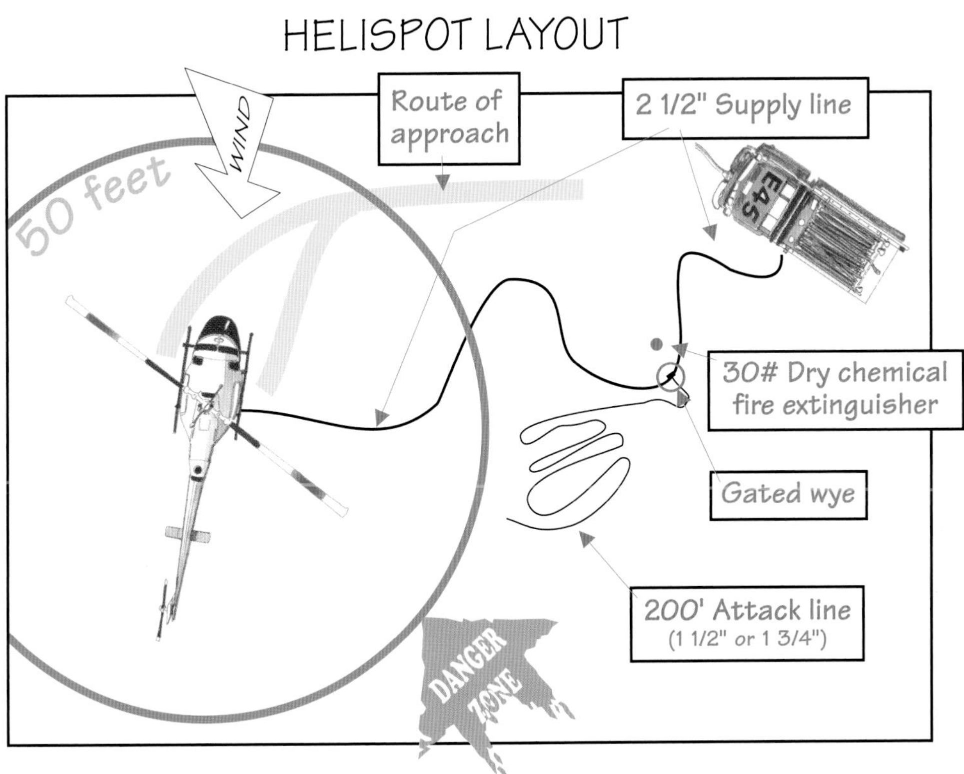

Figure 7.49 It is dangerous working around a helicopter.

helicopter rotor system. Dust control is critical; however, do not wet the area during takeoffs and landings.

- Place the engine at least 100 feet from the pad. Lay a supply line (2½-inch) to a wye with shut-offs. From one leg of the wye, lay the supply line to the pad. From the other leg, connect 200 feet of 1¾- (or 1½-) inch hose for a protection line. Remember: ONLY HELITACK PERSONNEL should fill the helicopter tank.

- Place a dry chemical fire extinguisher near the gated wye, about 50 feet from the pad.

- NEVER approach the helicopter until directed to do so.

Air Attack

Tell the air attack when you require status reports.

Air attack aircraft provide a platform for the air attack officer to control and coordinate the air operations on an incident. The primary function of an air attack officer is air operations safety (Figure 7.50). The air attack officer also has a tremendous view of the situation. The air attack officer can see the "big picture" better than anyone else on the incident. The air attack officer should work the aircraft in close coordination with ground forces.

Figure 7.50 The air attack officer is like a referee or traffic cop. They are in charge of all firefighting assets that fly. They also have the best view of the fire.

When the aircraft arrives over the fire, ask for an assessment of the situation and outline your objectives to the pilot. If you have to, orient the air attack officer or pilot to your location and where you want the drop. Use simple language and consider the pilot's point of view. To guide the aircraft: use known topographic features; reference the part of the fire; and use "clock orientation." (Place an imaginary clock over the aircraft, with the 12 o'clock position at the nose and the 6 o'clock at the tail.) As you see the heading of the aircraft, you say, "I'm at your 9 o'clock" if you are off the left wing.

If an air attack aircraft is not "at scene" (or if your organization does not use them) when the first airtanker arrives, you will have to provide direction to the airtanker. Discuss with the pilot your objectives and overall plan. Confirm your plans with the airtanker pilots and advise them of any safety hazards. *Lead planes* are used to provide a higher degree of safety for airtanker operations. They fly the course the airtanker is to follow, ensuring that the approach is clear of any hazards, assessing visibility and turbulence, and testing the escape route.

Jake says, "Air attack officers have the best seat in the house. Make good use of them."

Retardant drops can be used to:

- ***Hold*** (to allow time for ground forces to take action);
- ***Delay*** (to slow the advance so that the fire will hit barriers like ridges, highways or control lines after ground forces are in place);
- ***Herd*** (direct the head of the fire);
- ***Cool*** (reduce intensity of the fire so ground forces can place control lines);
- ***Control spots*** (keep the fire within lines);
- ***Reinforce*** (natural barriers, roads, or constructed fireline); and
- ***Protect*** (personnel and structures).

Use airtankers to slow the flanks and head of the fire and to catch spots. Follow-up with ground forces.

Confirm with the pilots the locations of poles, wires or extra-tall trees and snags.

Tactical Use of Aircraft

The incident commander MUST establish objectives and set priorities for the air operation. The incident commander must have a game plan and communicate it to the air attack officer and ground support personnel (Figure 7.51). Ensure that retardant drops will have follow-up by ground forces. Don't over-manage the air operation. Communicate to the air attack officer what you want, and require regular situation status updates.

Figure 7.51 When aircraft are involved in a firefighting operation, it is critical that it be well-coordinated. The incident commander must be in constant contact with air operations.

There are five general considerations for any kind of retardant-dropping operation:

- Are there ground forces available and in place to back up drops quickly and safely?
- Does the fire's behavior and potential justify airtanker drops? Will the drops be effective?

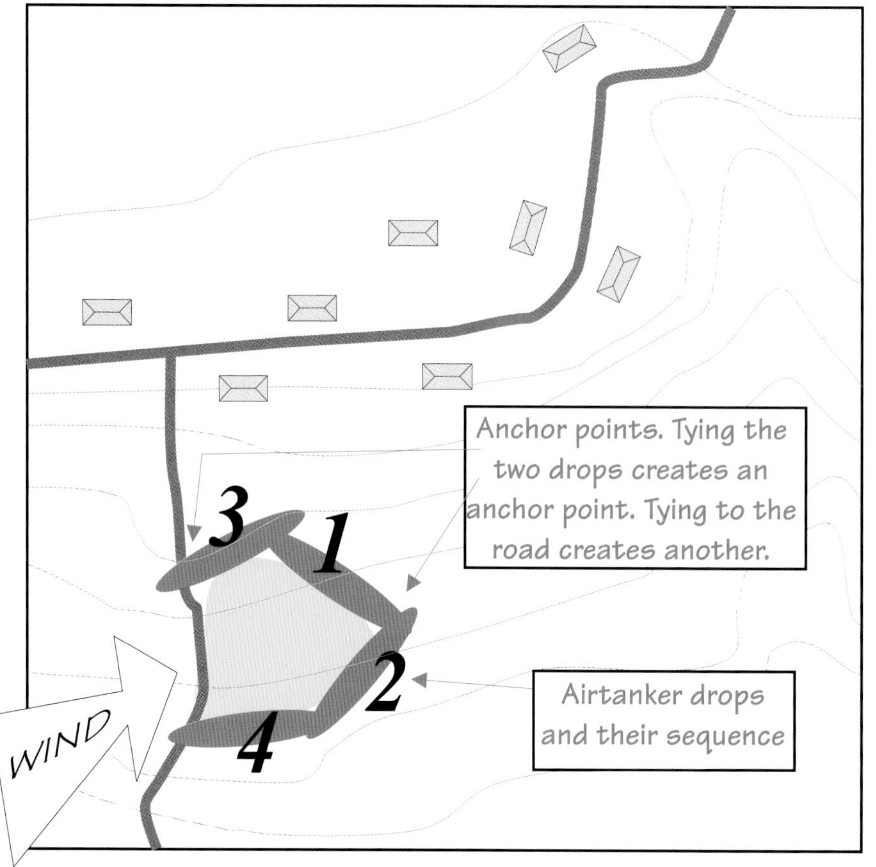
PLACEMENT OF AIRTANKER DROPS
ON A SMALL, SLOW-MOVING FIRE
Anchor points. Tying the two drops creates an anchor point. Tying to the road creates another.
Airtanker drops and their sequence
3
1
2
4
WIND

Figure 7.52 *On small, slow-moving fires, you can attack the head, then tie back to anchor points.*

- Do the values at risk justify the use of aircraft?
- Is there adequate time to complete the mission? Is there enough time and are conditions favorable for safe, effective drops?
- Is there another safer, less expensive way to achieve the same objective?
- Are the weather and light conditions conducive to safe and effective use of aircraft?

Unless the fire is small or structures are threatened, consider using a flanking action with the airtankers. In this way, ground forces can quickly follow-up on line construction. If the fire is spotting, use the tankers to hold the spot fires until ground forces can be positioned to control (Figures 7.52 through 7.54). If the fire is crowning, or if spotting is widespread, retardant drops may be ineffective. ***Always tie retardant lines to an anchor point.***

If an indirect attack is necessary, anticipate the fire's movement and place the drops accordingly. Pretreat fuels in preparation for backfiring; widen and strengthen control lines, natural fuel breaks and man-made barriers.

Cammeron Eck

Figure 7.53 *This heavy airtanker has placed several thousand gallons of retardant in support of this firing operation.*

Air operations personnel do use several terms that are unique to their program. Some of the terms that you should be familiar with:

- *Extend your previous drop* means overlap and lengthen.
- *Break left or right* means turn.
- *Early* means the drop was short of the target.
- *Late* means the drop was late and it overshot the target.
- *Low pass* means a low-level look at the target.
- *Move one wingspan left* is much better than move left 75 feet.

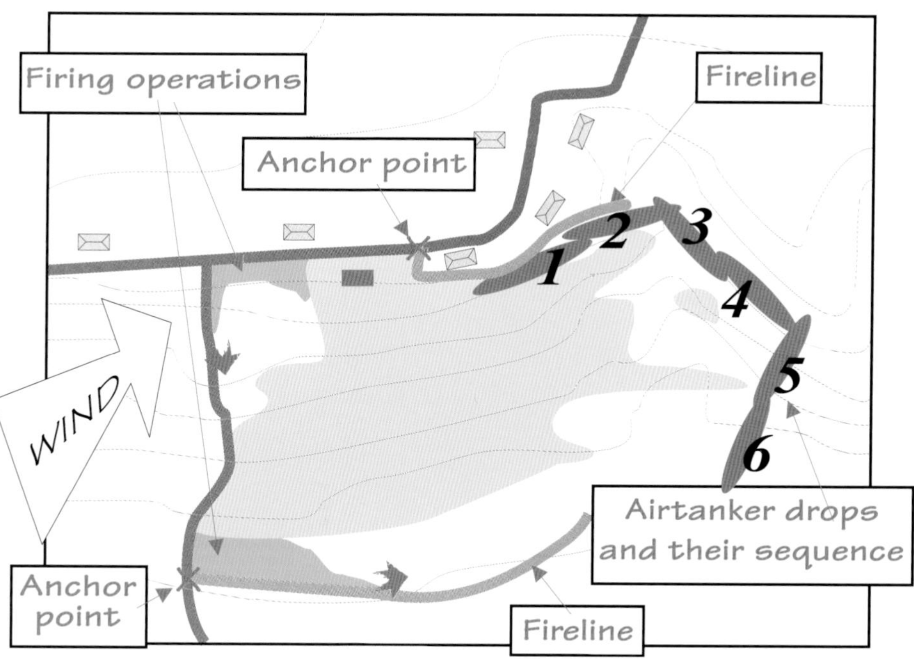

Figure 7.54 Fireline is constructed using indirect line behind the homes; using burning operations to clean up the line on the flanks; using airtankers on the left shoulder and across the head; and indirect line construction on the right flank.

Jake says, "Don't fly aircraft for public relations. If they are not being effective, shut them down!"

Once it has been determined that you will use aircraft, there are several principles that guide their use. Make a mental note of them when developing a plan.

- Determine tactics based on fire size-up and firefighting resources available.
- Establish an anchor point and work from it.
- Apply proper coverage levels (width and thickness of retardant line). Coverage levels are determined by the air attack supervisor or the lead plane pilot.
- Drop downhill and away from the sun when possible. The airtanker pilot normally determines how the drop run is made. Uphill drops are the exception.
- Drop into the wind for best accuracy.

- Honestly evaluate the drops; maintain effective communications with air operations personnel.

Don't over-manage air operations. Rely on the air attack officer to implement your plan.

- Use direct attack when ground forces are available.
- Plan drops so they can be extended or intersected effectively.
- Monitor retardant effectiveness and adjust its use accordingly. Don't be afraid to "shut down" the operation if the drops are not effective.

Use of Hand Crews and Bulldozers

The primary use of hand crews and bulldozers is to construct fireline. Hand crews usually work in direct attack, in steep and rugged terrain, or where bulldozers or tractor-plows cannot be used. Bulldozers are the most efficient at fireline construction, but they can also be the most damaging.

Hand Crews

When you work as a crew, you are working as a team. The team has its own dynamics. Some of the factors that a team brings to the fireline are:

- An established leadership and chain of command. Control of the team is maintained.

Hand crews can be used for more than just cutting line.

- Pre-planned assignments where a person's strengths are matched with a job.
- Teamwork, where the output is greater than the sum of the individual parts (firefighters).
- Each individual firefighter has his or her own responsibilities.
- There is reduced confusion over tasks and time is saved.
- A team spirit or esprit de corps is developed.

There are four "types" of hand crews. There are the "hotshot" crews operated by the state, tribal and federal agencies; inmate crews operated by state and local government organizations; the native American crews from the Southwest and Alaska; military personnel; and other crews organized by various private contractors or fire agencies.

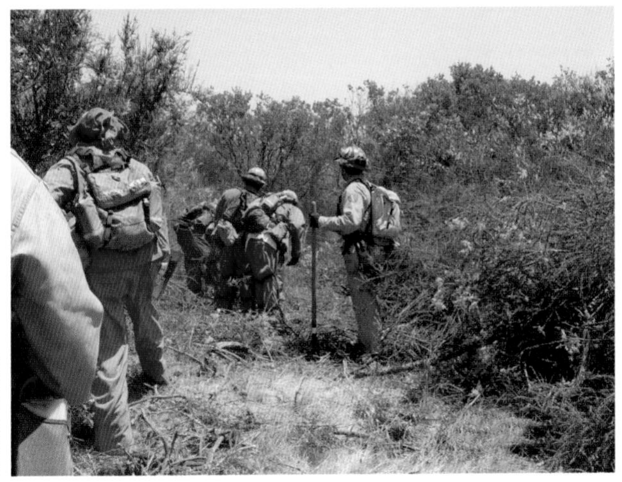

Trevor Wilson

Figure 7.55 This is an inmate crew constructing fireline in medium brush.

Trevor Wilson

Figure 7.56 This is the finished fireline. Note that the line is approximately 1½-times the height of the fuel, and it has been cut to mineral soil.

Hand crews are primarily used to construct fireline (Figures 7.55 and 7.56). But, they can be used to assist on hose lays, firing operations, protecting structures, mopping-up, cleaning up, and other logistical support functions. Match the task to the crew capabilities (Figure 7.57). Use the best, most experienced crews for the toughest jobs and hottest fireline. The effectiveness of the crews is controlled by these factors:

- ***Leadership*** - Good, competent leadership is key to the success of the crew.
- ***Training, physical fitness and experience*** - If the crew is well-trained, in top physical shape and has several fires under its belt, it will be very effective. If the crew has to be "trained" on the fireline, the amount of fireline cut will be much less.

The best hand crews are trained, fit, experienced, motivated, and well-led.

- ***Crew member turnover*** - If there are constant changes in the crew membership, the team aspects of the crew will suffer.
- ***Morale*** - If the crew is "down," its productivity will also be down. Morale is a tough thing to define, but if it is low, you will see it. One of the quickest ways to kill morale is the quality of the food or lack of it.
- ***Fatigue*** - If the crew members are exhausted, they will be ineffective and there is a much higher potential for accidents.
- ***Fuel, weather, topography and time of day*** - If the fuels are thick, the terrain is steep and it is the hottest time of the day, production will suffer. Also, working at night will reduce production rates.

ICS hand crew type

Type	Specifications	Examples (minimum crew size)
1	Highly trained (min. of 80 hours); no use restrictions; fully equipped and mobile; high experience levels with permanently assigned supervision.	Hotshot and some full-time state crews (18-20).
2 IA Capability	Basic firefighter training; squad bosses so crews can be used in small units; may need some logistical support.	Organized crews (18-20).
2	Some training and fireline restrictions; not fully equipped or mobile and without a full-time crew leader.	State and local inmate crews (18-20).
3	Basic training; will need transportation and tools.	Military crews (18-20).

Figure 7.57 This chart outlines the crew type by crew make up and leadership.

• ***Fire behavior*** - If the fire behavior is high to extreme, it will be very dangerous to work hand crews.

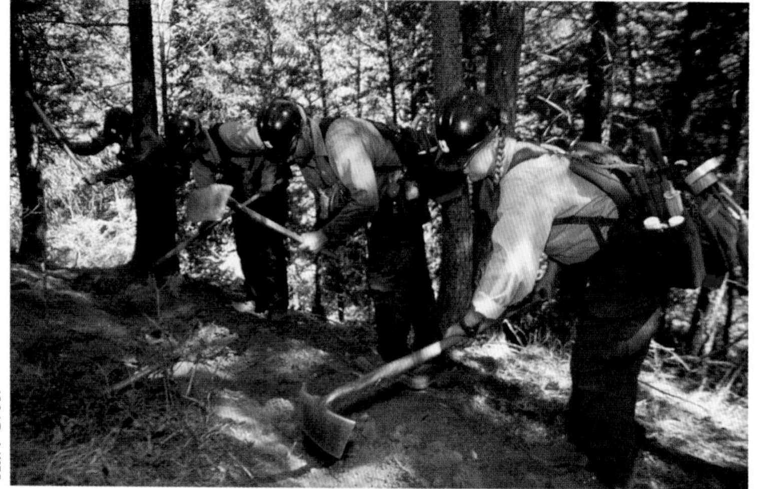
Kari Greer

Figure 7.58 If the hand crew has to construct undercut line, it is going to take a lot longer.

Production Rates

As you begin planning the attack, you must estimate the length and width of the line to be constructed, and the capabilities of the hand crew(s) in order to determine how many crews will be needed (Figures 7.58 and 7.59). Figure 7.60 shows some general production rate figures for a 15-person crew in various fuel types. These are average ideal rates that will be affected by the factors listed previously.

Towards the end of a shift, the production rates for most crews will decrease (Figure 7.61).

As a rule of thumb, a 15-person crew should be able to construct a 3-foot-wide fireline around a one-acre grass fire in one hour.

Kari Greer

Figure 7.59 If the line is "hot," this too will add to the time it takes to construct.

Extreme fire behavior is not the time for direct attack with hand crews.

Using Hand Crews on Hose Lays or Firing Operations

Even though hand crews are most often used to construct fireline, they can support progressive hose lays by carrying hose or using the lateral lines to begin mop-up (Figure 7.62). They can also be used to conduct firing operations and to burn-out or clean up shaky sections of line. Teaming hand crews with engine companies is a good strategy, especially when they have trained together.

Hand crew production rates

	Grass	Medium brush	Heavy brush	Very heavy brush
Line width	3 ft.	6 ft.	9 ft.	12 ft.
Length of line per crew member per hour	60 ft.	30 ft.	20 ft.	15 ft.
Length of line per hour (15 person crew)	900 ft.	450 ft.	300 ft.	225 ft.

Figure 7.60 This chart outlines estimated crew production rates for different fuel types.

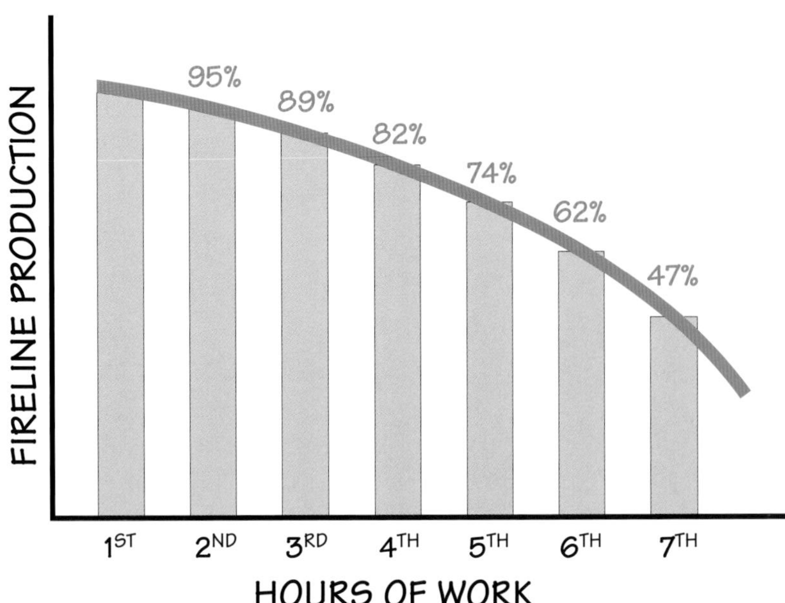

Figure 7.61 The rate of production of a hand crew decreases with time. Calculate this into your plan.

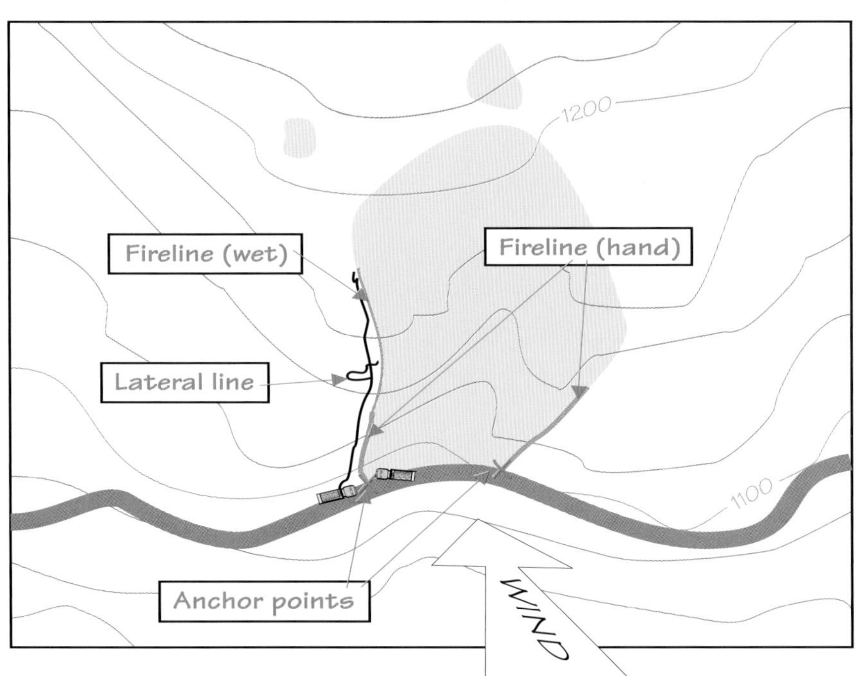

Figure 7.62 Hand crews can do more than just cut fireline. They can help on extensive hose lays, and with follow-up by constructing a fireline to mineral soil.

Cultural Differences

On large fires, it is common to have crews from all over the nation. There could be crews from the Southwest, and Alaska that have different requirements, such as food or dietary needs, based on cultural or ethnic backgrounds; housing; dress; and religion. There may be language barriers that could impact safety, and there may be historic rivalries that simply will not allow certain crews to work, sleep and eat in the same area. You should not mix male and female crews; free crews with inmate or wards; and any groups that may not get along.

These concerns are real and have to be accommodated. Be respectful and courteous, because your behavior will reflect on you and your crew.

Working with Inmate Crews

Some fire agencies utilize prison inmates or wards (wards are usually under 21 years of age) in organized hand crews. As firefighters, they can do anything other people can do. But if you are assigned to work with such a crew, there are several things you must keep in mind:

- Each crew has a supervisor and/or custodial officer. Contact with individual inmates should be kept to a minimum. Coordinate all work through the supervisors.

- Inmate/ward crews should not usually be split up. They should work as a group to facilitate security.

- Keep any contact with inmates/wards on a professional basis. Do not play cards, carry messages, or purchase items for them. Do not give or accept any gifts.

When a crew is tired and hungry, they can't be very effective.

- Inmates and wards are to be bedded down in separate areas from each other and from other personnel. They also shower at different times.

Treat them with respect, but always keep in mind that they are still in prison and thus do not have all of the rights and privileges you have. If you have any problems, contact their supervisor.

Working with Military Crews

When the regular forces are drawn down, agencies turn to the military as a source of personnel and equipment. The military is trained for field operations, but it is not specifically equipped or trained for firefighting. Before the military is committed to the fireline, several things have to occur:

- Each person should be provided with personal safety gear.

- Training should be provided in the basics of wildland firefighting and safety.

Military crews need someone to provide fire situation awareness.

- Each squad should have a qualified supervisor assigned, not as the leader of the squad, but as an advisor and safety officer.

The military will provide its own leadership. It is given missions, and the squads will be assigned. The firefighter assigned to each squad is there to provide technical assistance and to generally keep the squad from getting into an unsafe situation.

Kari Greer

Figure 7.63 Bulldozers are good news—bad news. If you want a lot of line constructed, they are the tool of choice. But they can also do a lot of damage that will take years for mother nature to repair. Weigh the potential good with the potential bad, and most of all, manage the dozer operations closely.

Bulldozers

Bulldozers are very efficient and effective in removing fuels along a fireline. In some of the heavier fuels like slash, they are the only tool that will work (Figures 7.63 through 7.66). The downside of the use of bulldozers is the damage they can do to the soil, fences, etc. Like any tool, if managed properly, a balance can be maintained between the good the bulldozer can do and the damage. Several factors influence the performance of bulldozers:

- ***Fire behavior*** - If the fire intensity is low, the bulldozer can work directly on the fireline. As the intensity increases, so will the distance away from the fire the bulldozer will have to work.

- ***Time of day*** - As night falls, the production rate will decrease. Never work a bulldozer at night without adequate lights.

- ***Terrain, soil type and number of rocks*** - Production rates are directly affected by the steepness of the slope, whether the line production is uphill or down, and whether the soil is hard, loose or sandy. If there are rocks, line production will be slower, especially if the rocks are large.

- ***Fuel type and density*** - The heavier the fuels, the slower the line will be cut.

- ***Size, age and condition of the bulldozer*** - The larger and newer bulldozer will usually produce more fireline per hour. If the bulldozer is older and "well used," it will work much slower.

- ***The accessories on the bulldozer*** - If the bulldozer is equipped with hydraulic blade angle controls, hydrostatic transmission, and an environmental protection cab, it will usually work faster.

Cammeron Eck

Figure 7.64 These two bulldozers are going to take this line to the creek and up toward the fire on the other ridge. The line is being fired as it is being constructed.

- ***The experience, skill and motivation of the operator*** - The main factor controlling the production rate is the operator. Having a "fire-wise" operator who is able to work with minimal supervision makes all the difference in the world. If the operator is right off the local highway construction job, you will have to closely supervise the operation. Don't let an inexperienced bulldozer operator work independently.

Kari Greer

Figure 7.65 This bulldozer is being guided by a "dozer boss."

Classification of Bulldozers

Some wildland agencies own and operate their own fleet of bulldozers. There are three general classifications of bulldozers: heavies (D-7s, D-8s, D-9s and TD25s); mediums (D-6s and HD-11s); and lights (D-4s and HD-6s). Avoid using the newer D-8s or similar bulldozers because they are too big for most fireline construction. They can be used to construct roads, remove large trees or build fireline on broad ridge tops. The lighter tractors can be used in lighter fuels. The most versatile bulldozer for wildland firefighting is the medium bulldozer equipped with an angle blade. A straight blade is better when working in and around timber. See Figure 7.67 for the ICS type specifications for bulldozers and tractor-plows.

Wes Schultz

Figure 7.66 This D-6 bulldozer is classified as a medium bulldozer.

The dozer boss is key to efficient, safe use of contract bulldozers.

How a Bulldozer Can be Used

The primary use of bulldozers on wildland fires is line construction. But they can also be used in the construction of escape routes and safety zones; opening up roads to provide access; mop-up; construction of helipads; and rehabilitation (water bars, scattering "dozer piles," and pulling berms).

Inspect the bulldozer before you put it to work.

ICS specifications for water tenders, bulldozers and tractor-plows

Type	Water tenders	Bulldozers	Tractor-plows
1	Pump - 300 gpm Tank - 5,000 gals.	**Heavy** Horsepower - 200+ (D-8/7, JD-850)	Horsepower - 165+ or more (D-7. TD-20)
2	Pump - 200 gpm Tank - 2,500 gals.	**Medium** Horsepower - 100+ (D-5, JD-650)	Horsepower - 140+ (D-6, TD-15)
3	Pump - 200 gpm Tank - 1,000 gals.	**Light** Horsepower - 50+ (D-4, JD-550)	Horsepower - 120+ (D-5, TD-12)
4	**	**	Horsepower - 90+ (D-4, TD-9)

Figure 7.67 Water tenders, bulldozers and tractor-plows are grouped by production potential into ICS types.

Bulldozers can be effective without damaging the environment.

The best place for bulldozer-constructed fireline is on ridge tops (Figure 7.68). This will reduce a lot of the damage that can occur when working sidehills or the bottom of a drainage. Avoid working in stream bottoms. If you must cross a stream, do it where the slope is gentle, but never perpendicular to the stream bed. In some areas, working in and around stream beds is restricted by law because of the potential damage to fisheries and riparian areas. When constructing line, always keep in mind the need for water-barring to prevent erosion. (A water bar is a ditch placed in a road or trail at regular intervals that carries the water off the road or trail. This helps prevent soil erosion.)

Figure 7.68 Bulldozers are more productive when grouped, with one dozer pioneering the line and the other finishing it.

If using a ridge top for your fireline, build it just off the top on the slope away from the fire. This will help

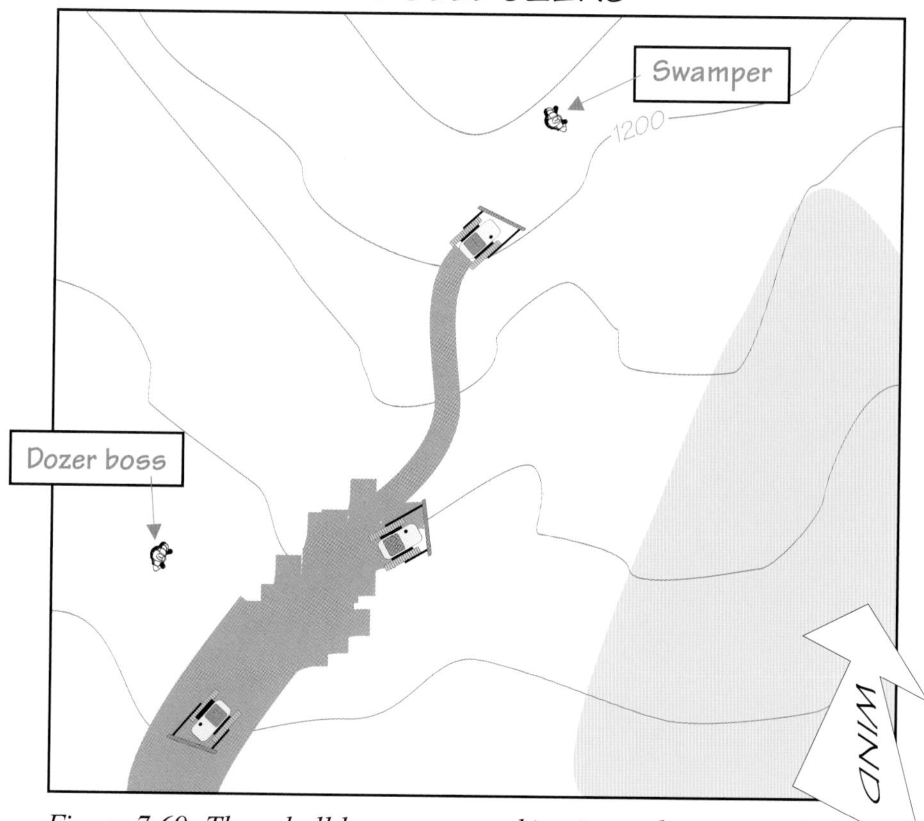

Figure 7.69 Three bulldozers are working in tandem on a ridge. The first bulldozer is pioneering the line; the second widening it and the third is cleaning it up. Note that a swamper is out in front of the lead bulldozer spotting the fireline. The dozer boss is back inspecting the final line construction.

Bulldozers can produce fireline very efficiently under the right conditions.

in firing operations. Also avoid sharp turns in the fireline. Use wide turns when possible. Unburned fuel should be cast away from the fire, burned fuels toward the fire. Try not to produce piles near the fireline. Spread all material when possible.

If your plan calls for the firing of the dozer line, make sure that the bulldozer operator is familiar with this practice and that you fire the line as soon as possible. Much properly constructed bulldozer line has been lost because it was not fired in a timely manner.

It is not a good practice to assign bulldozers to work independently. They should be worked in tandem and at least in pairs, AND with some firefighting support (Figures 7.69 through 7.71). This support can be from engine companies, hand crews, helicopters, airtankers, etc.

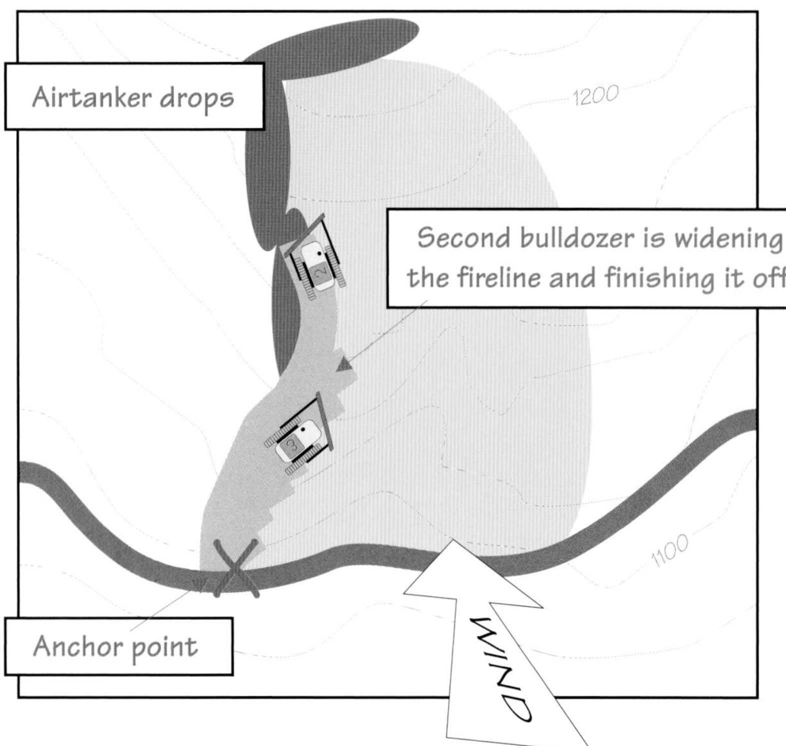

Figure 7.70 These bulldozers are following-up on airtanker retardant line.

If you have fire department owned and operated bulldozers working on your fire, pair them with contract bulldozers so that they can work in tandem. The department bulldozer can lead, and the contract dozer follow-up, cleaning and widening the line. In this situation, the operator of the department bulldozer may be able to function as the dozer boss for this particular fireline.

The Damage a Bulldozer Can Do

There has long been concern that bulldozers can cause more damage trying to control a fire than the fire would cause. Some jurisdictions require special permission before bulldozers can be used. This is especially true where permafrost is present, in national parks, or in designated wilderness areas.

A bulldozer can damage the environment in several ways. Damage can result from erosion and even gully formation along firelines. Soil compaction in sensitive areas is a concern. There can also be damage to archeological sites, fences, road pavement, and underground pipes or wiring.

Bulldozers can put in a lot of line, but they can also do a lot of damage.

The desired use of the bulldozer is to remove the fuel and leave the soil intact. Attempt to keep the natural slope and "build" only enough road for access for other firefighting resources. Leave stumps and root-balls, as they will help stabilize the soil and prevent erosion.

All these concerns can be alleviated if the fire management team and individual bulldozer operators are sensitive to the damage they may cause. You can avoid most of these problem areas by moving firelines away from sensitive areas or to a more gentle slope. There will be times when alternative fireline construction methods (hand crews, hose lays, retardant lines) will have to be used.

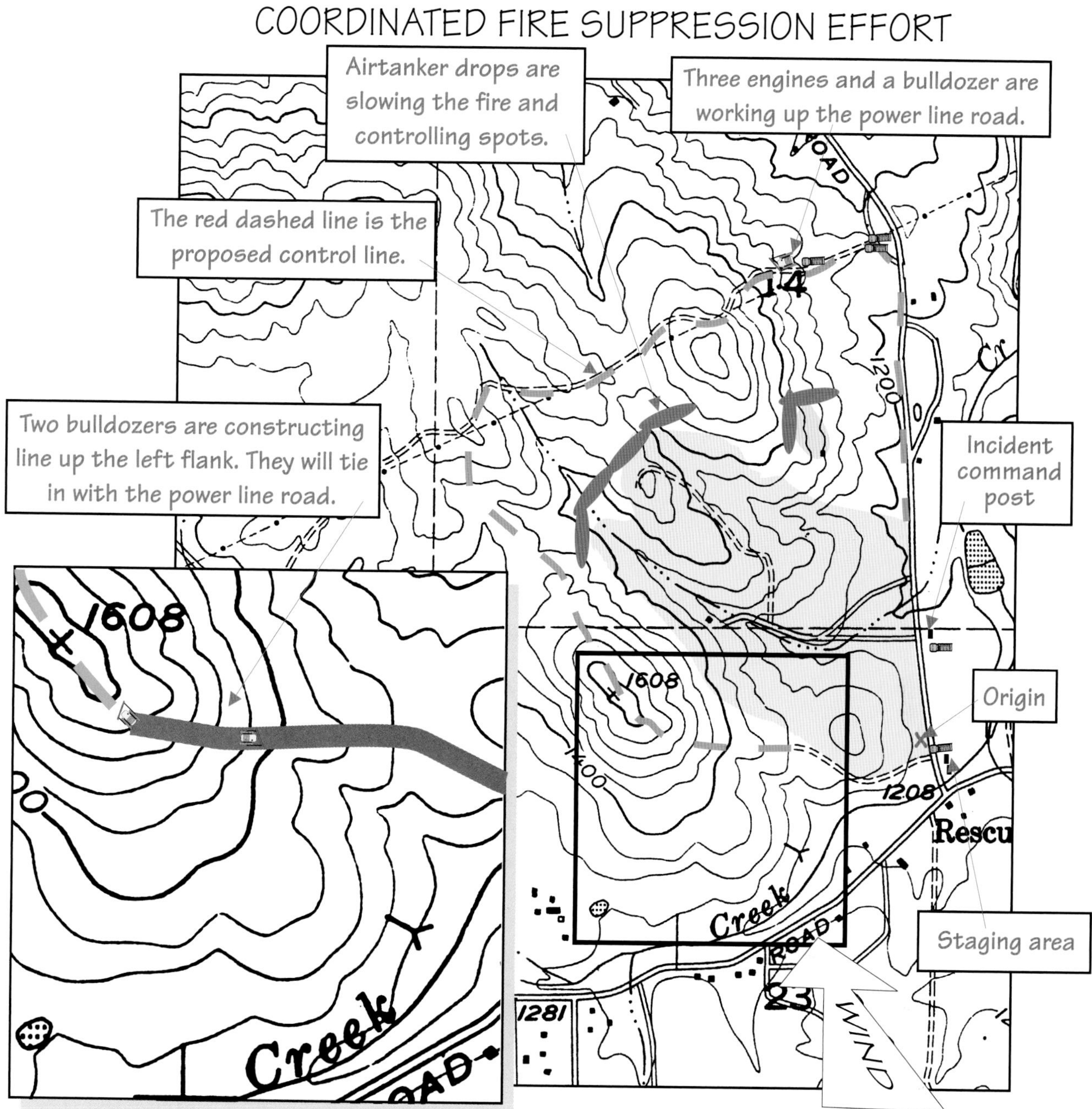
COORDINATED FIRE SUPPRESSION EFFORT
Airtanker drops are slowing the fire and controlling spots.
Three engines and a bulldozer are working up the power line road.
The red dashed line is the proposed control line.
Two bulldozers are constructing line up the left flank. They will tie in with the power line road.
Incident command post
Origin
Staging area
WIND
1608
Creek
ROAD
1208
1281
1400
1200
Rescu

Figure 7.71 The fire is being attacked by engines, bulldozers, hand crews and airtankers. The plan is to contain the fire within the red dashed line. Bulldozers are constructing line along the left flank. A bulldozer is opening up the power line road on the north flank. Three engines will then fire that line back toward the origin. Hand crews will fire the left flank from the power line road back toward the origin. The airtanker drops will hold the fire down until the lines are constructed and fired.

Management of Bulldozers

The level of management required is inversely proportional to the level of inexperience of the operator. If you are using a fire department bulldozer company, you normally can give the operator an assignment and it will be completed. But, in many situations, you will be using contract bulldozers; this changes the whole situation.

Even if there is only one hired bulldozer on your fire, you should have a qualified bulldozer boss assigned to it. This person will coordinate the use and operation of the bulldozer. The dozer boss can assist in locating the fireline to ensure the line is properly placed, as well as reduce the damage to the environment. He/she can also maintain the time records and provide communications and logistical support for the bulldozer.

The bulldozer is only as good as its operator.

Before a contract bulldozer is put on the fireline, it should be inspected by a qualified person to ensure that it is mechanically capable of the assigned work. The operator should also be checked for the proper safety gear. This includes protective clothing, helmet, gloves, boots and fire shelter.

The ideal accessories for a bulldozer assigned to wildland firefighting include:

- An environmental safety cab. This will allow the operator to work in hot, dusty conditions in relative comfort. The safety cab will also provide a greater margin of safety if the bulldozer is overrun by the fire.
- Adequate lights so that the bulldozer can operate safely at night. If the bulldozer is not equipped with lights, park it at night.
- Filtered air system. If the bulldozer is not equipped with a safety cab, the operator should have a system that will provide clean, filtered air. A short-duration self-contained breathing apparatus should be available for use in extreme situations.
- Fire blankets are necessary to provide protection for the operator if the bulldozer is overrun.
- Fire extinguisher. Belly-pan fires are commonplace on a wildland fire. A pressure-water type extinguisher is a must.
- A radio for fireline communications is preferred. If this is not possible, then the dozer boss must have reliable radio communication capabilities.

- Strobe light. The strobe light is mounted on the top of the canopy of the bulldozer and is activated as an identification beacon primarily for use by aircraft. Strobes are not normally used all the time, as they tend to "blind" those working near the bulldozer. They are used only when required to identify the specific location of fireline or equipment.

One very experienced dozer manager said, "My rule of thumb is to pretend that the equipment operator hates me and will use his machine to it utmost capability to kill me. He will drop trees on me, roll rocks on me, drive straight for me at high speeds and, in short, have no regard for my personal well-being. If you are always ready for this, you will never be surprised when something life-threatening actually occurs. I have been working around equipment for over 20 years with this approach, and they haven't got me yet!"

Jake says, "Working with heavy equipment is a lot like wrestling with your older brother, who doesn't know his own strength. It can be fun, but it is easy to get hurt!"

Safety

Bulldozers are very powerful, heavy and unforgiving. They are dangerous to be around and below in hilly terrain (Figure 7.72). If you are within 50 feet of a bulldozer, consider yourself in a danger zone (Figure 7.73). If you must talk to the operator, approach the bulldozer ONLY when you are told to and only if the operator sees you. The bulldozer always has the right of way. Be careful climbing up on the machine; the steel

Figure 7.72 Bulldozers are very effective, but they are dangerous to be around. Be especially aware if they are working above you.

Avoid working downslope from a bulldozer.

The bulldozer always has the right of way.

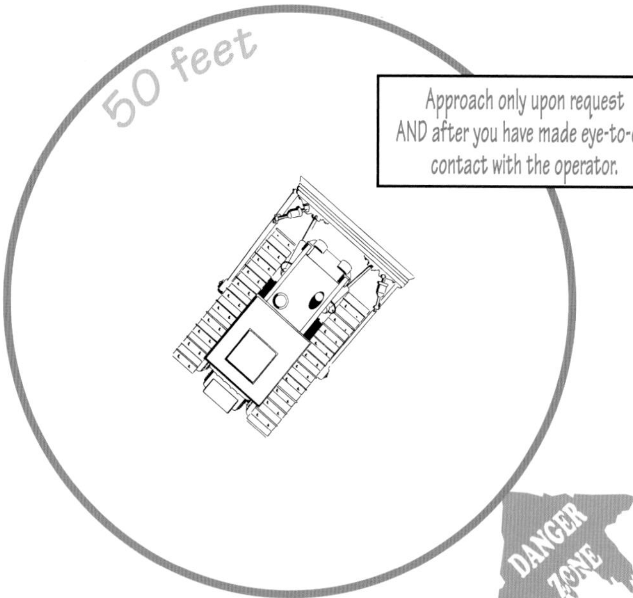
BULLDOZER SAFETY ZONE
50 feet
Approach only upon request AND after you have made eye-to-eye contact with the operator.
DANGER ZONE

Figure 7.73 Stay a minimum of 50 feet away from a working bulldozer. Try to maintain eye-to-eye contact with the operator.

can be very slippery. Use the handholds, not the levers. NEVER approach a bulldozer when it is moving or if the operator doesn't see you. It is a sure bet the operator can't hear you, so be sure he/she can see you. Always leave your headlamp on at night. The only time it is okay to ride on a bulldozer is if your life depends on it.

Be extremely cautious if a bulldozer is working above you. They tend to dislodge rocks or other things that roll.

Tractor-plows

Tractor-plows are used in the south and east to construct fireline (Figure 7.74). They are very effective in deep soils or boggy areas. The terrain has to be relatively flat and free of large rocks. All of the strategies, tactics and safety concerns used by bulldozers apply equally to tractor-plow operations.

Kari Greer

Figure 7.74 The tractor-plow is moving toward a spot fire. This is the primary piece of equipment used to control wildland fires in the south and east.

Skidders and "Skidgines"

You might ask yourself how a skidder can be used to fight fire—it doesn't move dirt or push over brush. Skidders are for moving logs, but they do have a small blade that can be used to cut fireline in lighter fuels. Since they cannot move a lot of dirt and are rubber-tired, they can be considered lower impact.

If you were to put a water tank and pump on the skidder, you then have a "skidgine" (Figure 7.75). Some skidgines can carry 500 gallons of water, and can meet the requirements of a Type 6 engine.

Tractor-plows work best on relatively flat terrain, where there are not big rocks.

Other "Big Iron" Firefighting Tools

In big timber areas, there are a lot of firefighting tools that may be available. A bigger bulldozer may not be the best tool for the job, especially if you want to be sensitive to the long-term environmental impacts. If the task at hand is to remove timber, you may need something like a feller-buncher, excavator, grander/ chipper or masticator.

Scott Kuehn

Figure 7.75 A "skidgine" can be an effective firefighting tool. It can cut line in lighter fuels, and because it uses tires, it is easier on the land.

- ***Feller-buncher*** - A feller-buncher can be used to clear standing timber or snags (Figure 7.76). They are safer, more efficient and more environmentally friendly than bulldozers or hand line.

- ***Excavator*** - You can use an excavator to pile slash and construct fireline. It can be especially handy if they have a "thumb" on the bucket. This is a hydraulic arm that acts just like your thumb. Excavators have flatter street pads, so they are lighter on the land, but don't ask them to climb mountains.

- ***Grinder/chipper*** - These machines can be used to reduce trees, brush and limbs to chips. Grinders use teeth or hammers to do their work, whereas chippers use sharp blades. Grinders tolerate some dirt and small rocks, but a chipper cannot.

Scott Kuehn

Figure 7.76 A feller-buncher, with "hot saw" can be used to cut and pile standing timber.

Scott Kuehn

Figure 7.77 Big equipment like this needs a very skilled operator.

• ***Masticators*** - A masticator is a grinding head on wheels or tracks. They move through brush and just grind it up.

The key to using any heavy equipment is to match the machine with the work at hand, and ensure the operator knows what they are doing. Something as big as the machine shown in Figure 7.77 requires some real skill to operate.

8

Wildland/Urban Firefighting Strategy and Tactics

Wildland fires have been destroying dwellings for centuries, but it wasn't until recently that the terms *wildland/urban interface fires* and *structural wildland interzone* were coined. This problem does not occur just in California. Major structural losses have occurred in all of the western states, in Florida, and countries like Australia, China, and South Africa (Figure 8.1).

Fire departments routinely respond to wildland fires and commonly protect exposed structures. In some areas this is a daily occurrence. However, this chapter is directed toward the large fire that threatens hundreds of homes at once, taxing the limited firefighting resources "at scene." Some examples are fires with names like Bel Air, Malibu, the Tunnel Fire in the Oakland Hills, Painted Cave in Santa Barbara, Black Tiger in Colorado, Cerro

Kurk Taylor

Figure 8.1 This fire is moving into the community of Devore. These spot fires are the result of shifting Santa Ana winds.

Grande in New Mexico, Old and Cedar in California. These are special situations, that require special strategies and tactics. A wildland/urban interface fire is a fire that is moving out of the wildland vegetation into a developed area. The fuel type is changing from timber or brush to structures and ornamental vegetation.

A wildland/urban fire presents conditions that require special strategies and tactics.

The wildland/urban interface problem has existed for years. In the last 50 years, the problem has grown more complex as more development has occurred in or adjacent to wildlands. Before we get too far into this chapter, we need to discuss some important definitions:

Normal strategies are too rigid.

- ***Structural Fire Protection*** – A service with the primary responsibility to protect the structure from interior and exterior fire ignition sources. This fire protection service is normally provided by local government fire departments, with specially trained and equipped personnel. After life safety, the priority is to keep the fire from leaving the area of origin. It includes the responsibility of protecting the structure from an advancing wildland fire.

- ***Structure Protection*** – To protect the structure from the threat of damage from an encroaching wildland fire. This normally does not include an interior attack of fire that is inside the structure. It involves taking special efforts to construct fireline to protect structures, and/or the extinguishment of spot fires near or on the structure. This protection can be provided by both the structural fire protection firefighter and the wildland fire protection firefighter.

- ***Wildland Fire Protection*** – A service with the primary responsibility of protecting natural resources and watersheds from damage by wildfires. State and federal forestry or land management agencies normally provide wildland fire protection with specially trained and equipped personnel.

- ***Wildland/Urban Interface*** – Where humans and their development meet or are intermixed with wildland fuels. There are four different wildland/urban conditions:

-Interface condition – is a situation, normally within a city or urban area, where structures abut wildland fuels

Orange County Register/Bruce Chambers

Figure 8.2 This is a classic interface fire. Note the clean lines between the wildland and the community. Don't be fooled that the only homes you need to protect are on the perimeter. Spotting will carry fire well into the community. A high percentage of the homes will be lost from house-to-house spread.

(Figure 8.2). There is a clear line of demarcation between the structures and the wildland fuels along roads or back fences. Wildland fuels do not continue into the developed area. The development density for an interface condition is usually three or more structures per acre. Fire protection is normally provided by a local government fire department with the responsibility to protect the structure from both an interior fire and an encroaching wildland fire (unless the line of demarcation is also a jurisdictional boundary).

-Intermix condition – is a condition, normally not in an urban city, where structures are scattered throughout a wildland area (Figure 8.3). There is no clear line of demarcation; the wildland fuels are continuous outside of and within the developed area. In some areas, ornamental plants replace the natural fuels. The development density in the intermix ranges from two structures per acre to one structure per 40 acres. Fire protection districts funded by various taxing authorities normally provide fire protection.

Figure 8.3 This is an intermix condition. The homes are scattered in the wildland areas. If the homes are not constructed with fire resistant materials, each house will have to be defended, one at a time.

-Occluded condition – is a situation, normally within a city, where structures abut an island of wildland fuels (park or open space). There is a clear line of demarcation between the structures and the wildland fuels along roads or back fences. The development density for an occluded condition is usually three or more structures per acre, and the area is less than 1,000 acres in size. Fire protection is normally provided by a local government fire department.

-Rural condition – is a situation, normally found in a rural or forest setting, where small clusters of structures (ranches, farms, resorts, or summer cabins) are exposed to wildland fuels (Figure 8.4). There may be miles between these clusters. Structural fire protection service may not be available.

Figure 8.4 The vast majority of the rural areas of the west look like this; scattered farms, resorts, etc. If fire visits this landscape, they will pretty much have to defend themselves.

Jake says, "These definitions are fine, but it comes down to one thing: are structures threatened or not? If they are, you have a new problem to deal with."

It is important to remember that when structures are threatened, regardless of their "arrangement," the whole dynamics of the firefight changes. Firefighting resources dispatched to protect structures usually are not actively involved in fireline construction. When structures are threatened, fires generally get bigger, and the dangers to firefighters increase.

Special Conditions

There are several conditions that may be present during a wildland/urban interface fire that mandate the use of ***special strategies and tactics***. They are:

- Numerous structures already involved or threatened (Figure 8.5). The number of homes at risk exceeds the number of available engine companies. This situation dictates that ***you must remain mobile*** and will have to triage the structures. You will not be able to protect them all.

Orange County Register/Bruce Chambers

Figure 8.5 Most of the homes on the block are lost. The one in the upper left may survive without intervention. The one in the lower center is just starting to burn along the eaves. It could be saved with very little effort.

- Fire is "driving" the incident. Firefighters are in a defensive mode, reacting only to the spread of the fire. This situation requires that you constantly monitor the fire behavior. Since "fire" is driving the incident, you've got to detect the subtle changes in the weather and fire behavior. You may want to use lighter hose to prevent fatigue. You may do more firing to create an expedient fireline.

- No time to plan and organize an attack. This dictates that you work with some independence, but with the overall plan in mind. Make sure that your actions ***contribute*** to the effort to control the fire. Listen to get a sense of what is happening; adjust your actions to complement what you hear; share intelligence with your supervisor and others on the fire.

- Water is limited. This dictates that you must use water wisely. By wasting water, you may be robbing someone of vitally needed water.

- Numerous non-fire problems, evacuations, and downed power lines. They must be dealt with. Either you do it, or get someone else to do it.

ORGANIZED INDEPENDENCE is one key to being successful.

Under these conditions, normal structural firefighting strategies will not work. They are too rigid. Firefighting forces will have to be flexible and remain mobile (not connecting to hydrants), moving from house-to-house, protecting as many structures as possible. ***"Organized independence"*** is one key to being successful under these conditions.

Jake says, "You must look at the bigger picture. You may not be able to save all of the homes...just save as many as possible, without risking lives."

The best way to protect structures from an advancing fire is to control the fire front before it reaches them. If you have time and the firefighting resources, attack the fire. If you don't have time or adequate firefighting resources, protect as many structures as you can. Or, you might try to protect the structures and construct some fireline. In any case, you will have to size-up the situation, establish a plan, and take action.

Sizing-up must be done. With this information, you can select the right strategy and tactic.

Size-Up

Your primary considerations as you arrive at the fire are firefighter safety, life threat, potential fire behavior, access, nature of the threat to structures, and water supply. Some of the factors that should be considered in the size-up are:

- ***Fire history*** - If you have knowledge of previous fires in the area and the conditions under which they burned, use that information in your size-up. Wildland fires tend to follow historic patterns under similar weather conditions. The "old-timers" can tell you "what happened last time" the area burned.

What was the prediction? Does it match what is happening?

- ***Weather conditions*** - What are the temperature, humidity, and wind speed and direction? Do they match the predictions? Is the air stable or unstable? What is the time of day? Are you entering the time for high fuel flammability? Are there local winds that are or will be a factor? You can't make a reliable forecast of what the fire is going to do without a good idea of what the weather is going to do. If you haven't heard a recent forecast, ask dispatch for one.

What type of fuels are burning?

Figure 8.6 The glow behind this home tells you the fire is close and that it would be unsafe for firefighters to attempt to save it. It is doomed!

- ***Fuels*** - Are they heavy or light? Are the fuels old and full of dead material? What is the fuel moisture? What is the fuel loading? The type of fuels will influence the tactics and type of firefighting equipment that you can use. Grass will be consumed very quickly, thus the threat will normally be short. Timber or slash fuels will take much longer to burn, thus they pose a much longer threat.

- ***Topography*** - Are there canyons or ridges, or is it relatively flat? Are there natural or man-made barriers that can be used as fireline? What is the elevation? The terrain will control the access and how you will fight the fire. Observe the terrain that is close at hand, and also observe what is out in front of the fire.

Firebrands will be the first "assault" on the structure.

- ***Fire behavior*** - Is the fire spotting, crowning, or just creeping? How much time before structures are threatened, given the fire behavior and weather conditions (Figure 8.6)? If the fire behavior is such that you can take direct action and reduce the threat, that should be the action of choice. If you don't have the time or the necessary resources to control the perimeter of the fire, take up defensive positions to protect the structures. The three fire behavior conditions that you must track all the time are rate of spread, number of firebrands, and length of heat (how long the fuels will continue to burn).

  - ***Firebrands*** - Firebrands are going to be a primary threat. The fire front does not have to be close for firebrands to be a problem.

  - ***Number of structures being threatened*** - What is their construction and density? Do they have flammable roofs and siding? Is there defensible space? Are there LPG tanks? Are there overhead power lines? Try to determine how many structures are directly threatened and when they will be threatened. The ***direct threat zone*** is that area which the fire is expected to threaten, not the whole community. Be realistic in estimating the threat (Figure 8.7).

Jake says, "Spot fires will be the first thing that will hit you. Don't be surprised at how many you will get—Situation awarness!"

- ***Access*** - Are the roads narrow, with many dead ends? Is there adequate width for two-way traffic? Are the roads cleared of brush, or are the fuels right up to the edge of the road, or worse—overhanging it? Will all of the bridges support heavy fire apparatus? Can you use the roads as fireline? Where can you stage arriving equipment?

Northtree Fire International

Figure 8.7 There are hundreds of homes in this picture. All could be potentially lost if a fire was to move over this ridge.

- ***Water sources*** - Are there hydrants? What other water sources are there? In a mobile attack situation the engine tank is the primary source of water. Fire hydrants, ponds, swimming pools and other fixed sources of water are only for refilling. DO NOT tie into a fixed source; if you do, you are no longer mobile. Floatable pumps are extremely useful. They can be used to refill engines from swimming pools, ponds or creeks. In addition, a firefighter with some fire hose, a nozzle, a fire hydrant, or floatable pump and a swimming pool can fight a lot of fire. Water tenders with drop tanks can create water sources. As with engines, water tenders must remain mobile, supplying water to the incident.

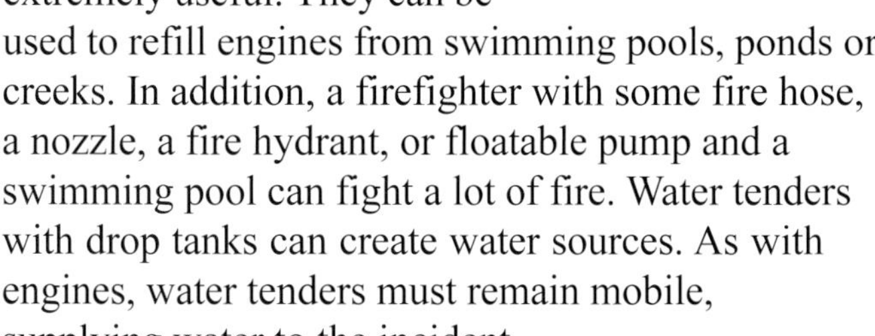

Jake says, "You will always need more water than you have...so don't waste it!"

- ***Evacuation*** - Will you have to evacuate people and animals? The decision to evacuate an area is a difficult one. If you expect extreme fire behavior AND you have time, initiate a request for evacuation (Figure 8.8). Only law enforcement personnel have the authority to ORDER an evacuation. In most areas, firefighters can advise people to leave, not order them. Remember that ingress of fire equipment and egress of residents at the same time are going to add to your traffic problem. If the residents are going to stay, turn them into an asset. Tell them what to do and

Figure 8.8 One of the problems with evacuations is when you want to move fire equipment into an area, and the same roads are being used to move the people out. Try not to allow this to happen.

get information from them on other structure locations, access, water sources, etc.

- ***Special hazards*** - Is there the potential for hazardous materials or explosives in outbuildings? Are there high-voltage transmission lines in the area? Are there above-ground fuel storage tanks?

Picture where the fire will be in one hour blocks. Constantly review the situation and your plan.

- ***Available firefighting resources*** - What firefighting resources have been dispatched? How long will it take them to get there?

Initial Actions

Based on your size-up and ***continuing observations,*** predict where the fire will be in one hour increments. What will the threat be for each of these periods? Consider the time of day, fuels, topography and weather conditions. What will the fire behavior be? What will the threat be to the structures (flying brands, radiation or direct flame contact)? When will your assistance be arriving? Now set your strategy, matching the anticipated threat with the firefighting resources that will be available at the time.

Establish a staging area as soon as you can.

Establish a staging area, a place for all incoming resources to report when "at scene" ***as soon as you can***. Make it easy to find, and pick a spot that is large enough to accommodate the anticipated resources. Be sure to notify dispatch and all the forces "at scene" of the location. On fast moving fires, the staging area will be very active and "dynamic." People and resources will be constantly moving in and out of the area.

You need help! What will you order?

Other Firefighting Resources

Don't put all of your efforts into protecting the threatened structures. Remember that you may have miles of fireline to construct ***and*** hold. You will need to establish a command and logistics structure right from the beginning. You could have all the engines on earth, but, if they aren't managed, they only add to the confusion; not a very efficient way of doing things.

Don't overlook the need for support personnel: law enforcement, animal control, utility companies, etc.

Some of the other resources that may be of some use:

- ***Bulldozers*** - can be very effective in fireline construction and opening up access. They can be used to construct fireline around structures, but use them with caution; they can cause a lot of long-term damage.

- ***Airtankers and helicopters*** - can be effective in protecting structures if the smoke and wind conditions don't restrict their use. Aircraft are most effective if they can place their drops just ahead of the flame front as it approaches the structure.

- ***Hand crews*** - can be used to construct fireline, to create defensible space around structures, and in firing operations.

Notify the various utility companies early into the incident. They can be of assistance.

- ***Law enforcement*** - If evacuations (recommended or mandatory) are necessary, local law enforcement personnel can provide an excellent service. Also use them for roadblocks and general traffic control. If a mandatory evacuation is ordered, there is an obligation to ensure that the area is secured. This is a role for law enforcement.

- ***Animal control*** - If there are lots of horses and livestock in the area, animal control personnel can be used to move and contain them.

- ***Utility companies*** - This includes water, electrical power and telephone companies. Make contact with them early into the incident. Let them know what is happening, and what you need. Understanding the water system may be critical to firefighting operations. Shutting down power lines may improve the safety in the area, but it may also cut off power to pumping stations vital to the water providers.

- ***Road departments*** - can close roads and provide heavy equipment and water tenders.

- ***Disaster relief organizations*** - such as the Salvation Army and American Red Cross can be useful in operating shelters for evacuees. They can also provide counseling for those who have lost their homes.

- ***Media*** - Use the broadcast media to inform the general public of what is happening and what they should do. Consider using the local radio and television stations if large-scale evacuations are necessary. Assign a public information officer as soon as you can. If not, operations may be hampered by press inquires.

Jake says, "Don't overlook the media. They can help keep the general public informed. Use them to your advantage. Advise them to focus on facts...not sensationalism."

Structure Triage

On a major wildland fire there are usually not enough engines to attack the spread of the fire and protect every structure threatened. The decision as to which structures will be protected is based on "sorting" or "prioritizing"—this is called structure triage. When you begin to triage the structures, your goal is to protect and save as many as possible with the firefighting resources you have at hand.

Three types of structures: those threatened, those not threatened, and those that are lost or can't be protected.

From a triage point of view, there are three types of structures (Figure 8.9):

- ***Those that need little or no attention*** - They are constructed of materials that will not easily ignite, and they have proper clearance.

- ***Those that need protection*** - These structures are constructed of flammable materials and may not have the needed clearance, but with an effort by firefighting personnel, they probably can be saved.

- ***Those that are lost*** - Even though they may not yet be on fire, these structures cannot be saved. They are primarily those structures that are constructed of flammable material and that don't have the proper clearance, or where the ingress/egress is such that it would be dangerous to attempt to protect them. The structures in this category generally have other problems such as wood or debris piles near the structure.

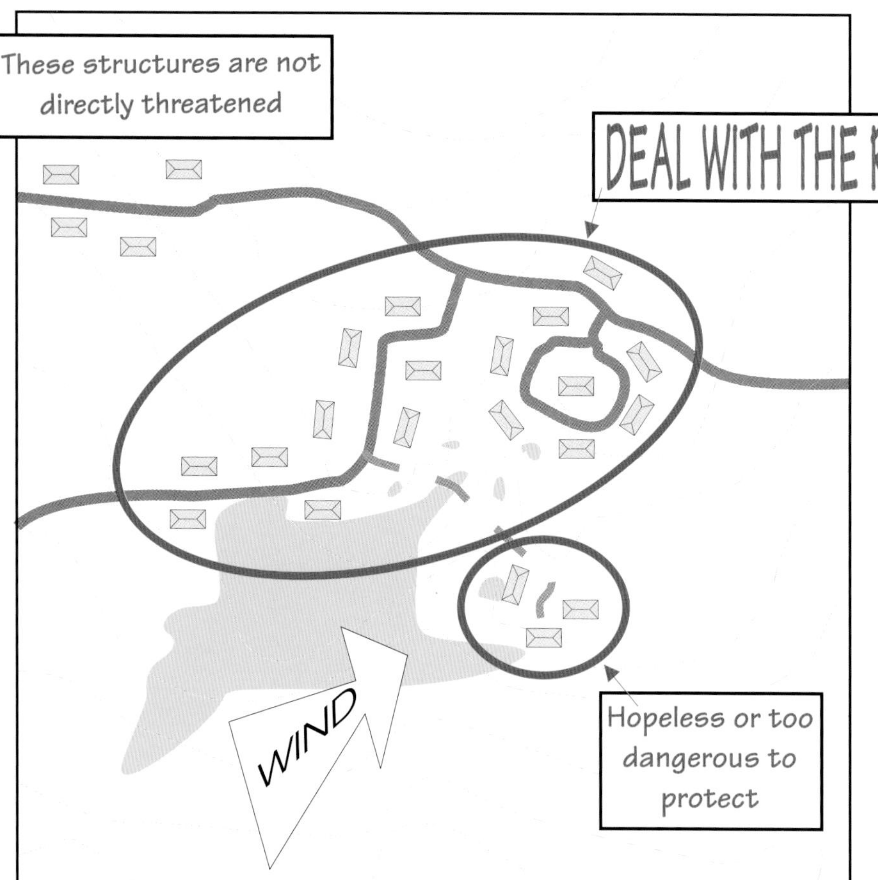

Figure 8.9 The first sorting identifies those structures that are lost or unsafe to protect, and those that are, are not, or will not be threatened. The remaining are the ones you should concentrate on protecting.

In your effort to select the structures that you have the best chance of saving, you have to consider multiple factors. The primary considerations are as follows:

Save what you can!

- ***Structure*** - How combustible are the roof and siding? What is the pitch of the roof, and how much debris is on top of it? Are there "heat traps" like open gables, unscreened vents, or overhanging decks? Are there large windows that could easily be cracked by the heat of the fire? What is the size and shape of the building? What is its position on the slope? (Figures 8.10 and 8.11).

Figure 8.10 The litter on this roof acts like a "magnet" for burning embers.

- ***Fuels*** - It is estimated that 90 percent of the structures that are lost to an advancing wildland fire are lost because of a lack of adequate clearance from flammable fuels. You need to analyze the fuels that are near the structure as well as those fuels that may produce firebrands or radiant heat (Figure 8.12). You should look at the size and arrangement of the fuels, their age, proximity, loading, and type. How long will the fuels burn? What is the landscaping like, and is there defensible space? Is there an accumulation of "junk" or stacked wood in the yard? Are there above-ground fuel tanks that may present a problem? (Figure 8.13).

Figure 8.11 If there are flying embers, they will find these openings to the attic. Some vent screens are made of plastic and melt when hit with a blast of hot air. Of course, they fail.

- ***Fire behavior*** - The level of fire behavior will give you some idea of whether you will be able to protect a structure. If the fire is just creeping through the ground litter, you have a different problem than a situation where the fire is crowning (Figures 8.14 and 8.15). When analyzing the fire behavior, study the

Figure 8.12 Unless someone takes some action soon, both structures will be lost.

Figure 8.13 This house is just another fuel. It would take hours to clear a defensible space.

rate of spread, topographical and weather-related influences, flame length, spotting potential, and timing. Are there any barriers that you may use to your advantage?

• ***Firefighting resources*** - What kind of firefighting resources do you have to fight the fire? What are their kind and type? Are the firefighters experienced in this type of firefight? How much water do you have available? How long will it take to get these resources in place and into action?

• ***Firefighter safety*** - The safety of all firefighters is paramount. There isn't a house or property anywhere that is worth the injury or death of a firefighter! Is the route in and out safe? Is the fire behavior too extreme? Are there overhead power lines that you should concern yourself with? Are there LPG tanks, stored hazardous materials, or other special concerns (Figures 8.16 and 8.17)?

One of the difficult decisions a firefighter will have to make is determining that a house cannot be saved, especially with the owner standing next to you. If there is time, take a second to show empathy. Some owners will be understandably angry no matter how obvious the reason for your decision. Figure 8.18 outlines the factors you should consider during structure triage.

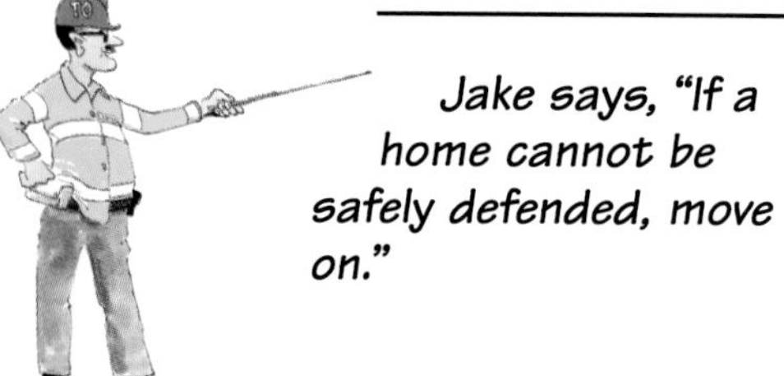

Jake says, "If a home cannot be safely defended, move on."

There is no simple rule regarding when you should abandon the effort to protect a structure. Rely on your judgment to guide you. Err on the side of safety. The following is a list of factors or conditions that may indicate that it is time to move on:

• The fire is making significant runs (not just isolated flare-ups) in the standing live fuels, and the structure is less than two flame lengths from those fuels.

- Spot fires are igniting around the structure or on the roof faster than you can put them out.

- Your water supply is insufficient to allow you to continue firefighting until such time as the threat subsides.

- You cannot safely remain at the structure and your escape route could become unusable.

- The roof is more than one fourth involved in windy conditions, and other structures are threatened or involved.

Guy M. Delaney

Figure 8.14 The house in the white circle is in jeopardy. There are two engine companies positioned to defend it. The fire appears to be burning away from it. It is the spot fires that will threaten and destroy it.

Kurt Taylor (1-4)

Figure 8.15 This sequence of pictures covers 8:04 minutes. Even though there were firefighters there, they were not able to save it. In the lower left corner of picture #1, you can see the only access to several houses on this hillside. The real lesson here is how fast the situation changes!

Figure 8.16 Will it be safe to park your engine near this home and attempt to protect it? Probably not in most any burning conditions.

Figure 8.17 This is not a road you want to travel on if there is a fire in the area. There is no clearance whatsoever. Imagine trying to move down this road at night, in the smoke, with fire on your tail! This is a dangerous situation.

• Interior rooms are involved and windows are broken. It is windy, and other structures are threatened or involved.

Evacuation

When do you attempt to evacuate an area? As a firefighter, this may be one of your tougher calls. Sometimes you will not have to say anything to the residents. At the first sign of smoke, they will choose to leave. But there are people who will not leave under any circumstances. Fire personnel can suggest and recommend evacuations, but it is the responsibility of law enforcement personnel to move people against their will.

There are two basic reasons to evacuate an area: to move the residents out of harm's way, and to clear the area so that firefighting personnel can move about more freely. The first of these reasons is obviously the most important.

The decision to evacuate must be made as early in an event as possible, by an official with the authority to carry it out, and in concert with the local law enforcement agency. The following factors must be considered:

Deciding when to evacuate is a tough call.

- ***Fire behavior*** - What are the rate of spread, spotting conditions, and the overall intensity of the fire? Does the fire behavior pose a real threat to the residents? How reliable is the information about the fire?

- ***Timing*** - When will the fire be in the area? Is there enough time to develop an evacuation plan, notify the appropriate

Factors to consider during structure triage

Factor	Why
Proximity of the fuels	If there are fuels (wildland or ornamental) right up to the structure there is NO DEFENSIBLE SPACE. If the fire is burning with any intensity at all, it will be difficult to save.
Fire behavior and intensity..spotting	The intensity of the fire will dictate how much DEFENSIBLE SPACE will be necessary. The greater the intensity, the wider the break in fuel will have to be. If the fire is spotting, control all of the spots within the "spotting zone".
Flammability of roofing and siding	If the structure has flammable siding and roof covering, it will be difficult to save. If there is also no clearance, it may be impossible to save.
Timing and firefighting resource availability	If you don't have time to set up or there are no available firefighting resources, you have very few options.
Safety	DO NOT put firefighters at undue risk to save a structure. Safety first. The house can be rebuilt.

Figure 8.18 *These factors should be considered when selecting which structure can be protected. Remember, no structure is worth getting hurt over. If your gut says go, go!*

Can you evacuate an area safely?

Is there time?

Do not call for an evacuation if you don't have a plan.

authorities, communicate to the residents that they must/should leave, and then move them from the area? There may be times when it would be best to have them stay put rather than risk being caught by fire while trying to evacuate.

Only law enforcement personnel are authorized to evacuate people against their will.

- ***The threatened population*** - How many people have to be moved? What are their ages? Are there any special problems or considerations? Will people be forced to leave if they don't want to go?

- ***Roads*** - Are the evacuation routes capable of handling the anticipated traffic loads? Are these the same routes responding fire apparatus will be using? Are the escape routes safe to use with fire in the area?

- ***Security*** - How will the area be protected from the possibility of looters? What security measures will have to be put in place? Who will do this?

Make sure everyone knows the plan.

In most cases, when an evacuation is implemented, it is decided to do so quickly and without a formal plan. But, if time allows, develop a formal plan that addresses the timing of the evacuation (when people will be asked to move), the routes they will take (hopefully not the same routes fire apparatus are using), and the locations to where people will be moved. You will also have to involve the Red Cross or another appropriate agency to arrange food and shelter for evacuees.

Homeowners can be motivated helpers when their property is threatened. Allow them to assist when there is no danger of injury.

If, on the other hand, you decide not to evacuate, how do you turn the people who stay into an asset? Residents can provide information on the location of all the structures in the area, sources of water, the road system, and any special hazards.

They can also help prepare structures for defense by moving flammable materials away from the structure, clearing the roof and gutters of litter, closing windows and curtains, and so on. They can sometimes assist fire department personnel with the actual defense of a structure. However, be very careful that residents are not put in any situation that may get them injured.

Action Plan

You must develop an action plan, even if it is just a couple of notes on a slip of paper. Use the information that you have gathered during the size-up, the review of the structure protection problem, and your knowledge of the firefighting resources committed to the incident. The fire behavior and fuel type will dictate the attack mode you will be able to use. Figure 8.19 outlines the three attack modes (Figures 8.20 thru 8.23). If conditions allow, take the offense, but keep safety first in mind. Develop a plan that predicts where the fire will be at certain times, and outline what the firefighting resources that are "at scene" will be assigned to do. ***Look at the big picture.***

You may be defensive on one division and offensive on another.

Structure protection attack modes

Attack mode	Situation	Action
Defensive attack	The main fire cannot be controlled in time to eliminate the threat.	Concentrate on saving as many structures as you can.
Offensive attack	Main fire can be controlled before the structures are threatened.	Attack the main fire.
Combination attack	Part of the main fire can be controlled and/or defensive action will create control lines.	Attack the main fire. Protect structures as they become threatened. Extend perimeter control from locations where structure defense has given you an anchor point.

Figure 8.19 There are three general attack modes when dealing with structure protection. Use the one that is appropriate.

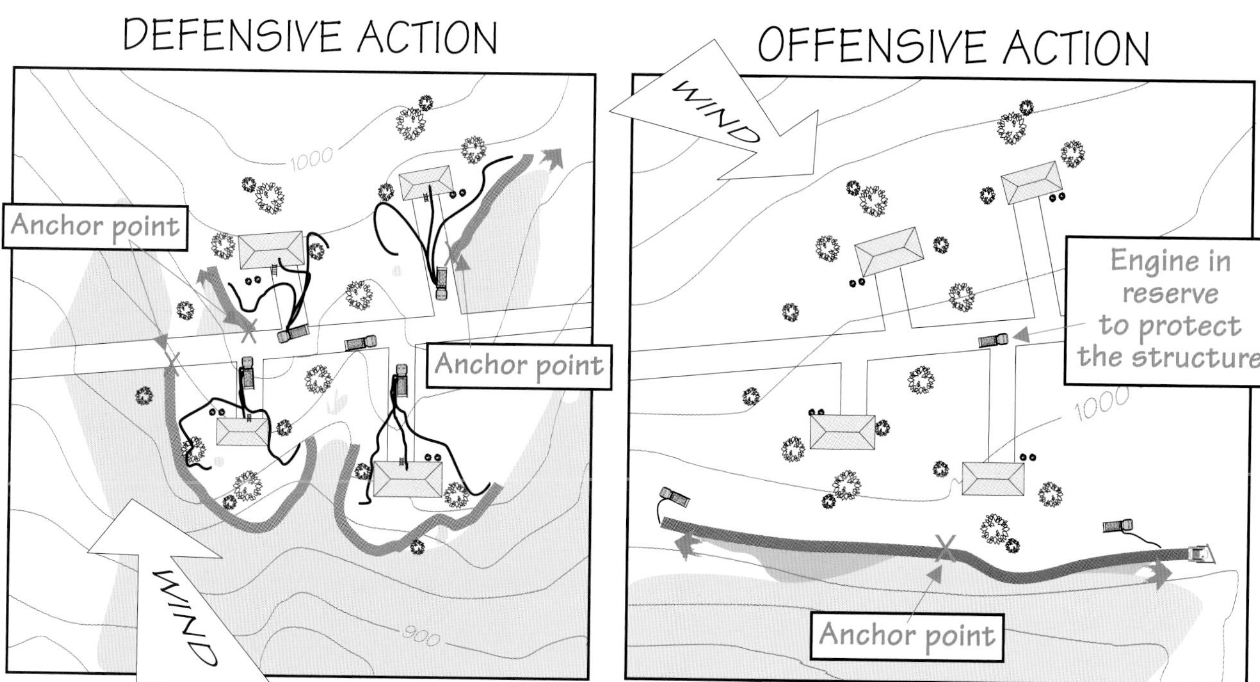

Figure 8.20 Take defensive action to save as many structures as you can. Protect the structures as the fire passes. Move when the structure is secure. Perimeter control will have to come later.

Figure 8.21 If conditions permit, take offensive action to contain the fire. Two engines and a bulldozer are constructing fireline and firing it out as they go. A third engine is positioned to protect the structures if needed.

COMBINATION ATTACK

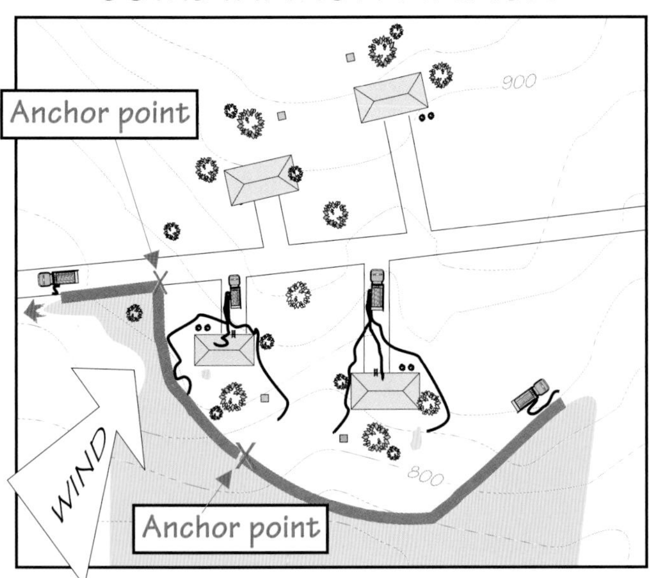

Figure 8.22 A combination attack strategy is used when some structures must receive direct defensive protection, but it is also possible to directly attack the fire. Once the line is secure around these structures, the two engines protecting them can be used elsewhere.

Consider what you can do with the firefighting forces and water at hand.

As a rule of thumb, you should assign one engine per structure in the "structure threat zone," with one additional engine for every four to be used to patrol the area and remain "flexible." If the structures are close together (50 feet or less), one engine company may be able to protect two structures. Large commercial structures or structures already involved will require the assignment of additional engine companies. These are idealistic numbers; in most cases, you will not have the luxury of having all the apparatus you think you need (Figures 8.24 and 8.25).

The time that will be required to actively protect the structure will depend on the fuel type and fire intensity. As a rule of thumb, in light fuel (such as grass and sagebrush) when flying firebrands are not too much of a problem, you may have to stay with the structure for 20 to 40 minutes. In brush, with passage of a single fire front, you may have to stay approximately 30 to 60 minutes. If the fuels are heavier and spot fires are a problem, the time will be increased. You may have to protect the structure for several hours in timbered areas, especially if the fuel buildup around the structure will expose it to radiated heat or firebrands for long periods of time.

Do not waste water.

Water requirements depend on the fuel type, fire intensity and the number of flying brands. You may need to expend only 200 gallons of "wisely used" water for the protection of a structure in light fuels. In an active firefight, plan on using an average of 400 to 800 gallons of water per engine per hour. Request and assign water tenders to support the engines.

It will be tough to write-off a structure, but you may have to do it to save more homes.

You must communicate your overall strategy to your dispatch office or supervisor. Give an update on the fire's behavior and progress. Describe the threat to structures and your plan of attack. Order additional resources if necessary.

Figure 8.23 This series of pictures span about two hours. The fire is being driven by Santa Ana winds as it moves into this small community. Several engine companies have been deployed to protect the homes. Two large helicopters have been assigned to support this defensive action. The homes are not being threatened by a full frontal assault, but rather by numerous spot fires.

ESTIMATING FIREFIGHTING RESOURCE NEEDS

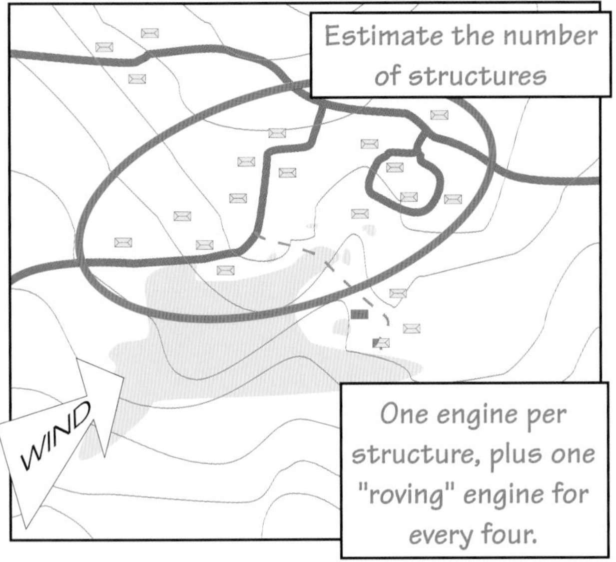

Figure 8.24 Count structures or mailboxes to determine the number of structures threatened. A general rule of thumb is one engine per structure; if the structures are within 50 feet of each other, one engine per two structures. For every 4 engines, order one additional as a reserve or "roving" engine. Be sure to order the appropriate supervisory personnel. These numbers represent ideal situations, which rarely occur in real life.

Figure 8.25 To get an idea of how many structures may be threatened in an area, count the mailboxes. Now you have to find them and determine if they can be safely protected.

Communicate the plan to the firefighting resources already on the fire. Outline the plan, the access, hazards, safety zones and escape routes. Identify the staging area, the location of the incident command post, and water sources.

The key is to match your firefighting resources to the predicted fire problem. This includes the type of engine companies (Type 1 or 3), water tenders, adequate management, and supervisory and logistical support personnel to ensure the operation functions in an effective and efficient manner.

Dealing with Homeowners

Where there are homes, there are going to be people. If their homes are being threatened, the homeowners are concerned, angry, and excited. Most will want to evacuate the area on their own. If possible, instruct them to leave via a route that will not interfere with responding firefighting apparatus. Homeowners and other volunteers can save homes.

Those that want to stay and help should be instructed in what to do. Figure 8.26 outlines actions they can take to protect their home. Be sure to ask the locals about water sources, hazards, and access. This information could be vital, as the situation gets worse.

Firefighting Tactical Situations

There are four firefighting tactical situations associated with protecting structures from encroaching wildfire. Figure 8.27 describes the four levels: ***spotting***, ***full control, partial control*** and ***no control***. If the fire is burning with such an intensity that it cannot be attacked, seek refuge in the structure or cab of the engine. Wait until the conditions improve, then resume

What can homeowners do to save their homes?

Task	Specific actions and reasons
Evacuate family members and pets	Move those people (and pets) that will not be taking an active part in protecting the structure. This reduces the number of people exposed to the risk.
Remove combustibles	Move all flammable materials (woodpiles, tarps, leaves, etc.) away from structures. Store the lawn furniture IN the pool. This removes fuel loading around the structure.
Cover/close openings	Close or cover all outside attic, eave or basement vents. Cover windows and pet doors. Remove lightweight drapes and curtains. Close all shutters, blinds, or heavy window coverings. This reduces the potential of the fire entering through an opening or via radiant heat through a window.
Close all doors INSIDE the structure	This will reduce drafts and help confine a fire if it does get into the structure.
Shut off the gas	Shut off any source of gas (LPG, natural gas, or fuel oil). This reduces and isolates the danger.
Water for firefighting	Connect garden hoses. Fill the pool or hot tub, garbage cans, tubs or other large containers with water. Get several buckets. The availability of water may be vital to the survival of the structure.
Ladder the structure	Place a ladder at a safe place for use by firefighters to get on the roof and extinguish spot fires.
Pumps or generators	Fuel and get ready any gas-powered pumps or generators. These may help in the firefighting.
Escape	Back your car into the driveway and roll up the windows. This may be your way out. Place your valuables in the car.
Doors and lights	Disconnect the garage door opening system. Leave your door unlocked. Turn on the porch light. This will help firefighters find your home and assist them in making entry if needed.

Figure 8.26 Here are some things homeowners can do to save their homes. Turn a potential problem (residents that stay) into an asset.

The four structure protection situations

Situation	Conditions/Actions
Spotting	Airborne firebrands are the immediate problem. This threat may last for hours. Remain mobile; survey all assigned areas (use lookouts or patrols); lay lines only when necessary; attack spots as quickly as possible (ensure they will not rekindle); under extreme conditions (many spots in heavy fuels) and not enough water, move out.
Full control	Perimeter control is possible, you pick the location for the control lines. Take action to stop the fire at the edge of the yard; construct a line that will hold and not rekindle; fire-out the line if time is short; connect your line with other control lines or anchor points.
Partial control	The location of control lines is not your choice. Time is limited or fire intensity prevents full control. Knock down the fire front moving at the structure; lead the fire around the structure with your hose lines; check the structure for ignitions.
No control	The fire has passed you by or is too hot to control. Protect firefighting and any civilian personnel; protect the apparatus; attack fire in/on structure if safe; use breathing apparatus and hose lines as necessary; move to safety zone or retreat if necessary. All is not necessarily lost after the fire moves past a structure. Move back in and take action.

Figure 8.27 As the fire moves toward structures and you are set up to protect them, you will find there are four situations you may find yourself in. This chart describes them.

firefighting. This usually happens in a wind-driven fire when conditions are extreme but short-lived.

Spotting will be first.

If you find a roof fire that does not involve more than one fourth of the roof, you may be able to save the structure. Use both exterior and interior attack lines. Pull the ceiling to gain access to the attic. Any time firefighters are actively fighting fire in the interior of structures, they must be in full structural personal protective equipment. If knockdown is not accomplished quickly, the chances of success are doubtful. Move on.

Consider using chain saws in mop-up/overhaul when time is critical. They are a quick way of removing burned material from a home or deck.

Staying Mobile

You must remain mobile; ready to move as the situation changes. You may have to move because the fire is too intense, or someone needs immediate help. Or you move because the threat has passed and it is time to find another structure to defend (see Figures 8.28 and 8.29). Whatever the reason, make your move deliberate and purposeful. Some key points to remember when setting up and moving:

Stay mobile and flexible.

- Keep all hose lines as short and as few as possible. You don't want to pick up any more hose than necessary. Do not place any lines in front of the engine.

Connect to a hydrant to get water, not pump from it.

- Do not connect to a hydrant. Use it as a water source, but not as an "anchor" to one spot. Use a rapid refill evolution from either side of the engine.
- When reloading and time allows, drain the hose and fold it in an accordion lay on the hose bed, or hang it on racks, the ladders or any other convenient place. Keep the hose connected to the pump discharge and the nozzle in place. You are now ready to quickly redeploy the line at your next stop.

Be prepared to "load and go!"

- If you must move quickly and abandon the lines, shut down the lines, disconnect from the engine and retrieve the nozzle. Leave the hose lines so they can be seen; they may be needed later. Don't forget your ladder, if it was used. This whole operation should take less than two minutes.

If you have to move-out quickly, just "cut and run!"

- If time allows, make one more search for hidden fire. There are many places where firebrands may have entered the structure. You don't want to leave "a save" only to find a loss later. It is very important to have someone check the structure later.
- When moving, constantly check the progress of the fire, and be alert for hazards like downed trees and power lines, rolling debris in the road, and burned-out bridges. Be very careful if the route is smoky.

STRUCTURE PROTECTION WITH LIMITED FIREFIGHTING RESOURCES

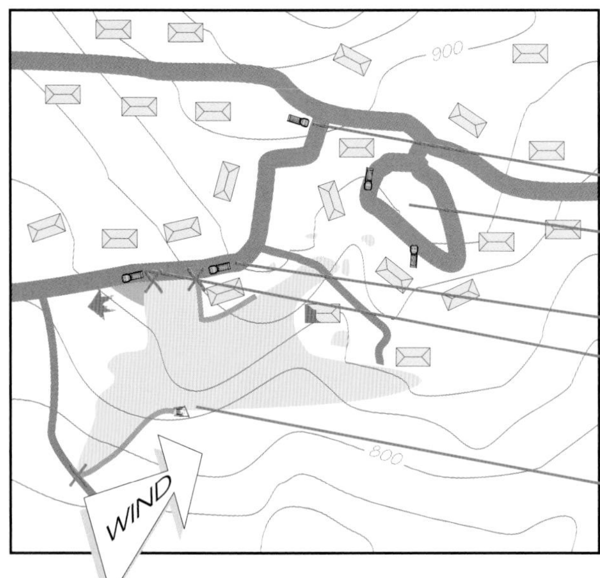

This fire is burning with a southwest wind into a rural community. Under ideal conditions, you would have ordered 15 engines to protect the structures. They aren't available. You only have 5 engines and a bulldozer.

HOUR 1

Engine 1 is instructed to protect structures. Engine 2 is to start firing from Engine 1 toward the origin. Dozer 1 is to start constructing line from the origin up the right flank. Engines 3 and 4 are to protect structures on the Loop Road. Engine 5 is held in reserve. No attempt will be made to protect the two homes on Dead End Road.

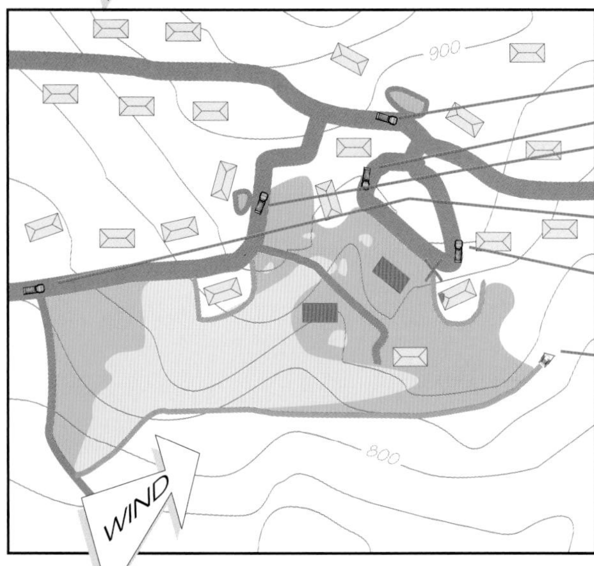

HOUR 2

Engine 1 protects the home as the fire passes, then moves to control a spot fire. Engine 2 completes firing and secures the line. Engine 3 continues to protect a home and Engine 4 has moved to a home that can be saved. Engine 5 picks up a spot across the main road. Dozer 1 continues constructing line up the right flank.

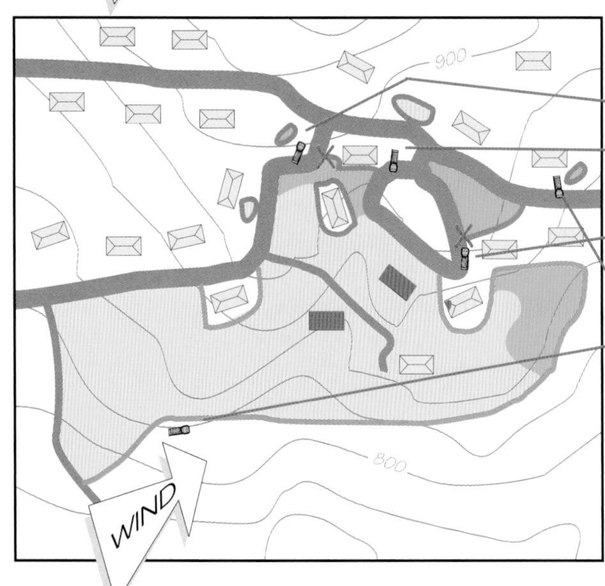

HOUR 3

Engine 1 moves to control another spot and assist Engine 3 in protecting two homes. Engine 2 begins to patrol the right flank. Engine 4 protects two homes and constructs line around them. Engine 5 picks up another spot and assists Engine 4 in constructing line. Dozer 1 finishes constructing line on the right flank.

Figure 8.28 Fighting fire with limited resources requires "being dynamic." Stay mobile; protect what you can.

Water Usage

Use water very sparingly, it will usually be in short supply. If there is a hydrant system in the area, learn something about it. Determine the pressures, available flow, gallons in storage, and if it requires electrical power to function. Systems often fail, once power is lost. If you must rely on other water sources, locate them and determine how to access them. Let other companies know if you find a source of water. Again, DO NOT hook up to a hydrant, except to refill. You will tend to restrict your mobility and waste water (Figure 8.30).

Avoid wetting down the area. It wastes water and will make it hard for you to fire the area later if that is necessary. Keep fire from getting into heavy fuels. ***Extinguish fire at its lowest intensity,*** not when it is flaring up. Wait until the fire moves into light fuels. Avoid wasting water on the fire at the peak of its "heat wave." Use water or foam to cool the sides of structures exposed to extensive radiated heat, but avoid hitting windows, especially if the glass is hot. ***Apply water only if it controls fire spread or significantly reduces heating of the structure.***

Figure 8.29 A major fire under north winds is entering a highly developed area. Engines are positioned to protect the homes. One structure has been lost. Two others damaged. Two engines are starting to fire-out the main highway. Engine companies are controlling the various spot fires. Three engines are being held in reserve.

Figure 8.30 Don't waste water. You may be robbing it from an area that really needs it.

Foams and Gels

Class A foam and gels can be a big help in structure protection situations. Like other tools, foam has to be used at the right time and in the right way to be effective.

Class A foam and gels can be applied to a roof or wall of a structure as protection from firebrands or radiant heat. The

Figure 8.31 Class A foam or gel can be of tremendous value. Each in its own way can defend a structure. Don't overlook this valuable tool.

foam blanket may last a couple of hours in cool, calm conditions, but more likely 15-30 minutes in most situations. Gels can last for hours (Figure 8.31).

Wait until the last minute to apply the foam blanket. If you have to vacate the area after foaming, come back and check (if safe) after the fire front has passed. Gels can be applied some time before the fire will be there. Often, the foam or gel blanket will give you a second chance to save a structure.

Class A foam or a gel can be of tremendous help.

If you have Class A foam or gel capabilities, you may want to apply either one to the exposed structures and move out of the area while the fire moves through. Then return and check for fire in each of the structures. This is a useful tactic when the fire is moving fast and with a high intensity, and making a stand to protect structures is too dangerous for firefighting personnel. See Chapter 5 for more information on Class A foams and gels.

Gels have to be applied by someone that has been trained.

Common Ignition Points

The fire may be some distance from the structure, but a flying brand may have landed on the roof. Flammable roof material and debris are the first areas to check. As the fire draws closer, the eaves or gable roofs will hold heat. Flammable siding will also be a problem if you have limited clearance.

Flammable roofs and lack of clearance are your greatest problems.

Openings like unscreened vents, windows, holes in roofing materials (like tile openings), open doors and breezeways, and crawl spaces or openings in mobile home skirting are also places the fire can enter a structure. ***In windy conditions, firebrands can enter virtually any opening, even under roof tiles and tightly fit siding.***

Wooden patio decks, lawn furniture, stacks of firewood, and trash piles can be sources of radiant heat (Figure 8.32). ***Don't underestimate the effects of radiant heat. It tends to be cumulative.***

Yard litter or junk (old cars, lumber piles, etc.) can also be a source of heat. They may also be difficult to work around and may be hazardous (Figure 8.33). If there are a lot of these types of exposures to protect, the time commitment will increase.

Radiant heat can also be a big problem, especially if there is a high concentration of heavy fuels nearby.

Structure and Site Preparations

If time allows, there are several things you can do to make your job of protecting the structure easier. Conditions to be considered are:

- ***Defensible space*** - Remove flammable material from the area. If the yard is cluttered with flammable material (firewood, scrap lumber, trash, etc.), move it well away from the structure. If there is grass or brush close to the structure, cut it back. Limb up trees, reducing the ladder fuels. Do whatever is reasonably necessary to give yourself some working room and to reduce the potential for convective and radiated heat to ignite the structure. The amount of defensible space needed depends on the anticipated fire behavior. Clearance should be at least two times the flame length—three times, if possible. If the fire will be approaching from below, the steepness of the slope will dictate even greater clearance.

Figure 8.32 The fire in this home probably first started on the deck. It is now on the roof and is lost.

Jake says,
"Sometimes it will be impossible to adequately prepare a structure. Move on, put your efforts in a potential winner. Firefighter safety comes first."

Figure 8.33 Even though this structure has metal siding, it will be lost because of the heat that will be generated from all of the materials placed next to it.

• ***The structure's exterior*** - Plug holes and unscreened vents. Clear the roof and rain gutters of litter. Cover evaporative coolers; turn off the blower. Turn off the gas at the meter or at the LPG tank. Leave the electricity on unless there is fire inside the structure and you will be attacking it. Close the garage door.

• ***The structure's interior*** - Close the doors and windows; close the drapes if they are of a heavy material or will not easily ignite. Turn off fans, coolers, air conditioners, etc. Turn on the porch light. Make sure external doors are unlocked, but closed.

• ***Hazards*** - Some of the more common hazards are LPG tanks, above-ground fuel storage tanks, junk cars, and miscellaneous car parts (shocks, fuel or air tanks, mounted tires, drivelines, etc.). Clear flammable fuels from around these items.

Above ground fuel storage can be a real problem.

• ***Pets and livestock*** - If you will have to deal with a lot of animals, have animal control personnel called out. You may have to free penned or chained animals; most of the time they can avoid being hurt if they can move freely. Household pets may have to be confined to the house or garage. If an evacuation has been initiated, instruct homeowners to take their pets, if possible.

If you have time, improve the defensible space.

• ***Fireline construction*** - There may be times that you need to construct fireline just to protect a structure or group of structures. The placement of this fireline is critical. The fireline should be placed well away from the structures so that radiated and convected heat cannot ignite them. The slope, wind, and type of fuel will dictate the necessary fireline width. You may allow the fire to move to this line; however, most often you will fire it out, widening the buffer between the fireline (and structures) and the main fire.

• ***On-site protection*** - Ladder the building, using the owner's ladder when possible. Keep your ladder on the engine so that if you have to leave quickly, it isn't something else you have to collect. Place the ladder on the side of the structure away from the heat of the fire, in a visible location, and close to where your hose lines will be positioned. Move private vehicles out of your way and to a sheltered location away from the heat and firebrands. Park them headed out, unlocked, with the keys in the ignition and the windows up. Be sure they are not parked on flammable fuels.

Defending the Structure

The job of protecting structures from an advancing wildland fire is usually assigned to engine companies. In this process, the engine crew stays with the structure until the fire passes (unless it is prudent to leave sooner). Once the structure is safe, the engine company may be moved to protect another threatened structure.

Jake says, "When moving into an area, plan your escape...walk around the structure and lay out your plan. Thinking ahead saves lives."

Engine Operations and Hose Lay Evolutions

Your whole operation will center on the engine. It is important that you position the engine for ease of operations and the ability to escape a dangerous situation, if necessary.

There are several key things to remember when making access to a structure and positioning the engine (Figure 8.34):

- Note landmarks and hazards along the way. You may have to leave the area in a hurry and under very smoky conditions. Make note of any potential safety zones.

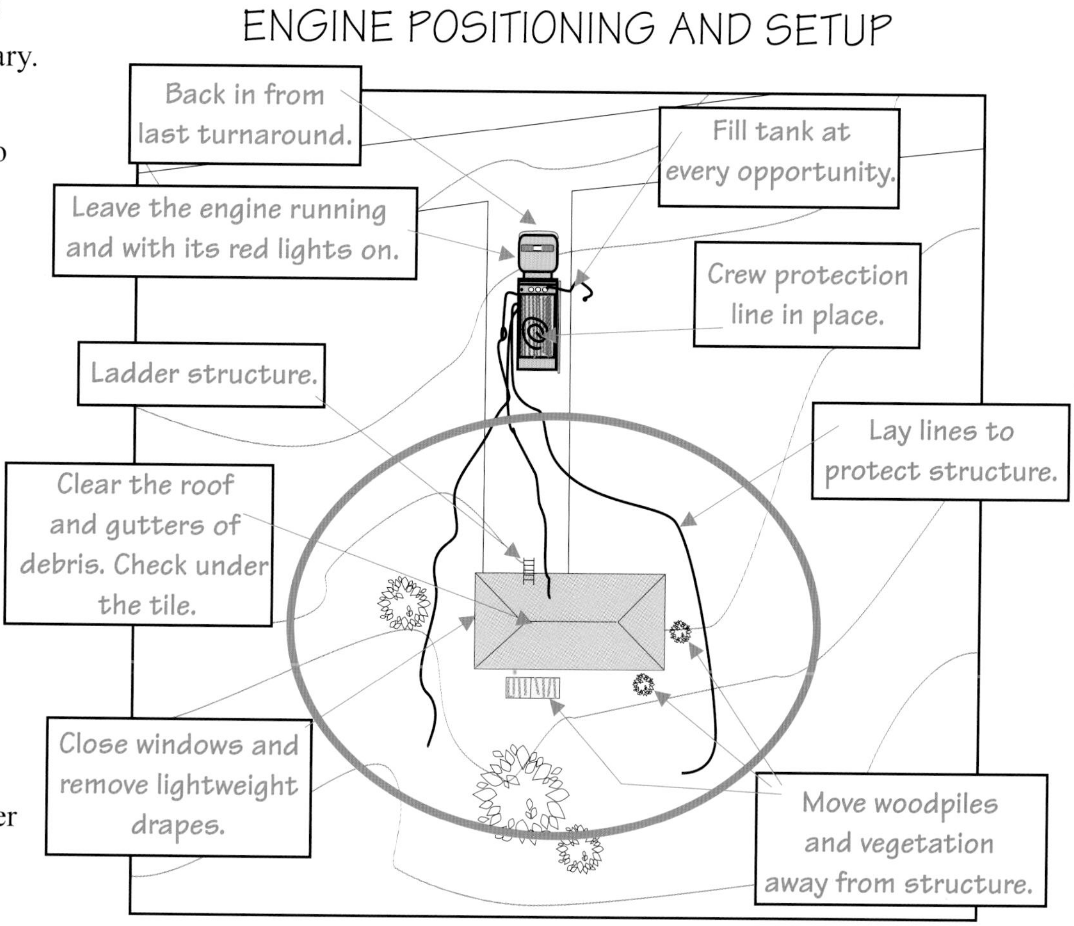

ENGINE POSITIONING AND SETUP

Back in from last turnaround.
Fill tank at every opportunity.
Leave the engine running and with its red lights on.
Crew protection line in place.
Ladder structure.
Lay lines to protect structure.
Clear the roof and gutters of debris. Check under the tile.
Close windows and remove lightweight drapes.
Move woodpiles and vegetation away from structure.

Figure 8.34 It is critical that you position you, your personnel and apparatus in positions to protect the structure, but also so that you can make a quick move, if necessary. Prepare the structure and lay out protection line(s).

- Back in from the last turnaround. If the driveway is short (less than 50 feet), you may want to park in the street.

Stay mobile. Don't hook to hydrants.

- Park the engine so that it is NOT blocking other traffic. Be careful not to park in flammable vegetation, under power lines, near sources of intense heat or LPG tanks. Park close enough that you don't have to lay extensive hose lines. Be sure to close the doors and windows on the engine. Don't leave flammable items (hose packs, personal gear, sleeping bags) exposed on the top of the engine.

- Survey the area for hazards like holes, drop-offs, septic tanks, wire fences, low service drops or other wires, agitated or protective animals, fuel storage tanks, LPG tanks, or pesticides.

- Locate a garden hose and place it in your tank filler. Seize every opportunity to top off your tank. You will need every gallon of water you can find.

- ***Exterior preparation*** - Once the engine is positioned in a safe place, it is time to lay the protection lines. Walk around the structure and survey the situation. Check the clearance and condition of the fuels. What is the vulnerability of the structure? Look for on-site resources, such as a water source, ladder, hoses and other tools that can be used instead of taking them off your engine. Shut off the gas, but keep the electricity on so that any on-site water pumps will continue to work, and so that the home can be easily found at night.

Know what attack strategy is being used. Work for perimeter control when possible.

Establish the perimeter protection. Use 1½- or 1¾-inch lines. Smaller lines may be used in light fuels or on firing operations. Combination nozzles work best, although you may have to change to straight-stream nozzles in high-wind conditions. Use single-jacket hose if available; it is easier to handle. The most common hose lay is two lines, one on each side of the structure. If the structures are closely spaced, lay the two lines between the structures. If a portion of the perimeter is not exposed, use a single line on the threat side. ***Don't deploy more lines than you need or can effectively utilize.*** Cover the full perimeter. Attach all lines on the same side of the engine. This will facilitate quick disconnecting, if that becomes necessary.

If roof protection is needed, ladder the structure and lay a line to the roof. Use the owner's ladder if possible. Remember, even a roof constructed of a fire-resistant material may need to be protected.

Coil a crew and engine protection line on the hose bed. Charge the line and check it. Place this line so it is easy to reach. Use it only for the protection of the crew or the engine.

As the fire approaches, get in or behind the structure for protection.

If it becomes necessary to attack an interior or roof fire, use $1\frac{1}{2}$- or $1\frac{3}{4}$-inch lines. Don't use larger lines unless you have the water and are convinced they will provide a "quick knockdown." If you don't have the water, stay with the smaller diameter hose.

Interior attack - All interior attacks must be made by firefighters who are in full turnouts and breathing apparatus. Work to suppress any fire on the interior of the structure. If it is well-seated and you don't have a good water supply, the structure, again, may be lost. If you suspect fire has moved through vents into an attic, check it out. Many a structure has been lost because a small fire in the attic space was overlooked.

Creating a spotting zone - Create a spotting zone around the structure within which you will attack spot fires. Spot fires usually start well before the main fire front arrives. In fact, the main fire may not arrive for hours. Be sure that your hose lines cover all of the area you consider the spotting zone. Move around, looking for and suppressing the spots as they occur (Figures 8.35 and 8.36).

Reducing the heat - As the main fire approaches, reduce the heat from hotspots if you can. Use the hose lines to knock down flare-ups as they occur. Don't just stand there wasting water by launching it into the air.

SPOTTING ZONE

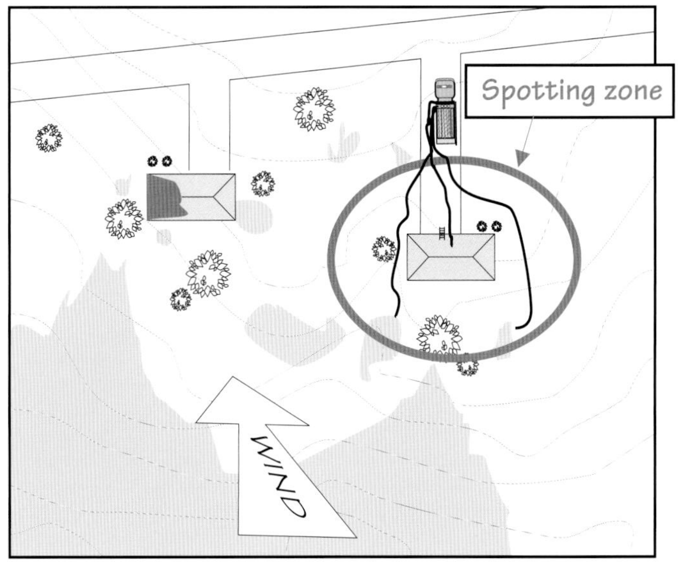

Figure 8.35 The spotting zone is the area around a structure that you will defend. Control all spot fires within this area. Note the structure to the left is not being defended. More than $\frac{1}{4}$ of the roof is involved. It is lost.

Kurk Taylor

Figure 8.36 This spot fire is within the spotting zone. Action will be taken from the air as well as from the ground. Two firefighters (in the circle) are protecting the structure.

Wetting down an area is a waste of time and water. Use the limited water you have to actually suppress the fire. Use the water to reduce or limit the buildup of heat. Use it to prevent the fire from getting into heavy fuels (patches of brush, wood piles, lumber), and use it to knock down surface fires so that the fire doesn't move into the aerial fuels.

Study the area...make sure you are not in a hidden chimney.

The heat wave - As the fire moves near your location, heat will build up. The duration of this buildup is controlled by the fuels that are burning, as well as the overall fire behavior. In light fuels, the flame front may pass by very quickly, often in minutes. In brush, the burn-out time may take 10 to 15 minutes (Figure 8.37). If you are protecting a structure located in or near heavy fuels, the duration of the heat may be hours. Many a structure has been lost because of the accumulated effect of radiated heat over a long period of time. As the heat peaks, you may have to seek refuge behind the structure or even inside. Don't launch water into the flame front. As previously noted, you will be wasting water. Move back until the fire has passed, and then take whatever action you can.

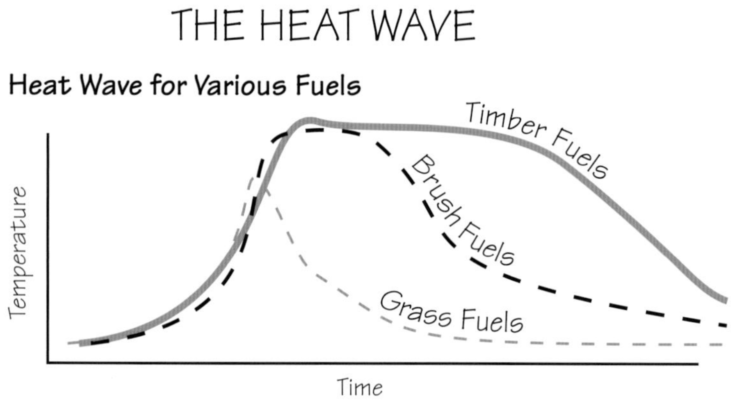

Figure 8.37 As a fire moves closer to the structure, the temperature will increase. This is called the "heat wave." The duration of the heat wave is tied to the type of fuels and the level of fire behavior. Timber or heavy fuels will generate heat for very long periods. Grass fuels, on the other hand, will burn very hot, but for only a short period of time.

When the fire is controllable, limit the heat build-up by keeping fire out of heavier fuels. Work on the fire where it has moved into lighter fuels. At the other extreme, wait until the worst of the heat wave has passed, then put water on the structure or on threatened fuels. In between these extremes, apply water only if it significantly reduces direct heat impinging on the structure.

Firing-out - It is not uncommon for firefighters to be reluctant to fire-out around a structure. However, this may be the only way to safely protect a structure. If an area is going to burn, why not burn it on your terms? If your plan is to burn-out around a structure, don't wait too long. Do it early enough so that your fire can widen your control lines in time to do some good. Coordinate your burning operation with others. You don't want to start burning in an area that is planned to stay unburned. Also, you don't want to start a fire which will cause problems for other firefighters. Consider firing-out if:

Establish an area you will protect.

- Control lines will not hold if the full force of the fire hits them.
- Main fire's intensity would threaten structures.
- There is no time to wait for the main fire.
- Lower intensity fire will help prevent a crown fire.
- ***Control lines must be in place before you start firing.*** Use existing barriers, like lawns, roads, and driveways, or use line constructed with hand tools or bulldozers. Wet line will also work in light fuels. Construct your line just wide enough to hold "your" fire. You must time and coordinate the firing with adjacent forces and the division/group supervisor. Air operations must also be told of the firing. Many firing operations have been put out by uninformed, well-meaning air operations personnel.

If you are going to use fire, be sure everyone in the area knows it.

- ***Fire as soon as the line is constructed.*** Don't wait around for "indrafts." They will rarely help. Take the fire along the line AGAINST THE WIND and DOWNSLOPE. Use simple edge firing if the conditions are favorable. If a wider line is needed quickly, use strip firing. Burn in segments to take advantage of varying slopes and wind conditions. Fire a line no faster than your forces can hold it. Knock down flare-ups that threaten the line. Watch for spot fires.

Have the lines in before you begin firing operations.

Staying Mobile **-** Again, you have to remain mobile, not tied down to one spot for any length of time. There usually aren't going to be enough engines to protect all of the structures being threatened. As soon as one structure is saved, move to the next one. Try not to get tied to fire hydrants or large diameter hose lines.

"Is it time to cut and run?" - There may be times when the situation is deteriorating and you have to "cut your hose lines" and retreat to safety. If you have to move out, be sure to watch for hazards like downed powerlines, burning trees and snags, debris on the road (e.g., logs and rocks), smoky conditions, weakened bridges, or cattle guards. Move to a safety zone and wait out the fire,

Keep your options open to move out of an area if safety becomes a concern.

notifying your supervisor of your move. Then move back into the area to see what can be salvaged.

Again, stay mobile!

Entering private property - If you are protecting a structure, you usually have access to the building. This means you may need to enter it to check for fire. Be sure that you and your people treat the residents' property as you would want yours to be treated. If you do enter a structure, follow your agency's procedure for documenting what actions were taken, and informing homeowners.

Be sure to work with all agencies that have a stake in the fire.

Dealing with Multi-Jurisdictions

A lot of the time, there may be more than one agency with jurisdiction. The local fire department will be responsible for protecting the structures and a wildland fire agency is responsible for controlling the wildland fire. You cannot have two operations. The only answer is a unified command. Find your partner and "get married to them." Ensure that you have common radio frequencies. Set up a joint command post, and don't forget law enforcement.

9

Incident Command System

The Incident Command System (ICS) is an "all-risk" operating system that can be adapted to fit any type and size of incident. The system can expand or contract as the situation changes. Think of ICS as you would a computer operating system, like Mircosoft Windows. "Windows" and ICS are sets of rules that define relationships between, and responsibilities of, various parts of a "machine." Windows does not "know" a thing about word processing, spread sheets, or data bases. Likewise, the Incident Command System doesn't know a thing about dealing with fires, floods or earthquakes. Just as computers rely on people to operate the software programs, the ICS depends on people who bring their knowledge bank of experience and expertise on a particular subject to make the system work logically and efficiently.

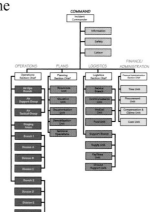

ICS doesn't know how to fight fire...it is the people who implement ICS that make it work logically.

Jake says, "Don't let this chart scare you. The only position that must be filled is that of the Incident Commander."

Background

During the summer of 1970, there were disastrous fires in Southern California. One of the conclusions developed in the after action report was that the major players in Southern California (USDA Forest Service, California Department of Forestry and Fire Protection, Los Angeles

Figure 9.1 Without a commonly used, functioning organizational structure, an incident may unnecessarily become a disaster. Before the 1980s, no such system existed, and there were a lot of unnecessary disasters simply because firefighting agencies had difficulties working together.

County Fire Department and the Los Angles City Fire Department) all used different major fire organization schemes, and different radio codes, etc. In other words, they simply were not setup to work together on a major fire (Figure 9.1). These four agencies (along with the California Office of Emergency Services, using a federal grant), formed a working team under the title of ***"FIRESCOPE."*** The Incident Command System was one of the "products" developed by this group. Originally, ICS was strictly for wildland fires, but over time it has evolved into an all-risk incident management system.

The Incident Command System is an organizational scheme that can be used to manage any type of emergency. It outlines the chain of command, lines of authority and responsibility, and the span of control (Figure 9.2).

The vast majority of wildland fires are controlled in their initial attack phase, so why

There are five elements to ICS: Command; Operations; Planning; Logistics and Finance/Administration.

ELEMENTS OF THE INCIDENT COMMAND SYSTEM

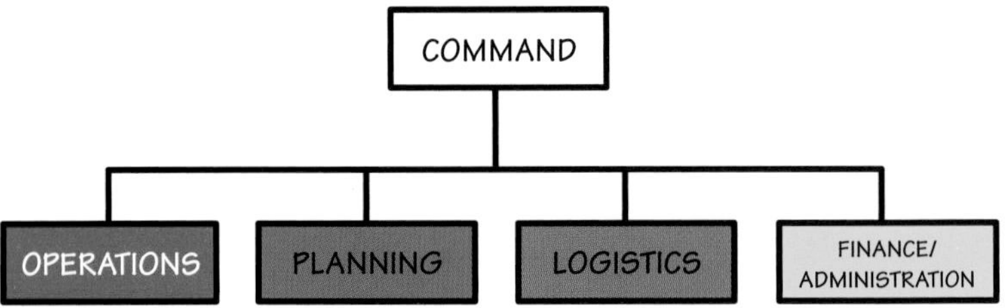

Figure 9.2 There are five organizational elements under the Incident Command System: Command (includes the person(s) responsible, information, safety and liaison); Operations (ground and air forces); Planning (incident and resource status, and technical specialists); Logistics (service, communications, medical, food, supplies and facilities); and Finance/Administration (time records, cost, claims and procurement).

Jake says,
"Once these large and very proud organizations began to work together, they found they had a lot more in common than they thought!"

do we need a "whole system" to manage these fires? Why not have a system for large incidents, and another to use when the fire is small? The key to any organizational scheme is that you use the same one all of the time, and the organization grows larger as the incident or fire grows larger. Hopefully, the organization is ahead of the fire's growth, but this very seldom happens, if ever.

It doesn't take hundreds of firefighters before you use ICS. You need only ONE.

As mentioned before, ICS is an organizational scheme that can be used for all types of emergencies: fires of all types, floods, search and rescues, hazardous material spills, any event.

Most people are introduced to ICS by showing them a large organization chart, and their first impression is, "Where am I going to get the people to fill all of those boxes?" ICS requires only ONE POSITION to be filled, that of an Incident Commander (IC). The IC fills all of the boxes (responsibilities) until he or she assigns the responsibility to someone else for a particular function (Figure 9.3). ***Each box is filled only if and when necessary.***

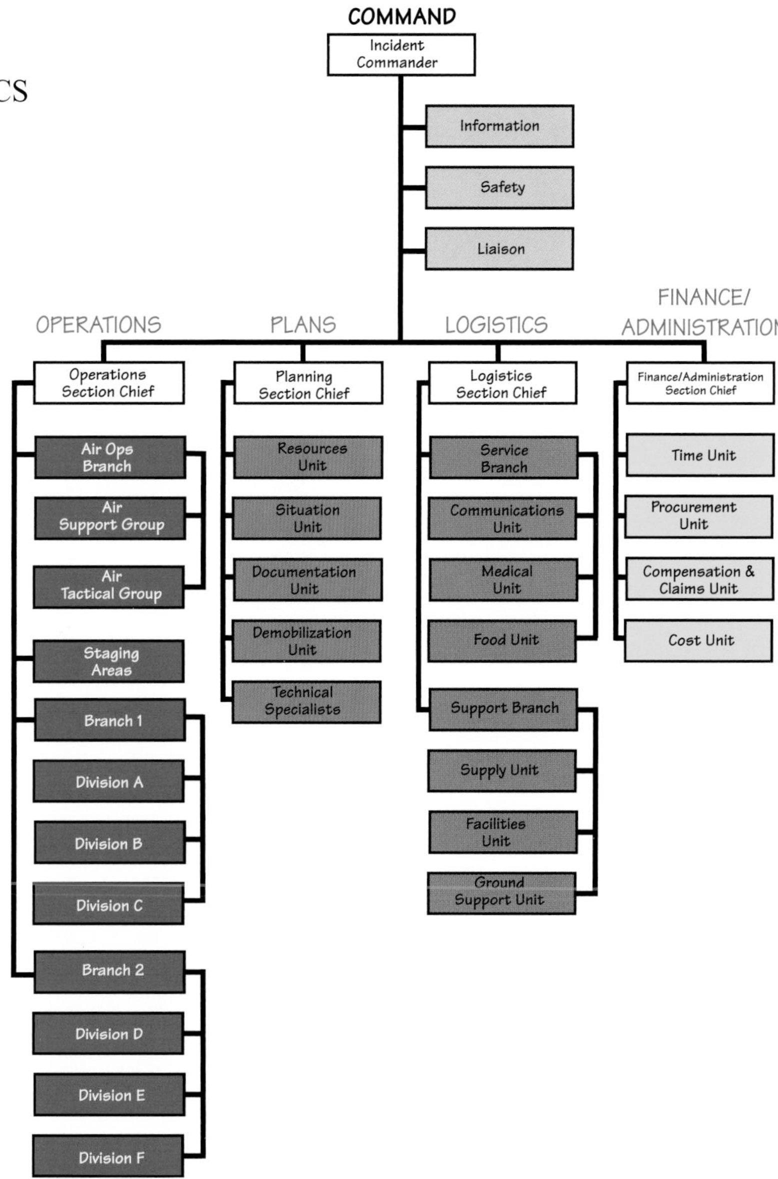

Figure 9.3 This is the organizational chart for a major wildland fire. Fill only the positions that are needed. Don't let the system run you—you run the system.

Components of ICS

ICS has nine critical components that make it work:

- ***Common Terminology*** - Everyone speaks the same language. Major functions are pre-designated and named, and the various organizational elements are standardized and consistent. Firefighting resources and facilities are also defined.

- ***Modular Organization*** - The size and complexity of the incident dictates how the various functions will be staffed. The organization expands and contracts as necessary.

Fill only the positions you NEED. Don't let the system run you.

- ***Integrated Communications*** - There is one communications plan for the incident. The plan outlines the use of various radio networks, and may include a command net, several tactical nets, a support net, air-to-ground net and air tactics net. Various locally used 10 codes are replaced with "clear text," or plain language.

When the fire crosses jurisdiction boundaries, you still have one action plan.

- ***Unified Command Structure*** - Most disasters involve several jurisdictions. ICS allows for a unified command structure where overall incident objectives and strategies are set. Unified command also ensures that there is one operational plan and integrated tactical operations are conducted, making maximum use of firefighting resources. Under a unified command structure, the incident action plan is implemented under the direction of one individual, the Operations Section Chief.

- ***Consolidated Action Plan*** - The incident action plan establishes the objectives and strategies that will be used to control the emergency. It covers all tactical and support activities for the incident.

Manageable span of control.

- ***Span of Control*** - The span of control (the number of people one supervises) for emergency management personnel typically ranges from three to seven.

Facilities to meet the need.

- ***Designated Incident Facilities*** - ICS designates several facilities that may be needed to manage the incident.
  - *Command Post* (location from which the incident is managed);
  - *Incident Base* (location where most of the support activities are performed);
  - *Camps* (sites where firefighting resources may be fed, housed, or parked);
  - *Staging Areas* (temporary parking areas or "storage areas" used by the Operations Section Area to hold resources);
  - *Helibases* (helicopters are operated and supported from helibases); and
  - *Helispots* (landing sites for helicopters).

- ***Firefighting Resources*** - Resources are categorized as either single resources, task forces or strike teams. A single resource is an engine company, bulldozer unit, hand crew, etc. ***A task force is any combination of resources, grouped to***

accomplish a specific task. A task force leader is assigned. ***A strike team is a set number and type of resources.*** There are very specific types of strike teams. A strike team leader is assigned.

Groupings of resources to help in tracking and operations.

- ***Resource Management*** - Accurate status of all the resources assigned to an incident is critical. There are three status conditions: assigned, available, and out-of-service.

The ICS Organization

Modular format fits any size incident.

The Incident Command System outlines five basic organizational functions: ***COMMAND***, ***OPERATIONS***, ***LOGISTICS***, ***PLANNING*** and ***FINANCE/ ADMINISTRATION.*** All aspects of incident management fit under one of these five functions.

On a small incident, these functions may be filled by the Incident Commander. As the incident grows in complexity, the ICS organization should grow with it. As you can see in Figure 9.3, each of the primary organization sections has sub-elements.

Until a sub-element position is filled, the person higher up on the organization chart is responsible for that function. In Figure 9.4, you can see that the Incident Commander has retained the responsibility for Information, Safety and Liaison, as well as for Finance/Administration and all of that section's responsibilities. The Planning Section Chief has not filled any of the sub-elements. The Logistics Section Chief has filled the Service and Support Branch positions. Neither of the branches have filled positions below them on the chart.

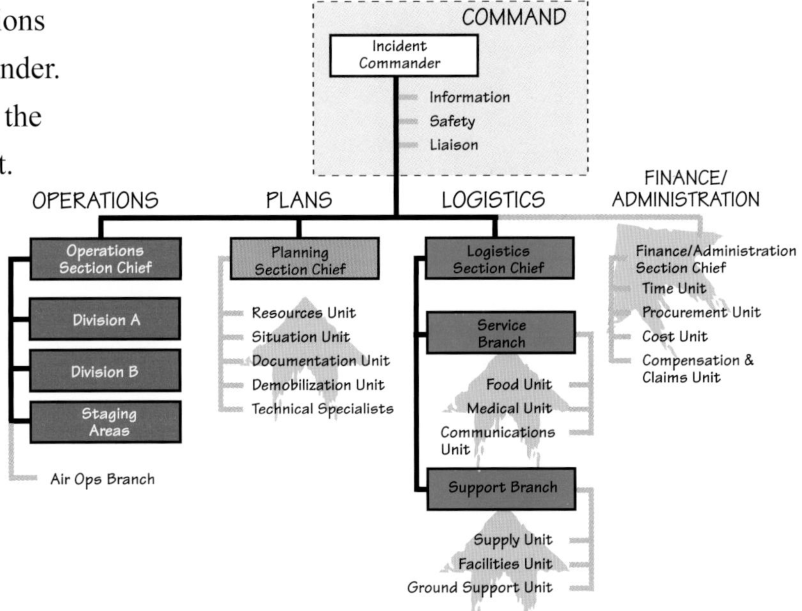

Figure 9.4 This incident is being managed by nine positions. Until a position is filled, the position above it retains the responsibility.

If a position is not filled, the supervisor of that position has the responsibility to fulfill the requirements of the position.

Command

There are four positions assigned to incident command: the Incident Commander, Information Officer, Safety Officer, and Liaison Officer.

The ***Incident Commander*** is the person in charge of and responsible for the management of the incident. The person filling this very important position has overall responsibility for the incident (Figure 9.5). The Incident Commander is commonly referred to on the radio as the "IC," preceded by the incident name: "Highway 65 IC." (Note: ICS position titles are used instead of agency radio call signs or personal names.) The IC draws his or her authority from the agency responsible for the incident, and is directly responsible to the agency line officer. The agency representative may be the Forest Supervisor, Fire Chief, Mayor, etc.

Jake says, "The IC is in charge, but it is the Division/Group Supervisors, that really coordinate the attack."

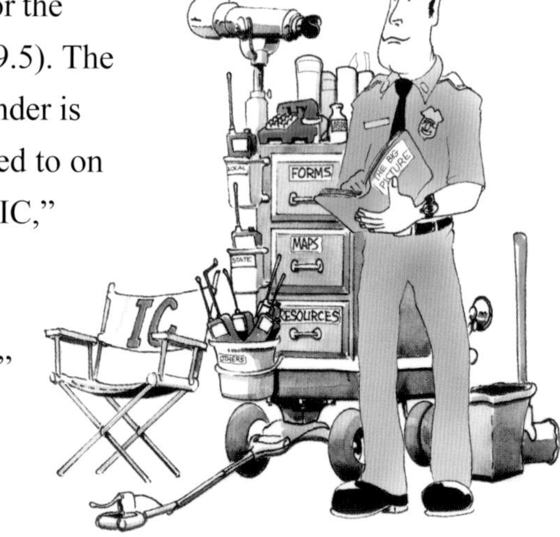

Figure 9.5 The Incident Commander has overall responsibility for the fire. This is also the only position that has to be filled; all others support the IC.

ICS is for managing any field incident.

As stated before, this position is the only one that is required to be filled under the Incident Command System. An initial organizational scheme may be quite simple; an Incident Commander and several resources (Figure 9.6). The Incident Commander fills positions in the organization chart, as necessary. It is very important that the IC be qualified for that position. It is common that as the incident grows in complexity, the person filling the IC's role may change. As this change in command takes place, the new IC must be fully briefed on the status of

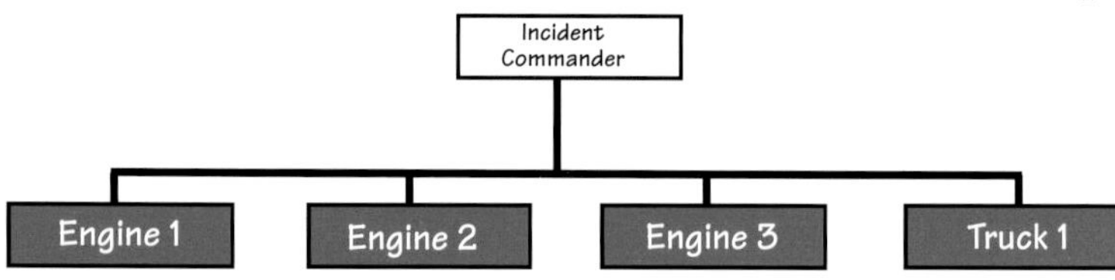

Figure 9.6 A first alarm assignment or the initial attack force may all answer to the Incident Commander. A need for other positions has not yet been identified, so the positions have not been filled.

the incident. The change in command is also announced over the radio so that everyone knows who is currently in charge.

Besides the four section chiefs, there are three other positions that answer directly to the Incident Commander. These three officers constitute the ***Command Staff*** (Figure 9.7). None of these positions are authorized deputies, but they can have assistance assigned as needed. (Note: A deputy is an individual that is fully qualified to manage the function of his/her superior. An assistant is a helper or subordinate.)

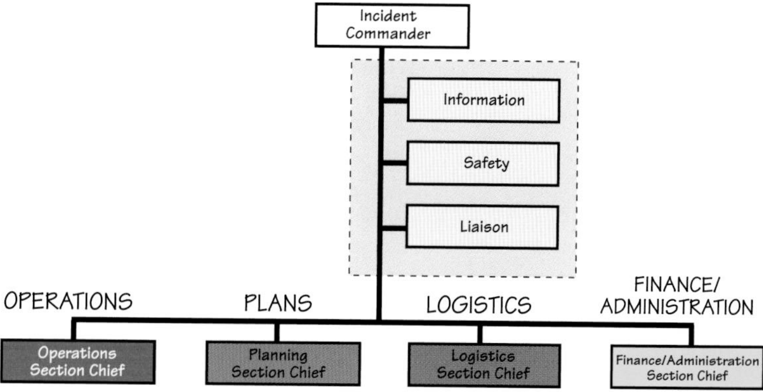

Figure 9.7 The Command Staff is made up of three positions: Information Officer, Safety Officer, and Liaison Officer. These positions are usually filled as the incident grows in complexity. There may be a need to assign assistants for each of the officers.

- ***Information Officer*** - The Information Officer is the central focus and contact point for the media or other organizations seeking information related to the incident. Only one Information Officer is assigned to an incident; this cuts down on confusion and conflicting information. On a more complex incident, the Information Officer may have several Assistant Information Officers.

Command includes safety, information and liaison.

- ***Safety Officer*** - The Safety Officer monitors the overall incident operations to ensure that safety is not being compromised. Only one Safety Officer is assigned to an incident, but this position may have several assistants who are assigned to monitor operations in the various functions or operating divisions. Even though the Safety Officer is a staff function, he/she may exercise emergency authority to directly stop unsafe acts if personnel are in imminent, life-threatening danger.

When a fire involves more than one agency, unified command is established.

- ***Liaison Officer*** - The Liaison Officer is responsible for coordinating involvement with representatives from assisting agencies. Other agencies or jurisdictions will often send resources to assist at an incident. These assisting agencies may also send an ***Agency Representative*** to work with the incident management team as a coordinator. The Liaison Officer does not direct the operations of these other agencies, but rather serves as a liaison between the Incident Commander and the various Agency Representatives.

General Staff

The key staff functions are managed by Section Chiefs who answer directly to the Incident Commander. These four positions are referred to as the ***General Staff*** (Figure 9.8). They can either provide relief or be assigned specific functions. On large incidents, especially where multiple agencies or jurisdictions are involved, the use of personnel from the other agencies can greatly increase intra-agency coordination.

The General Staff are the "big five," the IC and the four Section Chiefs.

The Operations Section is where all the work to abate the hazard or fight the fire is done. Until an Operations Section Chief is assigned, the Incident Commander is responsible for managing the tactical field operations.

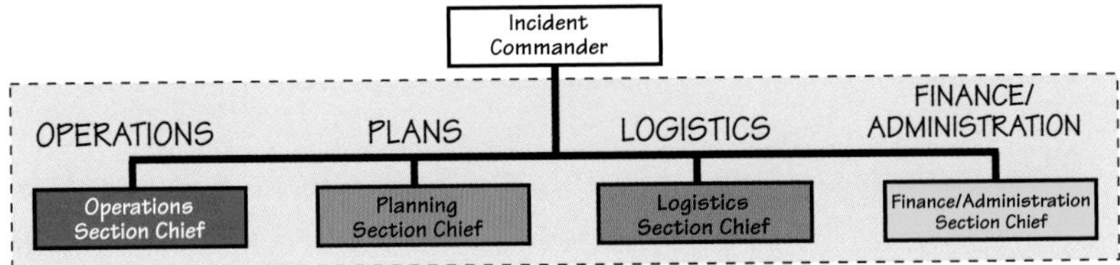

Figure 9.8 The General Staff is made up of the four Section Chiefs. Each of these key positions are authorized to have deputies to assist them in the performance of their responsibilities.

Operations Section

The ***Operations Section Chief*** answers directly to the Incident Commander and is responsible for the implementation of the incident's action plan. As a member of the General Staff, the Operations Section Chief is authorized to have a deputy.

Operations is where the fire is put out. All other functions support this element.

If the incident involves several jurisdictions, a representative from another agency can be assigned as a deputy to the Operations Section Chief. This will help with coordination and provide relief for this very important position.

Staging is under Operations.

One of the first actions that should be taken in organizing a wildland fire is to divide it into divisions and/or groups (Figure 9.9). This changes the traditional "pinwheel" shaped organizational scheme into one with a clear separation of responsibility (Figure 9.10). The number of divisions and/ or groups on an incident is dictated by the incident's size or complexity. If the incident grows to the point where more than five divisions are needed, branches become part of the organizational scheme. The goal is to keep the organization as simple and as streamlined as possible, while maintaining a reasonable span of control.

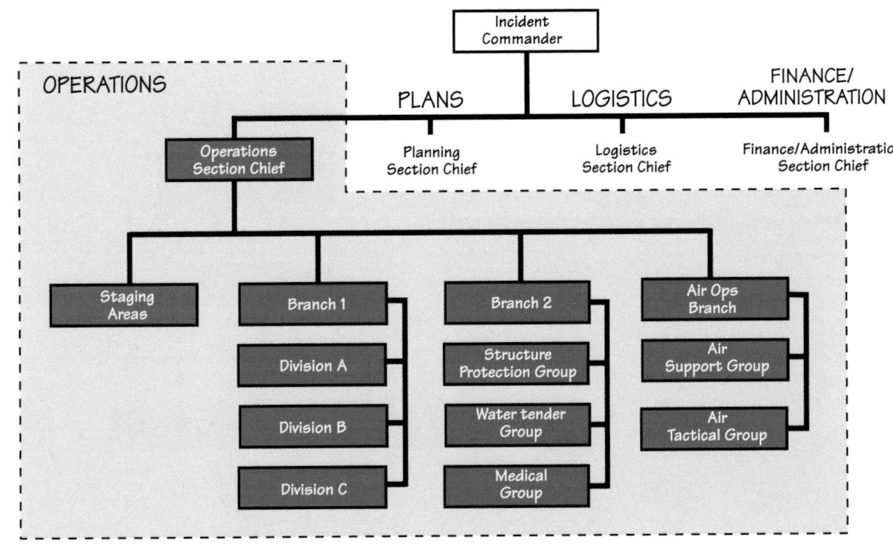

Figure 9.9 The Operations Section can include divisions, groups, and branches. These are used to organize the incident into geographic or functional areas. Air Operations and Staging are also managed by the Operations Section Chief.

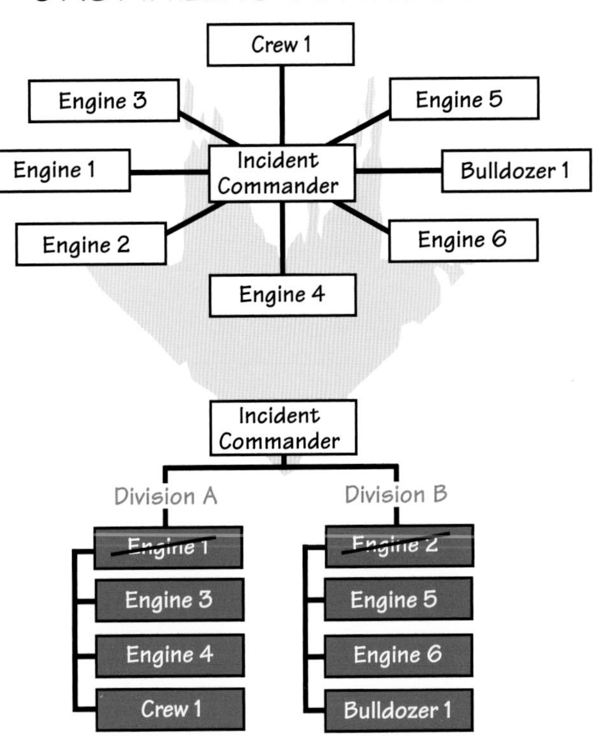

Figure 9.10 The establishment of divisions is the first step in organizing a developing wildland fire. In this case, the two engine company officers on Engines 1 and 2 are given division responsibilities. Their radio designation changes accordingly.

- ***Divisions*** - A division is managed by a Division/Group Supervisor. This position generally answers to the Operations Section Chief. However, if branches are established, the Division/Group Supervisor answers to the Branch Director. Divisions are used to divide an incident geographically. For example: by miles of fireline in a wildland fire, or by floors in a high-rise fire. If the division is an area of a wildland fire, it is given an alpha designator (start with "Division A," on one flank and "Division Z," on the other.). If it is a floor in a high-rise building, the floor number ("Division 14," "Division 42," etc.) is used.

- ***Groups*** - Groups are established to describe functional areas of the operation, like a "structure protection group" or a "medical group." They are managed by a Division/Group Supervisor and are on the same level in the organization as a division (Figure 9.11).

The operations organization is divided into branches and divisions.

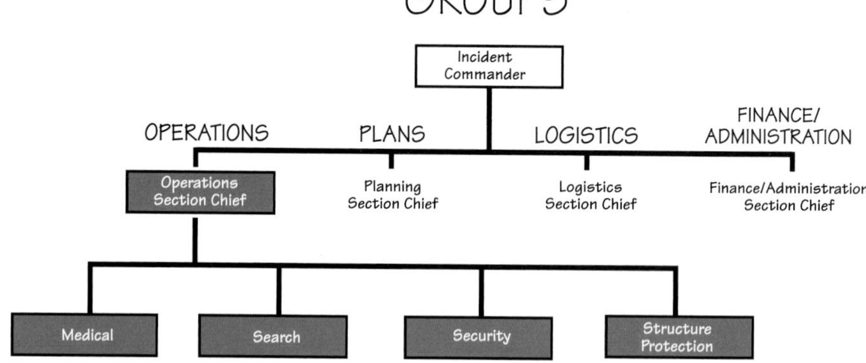

Figure 9.11 Groups are established for the management of functional activities such as structure protection, medical response and search and rescue.

Groups and divisions can function in the same area, but one does not supervise the other. When a functional group is working within a division, the two supervisors must coordinate their activities. Division and Group Supervisors report to the Incident Commander, or to the Operations Section Chief/Branch Director if these positions are filled. Deputies are not used at the division or group level.

Jake says, "One of the most important positions on any fire is the Division/ Group Supervisor. This is where the action really is!"

Some agencies have expanded the use of groups to manage resources that can be shared, like water tenders. A Water Tender Group Supervisor is given control of all the water tenders either on a single branch or on the entire incident. Division Supervisors who need a water tender can coordinate their needs with the Water Tender Group. This coordination can increase the efficiency and effectiveness of these types of resources.

Branches - As an incident develops, the number of divisions will also increase. If more than five divisions are required, you may need to establish branches. There are three reasons to establish branches:

Groups are for resources that can be shared between divisions.

- ***Span of Control*** - One person can effectively manage between three and seven reporting elements, depending on the complexity of the incident. The recommended span of control is one to five. The number of reporting elements falls outside the three to seven range, expansion or consolidation of the organization may be necessary.

- ***Functional Branch Structure*** - Some incidents involve multiple disciplines (e.g., structure protection, police, medical) and, as such, may require the use of a functional branch structure.

- ***Multi-jurisdictional Incident*** - If several jurisdictions or agencies are involved with an incident, it may be better to use branches for each of the involved entities.

Air Operations is under Operations.

Figure 9.12 Air Operations manages all air resources on an incident. The fire officer on the left is trying to explain why he didn't tell air ops about the wires.

Each branch that is activated will have a Branch Director assigned; deputies may be used at the branch level.

- ***Air Operations*** - Air Operations, if activated, functions at the branch level in the organization. The Air Operations Branch Director is responsible for the tactical operations of all aircraft assigned to the incident, as well as their logistical support (Figure 9.12).

Jake says, "Use staging areas to hold equipment in reserve. Establish a staging area early in the incident."

- ***Staging Areas*** - Staging Areas are established whenever it is necessary to position resources awaiting assignment. A Staging Area Manager is assigned to manage each of the staging areas. The manager answers to the operations side of the incident. All resources in the Staging Area are "assigned" to staging, and should be ready for deployment. There should be a three-minute response time for resources assigned to staging. Staging Areas should not be used to park out-of-service resources or for logistics functions. Staging should be one of the first elements of ICS that is established on a developing wildland fire.

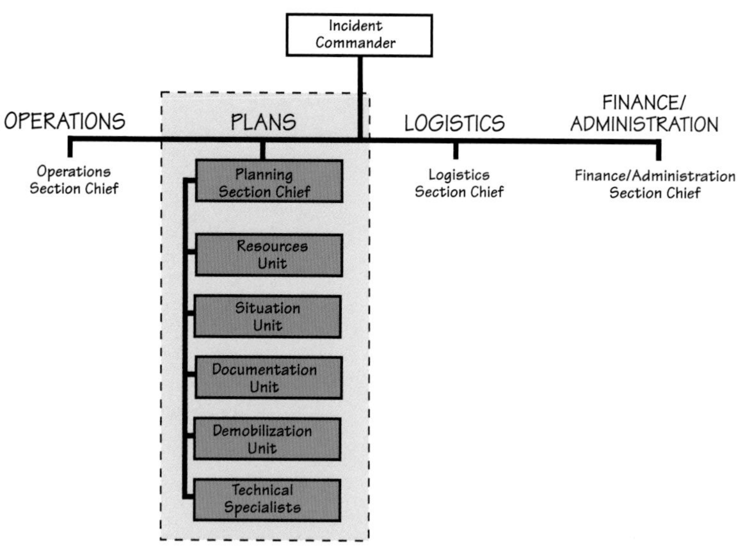

Figure 9.13 The Planning Section is responsible for maintaining and displaying resources and situation status, preparing the Incident Action Plan, developing the Demobilization Plan, and providing technical specialists as needed.

Planning Section

The Planning Section is managed by the ***Planning Section Chief***. The primary function of this section is to develop the Incident Action Plan for each operational period; keep track of the status of all assigned resources; and track the fire. There are several sub-elements within the Planning Section: the Resources Unit, Situation Unit, Documentation Unit, Demobilization Unit, and Technical Specialists (Figure 9.13).

The first responsibility of the Planning Section is to "check-in" all of the resources. Other tasks that the Planning Section Chief is responsible for include:

- Maintaining and displaying information on resources and situation status.
- Preparing the Incident Action Plan.
- Developing the Demobilization Plan.
- Providing technical specialists as needed.

Planning tracks the fire and resources. This section outlines the plans for the next shift.

One of the most important functions of the Planning Section is to anticipate potential problems and events that may come up beyond the current operational period.

- ***Resources Unit*** - Is responsible for all check-in activity, and for maintaining the status of all personnel and equipment resources assigned to the incident.

"Plans" must know where you are and your status.

- ***Situation Unit*** - Collects and processes information on the current situation, prepares situation status displays and situation summaries, and develops maps and projections.
- ***Documentation Unit*** - Prepares the Incident Action Plan, maintains all incident-related documentation, and provides duplication services.

• ***Demobilization Unit*** - On large, complex incidents, the Demobilization Unit assists in ensuring that an orderly, safe, and cost-effective release of personnel will be made when they are no longer required at the incident.

"Sit Stat" tracks the fire; where it is and what it is doing. "Plans" predicts where the fire is going to go.

• ***Technical Specialists*** - Technical specialists are advisors with special skills that may be required at the incident. Technical specialists initially report to the Planning Section. They may continue to work within the Planning Section, or may be reassigned to another part of the organization. Technical specialists can be in any discipline required (e.g., aviation, hand crew usage, environment, water supply, hazardous materials).

Logistics Section

The Logistics Section is responsible for all the services and support needs of an incident, including obtaining and maintaining essential personnel, facilities, equipment, and supplies (Figure 9.14). Logistics, services, and support are very important, and often under-appreciated. Early recognition of the need to establish logistics support can reduce time and costs.

Figure 9.14 The Logistic Chief is the "unsung hero" of an incident operation. This position is responsible for the feeding and other needs of the firefighters.

The Logistics Section is managed by a ***Logistics Section Chief*** who answers directly to the Incident Commander. A deputy position is authorized, and if the workload dictates, two branches can be established; one for ***support*** and the other for ***services***. There are six functional units within the Logistics Section: Communications Unit, Medical Unit, Food Unit, Supply Unit, Facilities Unit, and the Ground Support Unit (Figure 9.15).

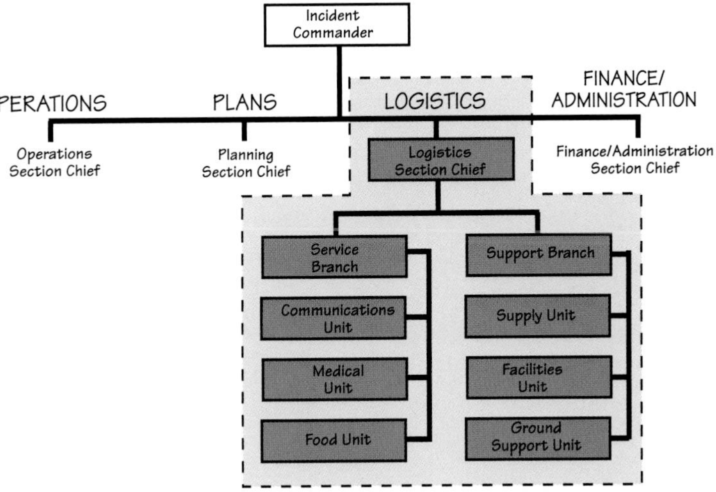

Figure 9.15 The Logistics Section has six functional units: Communications Unit, Medical Unit, Food Unit, Supply Unit, Facilities Unit, and the Ground Support Unit.

Logistics provides the material and services needed to keep the "machine running."

The following units are within the ***Service Branch***, if one is established. Otherwise, they report directly to the Logistics Section Chief.

- ***Communications Unit*** - Develops the Communications Plan, distributes and maintains all communications equipment, and manages the Incident Communications Center.

- ***Medical Unit*** - Develops the Medical Plan, and provides first-aid for personnel assigned to the incident. This unit also develops the Emergency Medical Transportation Plan (ground and/or air), and prepares medical reports.

Logistics includes service and support branches.

- ***Food Unit*** - Ensures that all personnel committed to the fire have adequate food and drinking water. The unit may prepare menus and food, provide food through catering services, or use some combination of both methods.

Jake says, "Logistics is the toughest job on the incident. Everyone is an expert on food and other support functions."

The following units are within the ***Support Branch***, if one is established. Otherwise, they report directly to the Logistics Section Chief.

- ***Supply Unit*** - Orders personnel, equipment, and supplies. The unit stores and maintains supplies, and services non-expendable equipment. All resource orders are placed through this unit. If this unit is not established, the responsibility for ordering rests with the Logistics Section Chief.

- ***Facilities Unit*** - Sets up and maintains whatever facilities may be required in support of the incident. It provides managers for the Incident Base and camps, and provides security support for the facilities and incident as required.

- ***Ground Support Unit*** - Provides transportation, and maintains and fuels vehicles assigned to the incident.

Finance/Administration tracks any item that may be a cost charged to the fire. This includes injury and damage claims.

Finance/Administration Section

The Finance/Administration Section is responsible for monitoring incident-related costs, and administering any necessary procurement contracts. The incident doesn't have to be large to require some level of finance/administrative support. If the Finance/Administration Section is established, there are four units that may be activated: Time Unit, Procurement Unit, Compensation/Claims Unit, and Cost Unit (Figures 9.16 and 9.17).

The ***Finance/Administration Section Chief*** is responsible for all cost-related activities on the incident. A deputy is authorized, if needed.

- ***Time Unit*** - Ensures that all personnel time on an incident is recorded.
- ***Procurement Unit*** - Processes administrative paperwork associated with equipment rental and supply contracts. Responsible for equipment time reporting.
- ***Compensation/Claims Unit*** - Is responsible for seeing that all documentation related to workers compensation is correctly completed. This unit also maintains files of injuries and/or illnesses associated with the incident. It also handles investigation of all claims involving damaged property associated with or involved in the incident.

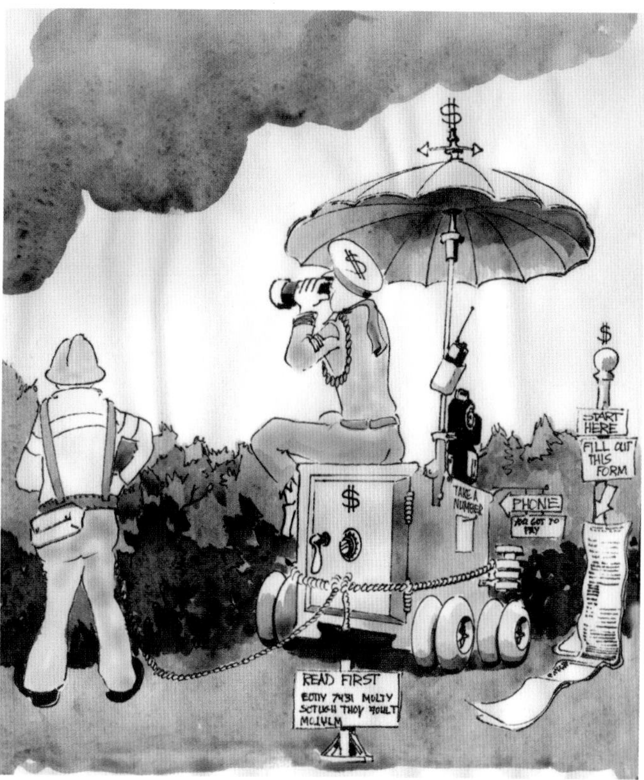

Figure 9.16 The Finance/Administration Section Chief is responsible for collecting all of the bills and paying them.

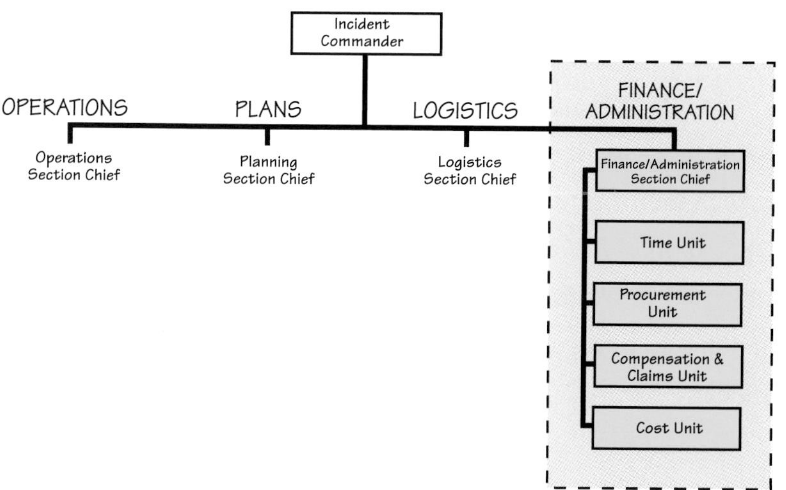

Time slips are mandatory if you expect to be paid.

Figure 9.17 The Finance/Administration Section is responsible for all cost-related and administrative services on the incident.

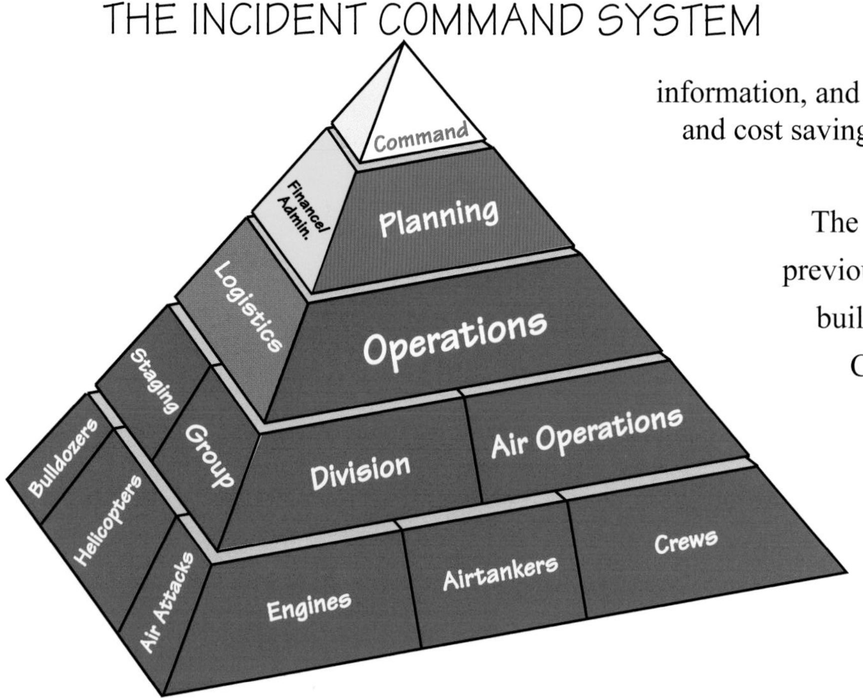

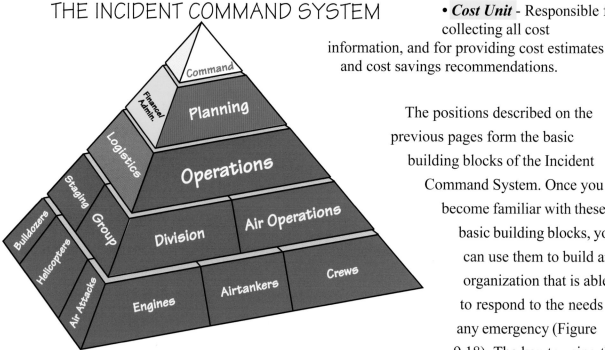

Figure 9.18 The Incident Command System "pyramid" of authority and responsibility. Be sure you tailor the size of the organization to the needs of the incident.

• ***Cost Unit*** - Responsible for collecting all cost information, and for providing cost estimates and cost savings recommendations.

The positions described on the previous pages form the basic building blocks of the Incident Command System. Once you become familiar with these basic building blocks, you can use them to build an organization that is able to respond to the needs of any emergency (Figure 9.18). The key to using the Incident Command System is to balance the needs dictated by the emergency with the size of the organization. It is easy to overstaff an organization to such a level that it will actually hinder operations.

As you can see, the primary functions at each level of the ICS organization have distinctive titles (Figure 9.19). When used on the radio, each position title is prefaced with the name of the incident (e.g., "Deer Valley Operations," "Point IC," etc.).

Incident Facilities

As an incident grows, several facilities or sites must be established and identified. The Incident Command System includes six such facilities: ***Incident Command Post***, ***Staging Areas***, ***Incident Base***, ***Camps***, ***Helibase***, and ***Helispots***. There is only one Incident Command Post and one base per incident. There can be as many of the other facilities as necessary for the smooth operation of the incident.

The ICP is where the incident commander is located.

Incident Command Post

The ***Incident Command Post*** (ICP) is where the Incident Commander and key staff functions work. More importantly, it is

ORGANIZATION TERMINOLOGY

Primary Position	Title	Support Position	Radio Call
Incident Commander	Incident Commander	Deputy	"Point IC"
Command Staff	Officer	Assistant	"Point Safety"
Section	Chief	Deputy	"Point Operations"
Branch	Director	Deputy	"Branch 1"
Division/Group	Supervisor	**	"Division D" "Structure Protection Group"
Strike Team/Task Force	Leader	**	"Strike Team 9441" "Task Force 6"
Unit	Leader	Manager	"Point Supply Unit Leader"
Single Resource	use unit ID	**	"ORCO Engine 22"

** A support or deputy position is not authorized or not applicable.

Figure 9.19 Each primary position within ICS has specific titles. Most upper-level positions are authorized to have support or deputy positions.

where these functions can be found. Since every incident will have an Incident Commander, every incident will have an ICP. This may be in the form of a vehicle, trailer, tent, or rented building.

Having one ICP is critical when the incident involves more than one agency or jurisdiction. If the various agencies and/or jurisdictions are separated, it is hard to have an effective Unified Command.

There should be only one command post. If there are two, it is a "watch out."

The ICP can be located with other incident functions. It should be close to the incident communications center and the planning function. The site should be large enough to accommodate the anticipated needs for the incident and ICP functions. It should be located away from noise and incident-generated traffic. It should be sheltered from the weather, and be secure from public traffic.

There are general guidelines for the establishment of an ICP. The ICP should be:

- Positioned away from the general noise and confusion associated with the incident.
- Positioned outside the present and potential hazard zone.
- Have the ability to expand as the incident grows.
- Have the ability to provide security, and to control access to the ICP as necessary.
- Identified by a distinctive banner or sign.

Announce ICP activation and location via radio or other communication so all appropriate personnel are notified.

Staging is a function under Operations.

Staging Areas

Staging Areas are locations where personnel and equipment are kept (or staged) while awaiting tactical assignments. They provide a point to which dispatched resources can report, and from which specific assignments can be made. Staging Areas are managed by a ***Staging Area Manager***. Staging Areas are within the Operations Section, thus the resources assigned to staging are under the control of the Operations Section Chief. As an incident grows, there may be a need to establish more than one Staging Area.

Some of the specific benefits of Staging Areas are that they:

Different staging areas may be established for different types of resources.

- Provide a safe location for resources to wait for assignments.
- Provide for greater accountability by having available personnel and resources together in one location.
- Prevent resources from free-lancing or doing their own thing.
- Minimize the need for excessive communications of resources calling for assignment.
- Control and assist in check-in of personnel who arrive at the incident via privately owned vehicles or other private means.
- Allow the Operations Section Chief (or IC) to properly plan for resource use, and to provide for contingencies.

- Reduce traffic congestion.

- Provide security by limiting or restricting public access.

A Staging Area may be in the same general area or adjacent to other incident facilities; however, it should have its own separate location and name. Separate staging areas may be established for different kinds of resources. For example: fire equipment and personnel in one area, and police-related resources in another. You want the areas to be easily found, but located off the main traffic routes so that the public isn't asking why all that equipment is just sitting there; rather than "fighting the fire."

Keep the staging areas close to the action.

The following are some other considerations for establishing staging areas:

- They should be close to the location of tactical assignments (hopefully within 5 minutes of the action).

- They must be located in a safe area (for example, upwind from a hazardous materials spill, or out of the path of a fast moving wildland fire).

- They should have separate ingress/egress routes. You don't want to create a traffic jam by having incoming and outgoing resources on the same road.

- They should be large enough to accommodate anticipated levels of resources, and should be located in an area where vehicles and personnel will do minimal environmental damage.

- They should be clearly marked.

- Personnel assigned to staging should be flexible. It may be necessary to relocate staging areas to facilitate the needs of the incident or to protect personnel from approaching danger.

Incident Base

The Incident Base is the location at which the primary service and support activities are performed (Figure 9.20). In the old days, it was called the "fire camp." Not all incidents require the establishment of an Incident Base. If a base is established, this is where the Logistics Section is located. There should be only one base for an incident. Normally, the Incident Base is where all uncommitted (out-of-service or resting) equipment and personnel are

Northtree Fire International

Figure 9.20 Some incident base operations are the size of small cities. School or fairground facilities make good locations for an incident base. They have lots of room and parking.

located. If the incident is large, additional logistic support sites, or camps, can be established.

The management of the Base falls under the Logistics Section. If an Incident Base is established, a Base Manager will be designated. The Base Manager in a fully activated ICS organization will be in the Facilities Unit.

Camps

Camps are locations which are equipped and staffed to provide sleeping, food, water, and sanitary services to personnel assigned to an incident. Camps are established when the travel time is too long or the number of personnel needing support is too great to be serviced at the Incident Base. All services and functional activities performed at a Base may also be located or performed at Camps.

There can be several camps.

Camp Managers are responsible for managing the Camp, and for providing non-technical coordination of all organizational units operating within the camp. The Camp Manager reports to the Facilities Unit Leader in the Logistics Section.

Northtree Fire International

Figure 9.21 The helibase should be close, but not too close to the Incident Base.

Helibase

The Helibase is the main location within the general incident area for parking, fueling, maintenance, and loading of helicopters (Figure 9.21). It is often located at or near the Incident Base, at a nearby airport, or other convenient site. If located near the Incident Base, ensure that the noise and dust will not interfere with the Base Operations.

The Helibase Manager reports to the Air Support Group Leader, the Air Operations Branch Director, or the Operations Section Chief, depending on the organizational structure.

Helispot

Helispots are temporary locations from which helicopters can safely operate. They can be used to load and off-load personnel, equipment, and supplies. Helispots are managed by Helispot Managers who report to the Helibase Manager. If an incident has no established air operations organization, but does have one or more helispots, the Helispot Manager reports to the Operations Section Chief.

Incident Planning

The incident objectives are broad statements of intent.

Incident planning actually starts with the arrival of the first firefighter on a fire. One of the first items of business for this firefighter, or Incident Commander, is to establish Incident Objectives—broad statements of intent. Some examples are:

"Keep the fire south of Highway 32;

East of the community of Forest Ranch;

North of Honey Run Road; and

Cut it off to the west when an opportunity presents itself."

These objectives then set the direction for the firefight until the fire is controlled, or the situation dictates that the objectives be modified.

The plan can be given as oral instructions, but it is better if it is written. Written plans are essential so that all levels of a growing incident and organization have a clear understanding of the tactical actions.

The Incident Briefing form is a tool to organize your thoughts and record important information.

Incident Briefing

The Incident Command System provides a planning document for the development of the plan, objectives, and the management of the incident in its initial attack period. It is the ICS Incident Briefing

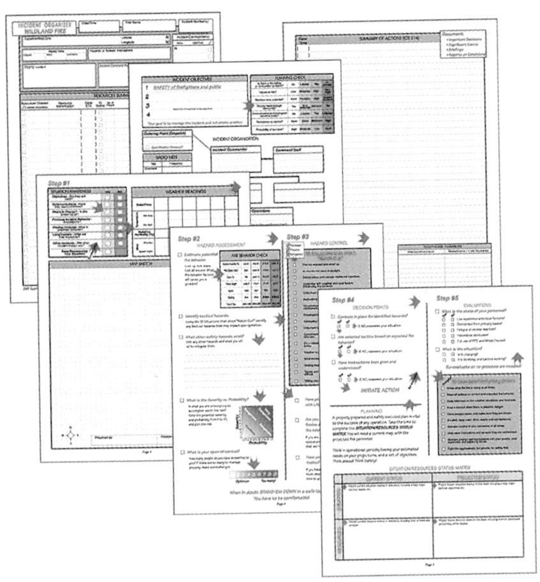

Figure 9.22 The Incident Briefing form has become an "organizer" for the incident commander. It not only is a place to record important information, new (commercial) versions include check lists and other reminders. See the Appendix for copies of one of the more widely used "Incident Organizer."

(Form 201). Over the years, the "201" has been expanded to include check lists and other key "reminders" (Figure 9.22).

The planning document must reflect the needs as identified by the Division/Group Supervisors and Operations Section Chief, and must be completed far enough in advance of each operational period to ensure that requested resources are available when needed. Large incidents should have a written Incident Action Plan. Several standardized forms have been developed to aid in the preparation of the plan.

Incident Action Plan

The Incident Action Plan is the document that provides all supervisory personnel with information about the incident and how it will be managed for a specific period of time, called the "operational period" (Figure 9.23). The written plan should provide a clear statement of the objectives and actions; the basis for measuring work effectiveness and cost effectiveness, and a way to measure progress and accountability. There are several important pieces to this plan: the Incident Objectives; the Incident Organization; the various assignment lists; weather and safety statements; and other supporting material, such as a Communications and Medical Plan, a Traffic Plan, etc. It is the responsibility of the Planning Section Chief and staff to prepare an Incident Action Plan for each operational period. The plan is then used as the script for each shift briefing.

> The incident action plan outlines who is to do what, when, and with whom. You've got to bring the "how" with you.

There may be times when it seems like the person who developed the plan was on another fire, or didn't know what was really going on with this one. It is vital that the planners have a good understanding of the situation. To meet this requirement, planners need to know:

- What has happened to date?
- What progress has been made?
- How good is the current plan?

- What is the incident growth potential?

- What is the present and future resource and organizational capability?

It is especially important that planners know ***in advance*** what the likelihood is of obtaining additional resources support from outside sources, for use in the next operational period. If there are readily available resources of the proper kind and type, then the planning process can encompass a wider variety of potential strategies than would be possible under very limited resource constraints.

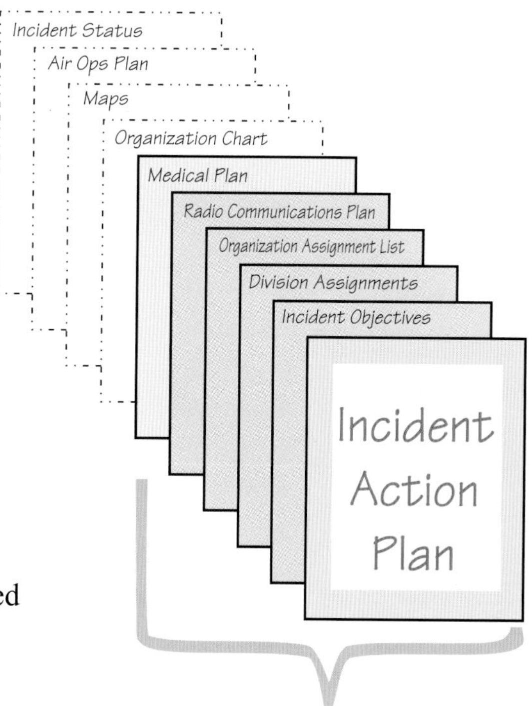

Figure 9.23 The Incident Action Plan outlines the operational perimeters for a given shift. On large fires, this can be quite a large document.

Resource Management

Resources are defined as any person, crew, team, piece of equipment, or grouping of people and/or equipment that is needed or used to abate an incident. This includes fire engines, bulldozers, hotshot crews, aircraft, division/group supervisors, etc. There are three general types of resources managed under ICS:

- ***Single Resources*** - A single resource is one person, one engine, or one of something else (Figure 9.24).

- ***Task Forces*** - A Task Force is a combination of single resources with common communications and a leader. A Task Force is assembled for a particular tactical need. This grouping can either be predetermined or formed at the incident.

Figure 9.24 This is a single resource positioned to protect these homes. It may have been ordered as a single resource, or part of a strike team that has been broken up.

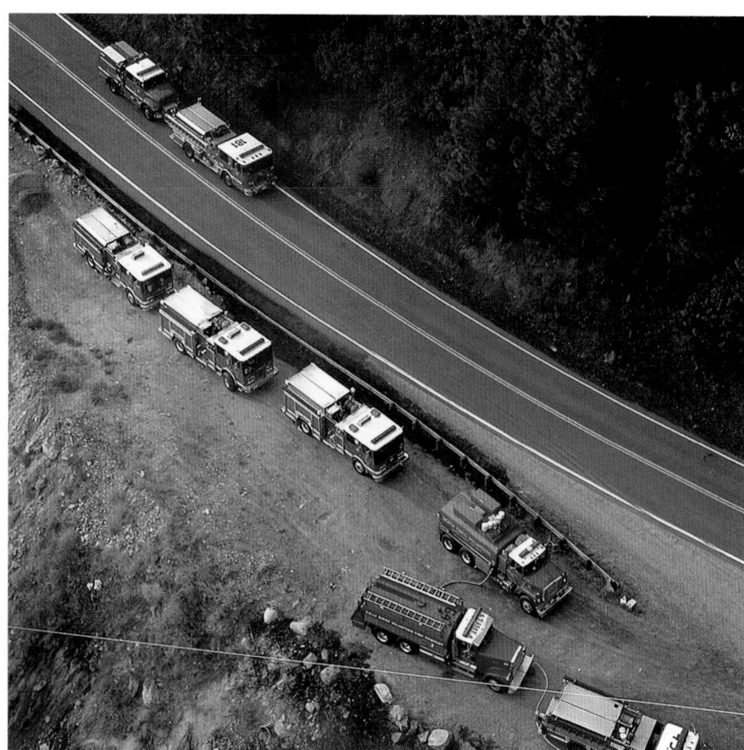

Figure 9.25 *A strike team is a grouping of resources, with very specific minimum requirements. The grouping can only be called a strike team if they meet these specifications. If they don't, it is a task force.*

• ***Strike Teams*** - A Strike Team is a combination of a ***designated number of the same kind and type of resources*** with common communications and a leader (Figure 9.25).

You would use a Task Force or Strike Team to increase the effectiveness of the resources (pre-designated teams work better together), reduce the span of control (the team leader answers for the team), and reduce communications traffic (each team has a common communications capability).

Originally, the development of Strike Teams was to facilitate ordering fire suppression resources, be it pumping capacity or line construction potential. It was also hoped that it would cut down on the information needed to be passed on through the dispatch or logistics channels. Unfortunately, it has turned out to be a simple way to order five engines with a leader, even if you don't need five engines. If the Strike Team concept is misused, it puts an unnecessary drain on the resource pool. More and more agencies are now stressing the ordering of single resources. If you need three engines, that is what you order; if you need a leader, that is what you order. Remember to tailor your resource orders to your actual needs; keep it simple!

Strike teams can be broken up if operational needs so dictate.

The advantages of using Task Forces or Strike Teams include:

- Enabling you to use resources more effectively. You can give the team leader an assignment and they can carry it out. You don't have to waste time instructing each resource on what you want or need.

- Providing an effective way of quickly ordering resources.

- Reducing radio traffic by channeling communications through a team leader, rather than to each single resource.

- Increasing the ability to expand the organization for large incident operations while maintaining good span of control.

- Providing close resource control and accountability.

Each resource or team will be in one of three status conditions, ***assigned***, ***available*** or ***out-of-service***:

- ***Assigned*** means that the resource is committed to its assigned task and is not available for another assignment. An engine pumping a hose lay is assigned.

- ***Available*** means a resource is assigned to an incident and is available for an assignment. An engine parked in staging is available.

- ***Out-of-Service*** means the resource is assigned to an incident but is neither assigned, nor available, for assignment to a task. The engine may be "mechanically challenged," there may be insufficient personnel to operate the equipment, or the crew may be off-shift and resting. In some cases, a piece of apparatus may be out-of-service because it cannot be used at night (helicopters and airtankers), or for fiscal reasons, such as no authorized overtime.

These are minimum requirements, not standards for staffing.

The maintenance of accurate status is important to the smooth functioning of any emergency operation. Under ICS, the only people who can change the status of a resource are those that control it. It could be an engine company captain saying the engine has a flat tire, or the Operations Section Chief moving an engine from the fireline so the crew can rest. The person making the status change must ensure that the change reaches the Resource Unit (the keeper of equipment status on major incidents) or whoever is keeping this information. It is the responsibility of each Division/Group Supervisor (or Branch Director or Operations Section Chief) to keep the Resources Unit informed of status changes for units assigned to field operations.

The Type 3 and 6 engines are the most widely used wildland firefighting engines.

How to Use ICS

The Incident Command System works only if all those assigned to the incident understand and use it. You should use ICS all the time, even on the routine small grass fire, a structure fire, or medical aid. This doesn't mean that you have to set up a 40-person command structure for a medical aid response. It does mean that you have to have an Incident Commander (someone is in charge). All other position needs are dictated by the complexity of the incident. Don't ever let the "system" run the incident. All too often, Incident Commanders will fill boxes "because they are there."

The only person that can change the status of a resource is the person in charge of it.

Use of ICS on a Small Incident

If you are out-of-service—Operations, Plans, and Logistics all need to know.

You have been dispatched to a small roadside fire. You are the first unit there. As you arrive, you see a small spot of grass burning very slowly. You feel that you can handle the situation and tell your dispatcher:

> *"Engine 45 at scene; can handle; small spot of grass; committed 20 minutes."*

You might not think you have used ICS, but you have. You have used just as much of ICS as you needed. You took charge (you are the IC) and reported the situation status and equipment needs to your dispatch office.

If the situation was a little different, let's say you needed traffic control and an investigator, you would add that information to your report on conditions. You have now functioned as the IC, Planning Section Chief, and Logistics Section Chief.

You should use ICS all the time. Don't wait until the "big one."

Use of ICS on an Expanding Situation

The small grass fire has turned into a larger brush and timber fire. You are the first unit "at scene," and realize that you will need assistance. Your report on conditions may sound like this:

> *"Engine 45 at scene; 20 acres of brush and timber; moderate rate of spread; moving up Hawkins Hill toward a couple of homes; continue first alarm assignment."*

The first person on the scene is the IC. They are responsible for Command, Operations, Plans, Logistics, and Finance/Administration.

The dispatcher should acknowledge your message and assign you an incident designator. In this case "Hawkins Hill" will be the name of the incident. From that point forward, the Incident Commander for this incident will be referred to as "Hawkins Hill IC." As long as you are IC, you will use "Hawkins Hill IC" as your radio call sign, not Engine 45. See Figure 9.26 on the next page.

As additional units arrive at the fire, they will call "Hawkins Hill IC" on the radio and ask for instructions. If a chief officer arrives and decides to take command of the incident, he or she becomes "Hawkins Hill IC." If an air attack is utilized, it will be referred to as "Hawkins Hill Air Attack."

If the IC needs to divide the fire and assign two division/group supervisors, they will be designated Divisions A and B. If a Staging Area is to be established, an appropriate site is selected, a manager is assigned and everyone is told of its location. The manager will be designated "Hawkins Hill Staging." See how easy it is to use just as much of ICS as you need?

The IC has the capability to organize the fire using a "special purpose Group" for structural protection (Figure 9.27). In this situation, the Structure Protection Group is positioned along the northern road, not involved in line production, but rather to protect the structures if the forces assigned to Divisions A and B cannot stop the fire. The structure protection responsibilities within Divisions A and B are the responsibility of each of the divisions.

Fill only the positions you NEED. Don't let the system run you.

Think divisions early on in the incident.

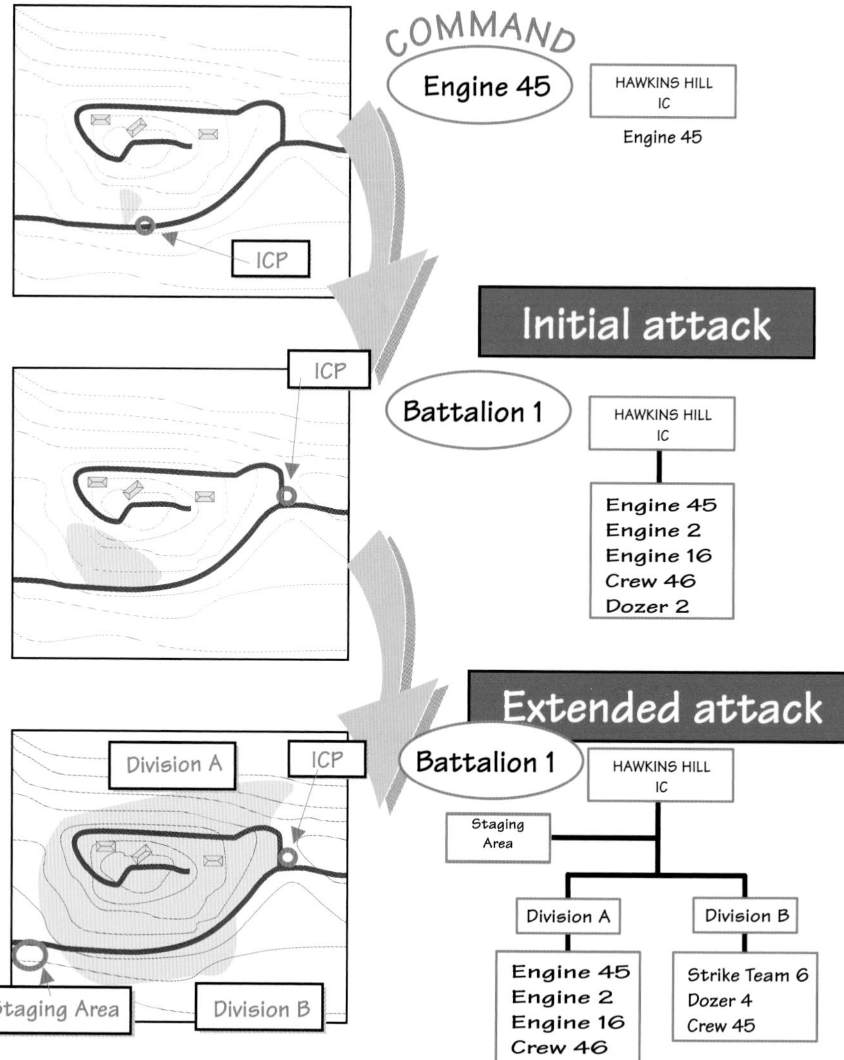

Figure 9.26 In this graphic there are three stages of the incident: When Engine 45 goes at scene and the engine captain assumes the role of "Hawkins IC." When Battalion 1 arrives, he/she becomes the IC. As the fire moves up the hill and a second alarm is requested, the fire moves into an extended attack operation, with an IC and two divisions.

Major Incident Management

Every so often, a fire or series of fires grows large enough to be called a major incident. You might think that there is an acreage threshold that turns a fire into a major event. However, acreage is only one factor that needs to be considered. What turns a fire into a major event is how the fire itself impacts people, property, or resources.

ORGANIZATIONAL STRUCTURE WITH TWO DIVISIONS, ONE GROUP AND A STAGING AREA

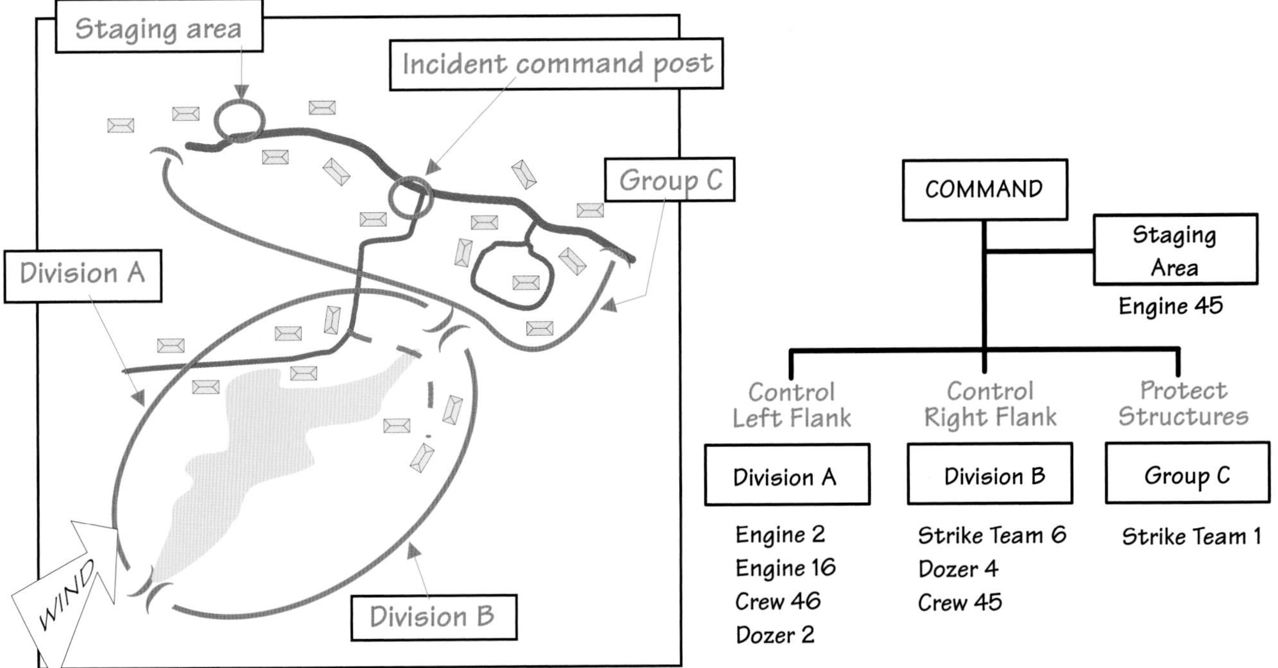

Figure 9.27 This fire is divided into two divisions (A and B), and a structure protection group (C). The ICP is located at the road intersection, and staging is located just down the road. Division A is responsible for controlling the left flank of the fire, and Division B has the right flank and its associated structures. Group C is responsible for the protection of the structures in front of the fire.

When a situation turns bad, several other things occur. More than one agency or jurisdiction becomes involved. The number of resources assigned usually increases, or resources become scarce and priorities have to be set. The Incident Command System also addresses these concerns with ***Unified Command***, ***Complexes***, ***Area Command***, and ***MultiAgency Coordination***.

Unified Command

Unified Command allows all agencies which have jurisdictional or functional responsibility for part of the incident to jointly develop a common set of incident objectives and strategies. This can be accomplished without the initial responding agency losing or giving up its authority, responsibility, or accountability. Unified Command is an important feature of ICS. It allows agencies having a legitimate responsibility at an incident to be part of the command structure. Specifically, Unified Command provides the following advantages:

Know and use the system to get what you need. You must know the right person to ask.

- The incident is managed under a single set of objectives, coordinated in the Incident Action Plan.
- There is one Operations Section Chief who is responsible for implementing the plan.
- There is one Incident Command Post.
- All agencies with responsibility for the incident have an understanding of one another's priorities and restrictions.
- No agency's authority or legal requirements will be compromised or neglected.
- The combined efforts of all agencies are optimized as they perform their respective assignments under a single plan.
- Duplication of effort is reduced or eliminated, thus reducing cost and chances for frustration and conflict.

Unified Command is one of the most important features of ICS.

The great thing about the ICS Unified Command setup is that it will accommodate almost any jurisdictional arrangement. The simplest form is when there are several fire agencies fighting a fire that crosses jurisdictional boundaries (Figure 9.28). Things get a little more complicated when the incident involves several different agencies (police, fire, medical, etc.) in one jurisdiction. The most complicated form is when the incident involves several different agencies from several different jurisdictions.

The command structure should accommodate all jurisdictions that have responsibility for part of the incident.

Under a Unified Command structure, the various jurisdictions/ agencies are blended together into an integrated team. The resulting organization may be a mix of personnel from several agencies, each performing functions as appropriate and working toward a common set of objectives.

The proper mix of participants will depend on:

Jake says, "All too often the 'chiefs' simply don't work together...what a shame!"

- The ***location of the incident***, which often determines the jurisdictions that will be involved.
- The ***kind of incident***, which dictates the agencies that will be involved.

One of the biggest hurdles in any multi-jurisdictional incident is determining who is going to pay for what. Smaller jurisdictions don't usually budget for major

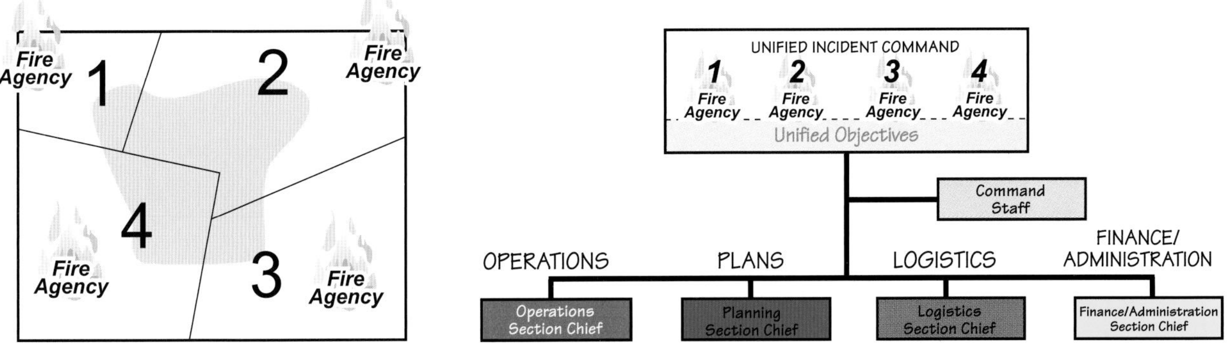

Figure 9.28 This Unified Command structure is one of the simplest in that it involves several fire service entities fighting a fire that is burning in several jurisdictions. Each of the fire departments may be from differing jurisdictions, but they are from one kind of operation (the fire service), and there is one common "enemy,"...the fire.

disasters. Larger state or federal agencies usually have emergency funds they can use, but in a lot of cases, the largest cost item (structure protection) really is not their responsibility. So, the early development of a cost-share plan is very important.

Fires don't follow maps, so they don't really care when they cross jurisdictional boundaries.

Another issue that will take some coordination is the establishment of a plan regarding who will order what resources through what channels. If everyone involved orders separately, it may facilitate payment (you ordered it, you pay for it), but it can lead to some costly duplication. Also, it is very difficult to order scarce resources, like airtankers and helicopters, if everyone is doing their own thing.

Ordering can be streamlined in two ways: designating a single ordering point (all orders for the incident go to one dispatch center for processing), or giving the Logistics Section direction on which kinds of resources will be ordered through which dispatch office. The issue of who pays is still there, but the main thing is that the needed resources are ordered and the orders get processed.

One of the biggest issue on a major fire is "who pays."

Some of the other issues the Unified Command will have to deal with are:

- Selection of an Operations Section Chief that they all feel comfortable with. This person should be highly qualified and experienced. There must be full agreement with the choice, and he/she must be given full authority to implement the plan.

- Designation of one spokesperson for Unified Command.
- Designation of the lead Information Officer.

Complexes

The term "complex" is used to describe a situation when two or more individual incidents located in the same general area are managed by a single incident management team. When a complex is established, the individual incidents (fires) become branches within the Operations Section of the ICS organization. If one of the fires within a complex grows too large, it may be wise to establish a separate ICS organization to manage it.

Establish a complex when several fires can be effectively managed by one command organization.

Complexes are most often established after a series of fires started by lightning. But that doesn't preclude the use of a complex in response to an earthquake, tornado, flood, or any event where several separate but common events are within proximity of one an other. A complex can also be established if a new fire starts near an existing fire, and the new fire is managed by the existing ICS organization.

Expanding the ICS Organization

There may be times when an incident grows so large that you may have to divide it into two or more separate incidents, each managed by separate organizations (Figure 9.29). This could happen if a flood covers several counties or states. If possible, retain a Unified Command structure. If that isn't possible, utilize an area command or establish a MultiAgency Coordination system. Both are discussed later in this chapter.

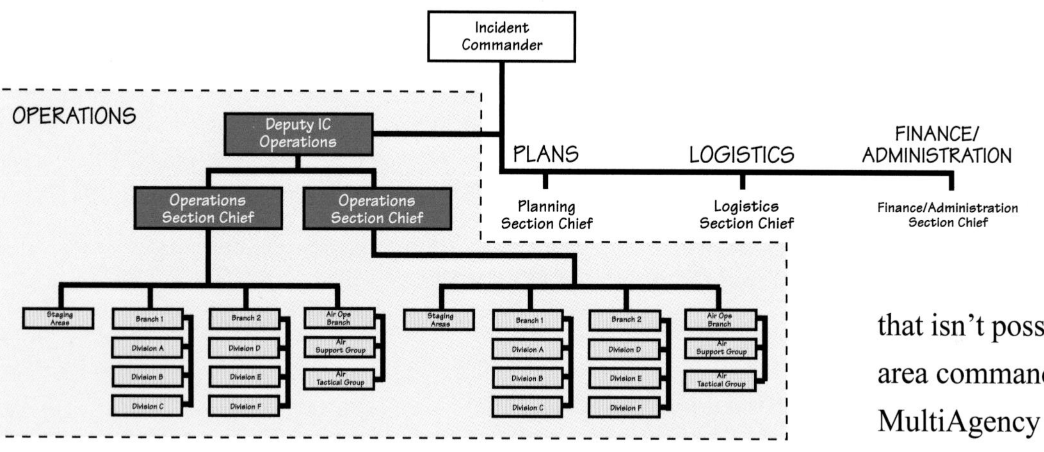

Figure 9.29 If the incident is of such a size that the Operations Section needs to be divided, establish the second Operations Section Chief and a Deputy Incident Commander-Operations position.

ICS is also designed to allow for multiple sections (for example, two Operations or Logistics Sections), if the incident is so large that the span of control is excessive. It is recommended that if two or more sections are overtaxed, the incident should be divided into two separate incidents.

Use the ICS to tailor the organization to match the complexity of the incident.

Another option is to conduct planning at the branch level, developing an Incident Action Plan for each branch. If it becomes necessary to do this, ensure that the overall objectives for the total incident are followed.

Area Command

An Area Command is established to oversee the management of several incidents in the same area, usually of the same type (each being managed by a separate ICS organization), or to manage a very large incident when several management teams are in place to direct it. If several jurisdictions are involved, a Unified Area Command organization should be established (Figure 9.30).

An Area Command is a new layer added to manage large, multiple, and complex events.

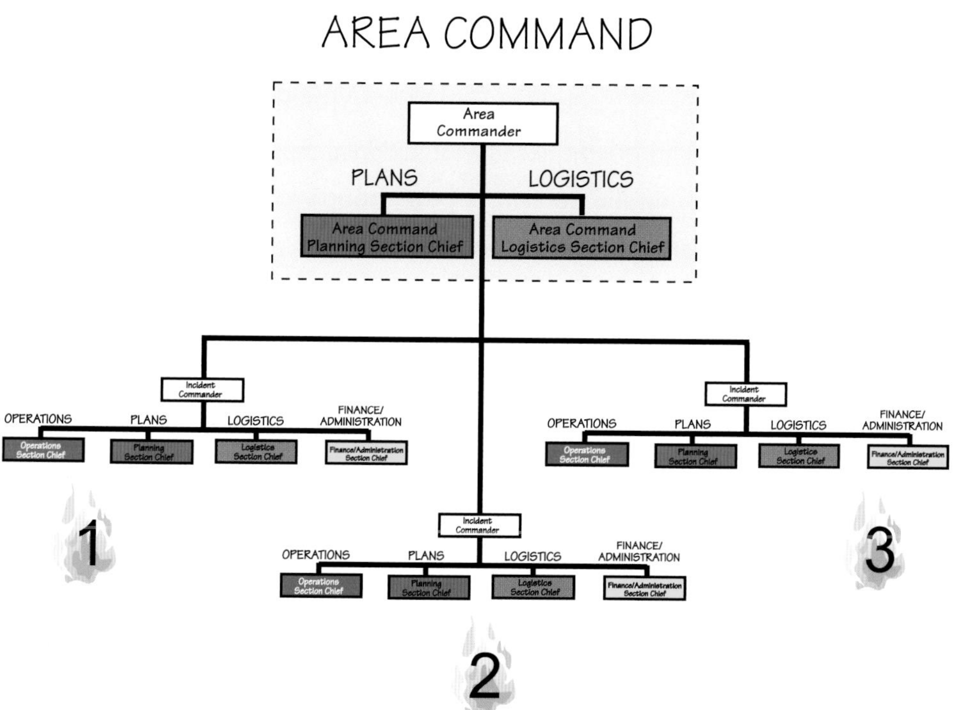

Figure 9.30 Area Command is established to manage a series of large fires in a geographic area. They are a layer between the Type 1 Incident Management Teams and the regional dispatch center.

The responsibilities of an Area Command organization are:

- Set overall agency incident-related priorities.
- Allocate critical resources based on these priorities.
- Ensure that the various incidents are properly managed.
- Ensure that the incidents' objectives are met and do not conflict with each other or with agency policy.

MultiAgency Coordination

The concept of MultiAgency Coordination (MAC) is quite simple. It is when the leadership of different agencies or jurisdictions meet to discuss and set common goals with regards to the management of emergency operations or any other topic. The term "coordination" is key to understanding the authority of a MAC. It is not a "command," thus does not have the authority attributed to that term. The power of a MAC is vested in the authority of the individuals who participate in the coordination meeting (Figure 9.31).

The key to the success of a MAC is the willingness of the participants to cooperate and compromise in the development of a common goal, and that the individual members have the authority to follow through on the decision of the group.

MultiAgency Coordination is not new. In its own way, the United Nations is a MAC. One example of a MAC specific to wildland firefighting is the National Wildfire Coordinating Group, which includes representatives from various federal wildland fire agencies and state forestry organizations, and which provides overall nationwide coordination. There are MACs at the state and local levels that coordinate disaster planning and responses.

A MAC functions outside the jurisdiction's line of authority or command and control system. The MAC does not communicate directly with Incident Commanders. However, it may communicate to Incident Commanders through other agency representatives.

Some of the functions of a MAC are:

- Development of priorities to be used in filling orders. This is especially important when resources are limited or scarce. This

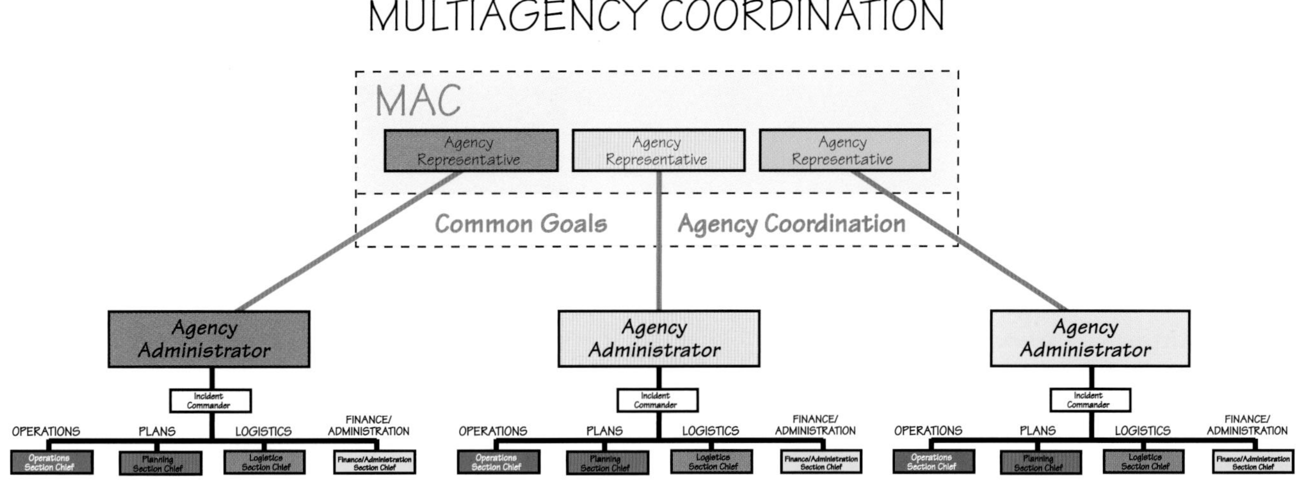

Figure 9.31 MultiAgency Coordination (MAC) is when representatives from different agencies and jurisdictions meet to develop a common set of goals and objectives. This example is for a series of large fires in three jurisdictions, but MACs meet regularly to improve interagency operations and coordination of other issues. For example, MAC groups manage the National Incident Management Teams.

includes the development of overall situation assessment. Normally, priorities are determined by the threat to life, property, and natural resources.

- Resource allocation. There may be times that resources are moved from one incident to another.
- Coordination of media information.
- Providing a common point for agency leadership contact. MAC can also provide assistance and information services to interested elected and appointed officials.
- Resolution of conflicts between different agencies.

The power and effectiveness of MAC lies in the authority of the individual members.

As noted, a MAC will not be able to function if the players don't have the authority to follow through with the collective decision, or if there is a lack of trust between the members. Trust is important, and trust is difficult to develop if the individuals are strangers. MACs should be established and functioning well before a disaster.

The difference between a MAC and Unified Command or Area Command is that MAC is a coordination function, and the two Commands have direct line authority.

Jake says, "A MAC is only as good at the people that are involved, and it is TRUST that makes it work."

National Incident Management System

On September 11, 2001, the world changed. There had been efforts made to have ICS adopted as the national incident management system, but the events of that day finally proved the need for such a system. On February 28, 2003, President Bush issued Homeland Security Presidential Directive-5. HSPD-5 directed the Secretary of Homeland Security to develop and administer a ***National Incident Management System*** (NIMS). NIMS, a direct spin-off of ICS, provides a consistent nationwide template to enable all government, private-sector, and nongovernmental organizations to work together during domestic incidents.

MACs operates above ICS. It coordinates (not commands) allocations of scarce firefighting resources and prioritizes incidents.

NIMS establishes standardized incident management processes, protocols, and procedures that all responders—Federal, state, tribal, and local—will use to coordinate and conduct response actions. With responders using the same standardized procedures, they will all share a common focus, and will be able to place full emphasis on incident management when a homeland security incident occurs—whether it be terrorism or a natural disaster. In addition, national preparedness and readiness in responding to and recovering from an incident is enhanced since all of the Nation's emergency teams and authorities are using a common language and set of procedures.

A MAC will be successful if the members are willing to participate, cooperate and compromise, and they are able to carry out the decisions.

NIMS incorporates incident management best practices developed and proven by thousands of responders and authorities across America. These practices, coupled with consistency and national standardization, will now be carried forward throughout all incident management processes: exercises, qualification and certification, communications inter-operability, doctrinal changes, training, publications, public affairs, equipping, evaluating, and incident management. All of these measures unify the response community as never before.

The key features of NIMS are:

- ***Incident Command System (ICS)*** - *NIMS establishes ICS as a standard incident management organization with five functional areas —* ***command****,* ***operations****,* ***planning****,* ***logistics****, and* ***finance/administration*** *— for management of all major incidents. To ensure further coordination, and during incidents involving multiple jurisdictions or agencies, the principle of unified command has been universally incorporated into NIMS. This unified command not only coordinates the efforts of many jurisdictions, but provides for and assures joint decisions on objectives, strategies, plans, priorities, and public communications.*

Part of the new NIMS is ICS.

- ***Communications and Information Management*** - *Standardized communications during an incident are essential, and NIMS prescribes inter-operable communications systems for both incident and information management. Responders and managers across all agencies and jurisdictions must have a common operating picture for a more efficient and effective incident response.*

- ***Preparedness*** - *Preparedness incorporates a range of measures, actions, and processes accomplished before an incident happens. NIMS preparedness measures include planning, training, exercises, qualification and certification, equipment acquisition and certification, and publication management. All of these serve to ensure that pre-incident actions are standardized and consistent with mutually-agreed upon doctrine. NIMS further places emphasis on mitigation activities to enhance preparedness. Mitigation includes public education and outreach, structural modifications to lessen the loss of life or destruction of property, code enforcement in support of zoning rules, land management and building codes, and flood insurance and property buy-out for frequently flooded areas.*

Jake says, "If you understand ICS, you will have no problem working with NIMS."

NIMS is much more than ICS.

- ***Joint Information System (JIS)*** - *NIMS organizational measures enhance the public communication effort. The Joint Information System provides the public with timely and accurate incident information and unified public messages. This system employs Joint Information Centers (JIC) and brings incident communicators together during an incident to develop, coordinate, and deliver a unified message. This will* *ensure that Federal, state, and local levels of government are releasing the same information during an incident.*

- ***NIMS Integration Center (NIC)*** - *To ensure that NIMS remains an accurate and effective management tool, the NIMS NIC will be established by the Secretary of Homeland Security to assess proposed changes to NIMS, capture, and evaluate lessons learned, and employ best practices. The NIC will provide strategic direction and oversight of the NIMS, supporting both routine maintenance and continuous refinement of the system and its components over the long term. The NIC will develop and facilitate national standards for NIMS education and training, first responder communications and equipment, typing of resources, qualification and credentialing of incident management and responder personnel, and standardization of equipment maintenance and resources. The NIC will continue to use the collaborative process of Federal, state, tribal, local, multi-discipline and private authorities to assess prospective changes and assure continuity and accuracy.*

Jake says, "ICS is now the nationally accepted incident management system."

NIMS is a national approach to managing major disasters in the United States.

These last five bullet points are right out of the Department of Homeland Security guidelines. You can see that the Incident Command System is National in scope and will be the organizational scheme used on any major event in the United States. Its "roots and heritage" are within the wildland firefighting community.

10

Firefighting Realities

In the material leading up to this chapter, we have discussed weather, fuels, equipment, methods, strategy, tactics, and safety. In this chapter, we will attempt to bring it all together: concepts that will lead to the suppression of a wildland fire, and some of the realities of life. We will discuss how small fires burn, verses large fires; how the philosophy of initial attack operations differs from that of extended attack; and how to plan and execute the selected firefighting strategy.

This chapter will deal with some of the "realities" of wildland firefighting.

This is a dangerous profession! And despite all of the rules and efforts of agency leadership, firefighters will be hurt. This is not to say we should accept this. It means that we must fight fire, always keeping the ever-present dangers in mind, and take a course of action that minimizes the risks. We want to reduce risks as much as possible, but there is a point at which reducing risk also greatly reduces the effectiveness of the firefight.

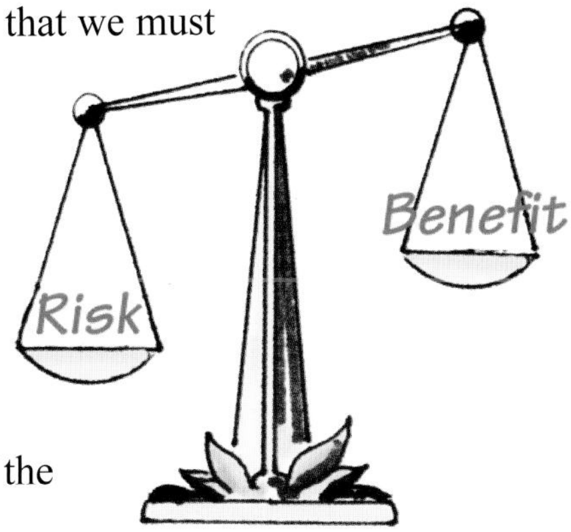

If risk and benefit are out of balance, reduce the risk.

Figure 10.1 You have to balance risk and values protected. When risk outweighs benefit, change your plan.

Balancing risk and the values protected must be considered in your planning (Figure 10.1). If you are operating under a plan that is going to cost more than the projected value of what is to be saved, one may want to rethink it. If the plan creates a higher risk to firefighter and public safety to protect property (that will re-grow or can be replaced), you must rethink it. There has to be some value to your actions! Impact on your ego must not be calculated into your equation…that is just the cost of doing business (Figure 10.2).

Figure 10.2 What is the risk-to-benefit ratio here? If that fire turns and runs on this home and that firefighter, we may read about it in the next edition of this handbook. The risk of injury is too high. The firefighter should move back to the corner of the house and shield himself from the heat that may be coming his way.

It is the planning, not the plan, that is important.

You cannot always put a dollar value on risk or benefit, but that is the scale most people use. The key is that you think through what you plan to do versus the potential benefit.

Dr. Gordon Graham, an internationally recognized expert in emergency safety issues and, specifically, the ***Six-Minutes for Safety Program*** that has been adopted nationally by the wildland agencies. He is credited with two very profound and important statements:

"If it is predictable, it is preventable."

"If it is predictable, it is preventable."

and

"We haven't found any new ways to kill firefighters in the last 50 years."

Think about these two statements and what they are saying. The first statement says, if you can predict certain events, then why do firefighters put themselves in jeopardy? It is not uncommon for firefighters to point to a section of line and say, "If it is going to get away, it will do it right there." Or, "If the fire gets below us, we are going to be in real trouble." Or, "If the helicopter doesn't come and get us, we are in real trouble." Are these three statements unrealistic, or are we just not hearing ourselves and saying, "It will not happen to me?"

The second statement is even worse...it says we haven't learned anything from all of the firefighters who have died over the last 50 years. We just keep on doing dumb things that get people killed! Why? Maybe the following will shed light on it.

Low frequency - high risk actions are very dangerous.

Risk management is also explained by Dr. Graham with a simple chart (Figure 10.3). This graphic demonstrates that you can develop a false sense of security through repetition. Even though it may involve risk, you tend to "let your guard down" when performing familiar or routine tasks. You have seen it before and know how to react to it. But, if you don't do something very often, like downhill line construction, or fighting a wildland fire, you don't have any experience to fall back on to help you know when you are in trouble. The "low frequency - high risk" tactics are the ones that will get you in real trouble. You don't see something coming and don't have time to get out of harm's way...South Canyon, Thirty-mile, Cramer and Cedar...you name it.

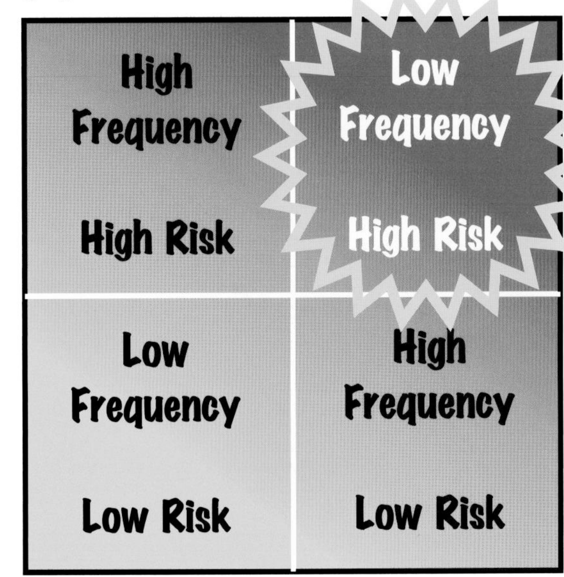

Figure 10.3 The "box" that will get you is the "low frequency - high risk" one. So, if you are doing something new, be especially careful.

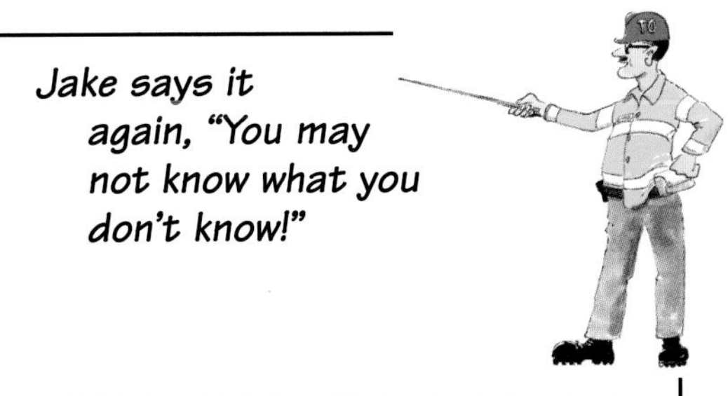

Jake says it again, "You may not know what you don't know!"

Situational Awareness

Situational awareness is a phrase that has been used in the aviation industry and the military for years. It is relatively new in the wildland firefighter's glossary of terms. Situational awareness is being aware of what is happening all around you…it is continuous "size-up" and "Look Up, Look Down, Look Around." This is easier said than done. Here I am working on a small segment of the incident and you want me to know what is going on?

Jake says, "There are few 'routine fires' any more."

"Sometimes the best strategy is to 'get out of Dodge'."

"Situational awareness can only come with experience."

"If fuels and weather conditions indicate you could have a bad day, automatically assume any fire you are on will be bad! Be justifiably paranoid!"

The following list of items spells out in broad terms what situational awareness is all about:

- Learned ***knowledge*** (aka training) from a handbook like this.
- ***Experience*** gained on the fire ground.
- ***Understanding*** the rules of engagement.
- Gathering new information (***intelligence***) about what is happening around you now.

All of this information needs to be combined into a "moving' picture" of what is happening in real-time, in an attempt to help you make timely and proper decisions. This assessment is not a onetime event, but a continuous process. Just as a pilot in an airplane must constantly assess the instruments and look out of the window for indicators, you as a firefighter must constantly assess your situation and measure it against what you are attempting to do.

If you don't know what is happening, who does? There are two views of a fire that are important. There is the view of the individual firefighters, concentrating on their segment of fireline, and the actions needed to fulfill their responsibility. The individual firefighter sees and hears things that are important to understanding what is going on. A firefighter is usually the first person to see a tree torch or to spot a small firewhirl. These may be the first indications that "things are changing." The other view is the broader picture of events. What is happening on the whole fire, or in the district, or region?

The point is that it is up to all of the individual firefighters to report things or events, which may be small at the time, but may have greater importance in the broader picture. If you as a firefighter see something that others may not have, report it to your supervisor.

- ***It's the Weather, Stupid!*** - The reality is that most accidents are the result of a change in weather that changed fire behavior. So, if you can track the weather to detect change, you are one step closer to predicting when fire behavior will change. Keep informed on fire weather forecasts.

- ***Think in four-hour blocks*** - One method to help organize your planning and decision making process is to think in manageable blocks of time. You can think in terms of operational periods or 12 hours, but it would be better to think in much shorter time periods. This will allow you to react to rapidly changing conditions. In fact, if things are changing fast, you may have to rethink actions minute by minute.

- ***Perception versus Reality*** - There is one major flaw in all humans. We sometimes don't see what really is going on. We are 'wired" in such a way that what we see with our eyes and hear with our ears is filtered through our brains before it is entered into memory. A lot of the time, ***we see what we want to see, not what is really happening***. This can be dangerous when your safety—and the safety of others—depends on what is accuately observed.

Jake says, "If things are really changing fast, maybe it would be best to back off a little and broaden your view of what is really happening."

"Anticipate the unexpected!"

"Planning is essential; plans are immediately obsolete."

"NEVER assume it can't happen to you! You could be dead wrong."

If you have ever been involved with an investigation, you will be amazed (and maybe alarmed) how differently witnesses to a single event report what "they saw." Sometimes you can have twelve versions of one event from ten witnesses! Be aware that this happens, and make every attempt to get the real story about what is going on.

- ***You may not know what you don't know!*** - Some would say this is the root cause of most accidents. The key is that you realize you might not have all the facts. Just consider that you may not have all the facts necessary to take an action that might prove dangerous. The best example may be attempting a downhill line construction.

- ***Use a checklist and an Organizer*** - When you are on the fireline is not the time to pull out your copy of the ***Firefighter's Handbook*** and read it. But, that doesn't mean you can't refer to your copy of the ***Fireline Handbook, Incident Response Pocket Guide*** or use an organizer to

Checklists are only as good as the one using it!

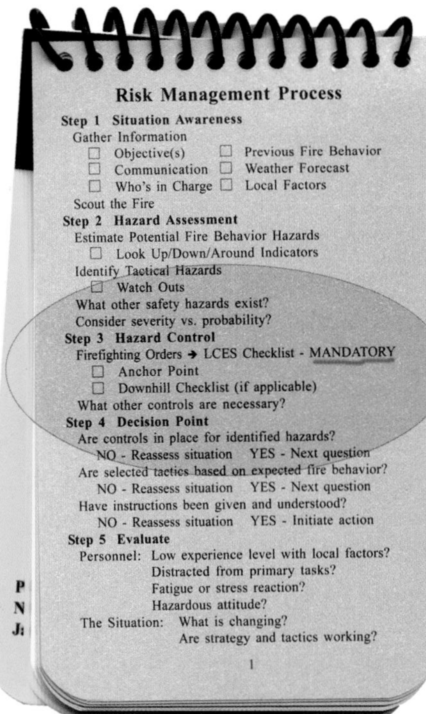

Risk Management Process

Step 1 Situation Awareness
Gather Information
☐ Objective(s) ☐ Previous Fire Behavior
☐ Communication ☐ Weather Forecast
☐ Who's in Charge ☐ Local Factors
Scout the Fire

Step 2 Hazard Assessment
Estimate Potential Fire Behavior Hazards
☐ Look Up/Down/Around Indicators
Identify Tactical Hazards
☐ Watch Outs
What other safety hazards exist?
Consider severity vs. probability?

Step 3 Hazard Control
Firefighting Orders → LCES Checklist - MANDATORY
☐ Anchor Point
☐ Downhill Checklist (if applicable)
What other controls are necessary?

Step 4 Decision Point
Are controls in place for identified hazards?
NO - Reassess situation YES - Next question
Are selected tactics based on expected fire behavior?
NO - Reassess situation YES - Next question
Have instructions been given and understood?
NO - Reassess situation YES - Initiate action

Step 5 Evaluate
Personnel: Low experience level with local factors?
Distracted from primary tasks?
Fatigue or stress reaction?
Hazardous attitude?
The Situation: What is changing?
Are strategy and tactics working?

1

Figure 10.4 If the firefighters on any one of the recent disasters had referred to this "process" and did what it required, would the disaster have occurred?

record information necessary to manage an incident, and record actions and events "for the record."

It is one thing to set and recite the 10 Standard Firefighting Orders and LACES, and it is another to say "we are violating them" and do something about it. Figure 10.4 is Page 1 of the ***NWCG Incident Response Pocket Guide***. If the firefighters on South Canyon, Thirtymile and Cramer had "whipped this out and analyzed their situation," would we be using them as an example? We would hope not.

Figure 10.5 is an example of an incident organizer used to capture thoughts, actions and important information related to an incident. A copy of this form is in the appendix. Pilots have used takeoff and landing checklists for decades...they work.

Jake says, "You know you're becoming 'experienced' when your guts are telling you much more than a checklist."

Trigger Points

It is important that you enter the term "trigger points" into you vocabulary. A trigger point is when you decide that once "something" is reached…the

Jake says, "Listen to your gut! Most of the time it only whispers."

Figure 10.5 This is an example of an incident organizer used on wildland fires.

humidity gets to a certain point; for example, you will do something…like move to a safety zone. You can use trigger points to think through options when you are not under pressure, so that you can react semiautomatically to an event when you may be a little more stressed.

Jake says, "Just because it was safe at 0900 hours, doesn't mean anything at 1400 hours, or when you move onto a south or southwest facing slope."

Some trigger points are instinctive. You are on the train tracks and see a train coming…you move off of the tracks. You hear a loud bang…and you duck. These are reactions to an external stimulus.

Most of the time trigger points are established for safety reasons…sometimes for operational reasons. Some of the more common trigger points for safety reasons are:

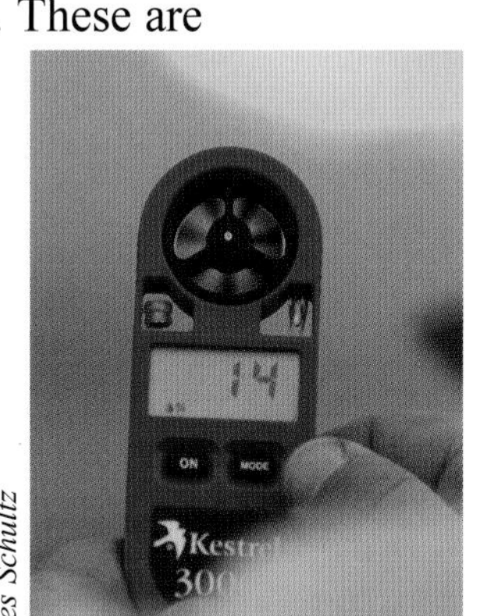

Figure 10.6 *If your trigger was 15 percent humidity, it is time to do what you had planned!*

Get familiar with the weather.

- When the relative humidity reaches a certain ***level*, *you do something*** (Figure 10.6).
- When the fire reaches a certain ***location***, like a canyon or ridge top, ***you do something.***
- When the wind reaches a certain ***speed*** or ***switches direction, you do something.***
- When fuel moisture, or the burning index reaches a certain ***level*, *you do something.***
- When you reach a certain ***time*** of day, ***you do something*** (Figure 10.7).

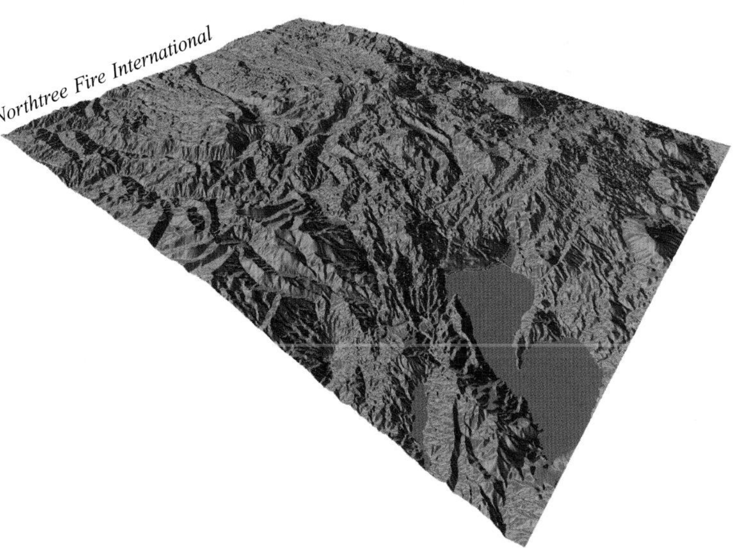

Figure 10.7 This is a computer-generated map that highlights the slopes that are south-southwest facing in red. If you have a trigger point that says you will not want to be on these slopes in the afternoon...and you are on one...move to safety.

All of these are trigger points. The key to the success in using trigger points is that the established parameter is realistic ***and*** you do what you have planned when the trigger point has been reached. Be sure that you have built enough margin…to do it without having to rush.

Different Rule of Engagement

You have learned that fire burns uphill faster than downhill, that wind is the main weather factor to be concerned with, and that airtankers don't put out fires! Now you will learn that you can't fight fire the same way, everywhere, every time.

Sometimes you can apply all of the firefighting resources available to you, and other times, you must tiptoe through the woods being very careful not to do damage to the environment. You can't always be a "bull in the china closet"; there are times you must be like a butterfly. This may sound a little trite, but it is meant to bring your attention to the fact that you can't fight fire the same way everywhere.

When you own the land you can manage it a little differently.

When you are sent in to fight fire, part of the initial size-up has to be determining on whose land are you fighting fire, and what are the rules of engagement? If you work for a government land management agency like the USDA Forest Service, USDI Bureau of Land Management, or the USDI National Park Service, your agency has some very strict rules on how you can engage and fight fire on lands in their jurisdiction. This is solely based on the rules on which the agency was established…each was formed for real different reasons.

When you are working for someone, you follow their rules.

If you are a member of a fire department or forestry agency that doesn't necessarily own the land it protects, but combats fire for the general public, you have a totally different set of rules. You can't arbitrarily manage the fire and let it burn an area because you think it should be burned. You have only one goal, and that is to protect life and property. That means putting out the fire as quickly, efficiently and safely as you can. All available resources are applied to the firefight, and you simply overwhelm the fire with the strength of your force.

The problem is brought to a head when you mix firefighters from multiple agencies in a firefight. No one agency can fight fire without asking for help. When help arrives, it may be from an agency with a whole different set of rules of engagement. If personnel from an agency that is heavy on initial attack and with a "pedal to the metal" approach to firefighting are asked to fight fire in a National Park, some real different kinds of sparks can develop.

National Parks and wilderness areas, for example, were designated for the preservation of the land and animals, and fire is a natural part of that land. In most cases, fire is what keeps the land healthy. You don't simply launch your bulldozers and cut line; you have to minimize the damage created by your presence and stop the fire when it wants to stop. A lot of time, you will be asked to fight fire very sporadically, allowing it to do its good work, only squirting water when it "looks real bad and you have to do something." The realities are such that you need to know, within the reasoning behind the establishment of the parks and wilderness areas, that fire can be beneficial, not always the enemy you have been taught it is.

When fire is properly used, it can be very beneficial.

Not all Fires are Bad

Another conflict that you may be confronted with is when you are asked to be involved in a prescribed fire…when you get to go out and light a fire on purpose. Setting fire for a specific goal is an awesome responsibility. Fire is not very smart, and it can't tell when it is supposed to be good, or when it has become very bad! Once out of the bottle, it is simply ***"fire,"*** and it does its own thing until you do your thing to put it out.

Jake says, "Ask the Chief about Hamlin Canyon next time you see him! Tee hee!"

If you are asked to burn an area for fuel reduction or other reasons, you have to do it when it will burn. But, you don't want to do it when it will burn too well, or you may get the dreaded "bonus acres." From experience, it is hard to feel like a professional firefighter when explaining how you "did set the fire" and yes, "it did get away!"

Note to the reader: Jake is poking a little fun at the author. In his list of career "successes" is a prescribed fire named Hamlin Canyon. It still holds the record in CDF for the number of "bonus landowners" at about fifty-two.

Fighting a wildfire is somewhat easier than planning and executing a successful prescribed fire. In a firefight, you study fire behavior for firefighter and public safety, and to predict where it will burn to and by when. Fire behavior prediction for a prescribed fire is more about how to build a fire that will burn sufficiently, but not so intensely that it jumps the control lines. All too often in our planning for a prescribed fire, we plan for fuel reduction without considering the damage that may be done to the soil…a living entity. You kill the soil, and you will live with the consequences for decades, not just years.

During a prescribed fire, your role is to keep the fire within the planned perimeter, and allow the fire to consume all of the dead and down material. So mop-up is different, really different. You don't "spread and put out," you "pile and let burn."

Let Burn Policy

The phrase "let burn" is old school language for today's "prescribed natural fire" or "wildland fire use"— the management of naturally ignited wildland fires to accomplish specific pre-stated resource management objectives.

If you own the property, you have more options.

What it means is that a land management agency allows naturally ignited fires (ones caused by lightning) to burn as they would "way back when." This policy is the norm for National Parks and in designated wilderness areas. A prescribed fire is any fire ignited by management actions to meet specific objectives.

The policy has merit, but it does not come without potential costs. When you make the decision in May to monitor a fire and allow it to burn, unchecked, you are accepting all of the weather patterns through the summer, the good, the bad and the ugly. Each year there are cycles of bad weather, where the winds blow, the temperatures rise and the relative humidity drops. You have to understand that, unless the fire goes out by itself, the weather will change the fire's behavior, and eventually you may have a problem later in the year. Smoke is another factor to consider over the long term. Fires in wilderness have been put out because the smoke was impacting the "view-shed" near by.

Jake says, "There are two suppression tactics: MIST and BIST...minimal impact suppression tactics or big iron suppression tactics. Each has its place."

The directives that establish this policy state that the fire can be put out if it poses dangers to private land or is creating more smoke than is desired, etc. The problem is that once the fire reaches a stage where you want to put it out, there are usually other fires burning, or the fire is too large to control with the available resources. Then there is the fact that in parks and wilderness areas, the tools you may use will be limited. In most cases, no mechanical tools can be utilized...no bulldozers, no chain saws, etc.

As stated, the policy has merit, but you must be aware of the potential economic, political and social consequences of deciding to "let a fire burn."

What are Your Objectives?

If the fire is small, you can see the whole thing, and you have the firefighting resources to put it out quickly; the need for a list of objectives is not too important. You just put the fire out and return to quarters.

But, if the fire is large and there are lots of firefighters involved, maybe some from other agencies, a set of overall objectives is necessary (Figure 10.8). You as a firefighter will not be establishing the objectives (that is for the incident commander), but you need to know what they are so you can do your part to accomplish them. If you or your supervisor don't know what they are, what Fire Order are you violating?

Figure 10.8 If no one knows what they are supposed to do, you can't expect them to do much in the way of production. People need to know what is wanted from them. Only in this way can a complex job, like firefighting, be accomplished in a safe and efficient manner.

Everyone needs to know what the plan is.

Group Decisions

Just because a group has discussed and made a "collective" decision, doesn't mean it is the best or right one to make. People have studied the dynamics of group decisions, and can show that if people don't speak up and state what they really want, but rather what they think the

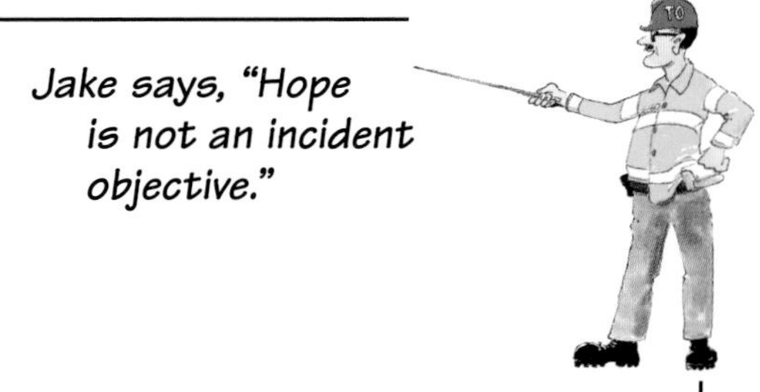

Jake says, "Hope is not an incident objective."

group wants, they are going to "Abilene." It is called the Abilene Paradox.

The book *The Abilene Paradox,* by Jerry Harvey, offers some interesting insights into how a group may go along with a decision that the individual members do not agree with. In *The Abilene Paradox,* an organization or group of individuals take an action contrary to the real desires of the group. One classic example is the South Canyon Fire disaster, in Colorado in 1994. On this fire, firefighters discussed the Standard Fire Orders and "Watch Outs" they were violating, but again, no one really spoke up. It is likely that they had similar fears to the Watergate group (e.g., fear of being branded not a team player, fear of being responsible for failure of the attack). In that disaster, 14 highly trained firefighters died!

If you don't speak up, you may take a ride to Abilene.

The author captures the essence of the paradox by the following short story:

> *There was a family of four who lived 60 miles from Abilene, Texas. One hot summer afternoon, a member of the family made an offhand suggestion that they should go to Abilene for dinner. It is important to note that he didn't really want to go, but thought the others might be bored. After a couple of minutes of discussion, each member of the family agreed that they should have dinner in Abilene that evening. After a long, hot drive and "not so good" meal, they returned to their home.*
>
> *Upon their arrival home, one family member stated, "It was a great trip wasn't it?" Another said, "To tell the truth I really didn't want to go, but you all pressured me to do it." The next person to speak said, "I didn't want to go either; I just went because all of you wanted to go." It was finally determined that none of the family members really wanted to go, but they did so because they thought that the others did.*

Jake says: "Don't go along for the ride with a group decision you believe is unsafe!"

Have you ever been to Abilene (Figure 10.9)?

Figure 10.9 These firefighters are "going to Abilene." Group decisions may be wrong if members of the group do not speak up and say what they really think. Many firefighters have gone to Abilene...with disastrous results.

What Jerry Harvey is trying to demonstrate with *The Abilene Paradox*, is that if members of a group do not voice their true feelings, or aren't allowed or encouraged to say what they really feel, an organization or group may develop a plan or pursue an action that is contrary to the true desires of the group members. We usually feel that conflict in an organization is one of the bigger issues a manager has to deal with. As Harvey points out, it is even more dangerous when an organization attempts to manage "agreement" that isn't really there.

Several places earlier in this handbook you have been told it is your responsibility to speak up. If you don't, maybe no one else will, and you all will go to Abilene. Remember South Canyon!!!!

Rank and Experience

There are situations when the best person for an assignment isn't the one with all the bugles on his or her collar. There are a lot of situations where a manager or chief fire officer has been able to surround him or herself with highly skilled and experienced people. This chief outranks subordinates, but is not as qualified or experienced as those under them. This isn't to say this chief is a bad person, but the realities are that others may be better suited for what is needed at the time.

Jake says, "Do not confuse years of service with experience."

The danger is that sometimes the chief doesn't want to own up to the fact that others would be better than he or she. Rank has that privilege, but is wrong when they exercise it. Don't go looking for all of the chiefs that you are better than, but realize that you may have to help some who don't realize they need help.

Too much Reliance

The primary and most important "tool" in fighting wildland fires is the firefighter! This is the person who swings the axe, operates the bulldozer, manages a part of the fireline, and commands the incident. But, today's firefighters and fire managers are placing too much reliance on several tools:

- ***Airtankers*** - It was mentioned earlier that airtankers don't put out the fires, they just help firefighters put out the fires; and, that airtankers are often not held on the ground when they are not being effective. Airtankers are most effective as an initial attack tool, but are being used more and more on large fires. Retardant drops should not be considered control line...they need to be followed up with some form of ground action. Airtankers have value, but don't base your strategy and tactics on having them available.

> *Jake says, "Don't base the success of your plan entirely based on the use of aircraft. They may not be available when you really need them."*

- ***Helicopters*** - don't carry a lot of fuel, so they have to return to base often. They also require lots of maintenance. This having been said, remember; they may not be available when you really need them. Like airtankers, do not base the success of your plan on them.

- ***Teams*** - Increasing numbers of agencies are relying more frequently on the use of teams to manage their fires. That is what they are for, but it does have a downside. If teams are assigned too quickly, the people they replace don't have the opportunity to learn by doing. Teams are good, but if not managed properly, or if the local personnel are not pulled into the team operation, some real learning opportunities will be lost.

- ***Others*** - Several pages were dedicated to LACES and the value and need for lookouts, but it is you who has to "Look Up, Look Down, and Look All Around" ***all the time***. Yes, lookouts are required when you can't see everything, but that doesn't free you from the requirement of keeping track of things yourself. You must constantly monitor, the weather, what aspect the terrain you are on is, where your other crew members are, etc. Have your own weather kit and compass. If you are on a south facing slope in the afternoon, what are you going to do?

Right to Refuse

Did you know you could say "NO!"? You have been told that fighting fire is a lot like fighting a war. There is an enemy; there are generals; there is a chain of command; and there are soldiers…that's you. The one big difference between the army and a fire agency, is in the fire service you can say no. You probably don't want to say no to washing dishes when it is your turn on the grounds (it will soften your hands), but you are expected to "turn down" an assignment if you don't think what you are told to do is safe.

A "**turn down**" is a situation where an individual has determined they cannot undertake an assignment as given, and they are unable to negotiate an alternative solution or mitigate the risk. The turn down of an assignment must be based on an assessment of risks, and the ability of the individual or organization to control those risks. Individuals may turn down an assignment as unsafe when:

You can say NO to an unsafe assignment.

- There is a violation of safe work practices;
- Environmental conditions make the work unsafe;
- They lack the necessary qualifications or experience; or
- Defective equipment is being used.

Small Fire Operations vs. Large Fire Operations

How fires burn dictates how you will fight them. Every fire is different in some way, but there are real differences in how small fires burn versus larger fires. There are also considerable differences on how you fight small fires versus large ones. There are five operational firefighting modes: initial attack; extended attack; large fire operations, mega-fire operations (e.g., Biscuit and Cedar); and recovery and rehabilitation. All five have very different philosophies.

Small Fires

Small fires tend to burn in two dimensions; the length and width of their surface areas, or a fire that burns in the surface fuels and does not generate enough heat to develop a strong plume. They are "driven" by the type of fuels, the topography and/or the weather. One or more of these factors may be dominant. When you get all three of them in alignment (a fire burning upslope with a strong upslope wind in dead or dying fuels), you may be heading for a fire that moves into the third dimension…vertical and plume-driven. The spread of a small fire usually can be easily predicted (Figure 10.10).

Small fires are only two-dimensional.

Tom McIsaac

Figure 10.10 This fire is only burning in two dimensions. If it moves into heavier fuels farther up the ridge, it may generate enough heat to move into the third dimension.

Small fires generally spread by direct flame contact, short-range spotting, or when rolling material (like pinecones) spread the fire down a slope. There may be fast runs upslope, with much slower burning downslope. The head of the fire will normally burn hotter and faster than the flanks. If suppression action stops the head, the two flanks may flare up and develop into two new heads. This is called "splitting the head."

You can have small fires that are the size of spots, or a small fire may be as large as thousands of acres. It isn't so much the size of the fire, but how it burns.

The organization of a wildfire incident and the philosophy of how you fight it can differ considerably. Generally, when you are involved in small fire operations, you are operating in a mode of "let's put it out and go home." Your organization is mostly tied up in direct suppression. The establishment of a formal Planning or Fiscal/ Administration Section is not necessary. Operations and some limited Logistics support, mainly food and drink are usually required.

95% of fires are controlled at less than 10 acres.

The vast majority of wildland fires, some 95 percent, are contained and controlled while they are small. It is the 5 percent that escape initial attack that become large.

Large Fires

A large fire does not necessarily have to be a big fire. The 1991 Tunnel Fire in Oakland Hills, California, was small by comparison. It was only 1,600 acres, but it burned 3,800 structures, killed 23 civilians, one fire officer and a police officer. It isn't the size, it is the "dimension" of the fire that counts (Figure 10.11).

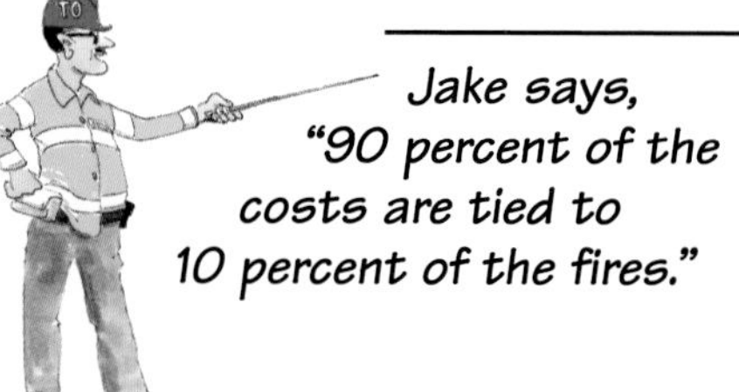

TO
Jake says,
"90 percent of the
costs are tied to
10 percent of the fires."

Large fires usually move into the "third dimension." They do not just move forward and broaden, they move vertically, bringing in air as the plume above the fire grows, sometimes to thousands of feet in the air. This brings the possibility of "collapsing thunderheads," long-range spotting and area ignition.

When fire moves into the category of "large," the attack philosophy usually changes. You anchor and start a flanking attack. All other actions are indirect, some distance in advance of the fire. On some sections of the fireline you'd best become an observer, because attempting to put it out would be wasted energy. If structures are involved, saving as many of them as possible will be the rule of the day.

The general rule with a large fire: be productive, but wait until the weather is favorable so you can safely get in and establish control lines.

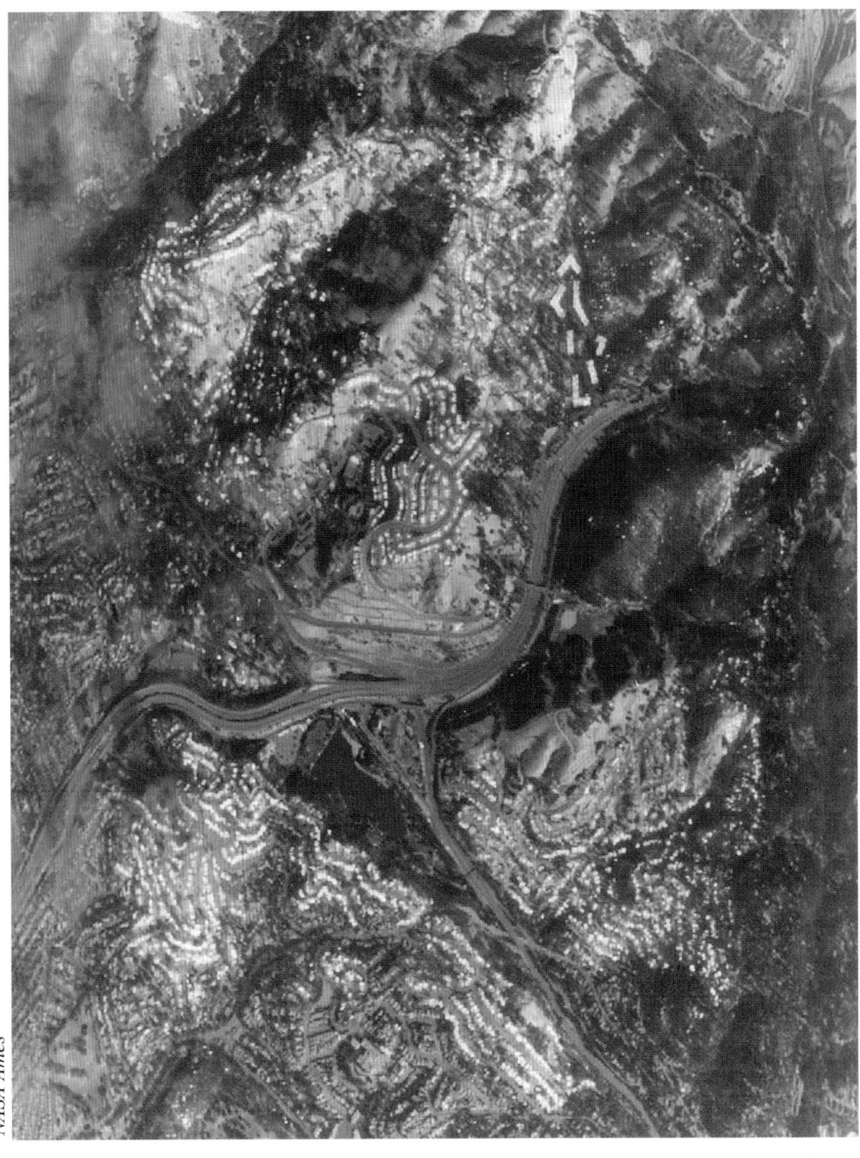

NASA Ames

Figure 10.11 This is an aerial view of the Tunnel Fire the day after its run. All the little orange dots are the hot foundations of 3,800 homes, apartments and condominiums.

The Really Big or "Mega-Fires"

You may only be involved in one "really big fire" in your career...but when you are, you will know it! The "really big fire" is one that is moving so fast and is so large it is beyond a human's ability to comprehend, let alone manage. Plans for such fires must be very broad in scope, but simple in message. The formal ***Incident Action Plan*** is close to worthless. Because events are changing so fast, the IAP is obsolete even before it is published.

Jake says, "The transition between IA and extended attack can be difficult and dangerous. Some say this should be a watch out!"

Some of the fires in California in October 2003, and Alaska in 2004 would fall into the category of "really big fires." If it wasn't for the fact that a lot of very experienced firefighters took independent, but coordinated action, more homes would have been lost and more people would have died!

Fires don't always do as you think they will. Yes, they do follow some "general rules," like normally burning uphill faster than down, etc., but they can and do violate these rules. Assume that your fire may violate the "rule" in an attempt to make you look bad. Learn to look for—and prepare for—the unexpected.

Think Big

Thinking big means looking a head and not allowing the fire "to be in charge."

Jake once "told" a class, ***"That he had the feeling sometimes, that the more engines you send to a fire, the bigger it gets!"*** Hopefully, this is not true, but it brings up a good point. It is safer to "think big" and put the fire out, than to think too small and "hope" you can put out the fire. ***If you think too small, "the fire will run you all over the country." Too many fires are "initial attacked" until they go out on their own.***

Here are some things that should be taken into consideration when developing the plan and the "mind set":

- ***Think about what really is going to happen*** - All through this handbook the word "efficient" has been used. This usually means smaller, less expensive, etc. Firefighters are taught that the "smallest" alternative should cost less, cause the least damage and pose the least risk to firefighting personnel. These perceptions can be very wrong. The fire may not let you think small. One sign an incident commander may be thinking too small is that he or she is ordering additional resources every couple of minutes. This means they have not projected where the fire will be and what it will take to put it out—no plan!

- ***Consider the consequences*** - If you fight fire for a land management agency where the agency "owns" the land, you can say, "so what if the fire gets big." If you work for a state or municipal fire department, where "you do not own the land," you do not have that option—you've got to put the fires out. You also may not have any impact on a fast moving fire. It is going to get big, regardless of what is thrown at it.

If lives and property are threatened, this must be raised to the top of your objectives. In developing alternatives, do some what ifs? What if: the wind changes direction or speed; the fire gets to a certain point (ridge, drainage, etc.); you lose aircraft support; the relative humidity drops; or the fire is heating up and spotting?

- ***What is the most reasonable alternative?*** - The first consideration is firefighter safety. Can what is proposed be done safely? If not, scrap this alternative. Pick the alternative that provides safety, keeps the damage to a minimum, and can be accomplished in an efficient and practical manner.

  The most acceptable attack may be establishing an anchor point and flanking the fire until it is contained.

- ***Success!*** - It is sometimes hard to accept the fact that most fires do what they want to do—based on the conditions of the fuel and the weather—until the fire behavior lessens to a point that firefighting efforts are effective.

The key is that every attempt be made to predict what the fire will do, determine what alternatives are available, and take all actions in a safe manner.

Making the Transition to an Extended Attack or Large Fire Mentality

There are a few fires each year that never require a transition from small to large fire mode. When you leave the barn, you know this is going to be "the big one." The fire is only minutes old and it already has a large smoke plume, and the organization is already in a large fire mode. The problem develops when the fire just stays one step ahead of the suppression effort. One more hour and you will have it contained. More equipment and personnel are ordered, but you just cannot catch it.

> *Jake says,*
> *"Managing the transition from initial attack to extended attack separates the leaders from the rest of the chiefs."*

The fire organization grows, a more experienced fire officer may even take charge, but no one has said, "We are out of initial attack, let's change our thinking on how we will fight this thing." Most organizations ensure that any change in command is well known…it is even announced over the radio for all to hear. But,

Northtree Fire International

Figure 10.12 This firefighter (in the circle) is responsible for "holding" this section of line. Later in the day, the fire did spot across the line, but they were able to catch it at less than an acre.

you very seldom hear a change in attack mode announced…maybe we should!

Hold the line already constructed

There isn't anything much worse than putting in the same fireline time after time. If a line is important enough to construct, then do what it takes to hold it (Figure 10.12). This is especially true when fighting a timber fire. Most old-timers will tell you, "Timber fires are fought at night…they are held during the day." What this means is, timber fires will normally lie down after dark when the humidity rises and the winds die down. Night is the safest time to construct fireline. The key is that during the day, the line is reinforced and held.

If a line is worth putting in, hold it.

In your plans, ensure that there are adequate forces to strengthen and hold the line you worked so hard to put in.

Fires usually are easier to put out at night.

Fighting Fire at Night

There seems to be less firefighting at night in recent years. This appears to be contrary to the fact that most fires usually quiet down at night. Not all, but most. Temperatures cool, humidity rises, winds are usually less. So why are more and more incident commanders tending to not take advantage of the fire being more easily controlled at night? Safety issues, such as large areas of snags or smoke inversions are legimate concerns at times (Figure 10.13).

Wes Schultz

Figure 10.13 Some fires are best fought at night when they usually burn with less intensity.

Not fighting fire at night means firefighters can sleep when nature intended them to sleep…it is dark and usually cooler. You don't have to find shade, it is provided. It also means that the fire is probably run on a 24-hour planning clock. This cuts in half the number of Incident Action Plans needed, the number of lunches needed, etc.

Fires rest at night, firefighters should take advantage of this.

Some swear by the 24-hour shift and its benefits, but fires usually don't burn as hot at night, and an opportunity may be lost. Again, evaluate the overall risks versus benefits before engaging or not engaging in night operations.

Jake says, "The day shift gets all of the glory, the night shift does all of the work."

Not all engines are created equal

It isn't the age, color, size or condition of the engine; it is the personnel that are assigned to operate it. The ***training***, ***experience***, ***leadership***, and ***motivation*** of the crew are what really count. It is the quality of the engine company—not how many you have—that is important. One well-trained, experienced and motivated engine crew can out-perform several inexperienced engine companies.

An engine is only as good as the people staffing it.

People under stress tend to revert to their familiar situations. An engine company highly trained and experienced in the art of fighting structure fires may not take an appropriate action to protect a home, but once that structure is on fire, they know exactly what to do. It isn't that they are bad wildland firefighters, they just are out of their element. They don't have the training—and, more importantly, the experience—necessary to take the appropriate action. The same holds true for the wildland firefighters that find themselves out of their element, asked to perform tasks they are not trained and experienced in. Again, situational awareness and knowing your limitations are key.

Training!
Experience!
Motivation!
Supervision!

Another thing with engines; in some places, you can mobilize more engines that you can possibly manage…California is famous for this. If you can't manage them, they sure aren't doing you any good, and they may be a long way from where they could be very useful…back home. Order what you need and release early!

Having a lot of engines means you have to manage them.

New fire environment

With year-round residents and human activities on the rise in many developing wildland areas, it is important for firefighters to become aware of the new fire environment in which they are increasingly involved. It is critical that they become familiar with the hazards they will encounter in the wildland/urban interface. Chapter 1 presented the Ten Standard Firefighting Orders, LACES, and the 18 Situations the Shout "Watch Out!" All of these must be remembered and adhered to. In addition, the wildland/urban interface presents unique hazards and situations that you must know.

The Wildland/Urban Watchout Situations are:

- Wooden construction and wood-shake roofs;
- Bridge load limits;
- Poor access and narrow, congested, one-way roads;
- Evacuation of public (panic response);
- Inadequate water supply;
- Natural fuels 30 feet or closer to structures; or
- Structures located in chimneys, narrow or box canyons, or on steep slopes of 30 percent or greater.

The Media

Jake says, "Don't get into a pissing contest with a skunk! You don't have the tools."

The media is the only industry protected by the United States' Constitution. And, they don't usually report good news. Good news doesn't sell...bad news does. Maybe that is more of an indictment of the general public, or the reporters, rather than us.

The best advice when dealing with the media...tell them the truth; don't talk about things you don't know about; don't get caught up in getting your "15 minutes of fame." You especially don't want to argue with the reporter. If they report something you know is not true, correct it. But, if you just don't like something...grin and bear it. Remember the old saying: "Don't argue with someone who orders ink by the barrel," or "you own your own reaction."

It is easier to learn something than to unlearn it.

Situation Avoidance

These basic firefighting tips may help you avoid a situation that could spell disaster:

- ***Anchor points*** - Take advantage of roads, creeks, trails, etc. Use breaks in fuel continuity to your advantage. Target rock outcroppings, landings, bare spots. Establish wet line in two directions.

- ***Mobile attack*** - Select drivable terrain and light fuels. Go "direct," one foot in the black, in "squirrelly" winds. Select appropriate hose and nozzle sizes. Use the attack method that suits the situation - ***Flanking - Tandem - Pincer***.

Avoid mid-slope and undercut line.

- ***Hose lays*** - Match hose and nozzle size to the fuel load. Calculate hydraulics; up and down. Use adequate, but safe nozzle pressure. Install laterals to expedite mop-up.

- ***Location and width of fireline*** - Keep away from snags and slash piles, etc. Use the leeward side of the ridge-top. Avoid mid-slope and undercut hand line. Utilize creeks, roads, trails, or other barriers. Match width of line to height of fuel (***one and a half times available fuel height***).

Mop-up is hard, dirty work, but it must be done.

- ***Bulldozers, excavators and tractor-plows*** - Provide a swamper to scout the line. Don't let people work on the slope below. More effective in pairs; match size to fuel. Pair contract dozers with a fire agency dozer or provide an experienced Dozer Boss. ***Keep under control to minimize damage.***

- ***Hand crews*** - Match production rates to fuel when ordering. Good for firing operations and mop-up. Can assist with hose lays. Use crew leader for your lookout and size-up. Maintain contact; check progress.

- ***Aircraft*** - Use airtankers and helicopters effectively. Locate and advise of hazards, such as power lines, tall trees, etc. Watch out for low drops and rotor wash. ***Follow-up retardant drops with ground forces...soon.*** Don't base safety on airtanker availability.

- ***Using fire*** - ***Backfire*** only on orders and with experienced personnel. ***Make sure everybody knows the firing plan.*** Timing is critical; watch the wind for change. ***Burn-out*** to reduce line length and mop-up. Utilize barriers as anchor points.

You are only as good as your training.

- ***Mop-up and patrol*** - Minimize property damage; repair fences, construct water bars, extinguish utility poles, etc. Deal with hazards (snags, slash and dozer piles, etc.) soon. Reinforce the control line, especially at corners. Patrol until you are sure it is out.

- ***Food and water*** – Food + Water + Rest = Maximum Productivity. ***Fresh crews are safe crews.***

- ***Rules of engagement*** – Most of all, do not violate any of the 10 Standard Firefighting Orders.

Training

You are only as good as your training! You were not born with the instincts to fight fire. You even had to be taught that touching something hot would burn you.

Is it possible to retrain an old dog...yes, with a lot of work.

Train as you fight fire; fight fire as you have been trained. It is very important that when you train, you do so in a realistic way, the way you will actually fight fire. If you don't, you may wind up doing something wrong (or deadly) when you are under pressure.

First Learned – There are three real good examples that point out the first learned concept of training.

For years, the California Highway Patrol required their officers to collect the brass (spent cartridges) after they finished shooting. This was the way they did it… as was the policy with most law enforcement agencies. Then came a tragic firefight between several CHP officers and some bad guys. The bad guys won this battle, and during the follow-up investigation, they found that several of the CHP officers had collected the brass and either had it in their pockets or in their hands. This may or may not have been the cause of the tragic loss of life, but it really changed how this critical training is now conducted.

Jake says, "I don't believe we are as smart as we'd like everyone to believe...especially when it comes to extreme fire behavior."

When the U. S. Air Force first introduced a new fighter to its fleet, it was designed to eject the pilot out the bottom of the aircraft when he was in trouble. So, their training included a drill that required them to invert the aircraft before they "punched out." Later models changed the ejection system to pop the pilot out the top of the aircraft. You guessed it. For those pilots that had been trained the old way, the first couple of ejections were fatal. The pilots instinctively inverted and punched out.

Several of the firefighters that lost their lives on Storm King Mountain were still carrying their hand tools and chain saws. They were trained that it was their responsibility to take care of their tools, and this kicked in when they were told to move off the line. It would have been failure if they had dropped them to lighten their load. Some feel that this contributed to the deaths of several of these brave firefighters. They were reacting to their training and it may have cost them their lives.

Training makes responses automatic.

The lessons to be learned from these three events:

- If you want a firefighter to react instinctively to danger, be sure the training is such that it builds-in the correct response to danger.
- If you have learned one way of doing something and now must change, the old way has to be unlearned first, and this is very tough to do, but "old dogs can learn new tricks."

This brings up a new term, that of "cognitive dissonance," or in "firefightereze," once you have learned something one way, you will resist learning it another way. This resistance to new information can be even harder to change if learning the first way was tough. So, the harder a lesson is to learn, the deeper it may be entrenched, and harder the relearning will be.

Jake says: "Train as you will fight; fight as you have been trained."

In a lot of cases where firefighters have been killed, some investigators feel that most of them were bewildered by the events that were taking place and were actually frozen by the quickness in which they occurred. You have probably been involved with something falling in front of you and you just didn't react quickly enough to stop it. Afterwards, you think to yourself, "I could have caught that," but for some reason, you just watched it fall. Or, perhaps something startled you, and in your response you froze—temporarily paralyzed while your brain tried to process the information and come up with an appropriate response. These precious moments of indecision and/or inaction can cost a firefighter his or her life.

Research has also found that when one is confronted with danger, your ability to think clearly is diminished greatly. A person not under stress can usually deal with more than one event at a time, but under stress, you will lose this ability to think. This is why the establishment of trigger points will help…you don't have to think, just react.

Jake says, "You can't get into too much trouble if you 'Anchor, Flank, Hold and Pinch.' These are the basics!'

The point of this whole discussion is that, when you train, build some realism into it. Use a windy day or power-fan to make it harder to deploy your shelters; when training how to react to an order to move to safety, drop your tools and move out fast; and if you are on the pistol range, leave the spent brass on the ground.

The Basics

The basics....anchor points and flanking action. That is about as basic as you can get!

There are times when we can get too fancy for our own good; there are times when we should return to the basics. We may attempt to use more and more firefighting resources, even though they are not making a difference. We wait for a team, or attempt to use some fancy tactic. As you get more experienced, you will find that it is better to return to the basics. You have the fundamentals, built on a foundation of safety. Knowledge is power. Now put it to work. Returning to the basics means:

Jake says, "It took me a long time to learn that the basics really do work! You don't have go be fancy to be effective!"

- ***Anchor your line*** - Pick a spot, like a road, old burn, river, etc., that will properly anchor your line and prevent the fire from hooking around and running loose.

- ***Flank the fire*** - Start moving along the fire's edge, keeping one foot in the black if you can. Use the black as a safety zone if you can. Move slow and deliberate, ensuring the fire is out as you go.

- ***Hold the line you have put in*** - It makes no sense to put in line you cannot or do not hold. If the line was worth the effort to construct, then hold it.

- ***Work at night*** - Normally, a fire will burn with less intensity at night. Use this to your advantage and "get on it" when it allows you to.

- ***Take care of yourself*** - Drink lots of water, eat healthy meals, and get rest when you can.

- ***Always have a safe place to move to*** - A good pilot always knows where the closest airport is in the event one is needed. You, too, must always know where you can go to be safe.

- ***Look Up, Look Down, Look All Around*** - Be your own lookout and safety officer. Know what is happening around you and why it is happening. Remember: your observations can make a real difference. Listen to your gut, and speak up when it tells you to.

Fire has a vote and it more than likely will vote against you.

Who has a Vote?

The realities of wildland firefighting are similar to those in combat...in that "the enemy has a vote." What this short phrase means, is that you are not totally in control of events. You can pick the place to fight, but the fire, with its three "henchmen" (fuels, topography and most of all weather), may attempt to foil your plans. And, if fuels, topography and weather all vote together, you have just become a spectator.

The human body is very strong, yet very vulnerable. It doesn't take a lot of heat to kill a person.

Don't ever underestimate how quickly a situation can change. Even if you follow all of the rules and have lookouts, and do everything right, fire can still flare up and strike you. A very tame, innocent-looking situation can turn deadly in a split second. Also, don't underestimate how little heat can kill you—one "gulp" of hot air can be fatal.

Jake says, "Don't fear fire, but always respect it!"

You must be able to recognize when you have to back off and rethink your plan of attack. There isn't an acre of this great country worth getting hurt over, so put things in true perspective, and attack the fire with proper respect. Fire has a vote, and it can bring serious injury and death to those whodon't respect it! Don't fear it, but always respect what it has the potential to do.

You have chosen a wonderful career, now go out there and get some experience and be the best wildland firefighter you can be!

Notes:

11

Firefighting Situations

This chapter of the handbook will present various firefighting situations and one or more practical approaches to them. You should remember that there normally is not just one right answer, but some approaches or solutions are better than others. Keep in mind all that you have learned. Apply your strategies and tactics.

Northtree Fire International

Figure 11.1 Last night's lightning storm started scores of fires in the area. In this picture, there are at least seven fires. A lot of wildland firefighting is about setting priorities; which fire to fight first; how to fight it; and more importantly, why?

There are times you will have more fires than resources. You will have to make some serious choices as to which of the fires will be worked on first. In Figure 11.1 there are at least seven lightning-caused fires; no telling how many there really are. Sometimes you will have to decide which fire you will attack first; which homes can or cannot be saved; and, a reported fire may turn into a medical aid call. The "a possible solution" is just one of many that would be correct and proper. How would you approach each situation?

Situation 1 - Several Small Fires

You have been dispatched, with an engine company of three, to a small fire on Smith Road. As you move into the area you see that there are at least three separate fires, scattered along the road for about a mile (Figure 11.2). The smoke columns are not uniform in size; one of the fires is significantly larger than the other two.

Problem

Which fire do you attack first? None of the fires poses an unusual situation, such as threatening structures.

A Possible Solution

Keep small fires small; worry about the big fire later.

The tendency would be to attack the larger fire first, leaving the smaller fires till later. ***You want to have as few large fires as you can.*** If you attack the larger fire first, the smaller fires will probably be larger by the time you get to them. Attack the two smaller fires while you have a good chance to control then, then tackle the larger fire.

If the fuels are heavy, you may need to leave one of the firefighters to mop-up. Move quickly; the other fire needs attention soon.

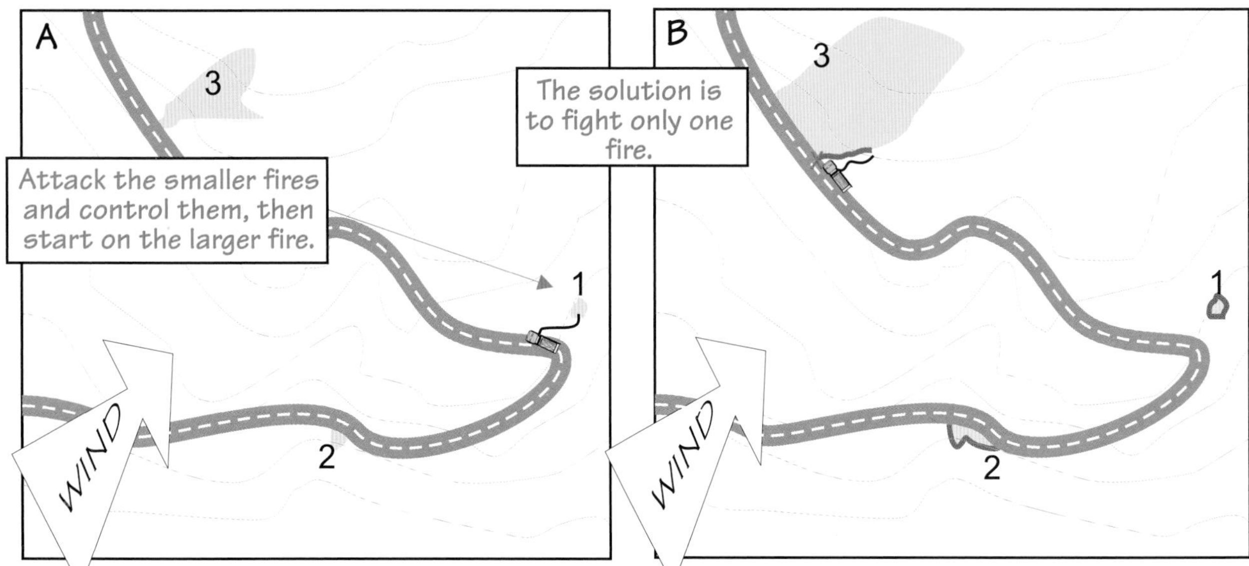

Figure 11.2 You have three fires scattered along a road. Attack the smaller ones first, then the larger one. As other forces arrive, assign them accordingly.

If one of the fires is threatening some high-value timber, a plantation or a structure, that fire takes priority; attack it first.

Alternative

If two of the fires are very close together, you might handle these as one, tying the firelines together.

Situation 2 - Fire on both sides of a canyon

You have been dispatched to a timber fire in a remote part of your district. From the dispatch information, you know your four-person engine company is part of a five-engine first alarm. You will be the first engine to arrive. As you approach the area, you see that the ***fire has spotted across the road and creek*** and is on both sides of the canyon. The main fire is about two acres and moving up the ridge with a moderate rate of spread. The portion below the road is one-quarter acre, and the slopover across the creek is small and not yet "running."

A fire on both sides of a drainage should be treated as two fires.

Problem

Which part of the fire do you attack? This situation is very similar to the one where there are several fires to attack. Fires like this—on two sides of a canyon—should be dealt with as if they were separate fires (Figure 11.3). If the canyon is narrow, the fires will

Anchor and flank.

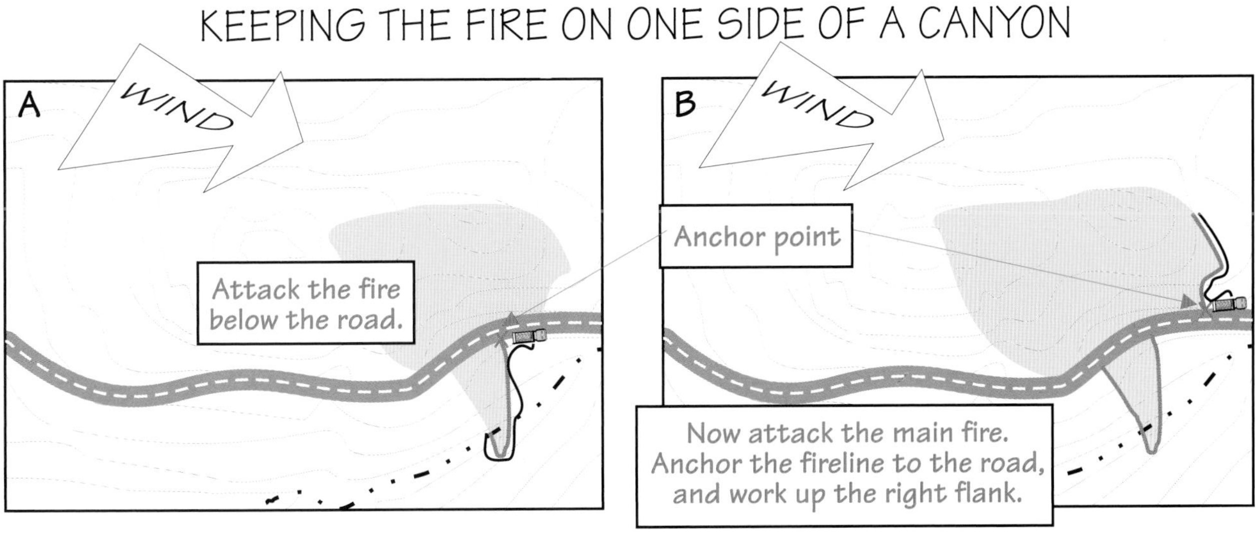

KEEPING THE FIRE ON ONE SIDE OF A CANYON

A

WIND

Attack the fire below the road.

B

WIND

Anchor point

Now attack the main fire. Anchor the fireline to the road, and work up the right flank.

Figure 11.3 You want to keep the fire to one side of the canyon. Once the fire below the creek and across the canyon is controlled, attack the main fire.

begin to influence each other. The heads will move in opposite directions, forcing you to split your forces.

A Possible Solution

You want to keep the fire on one side of the canyon. Attack the small fire below the road and control it, then start an action on the larger fire. Position the engine so as not to block access for the other forces. If there is water in the creek, consider using a portable pump to save tank water.

Fast-moving fires require quick decisions.

Situation 3 - Coordinated attack on fast moving fire

You have been dispatched as part of the five-engine first alarm to a fire just a couple of miles from the station. You are the officer of a four-person engine company. As you move down the road, you see a large column of smoke. The weather is very hot, dry and windy. It takes you about ten minutes to arrive at scene. The fire is already 20 acres in size, moving through grass, brush and into the timber. The terrain is such that a mobile attack is not possible. The fire appears to have started from the road. There is a small tract of summer cabins about a mile off to the right. There is heavy traffic along the road. You give a report on conditions to dispatch. You request a second alarm, traffic control, air support and two water tenders.

Protect exposures; establish an anchor point.

Problem

Where to set up and attack? There are no roads or streams to use as fireline in an indirect attack. The fire is too dangerous to attack with an indirect attack. You want to keep "one foot in the black."

Don't block the road and cut off your help.

A Possible Solution

Position the engine so that it does not block the road and will accommodate the safe positioning of additional engines and water tenders (Figure 11.4). Anchor the right flank at the road. Start up the fireline with a progressive hose lay. You will stay at the engine operating the pump and coordinating the assignment of the incoming firefighting resources until relieved of command. Put your most experienced firefighter on the nozzle. Provide a portable radio if

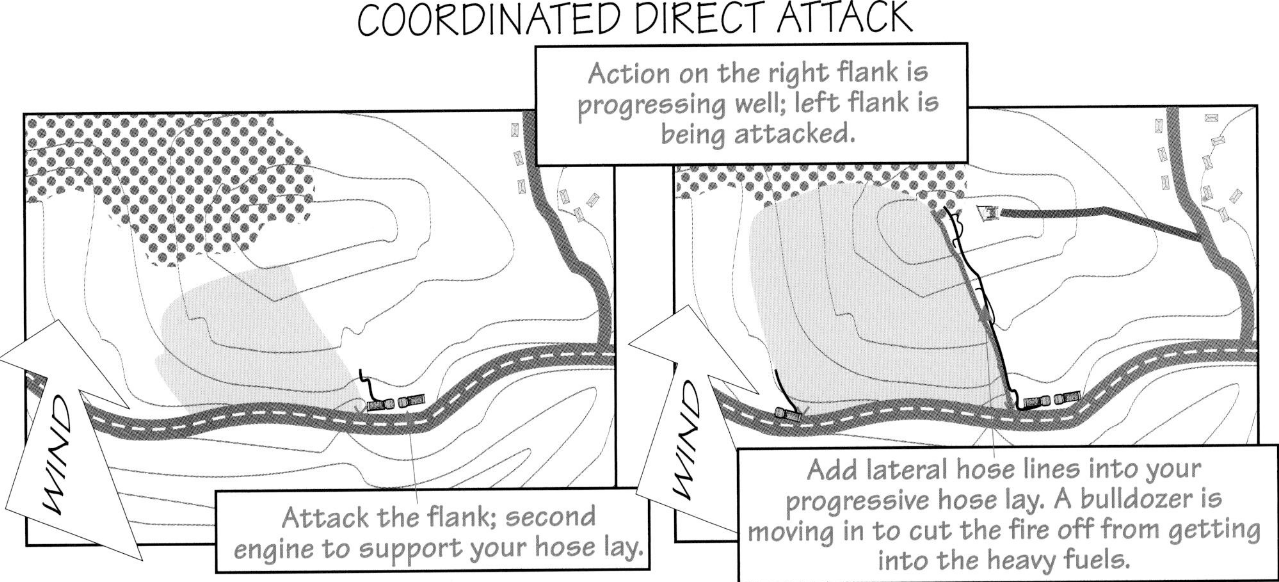

Figure 11.4 Attack the right flank, anchoring your fireline to the road. This flank is your highest priority because of the small residential community to the northeast. The second engine should support your hose lay. Good progress on the right flank. The third engine is starting up the left flank. Note the bulldozer moving from the houses, up the ridge toward the heavy fuels. It could be used to cut-off the fire before it gets into the heavy fuels and support the fireline construction along the right flank.

available. The other firefighters will be used to advance the hose line and pack hose.

Assign the other engines as they arrive. You will have to decide how the left flank and head will be controlled. One of the first engines should be assigned to assist your crew in its hose lay. You want to maintain the momentum of the hose lay and control the right flank as soon as possible. The bulldozer can be used to cut-off the fire before it gets into the heavy fuels.

Be prepared to change your plan when the situation changes.

Situation 4 - Fire moving toward structures

The wind has shifted in Situation 3. The fire is now moving toward the tract of homes. The fire is about one-half mile from the structures, spotting, with a moderate rate of spread. Several engines and the bulldozer from the first alarm are just arriving.

Move to safety zones when conditions warrant it.

Problem

With the change of wind, your hose lay is not being effective and the safety of the crew may be in doubt. The tract of homes is

Protect structures as necessary, but keep any available resources working on the perimeter where they can make the most progress.

now the most valuable resource that is threatened (Figures 11.5 through 11.7). How do you redeploy your forces?

A Possible Solution

It is imperative that you bring your crew to safety. Tell them to retrieve the nozzle(s) and brass and return to the engine. Abandon the hose. Instruct the incoming engines to move into the tract of

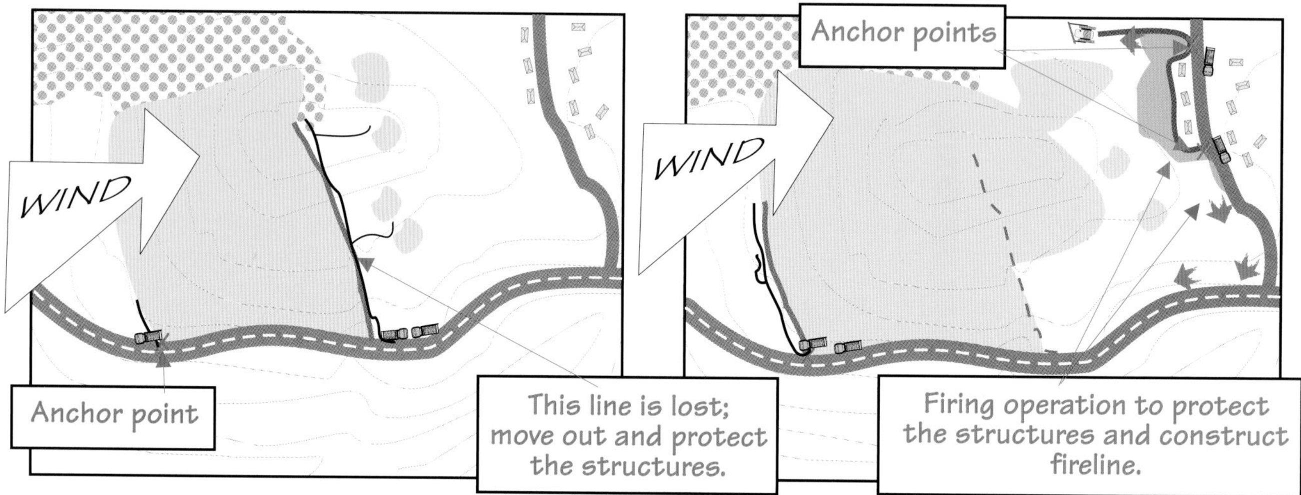

Figure 11.5 The wind has shifted and has caused the fire to spot across your fireline. The fire is now threatening the homes to the east. Move to the structures to protect them and use the access road as your new fireline. Continue to work on the left flank. Anchor the fireline to the road and fire-out around the structures. Carry the firing operation back toward the origin and behind the bulldozer as it moves west of the structures.

Figure 11.6 The bulldozer is cutting line along the fence line behind the homes. It will then take the line west toward the heavy fuels. The line will be burned-out as it is constructed.

Figure 11.7 This engine is positioned to protect this home as the dozer line is being burned. Once the line is secure, the unit can move to another home or begin mopping-up.

homes and prepare for the fire. If air support is available, use them to protect the homes. Don't abandon your left flank action. Continue that effort. Consider using the road as the fireline, burning it out as soon as the threat to the homes has passed. Determine the threat to the homes and the magnitude of the fire and order the appropriate equipment and overhead.

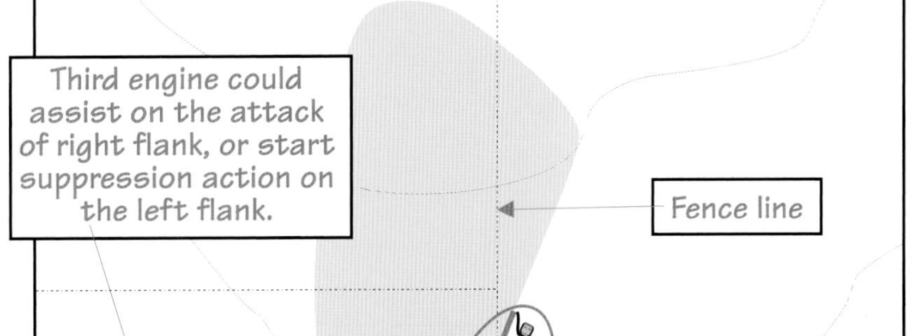

Figure 11.8 A grass fire is moving across flat country, being pushed by a strong wind. The plan is to use all three engines in a tandem mobile attack on the right flank. An option could be to use the third engine on the left flank.

Situation 5 - Grass fire is being pushed by strong wind

You have arrived at the scene of a grass fire being pushed by a 20-mile-per-hour wind.

Problem

The fire is between 40 and 50 acres. The terrain is fairly flat, with some fences and small draws. You are one of three engines dispatched on the first alarm (Figures 11.8 and 11.9). The fire appears to have started from a burn barrel.

A Possible Solution

This fire can be fought using a direct, mobile attack. Have a firefighter cut the fence and move through it. You have two firefighters charge a 1 inch line and start the attack. Utilize the Class A foam system in a very low concentration. This will aid in penetrating the grass cover and

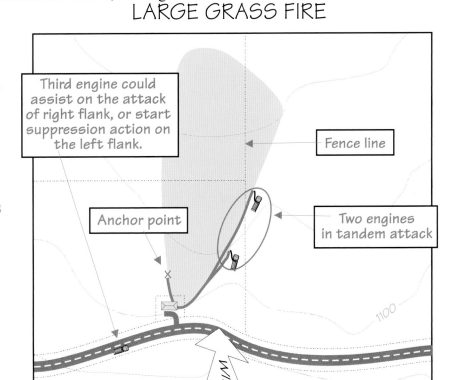

Bryan Day

Figure 11.9 This engine company is taking a flanking action. There is another engine company about 100 yards behind it.

Have somebody follow up to prevent flareups or slopovers.

reduce the risk of flare-ups. The third firefighter is assigned to patrol behind the engine with a backpump. The second engine is at scene; they are instructed to come to your position and take over the attack; you and your crew will follow behind in a tandem attack. The third engine is instructed to come to your location and take up your position, freeing you to find a water source and refill your tank. At the same time, you will be able to "recon" the other flank. Advise dispatch that the fire appears to have started from a burn barrel. Also, before you leave the fire, repair any fences you cut. That is your responsibility.

Alternative

If the fire is in light grass where the chance of flare-ups is less, the second engine could have started attacking the other flank in a pincer attack.

Estimate the progress of the fire.

Situation 6 - Structure protection in a rural setting.

You have been assigned to Group C on the Hardy Fire. The group supervisor has assigned you to protect the structures along Smith Lane. You have been told that there are approximately ten homes scattered throughout the area. The only water source is a pond at Smith Lane and Hardy Truck Trail. You will be working with two other engine companies (Engine 6 with four firefighters and Engine 726 with three firefighters). You have a total of five firefighters on Engine 45. There will be no additional help for several hours.

Tom Iraci

Figure 11.10 Rural areas like this are all very common. This looks green and nice, but it will burn.

Problem

As you move down Hardy Truck Trail toward Smith Lane with the two other engines, you see that there is very little clearance along the road (Figure 11.10). The fuel is heavy oak, scattered pines and grass. As you near the intersection, you see the pond and note the limited access to it. You also notice that the electrical power for the area is aboveground. This means that well pumps may lose their power source right away.

Smith Lane is a narrow two-lane gravel road that winds its way down the hill. The homes are scattered along this dead-end road for about a half-a-mile. You know that if the fire is active along the road, it will not be safe to use. You pass several cars as they evacuate the area; those fleeing look scared.

Identify safety zones.

You see the smoke from the fire off to the left. It is dark and boiling, and some of the trees are starting to torch. There is a five-mile-per-hour wind pushing the fire toward your location. You estimate that it will be about 45 minutes before the fire front reaches your location. It feels like spotting could start sooner.

A Possible Solution

Restricted communication capabilities shout, "Watch out!" Have a plan.

You stop to discuss the situation with the other engine companies. "Group C" has asked you to organize the protection effort along Smith Lane (Figure 11.11). You compare notes with the other firefighters. The Captain on Engine 6 noticed that one of the homes had a swimming pool in the backyard. The Captain on Engine 726 noted the wide area just down the road that could be used as a safety zone. The plan is to locate all of the structures and decide which ones can be protected, and how. Engine 6 is given the first four homes along the road to survey and protect, Engine 726 the next three, and you will go to the end of the road to check on the rest. You do not have radio communications between the units, so you will try to meet back at the safety zone in 15 minutes. (That meeting will not take place!)

Jake says,
"Be creative to solve communications woes; give them your portable; swap cell phone numbers; etc."

As you proceed down Smith Lane, you see that several of the homes have good clearance and nonflammable roofs. You count out the homes or driveways, as not all of the homes can be seen from the road. Your plan is to go to the end of the road and turn around. It takes you about four minutes to reach the end of the road. By that time you are sure that there are at least twelve homes, not ten. This means you will have to protect a total of five.

Make sure your firefighters know and understand the plan.

The last home is off the road about 100 feet. It is tucked into a thicket of brush, hardly visible from the road. There is no clearance, the roof is wood shake covered with pine needles. There is no way this structure can be defended. The next home has some clearance. The roof is wood shake, but it is clear of debris. There is an above-ground pool next to the house. The homeowner rushes to meet you and asks for help. You advise him that it would be wise for him to leave. He says, "I'm staying." You take a quick walk around the structure. You see that it can be defended, if the roof can be protected, and a woodpile is moved. You instruct one of your experienced firefighters to take the floto pump, fuel tank, five fusees, 200 feet of 1½-inch hose, nozzle, and shovel off the engine. You instruct the homeowner to move the woodpile and help the firefighter. Your instructions to the firefighter are: "Use the pump and hose to protect the roof and structure; ladder the front of the structure, using the homeowner's ladder; fire around the structure if you need to; control all spots within 50 feet of the structure. Be sure to keep the fire from the scattered woodpile. Seek refuge in the structure or pool. Stay put. I will come back for you. You have about 30 minutes to get ready. Be careful."

Turn the homeowner into a productive asset.

You look out through the trees to see where the fire is. You still feel you have at least 30 minutes before you will have to start fighting fire.

The next two homes are right next to each other. You feel that you can protect both of them with two lines if the engine is positioned next to home number 9. You drop off two firefighters to begin surveying the situation, while you check on the last home you are to protect. This home (#8) has good clearance and a steel roof that is clean. The homeowner is there. You instruct him on what he should do. You tell him that you will be just up the road. You note he already has several buckets of water and a shovel. You tell him that he should control all of the spot fires. If the smoke gets too much for him, he should go inside until the worst passes. He is not to enter the woods. He is to stay put. Once the fire is here he should not try to escape. You will be back to check on him. As you leave you see the crew from Engine 726 laying protection lines to protect the next home up the lane.

WATCH OUT! Starting to get spot fires.

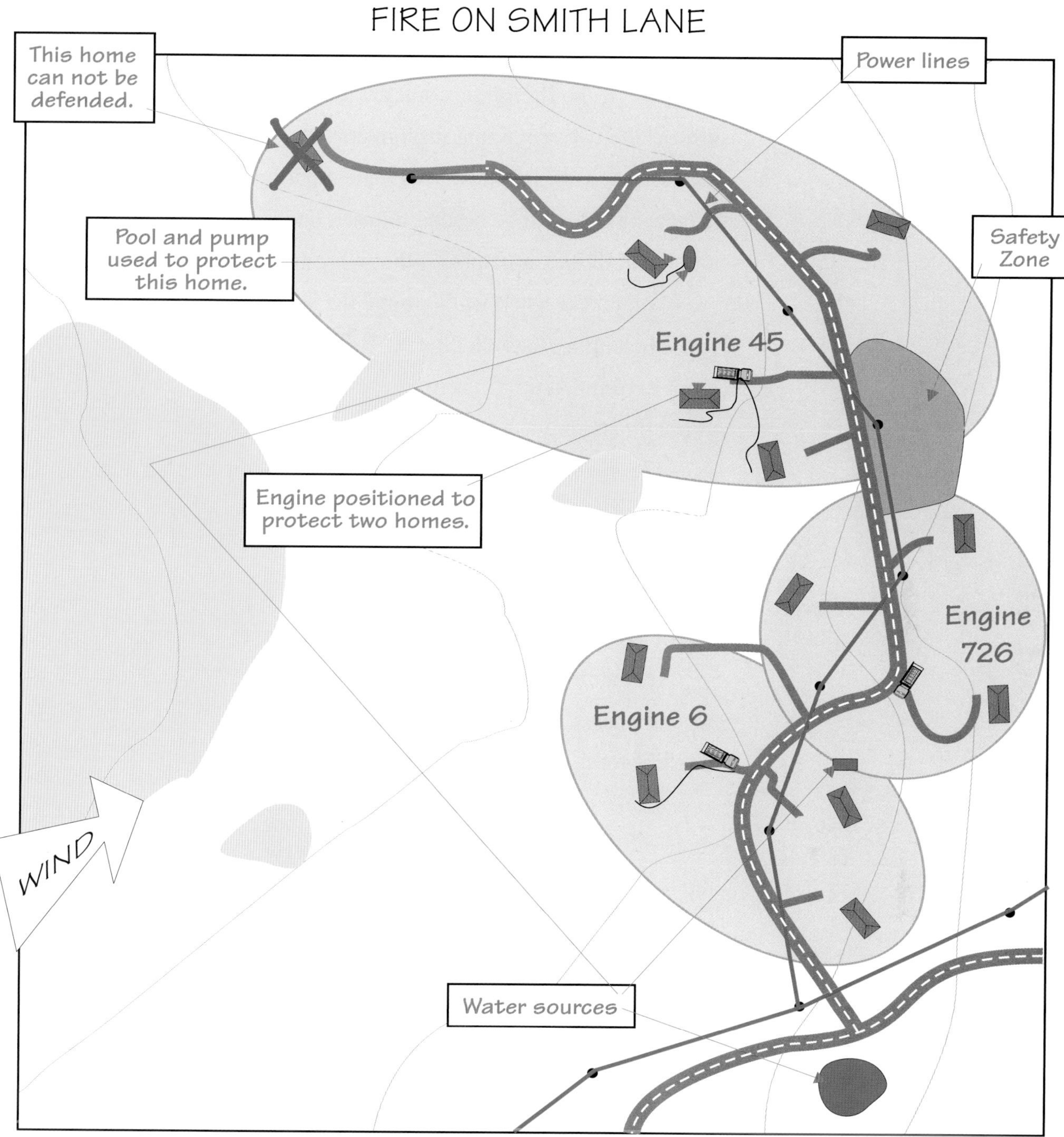

Figure 11.11 Three engines are assigned to protect 12 homes along Smith Lane. The last home on the lane is not defensible. Areas of responsibility have been identified. A safety zone and water sources have also been found. Once the fire passes through the area, one engine will remain in the area on patrol.

You move back down Smith Lane to where you dropped off the two firefighters. You back into the driveway, positioning the engine in the widest part of the drive. Both firefighters return to report what they found. Before they have time to report, you see a spot fire starting off to your left.

Prevent torching and crowning near structures.

One of the firefighters quickly moves to put out the spot. You instruct the two remaining firefighters to each lay a 1½-inch line between the two structures, to create a perimeter. Two 100-foot lengths of hose will be needed for each line. The third firefighter is told to ladder the structure on the right; its roof may be a problem.

You take a quick walk around the structures. You note that the fire is beginning to torch pine trees about 300 feet down the hill. You don't have much time. There is a garden hose near the engine. You turn it on and put it into the tank filler. Water spills out.

Improve your odds of saving the structure.

The fire is here. One of the firefighters shouts that he needs water. You run back to the engine and charge both lines. You hear the pump starting to supply water to the nozzle. You move toward the other firefighter. She is experienced, but has never been in a situation like this.

As the fire moves closer, the firebrands begin to ignite spots in several places. You stomp on a couple and put them out. The smoke is really getting thick.

It takes about ten minutes for things to really heat up. The main force of the fire is now upon you and your crew. You wonder how the firefighter with the floto pump is doing. He is experienced and should do fine.

You see the fire moving under several pine trees close to the house. You calmly instruct the firefighter to cool the fire before it has time to torch trees and move into the crowns. She covers the burning area with a mist of water, cooling the fire.

Keep mobile; check on the other engine companies.

The firefighter protecting the other home says he has smoke coming from the eaves of the roof: an attic fire. You grab an axe from the engine and move to assist him. You move into the home through the back door, bringing the firefighter and line with you. You find a place close to where the smoke was coming from, in the dining area. With an axe you begin pulling the ceiling. You are making a real mess, but you find the piece of insulation that is on fire. You wet it down and pull several more feet of ceiling to make sure you have it all; nothing else is found.

You feel the fire is moving on. Most of the lighter grass has been burned; trees are still burning. The scattered woodpile is burning; the family camper trailer and one pumphouse were lost. You feel you must check on the other firefighter and the homeowner just up the road. In ten minutes you should be able to move.

Make resources available for other assignments when you can.

Conditions are stabilizing; time to check on the others. One firefighter is left to patrol the area around the two homes. The lines are disconnected from the engine and left in place. You move down the road to check on the other firefighter and the homeowner. The home is still standing. There appears to be no damage. After a short discussion with the homeowner, the floto pump, hose and other equipment are loaded back on the engine. You move back toward the other engines. As you move back up the road, you see that the last home on the lane was lost. After checking with the other homeowner, you move back to where you left the last firefighter. You move around the structures checking for fire. None is found. By this time, all the equipment has been reloaded and a short note placed on the kitchen table in the house where the ceiling was pulled.

Watch out for holdover fires in abandoned vehicles, woodpiles, brush piles, etc.

As you move back up Smith Lane, you see that two more homes were lost. After meeting with the crews from the other two engines, it is decided that Engine 726 will remain in the area for the next two hours on patrol. You don't want to lose one of these houses now!

Engine 45 and Engine 6 move to the pond and top off the tanks. You then call your supervisor or move to the ICP to get the next assignment.

Don't let your engine become another exposure.

Situation 7 - A fire involving a LPG tank

You have been dispatched to a structure fire involving a liquid petroleum gas (LPG) tank. You will be the first engine at scene.

Problem

As you arrive at scene, you see an LPG tank burning from the top of the tank, and the roof and eaves of the structure next to it are on fire. There is another structure exposed to the radiant heat.

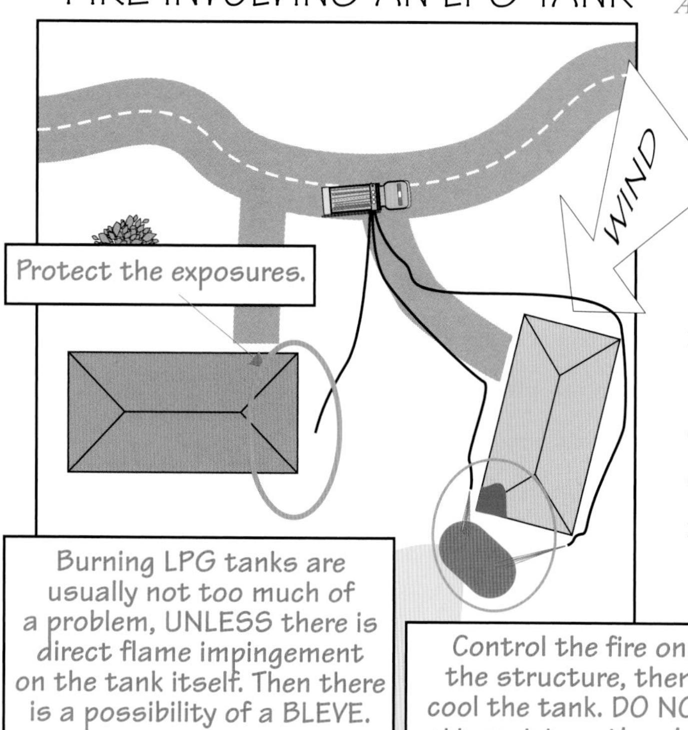
FIRE INVOLVING AN LPG TANK
WIND
Protect the exposures.
Burning LPG tanks are usually not too much of a problem, UNLESS there is direct flame impingement on the tank itself. Then there is a possibility of a BLEVE.
Control the fire on the structure, then cool the tank. DO NOT attempt to extinguish the venting tank.

Figure 11.12 The LPG tank is burning from the relief valve at the top of the tank. Radiant heat from the fire has ignited the roof of the structure. Attack the fire in the structure and begin cooling the tank. Since there is no direct impingement of flame on the tank, the chance of the tank rupturing is reduced. Clear the area of spectators. Cool the tank until the relief valve resets itself, or the gas is expended.

A Possible Solution

The first action is to position the engine so that it is not lined up with the ends of the tank. If a tank ruptures, the end-welds are usually the first to rupture (Figure 11.12). Then, evacuate the houses and clear the area of spectators. Instruct them to move back several hundred feet.

The next action should be to lay lines to extinguish the fire on the structure and to begin cooling the tank. Since there is no direct flame impingement on the tank itself, cool the top of the tank to slow the venting. Do not attempt to extinguish the fire on the tank. If you were to do that, the gas would continue to escape. LPG is heavier than air, so it would flow along the ground toward low spots. If it were to find a new ignition source, you would really have a problem on your hands.

Cool the top of the tank. The bottom of the tank is cooled by the liquid LPG. If you have the personnel, lay a line to protect the other structure. Don't use this line unless it is evident that the radiant heat from the burning tank is about to ignite the structure.

Watch out for BLEVE situations.

If you have direct flame impingement on the tank surface, you have an entirely different situation. If the impingement is above the liquid level (above the frost line) the heat may be weakening the metal. If this occurs, and the tank ruptures, you have a BLEVE condition. (A BLEVE is a boiling liquid expanding vapor explosion.) This type of explosion is very violent and dangerous. If you have direct flame impingement, clear the area of all spectators for several hundred yards. If you cannot apply water to the spot where the flame is hitting the tank or you don't have a water source to sustain a cooling operation, clear the area and let the tank burn.

Continue to cool the tank until the gas stops venting. Ask your dispatcher to send an expert on LPG tanks to the scene to assist you in securing the tank.

Situation 8 - Using legal descriptions and a map

You are at the scene of a small fire and you are asked for the legal description. Describe your location to a 40-acre block—this is ***Problem 1*** (Figure 11.13)

You have been dispatched to a new fire that is located in the SW¼, NE¼, of Section 26, Township 34 North and Range 1 West. Your are to proceed there immediately. Also, plot the route you would take from the nearest town—this is ***Problem 2***.

The answers can be found on page 414.

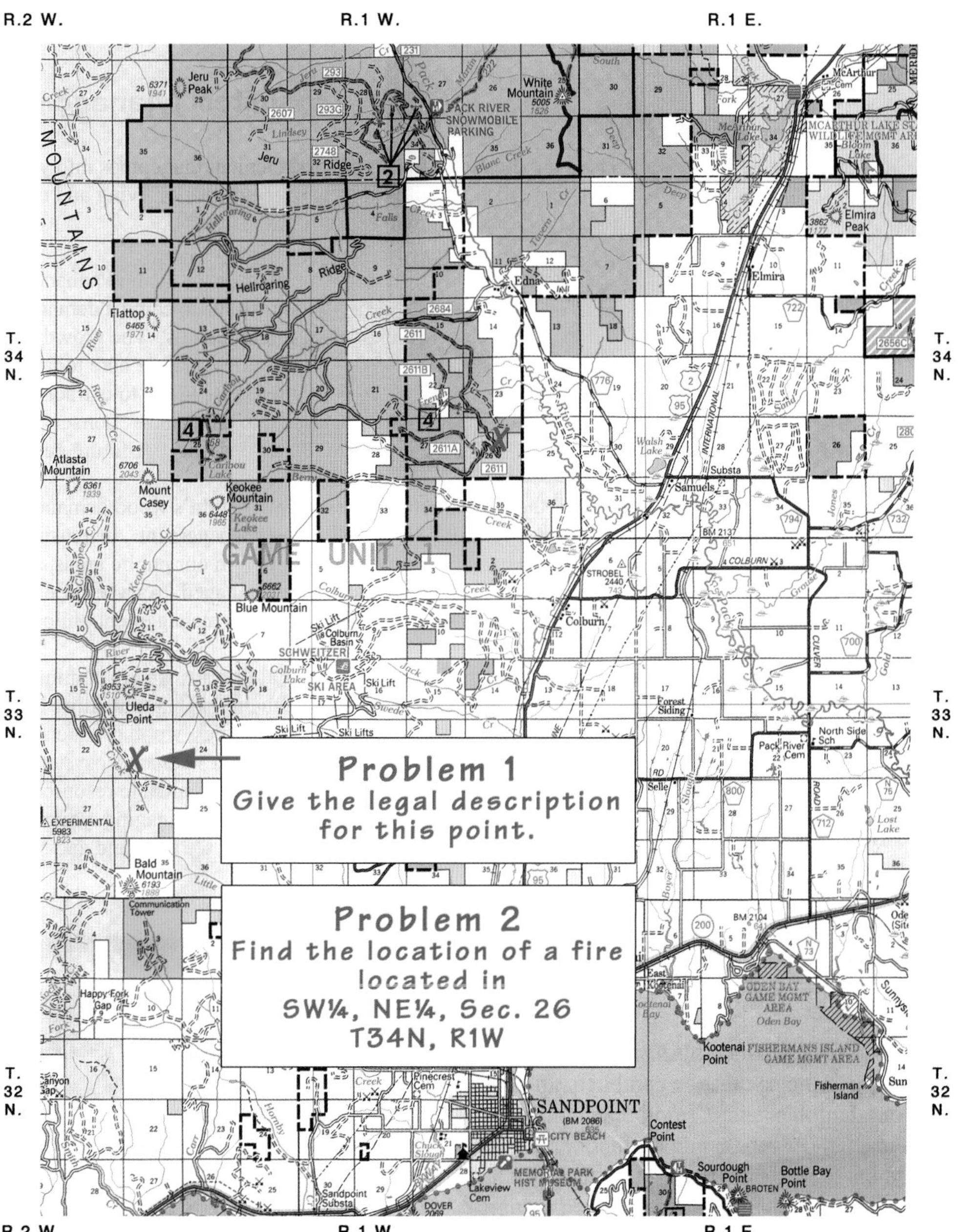

Figure 11.13 Here is a portion of a USDA Forest Service map for the Idaho Panhandle National Forest. Use it to describe in legal terms two locations.

Situation 9 - Fire moving toward heavy fuels

You are part of a first-alarm assignment to a brush fire. The fire is just down the road from the station. You are responding with five engines, a bulldozer, air attack, two airtankers and a chief officer. You will be first at scene.

Problem

The fire is about 10 acres and moving up a slope toward a knob. The wind is from left to right; it is hot and dry. There is an old road through the saddle behind the knob. There is no access to the knob. There will not be a threat to structures for some time. The right flank of the fire is burning toward a patch of very heavy fuel. The first airtanker has arrived and says there is a spot fire on the top of the knob. There will be no additional help.

Keep an engine in reserve for contingencies.

A Possible Solution

The spot fire needs attention (Figure 11.14a). Direct the airtanker to attack the spot fire and keep the fire from moving into the heavy fuels on the right flank. If the road on the right flank is passable, it might be used as part of the fireline. Establish a staging area and have dispatch instruct all responding resources to report to that location.

Attack spot fires while still small.

You develop an action plan that has the bulldozer and two engines move up the road ahead of the fire. Their assignment is to open up the road and fire it (Figure 11.14b). The objective is to keep the fire on the ridge and away from the heavy fuels. Two engines will be assigned to the left flank to begin a progressive hose lay. The fifth engine will provide water and assist either operation as needed. The air operations are told to keep the fire on the knob and to support the flanking operations upon request.

Keeping fire out of heavier fuels is usually a good strategy.

The road is opened, but the fire creeps through the retardant and moves down the back side of the hill. The road is fired-out, and it appears to be holding. The fire has not moved into the heavy fuels.

The fire is contained and the airtankers released. The engine crews are instructed to begin mop-up. The bulldozer is to cut line all the way around the fire.

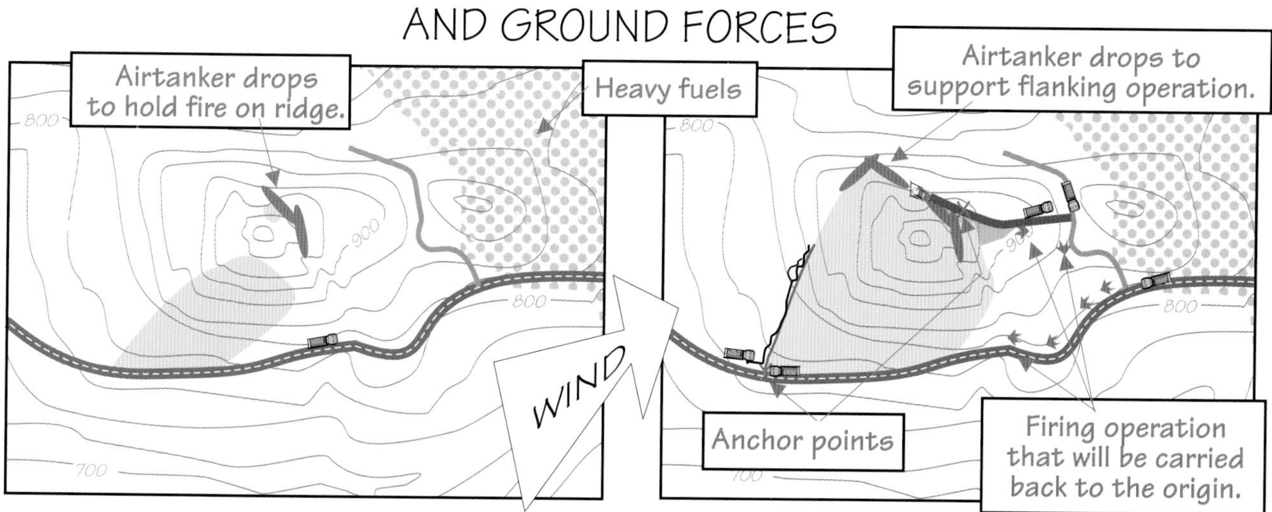

Figure 11.14a The first action is to try to keep the fire from moving into the heavy fuels on the right and over the ridge. The air attack officer is given that responsibility.

Figure 11.14b The bulldozer is working up the ridge, tying into the airtanker drops. This line is being fired from the top of the ridge, down the fireline, back toward the origin. The progressive hose lay is moving up the left flank. The airtanker drops are supporting both operations.

Situation 10 - Firing-out line

You are assigned as one of three engines to fire-out and hold a section of road.

Problem

The section of line you are assigned is along a road that climbs up a slope, around a switchback and through a saddle. There are no structure protection problems. The wind is pushing the fire up and over a hill, toward the proposed fireline.

Always fire-out going downhill, unless the wind dominates.

A Possible Solution

You want to burn INTO THE WIND and DOWNSLOPE, if you can. In firing this section of line, there are several key points: Start firing from the highest point, the saddle, and fire in both directions. Carry the fire down the road toward the southwest to cut off the head. This is the main firing operation and can be started by one engine company. The second engine should fire down the road the other way from the saddle, veering left down the canyon beyond the head. This is done to anchor the firing operations for the time being (Figure 11.15).

Burn-out the interior of switchbacks ahead of the main firing operation.

As you carry the firing toward the south, the third engine should fire the interior of the switchback before the main firing operation arrives. The switchback could be a difficult operation, so take it slow. As soon as the switchback is fired, continue moving down the road, toward the right flank of the fire.

This section of line can be fired with two engines, one doing the firing and the other patrolling. The third engine should stage at the top of the hill until it is time to fire the northeast corner of line. This

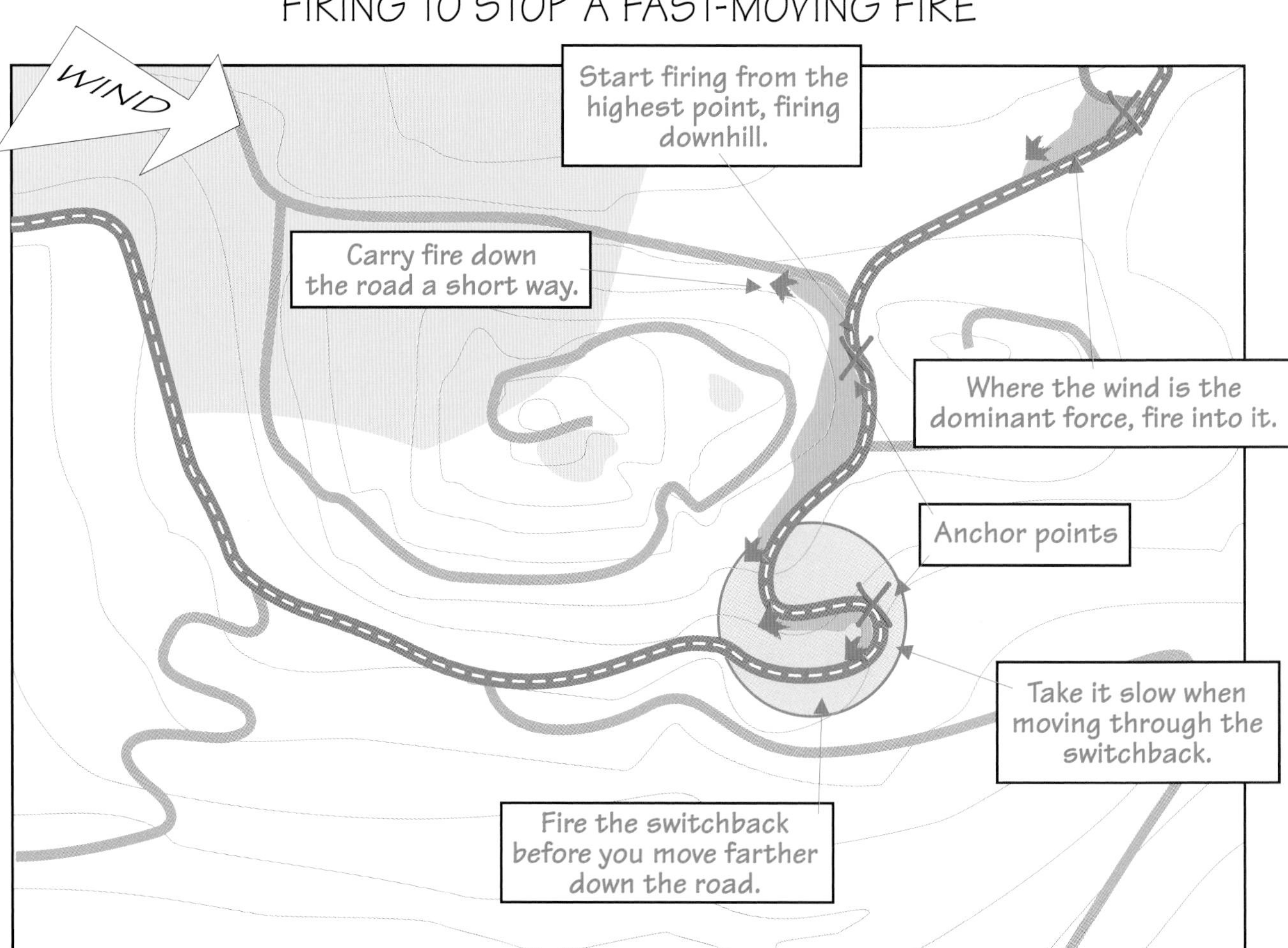
FIRING TO STOP A FAST-MOVING FIRE
WIND
Start firing from the highest point, firing downhill.
Carry fire down the road a short way.
Where the wind is the dominant force, fire into it.
Anchor points
Take it slow when moving through the switchback.
Fire the switchback before you move farther down the road.

Figure 11.15 Fire the road in several segments. Start at the saddle, firing down in both directions. Fire-out the road a short distance to anchor that section of line. When you reach the switchback, don't carry the main fire through it. Fire the interior of the switchback BEFORE you continue down the road. The switchback may be hard to hold. Fire the northeast section against the wind, back toward the saddle last.

section is also fired against the wind, even though it is slightly uphill. The wind is the dominant force.

Once the line is fired, the engines will be used to patrol, or part of the force can be reassigned.

Jake says, "You could shortcut the switchback with a scratch handline or wet line to fire from to avoid the time delay to fire the interior of the switchback."

Situation 11 - Firing against the wind

This is the same section of line that was used in Situation 10, except the wind is from the east, blowing toward the fire.

A Possible Solution

Fire downhill or into the wind, whichever is dominant.

The firing operation is conducted in the opposite direction, all against the wind (Figure 11.16). Again, the wind is the dominant

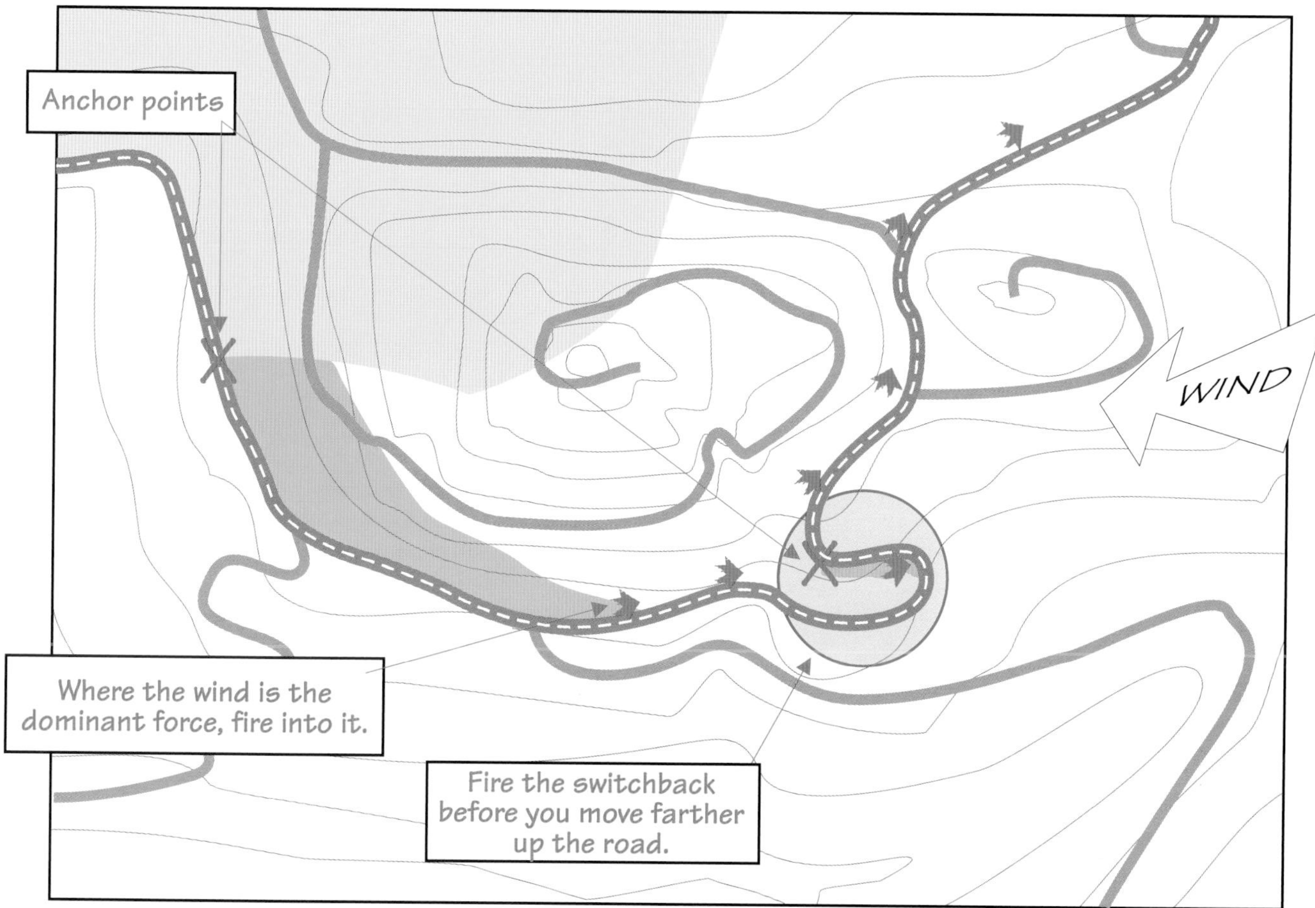
FIRING INTO THE WIND
Anchor points
WIND
Where the wind is the dominant force, fire into it.
Fire the switchback before you move farther up the road.

Figure 11.16 The wind is the dominant force; fire into it. Anchor your operation at the fire's edge, moving up the road. Fire-out the switchback before you proceed. Carry the firing operation to the end of your division.

force. If there is little or no wind, then the topography is the dominant force and you would fire accordingly.

All three engines can work as a team in this operation. Anchor the operation at the leading edge of the fire and begin firing uphill, into the wind. As the operation reaches the switchback, one engine should fire it, starting at the top turn, burning through the turn, back toward the main firing operation. Once the switchback is fired, proceed with the operation.

Coordination is mandatory.

Carry the firing up and through the saddle, to the end of the division. Watch out for wind eddies in the saddle. Establish an anchor point at the division boundary. Make sure the folks in the other division know you are bringing fire to them.

Situation 12 - You are assigned as a lookout

You are a member of a state forestry engine company. There are five of you on the engine—a special staffing pattern because of the weather—has you working today. It is a Type 3 wildland engine, with a crew leader and four firefighters. You are the most experienced firefighter (with three summers on this crew), with this crew leader. The crew leader has been told by the Division/Group Supervisor that two additional engine companies, a water tender and home Type 2 hand crew are en route to assist in the hose lay and line construction.

Figure 11.17 The crew assignment is to construct fireline up the ridge from "Point b" toward "Point a." You are the lookout and position yourself where you can see the ridge and where the crew will be working (near "Point c").

Problem

Your crew has been assigned to begin a hose lay to stop the fire from slopping over the ridge ("Point a") (Figure 11.17). The instructions were to anchor the hose lay in the creek at about "Point b," moving up the ridge. It is 1230 hours, in August. The summer has been dry and the temperature is in the mid-90s. The winds are variable, gusting to over ten miles-per-hour. Relative humidity is in the teens; flame length is over four feet in the grass and worse farther up the slope.

Because the ridge line cannot be seen from the bottom of the canyon, the crew leader assigns you to be the lookout. You have never had this assignment before, but you have attended two classes on LACES. All the crew leader told you was, "We're going to anchor to the creek and start a flanking action up the ridge line staying on edge of the fire. Find a spot where you can see the fire and us, and keep me informed on what is going on. Now move!" What do you do?

You must monitor the fire, the weather, the crew's progress, etc.

A Possible Solution

You have been on several fires that were burning like this, but this one just feels worse. You think to yourself, "It is going to be a long day." You think back to your training on LACES. What do I do, how, and where? You look around and spot a small knob above the engine on the opposite side of the canyon; you pick that as the spot from which you will watch the fire. You put on your PPE, make sure your portable radio is working and that you have an extra set of batteries. You pick up your McLeod and start up the hill.

It will take you about ten minutes to reach the knob from which you think you can see the fire and crew ("Point c"). Part way up the hill, you look toward the fire and see that it is still on the ridge, but is getting close to crossing near "Point a." You also see the crew laying hose up the ridge, moving slow because of the steepness of the terrain, fire behavior, and that there are only four of them. You hear on the radio that the other crews assigned to this division are just down the road.

Keep the crew informed...lookouts don't keep secrets!

You get to the knob, but notice that you can't really see the crew, so you move back down the hill a little, so you can see the ridge line, the fire and the crew at work. You take out your Kestrol weather instrument and take the weather, recording it in your notebook. You note that the wind is now steady at 12 miles-per-hour, and the relative humidity is at 14 percent. You also note that the fire is burning on a southwesterly facing slope. You radio the crew leader and tell him you are in place, and can see the ridge, the fire and the crew at work.

You see an airtanker approaching, and then start a wide circle. You switch frequencies on your portable radio and hear "Tanker 215" trying to contact your crew leader, "Engine 5466." (Unbeknownst to you, your crew leader had asked for some air support to help keep the fire from crossing the ridge. They were only able to get a SEATs, an Air Tractor-802 with 600 gallons of retardant...it may be small, but it can be very accurate.) Just then, your crew leader comes on the air and asks the pilot to drop on the ridgeline where the fire is about to cross (Figure 11.18).

Not all airtankers are created equal...but even a small one can be useful.

You have a minute now to assess your position. You know that if things go bad for the crew, they can move back down the ridge toward the river, or into the black. They are still working the fire in scattered brush and that would work. Your position is a little tougher. If fire gets below you, you have to get off this hill fast. If you go right, it will take you upcanyon and into some very thick stuff. Uphill is out…steep, rocky and worse. If you go left, you can very quickly be into light fuels and easily walking back to the road…that will be your escape route.

Figure 11.18 Airtanker 215 drops on the ridgeline just below where the fire has crested the ridge. Hopefully the drop will hold the fire until the hose lay reaches this point.

You study the fire for a minute and see what you think is a small puff of smoke down over the ridge, well into the green. You also note that the smoke on the main fire is beginning to swirl and move lazily back and forth. You make a note of this in your notebook, indicating that it is now 1320 hours. As you finish, you see a second, much larger airtanker making a run downcanyon…looks like he is going to try to lay a line down the ridge as a reinforcement to AT 215's drop. The drop appears to be long, most of the retardant falls into the black. You also see that just after the drop, the smoke behind the crew makes a real funny barrel-rolling motion…wing vortex. Just as quickly, you see a spot fire behind and below the crew (Figure 11.19).

Wing vortex can spread fire.

You are on the radio in a flash, warning the crew leader that they have fire across the ridge and the wind may be about to change directions. He says he will check it out.

Figure 11.19 You see that the fire has spotted over the ridge in two spots, and one tree is torching. It is time for the crew to be alerted.

You mention, "did he know he was on a south facing slope, and that the smoke column was starting to do some funny things?" You also state, that if they have to move to safety, move into the black and return to the engine. He acknowledges.

As soon as you see something is changing, let people know. Don't wait until it gets dangerous.

A lookout must also look after his/her personal safety.

You watch as the crew leader briefs the crew and see him move back down the line. The crew appears to be taking a break. The spot fires are growing in size. The crew leader comes on the radio and says they will have to abandon this line and they are retreating to the river. You see the crew disconnect the nozzle, pick up their gear and move down the ridge to safety. The crew leader tells you to stay there for a while, until he and the crew regroup and start the attack again. He says, "well done and thanks."

Note: Think about what you would have done in this situation. Lay out several different scenarios and discuss how you would react to them.

Situation 13 - Slopover below the road

You are part of a Type 2 hand crew assigned to a Division D on the Carson Fire. The crew assignment is to fire-out and hold a road. You may have some help, but not likely. You are firing-out a logging road and all appears to be going well.

Problem

The fire has spotted across the road and is running loose. Word is passed up the line to the crew leader that they have fire below the

Northtree Fire International

Figure 11.20 The fire has jumped the road. Your assignment is to pick up this slopover.

road (Figure 11.20). The crew leader stops the burning and heads back down the road. You and a couple of the crew arrive at the site of the slopover just as the crew leader gets there. The spot is about 50 by 200 feet in size and starting to move into the crowns of the trees. The crew leader looks at the three of you and says, "Go get it—and you are in charge." He points out that the fuel is heavier to the left. He wants you to keep him informed as to your progress. They will continue burning, but at a slower pace.

A Possible Solution

The three of us discuss our options. We could attack the spot from both ends, working toward the middle, or we could work together. We elect to anchor to the road on the left end of the slopover and take a flanking action to keep the fire from moving into the heavy fuels. We also decide that the line we put in will be our escape route and the road and black above the road would be our safety zone. We also discuss the fact we could move down below the fire and downcanyon to a road about 100 yards down the slope.

Anchor to the road and begin to flank the slopover.

We also discuss that I would also serve as the lookout. Since the spot is relatively small, one could see most of it at any one time. If I could keep moving and watching, there shouldn't be any surprises.

We tie to the road and start cutting direct line. It takes us about 20 minutes to cut the fire off from the heavier fuels, but the fire now covers about an acre and is heating up. We report in to our crew leader as to our progress and ask him to provide us some help—the slopover is heating up and will become a problem if we can't take some heat out of it. He says he will see what he can do.

We continue to cut line, every so often discussing the situation and our options if things get bad. We still feel our plan is sound, but we all agree we need water to take some of the heat out of the fire.

The crew leader radios that an engine and water tender will be there in 30 minutes. He says, "meet them on the road and line them out."

It is over an hour before the engine arrives and the slopover is now over five acres. They are instructed to take the right flank, anchor to the road and work toward us. Within 25 minutes we have a rough containment. The next step is to strengthen the line and then begin mop-up.

If you don't work quickly to contain spot fires, you may not get a second chance.

Situation 14 - An assignment you don't feel is safe

You are assigned to Division D on a couple-hundred-acre fire, burning in steep terrain and heavy fuels. You are part of crew that has been assigned to construct line down a ridge to a creek (Figure 11.21). You will tie in with line being constructed from below.

Problem

It is late morning when the crew arrives at Check Point 14, at the top of the ridge. You meet with "Division D," and he tells you he needs to construct line from the ridge toward "Crew 6" coming up from the bottom. The fire is some distance off to the left. You can see smoke off in the distance, but the fire is hidden by a ridgeline and vegetation. The weather forecast said that a cold front will be moving through the area later in the afternoon. You also note that this ridge runs from west to east, thus the fire is on a south-facing slope. You are not happy with this assignment.

Cammeron Eck

Figure 11.21 It is mid-afternoon and you are on a south-facing slope. The wind is switching directions and you feel that what you have been assigned to do is unsafe. What are you going to do about it?

A Possible Solution

You know the importance of this line to the overall plan to contain the fire, but...

You have the right and responsibility to decline an assignment if you feel it is dangerous.

As a crew, we are not happy with the situation. We have been asked to construct fireline downhill, above a fire on a south-facing slope. We would be on the ridge for most of the day, and a cold front may move through the area later in the day.

I point out to the Division/Group Supervisor our concerns, but he is insistent that we get started. I outline our concerns, stressing that it would be safer if the line was constructed that night. If not, we would have to decline the assignment.

The Division/Group Supervisor understands our concerns, and says, "let's discuss our options for holding this fire along this ridge." He informs the Operations Section Chief that there are too many safety concerns for the line to be constructed today, and suggests that no one be placed on that ridge today, and tonight "we hit it hard."

After some discussion, the plan is modified, the crew is told to rest, while further "recon" is conducted.

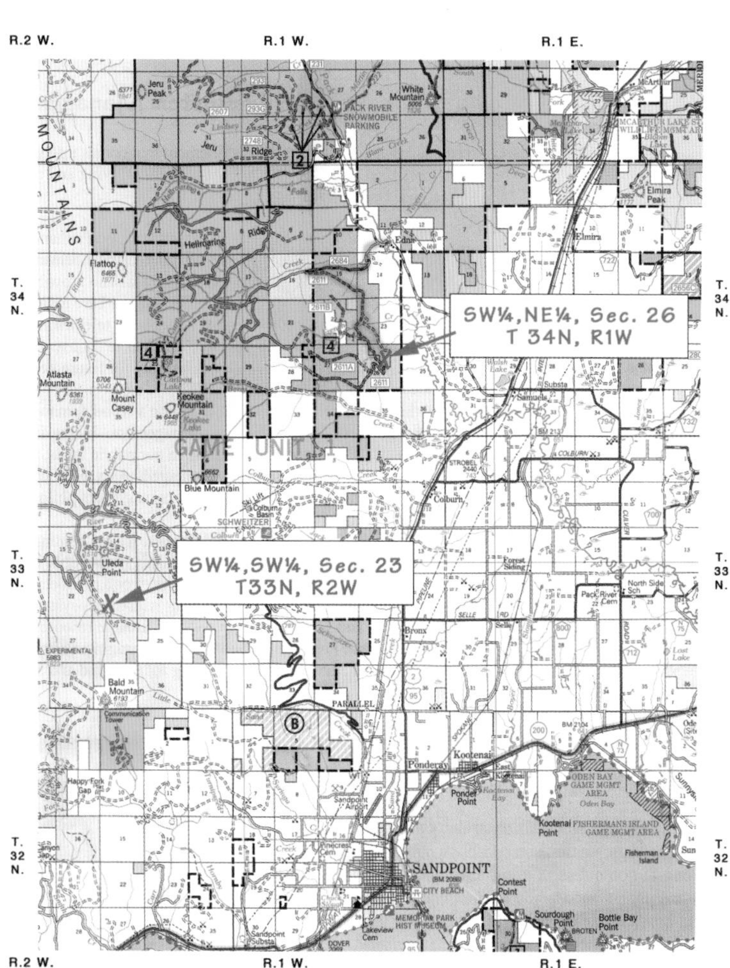

Figure 11.22 This is the solution to Problem 8 on Page 403.

An Alternative Situation

There is the possibility that the Division/Group Supervisor insists that your crew construct the fireline, as planned. You discuss with him your concerns, but he disagrees and insists we "get to work." I then tell him, "We decline the assignment because it will place the crew in a very dangerous situation."

The crew then moves off the line and returns to staging or reassignment. You make a note in your log about the "turn down" for later reference.

Appendix

Map Reading

The ability to read a map is an essential skill which must be acquired by all wildland firefighters. The importance of this skill is more readily appreciated if its labor-saving qualities are understood. Consider the value of the time saved during tactical and strategic planning when time is critical.

A map is a graphic representation of features on the ground.

Firefighting operations are seldom conducted in areas that are totally familiar to all the firefighters, and the incident commanders must base their decisions upon information obtained through ground or aerial reconnaissance. Although a map reconnaissance can seldom completely replace one on the ground, it can materially reduce the time required (Figure A1.1).

Figure A1.1 "Look for anything with a name on it."

Before proceeding to a detailed discussion of the methods and techniques of map reading, a formal definition of the word "map" is needed. The most generally accepted definition of a map is "a graphic representation of a portion of the earth's surface, drawn to scale upon a plane." The term *graphic* pertains to representation of physical features by lines, symbols, etc.

Maps may be classified according to the design of the map, according to the scale, and according to the use to which the map is put. Firefighters generally use two types of maps, road maps and topographic maps (Figures

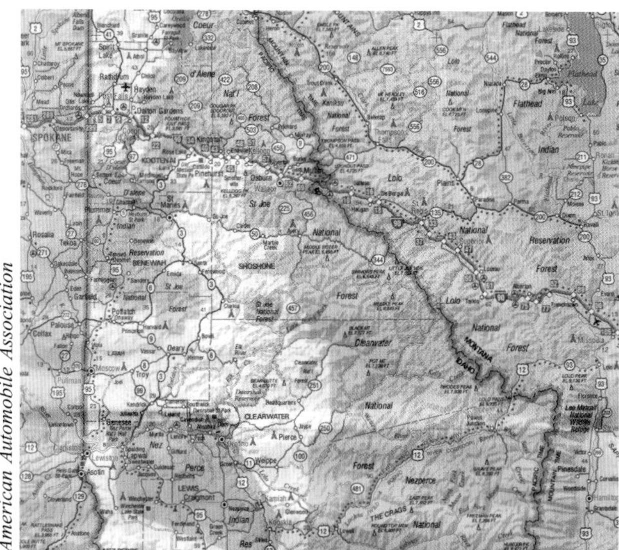

American Automobile Association

Figure A1.2 Road maps are classed as small scale. They primarily show towns and highways, and meet the needs of vehiculer travel, but are of little value to the wildland firefighter.

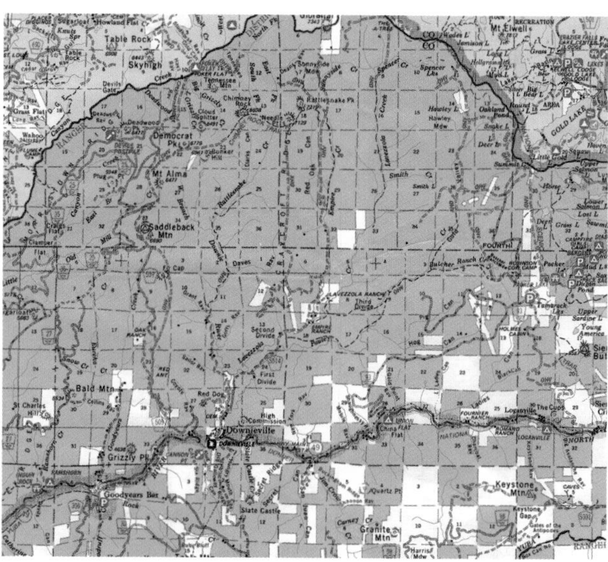

Figure A1.3 Forest maps are classed as medium scale. There is more detail on these maps than on road maps. They do include public land survey information.

A1.2 to A1.6). This section of the book will concentrate on the use of the topographic maps.

Map reading skills are critical to planning and operations.

Different map designs are used for different purposes.

The topographic map is a graphic representation of the three dimensions of the earth's surface; the vertical dimension being indicated by contour lines (a line on a map representing an imaginary line on the ground, all points of which have the same elevation). The topographic map is much more useful to the wildland

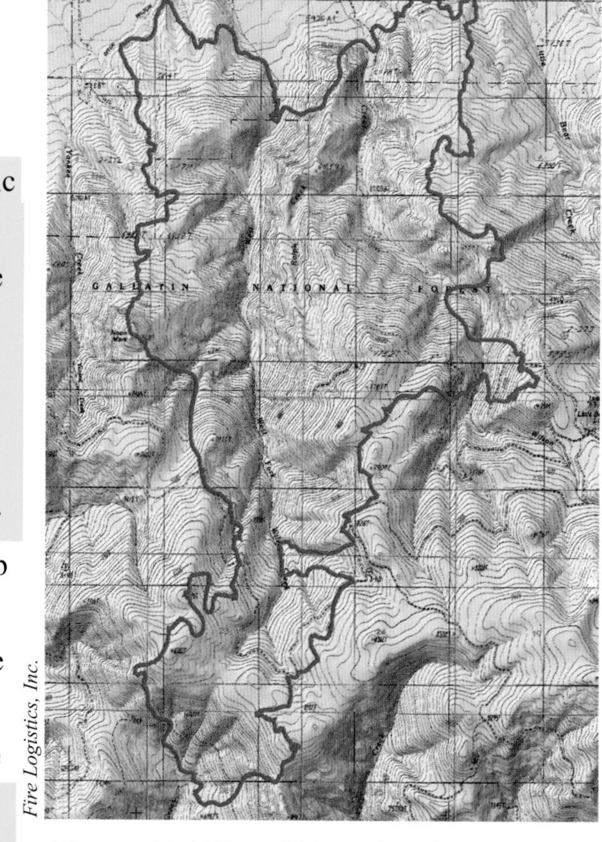

Fire Logistics, Inc.

Figure A1.4 This GIS produced map shows the fire perimeter, as well as topographic and other detailed information.

firefighter than a road map, because it depicts the terrain and is usually much more detailed.

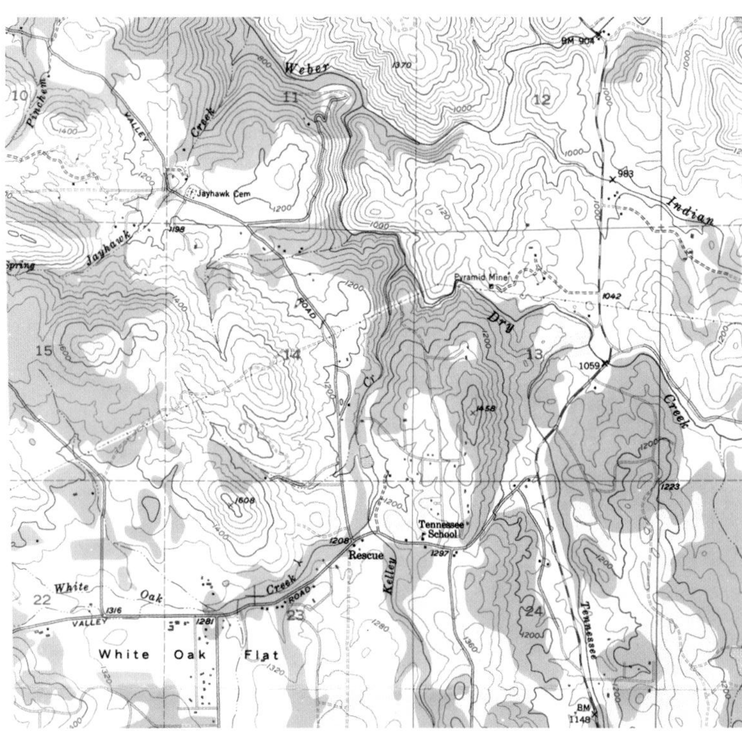

Figure A1.5 This topographic map (which is classed as large scale) provides road and stream information, and it also provides topographic or elevation information. This is the map type of choice for a wildland firefighter.

Geographic Coordinates

Visualize a map covered with roads, trails, contour lines, rivers and other symbols, and try to imagine how you would designate a particular feature so that another firefighter could understand precisely which feature you meant. The location of points or the indication of positions on a map or on the ground, may be accomplished by two methods: latitude and longitude coordinates, or public land survey data.

The oldest systematic method of location is that of geographic coordinates (latitude/longitude). This method was originally devised

Common map data

Type of map	Representative fraction (scale)	One inch represents	Map size (in lat/long dimensions)	Area covered by map
7.5-minute	1 : 24,000	2,000 feet	7.5 x 7.5 minutes	49 to 71 sq. miles
15-minute	1 : 62,000	about 1 mile	15 x 15 minutes	197 to 282 sq. miles
1 : 100,000 (US)	1 : 100,000	about 1.6 miles	1/2 x 1 degree	1,832 to 3,467 sq. miles
1 : 250,000 (US)	1 : 250,000	about 4 miles	1 x 2 degrees	4,580 to 8,669 sq. miles

Figure A1.6 Common map types and specifications for the maps you may be using.

LATITUDE AND LONGITUDE

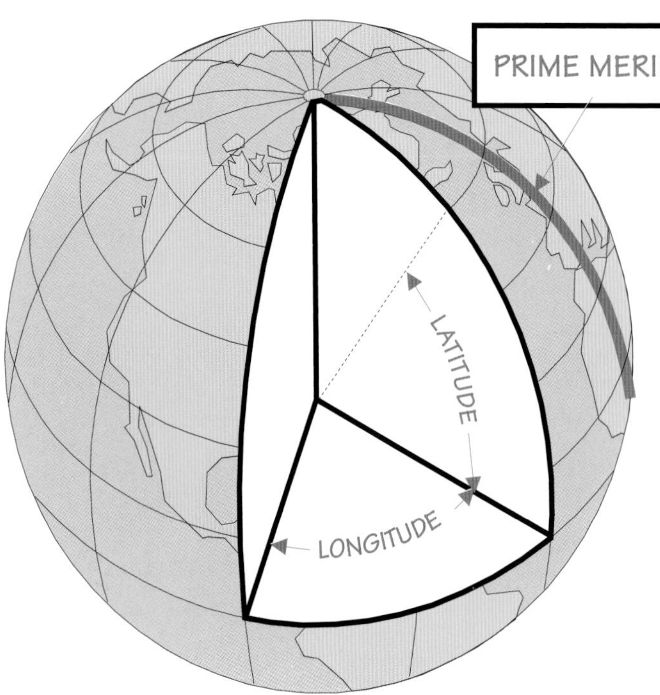

Figure A1.7 The use of latitude and longitude coordinates in wildland fire operations has become commonplace. You should know how to use this universal system of map reference. Global Positioning System (GPS) equipment is making the use of latitude and longitude much easier.

to permit the navigation of vessels out of sight of land by substituting the stars and planets for the landmarks of the coastline.

The position of any point on the earth's surface may be indicated by giving its distance north or south of the equator, and east or west of a line passing at right angles to the equator and through the poles.

Latitude and Longitude

This network of lines is the basis of the system of geographic coordinates. The distance a point is north or south of the equator is called its latitude, and the rings around the earth parallel to the equator are called ***parallels*** of latitude, or simply parallels (Figure A1.7). The distance a point is east or west of the line through the poles is called its longitude, and the lines converging at the poles are called ***meridians*** of longitude, or just meridians; the line from which measurements are made is called the prime meridian.

Latitude and Longitude are used worldwide.

- ***Latitude*** - Starting at the equator, the parallels of latitude are numbered from 0^0 to 90^0, both north and south. Thus the South Pole is at latitude 90^0 south, while the parallel halfway between the equator and the North Pole is latitude 45^0 north. The direction, whether north or south, must always be given. Because it is impossible to go farther north or south than the poles, no latitude value ever exceeds 90^0.

Without the "north" or "south" designations, the coordinates are not complete.

- ***Longitude*** - The prime meridian is the meridian that passes through Greenwich, England. From the prime meridian, longitude is measured both east and west around the world to the meridian exactly opposite. Lines east of the prime meridian are numbered from 0^0 to 180^0 and are called east longitude; lines west of the prime meridian are numbered from 0^0 to 180^0 and are called west longitude. The line directly opposite the prime meridian thus has a value of both 180^0 east and 180^0 west. No longitude measurement ever exceeds the value of

180°. (Note: Many foreign maps do not use the Greenwich meridian as the prime meridian; for example, French maps use the Paris meridian, and Italian maps the meridian passing through Rome.)

The values of latitude and longitude will mean more if they are compared with the linear values they represent. A degree of latitude represents 1/360th of the distance around the earth through the poles, or about 69 miles; one second of latitude is about 100 feet. The meridians of longitude converge from the equator toward the poles, so the linear distance of about 69 miles represented by a degree of longitude at the equator decreases to zero at the poles. One second of longitude represents about 100 feet at the equator; at the latitude of Rescue, California, it is about 79 feet.

The geographic coordinate of a point has two parts: latitude and longitude. When the coordinates are written, it is conventional to write the latitude value first. The coordinates of the post office in Rescue, California are approximately, Lat 38° 42′ 30″N, Long 120° 57′ 10″W. Remember, without the north and west designations, these numbers can also refer to three other places on the earth.

Before GPS equipment came on the market, the use of lat and long was impractical. But, now you can spot yourself anywhere on the ground to within a couple of feet. Figure A1.8 shows what the GPS coordinates look like.

Global Positioning System

The 24 GPS satellites are in orbit at 10,600 miles above the earth. The satellites are spaced so that from any point on earth, four satellites will be above the horizon. Each satellite contains a computer, an atomic clock, and a radio. With an understanding of its own orbit and the clock, the satellite continually broadcasts its changing position and time on the ground; any GPS receiver contains a computer that "triangulates" its own position by getting bearings from three of the four satellites. The result is provided in the form of a geographic position - longitude and latitude - for most receivers, to within a few feet.

Public Land Survey

The United States Public Land Survey (USPLS) started with the Land Ordinance of 1785 and covers all US land that was not settled by the time of the official government survey. It does not include the east coast states including Kentucky and Tennessee or Texas. Small areas of other states that were settled

Figure A1.8 GPS coordinates for the Rescue, CA Post Office. Note that you can also determine your elevation.

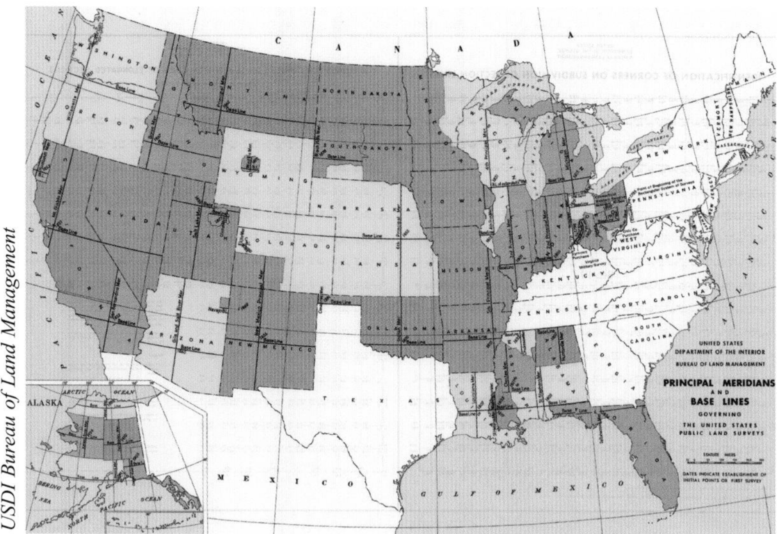

USDI Bureau of Land Management

Figure A1.9 The Public Land Survey established the primary meridian and base lines. They serve as the reference points for townships and ranges.

before the survey are also technically excluded. This explains the odd shapes of many mining claims in the west. Later title transfers followed USPLS lines.

Legal descriptions to one mile square sections have 4 parts: the Section Number, Township, Range and State/Meridian. Note that an official legal description always works from small to large areas, for example NW1/4, sec. 12, T.122N, R.71W, South Dakota, Fifth Principal Meridian.

- ***Initial Points*** - There are 35 initial points established in the United States. In a lot of cases, thire are prominent geographic features like mountain tops or hills. The line that goes north and south of the initial point is called the ***principal meridian***. The line that goes east and west through the initial point is called the ***base line*** (Figures A1.9 and A1.10).

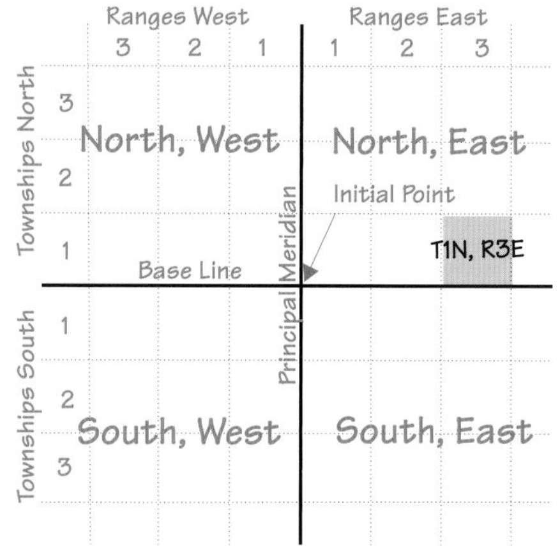

Figure A1.10 Townships are north and south of the base line; ranges (like mountain ranges) are east and west of the principal meridian. A township is 36 square miles in size.

- ***Township*** - These occur at 6-mile intervals north and south of the base line. Township values are normally whole numbers starting at 1, but some exceptions occur.

- ***Range*** - These occur at 6-mile intervals east and west of a principal meridian that is associated with each base line. The position of the base line for each principal meridian is also historical and arbitrary. Range values are normally whole numbers starting at 1, but some exceptions occur.

- ***Sections*** - The intersection of Range lines and Township lines define 6 by 6-mile squares called Townships, which are divided into 36 sections. These are normally 1 by 1 mile squares, but some are altered to correct for the spherical earth. Most of these departures are pushed to the northern and western tiers of sections before the next standard parallels and guide meridians. Others are altered by simple survey errors. Note that old survey errors

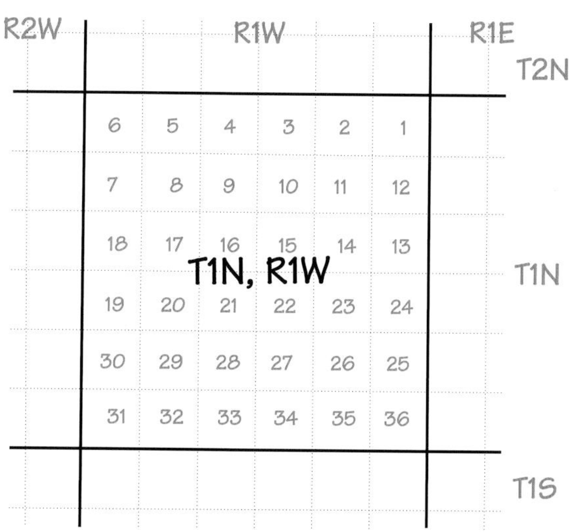

Figure A1.11 A township is six miles on a side. There are 36 sections in a township. Note how they are numbered.

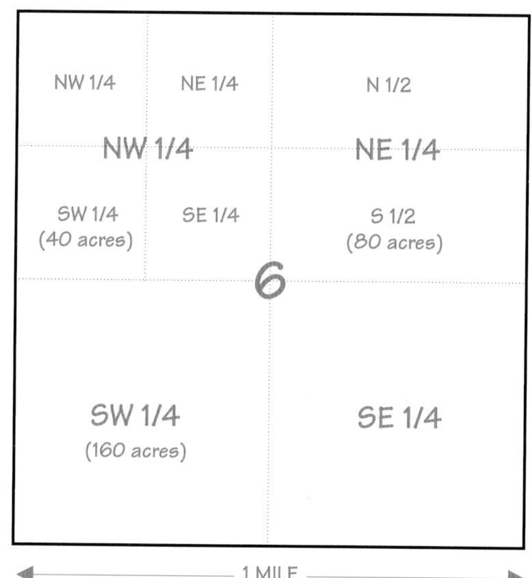

Figure A1.12 Sections can be divided into smaller areas.

have legal precedence over new survey results so the odd-looking lines stay put. Section numbers always range between 1 and 36 (Figures A1.11 and A1.12).

- ***Township and Range*** - The numbers in red along the sides in the form T 24 N and along the top and bottom in the form R 7 E give coordinates in the Public Land Survey. The Public Land Survey system in the U.S. is called the "township-range" system. Each 24 square mile area formed by intersecting latitude and longitude lines is broken into 16 township-range units, which in turn are broken into 36 one mile square sections. This system allows every part of the U.S. to have a uniquely defined location.

You can sometimes find public land survey markers on the ground.

When describing a location using public land survey reference points, you start with the smallest reference point - the section number - then township, range and meridian. Sections can be broken down into halves, quarters, quarters of quarters, etc. As an example, the public land survey location for the post office in Rescue, California is Center NE¼, Section 23, T10N, R9E, Mt. Diablo Meridian. This description places the post office within an area that is 160 acres in size, whereas a lat/long description is a precise point

Not all sections contain 640 acres.

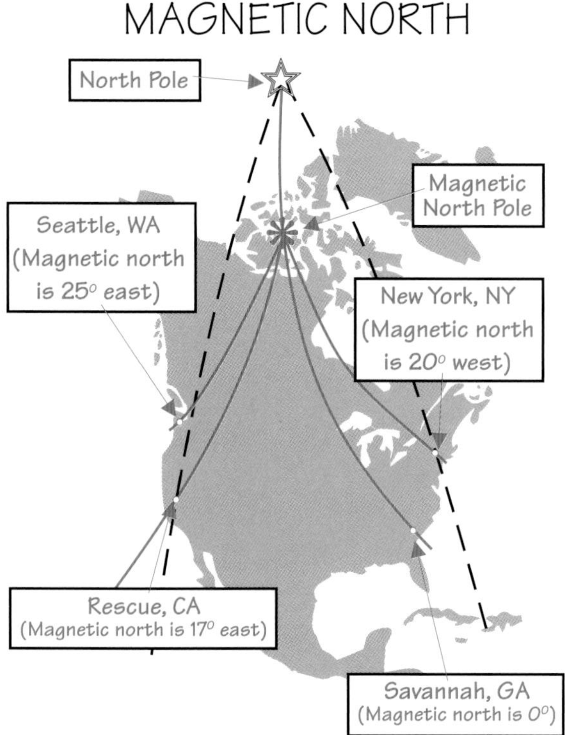

Figure A1.13 The magnetic center of the earth is not the same as true north. Note how the direction or declination to true north varies.

on the surface of the earth.

One unique feature of the public land survey is that it is the only geographic coordinate system that you can find on the ground. In most areas physical markers identifying the section corners can be located.

In some areas, especially the very rugged and mountainous areas or where two survey parties joined, the sections are not "exactly" one mile squares. Some sections may only contain 100 acres, where others may include well over 1,000 acres. Even though these are not "whole" sections, they are still referred to as sections.

Before the formal survey was begun in 1795, land ownership was identified using a system called "metes and bounds." This is where markers or prominent landmarks were used to identify the property lines. There was no formal grid-like system, you just came up to another person's property line and "met" it. Most of the states east of the Mississippi River use this system.

The declination changes based on your location.

Magnetic North

There is the North Pole, located at the top of the earth where all the longitude lines converge. And, there is the point to which all compasses point: the magnetic north pole. This point is located in northern Canada (see Figures A1.13 and A1.14). The correction used to locate the North Pole is called declination. Declination is an adjustment in readings between where the compass says north is and where it actually is. The declination changes by location. The proper declination is usually shown in the margin of the map. The declination for Rescue, California, is 17 degrees east. That is to say that the compass will point to the magnetic north, which from this location in California is 17 degrees east of the true north pole. If a particular map also displays a grid system, it may also show the declination for the grid.

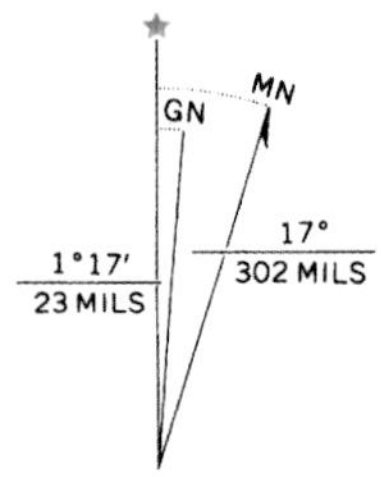

GN MN
1°17′ 23 MILS
17° 302 MILS
UTM GRID AND 1973 MAGNETIC NORTH DECLINATION AT CENTER OF SHEET

Figure A1.14 This shows the declination from north to magnetic north.

Scale and Distance

A map is a graphic representation of a portion of the earth's surface on a plane, drawn to scale. This means that there is always a definite relationship between distance on the map and the corresponding distance on the ground. There are three common methods of expressing this relationship: graphic scale; representative fraction; and words and figures.

Use the graphic scale to quickly determine distance.

Graphic Scale

Of the several methods used to determine or measure distances on a map, the graphic scale method is the easiest. A graphic scale is a ruled representation of ground distances, drawn to the scale of the map. The graphic scale is divided into two parts. The part of the scale to the right of the zero, marked off in full units of measure, is called the primary scale. The part to the left of the zero is divided into tenths of a unit, and is called the secondary or extension scale (Figure A1.15).

Knowing the scale of a map is necessary before you can fully utilize it.

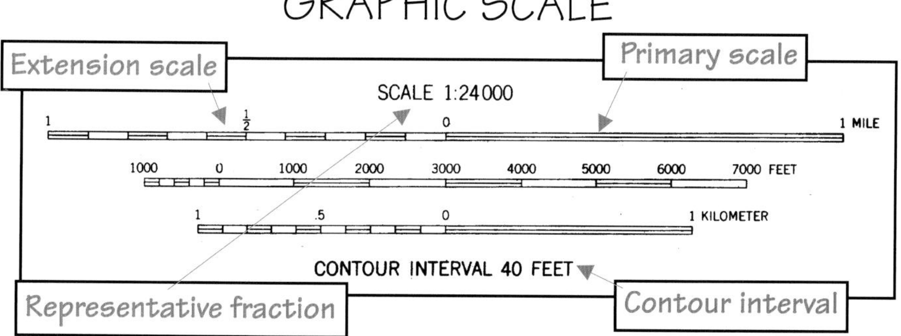

Figure A1.15 The graphic scale relates the distance on the ground to the distance on the map.

To measure the straight-line horizontal distance between two points on a map, lay a straight paper strip on the map so that the edge touches both points, and make a tick mark on the edge of the paper at each point. Then transfer the paper to the graphic scale of the map and read the horizontal distance between the points.

7.5-minute quadrangle maps utilize a scale of 1:24,000.

To measure distance along a winding road, a stream, or any other curved line, the paper strip method is again used. Make a tick mark near one end of the paper strip and place it at one end of the curved line to be measured. Align the paper strip along the first fairly straight segment or portion of the line, and make a tick mark at the other end of that segment, on both the paper strip and the map. Keeping both tick marks together, pivot the strip until the measurement is complete. Then place the paper strip on the graphic scale and read the horizontal distance measured.

Representative Fraction

The scale of a map is the ratio of any horizontal distance on the map to the corresponding distance on the ground. This may be expressed as a fraction, called the representative fraction or "RF." For easier use, the fraction is always reduced so the map distance is ONE.

The map distance and ground distance may be in any unit of measure, but both must be in the same terms; if the map distance is given in inches, the ground distance must also be in inches. Thus, a representative fraction of 1:10,000 means that one unit of measure on the map equals 10,000 of the same unit of measure on the ground.

Representative fraction is a ratio of distances on the ground and on the map.

Ground distance may be determined by measuring the distance between two points and multiplying this by the RF.

EXAMPLE: Map distance (MD) equals 5 inches; RF equals 1:24,000. What is the ground distance?

We know from the RF that one unit on the map equals 24,000 units on the ground. Therefore 5 units (inches) on the map represents 5 times 24,000 or 120,000 inches on the ground:

Map distance x RF = 5 inches x 24,000 = 120,000 inches.

Since it would be a nuisance to measure several thousand inches on the ground, you may convert this figure to feet, yards or miles:

120,000 / 12 = 10,000 feet
120,000 / 36 = 3,333 yards
120,000 / 63,360 = 1.89 miles

The reverse of the above situation may occur where you are given a certain ground distance and want to convert it to map distance. To find the map distance, divide the ground distance by the RF.

You are told to move a mile down the fireline and wish to plot this distance on your map. The scale of your map is 1:24,000.

EXAMPLE: Ground distance is one mile or 5,280 feet;
RF equals 1:24,000. What is the map distance?

The map distance is 5,280 feet / 24,000 or 0.22 foot. This equals 2.64 inches.

The representative fraction is used to convert map distance to ground distance or vice versa.

You may find a map on which the RF is not printed or has been torn off. You can determine the RF of the map by identifying two points on the map, measuring the distance between the same points on the ground, then using the formula, RF = ground distance / map distance.

EXAMPLE: Map distance = 4.33 inches; ground distance = 1.64 miles. What is the representative factor for the map?

The ground distance is 1.64 miles (or 103,910 inches) / 4.33 inches equals 23,998 or 1:24,000.

Words and Figures

Sometimes the scale of a map is represented by a statement in words and figures; this is common on road maps. The most common example is a phrase such as "1 inch equals 1 mile." This means that 1 inch on the map equals 1 mile (63,360 inches) on the ground or a RF of 1:63,360.

Elevation and Relief

Elevation is defined as the vertical distance a point lies above or below sea level. Relief means the difference in elevations in a local area, or the general shape of the terrain or landform. There are several methods to represent the third dimension of ground features, their elevation, on a flat map. The most commonly used method is the use of contour lines.

Relief means the difference in elevations in a local area.

A contour line is a line on a map representing an imaginary line on the ground, all points of which have the same elevation. In other words, it is a line of constant elevation. The line follows the land in and out of valleys, across ridges, and through streams, but always remains at the same height above sea level. Since each contour line has an assigned value (elevation) and lies a definite interval above or below other contour lines, this is the most exact method of representing elevation on a map. Contour lines are usually shown in brown. If the contour line is larger than the rest, it is an index contour. Somewhere along that line will be the elevation interval.

Characteristics of Contour Lines

Contour lines connect points of the same elevation.

The use of contour lines is the most accurate method we have of depicting relief on a map. It is the method used on almost all maps used in wildland firefighting. But in order to interpret topographic details, we must first have a sound knowledge of the basic characteristics of contour lines.

Each contour line has a definite elevation assigned to it, and it represents that elevation no matter where it wanders on the map. The difference between elevations of two adjacent contour lines is known as the contour interval. This is always a "round number" (e.g. 10, 20, 40, or 100 feet), and it is always given in the marginal information directly below the graphic scale. Thus, if the contour interval is 40 feet, then each contour line represents an elevation 40 feet higher (lower) than the next one. Contour measurements always start from sea level or zero elevation and, for convenience in reading, every fifth contour line starting from zero is made heavier, and its elevation is printed in occasional breaks.

Learn to picture the terrain "dimensionally" when you look at a contour map.

Contour lines are smooth curves, and tend to parallel each other, each approximating the shape of the ones above or below it. This reflects the fact that changes in ground form are usually gradual. The spacing of the contour lines is very informative:

- Contour lines evenly spaced and close together indicate a uniform, steep slope. The closer the contour lines are to each other, the steeper the slope (Figure A1.16a).

- Contour lines closely spaced at the top and widely spaced at the bottom indicate a concave slope (Figure A1.16b).

- Contour lines that are evenly spaced and wide apart indicate uniform, gentle slopes (Figure A1.16c).

- Contour lines widely spaced at the top and closely spaced at the bottom indicate a convex slope (Figure A1.16d).

THE SPACING OF THE CONTOUR LINES SHOWS THE RELIEF

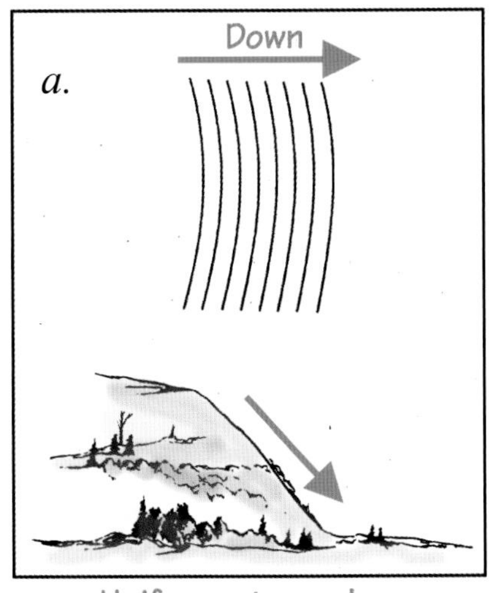

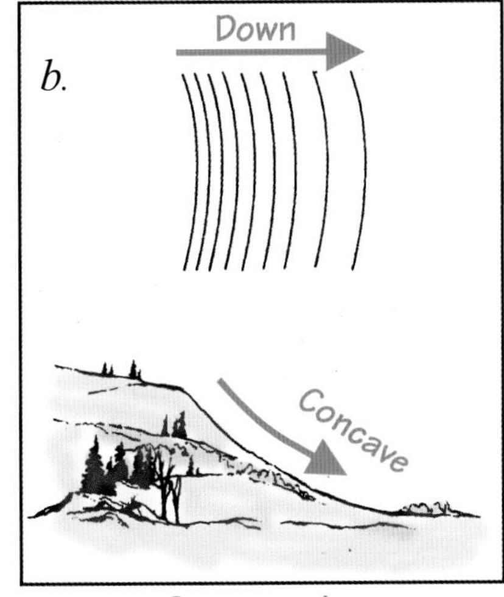

Concave slope

The closer the contour lines, the steeper the slope.

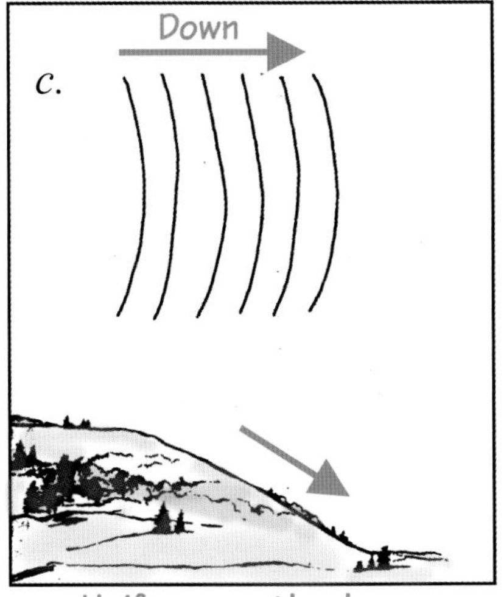

Uniform gentle slope

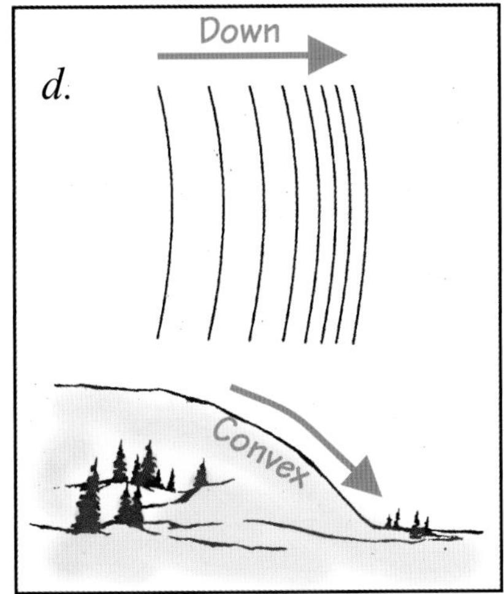

Convex slope

The steeper the slope, the faster the fire will burn up it.

Figure A1.16 The spacing of the contour lines will give a feel for the shape and steepness of the terrain.

A knowledge of the type of slope is important when firefighting. The steeper the slope, the faster the fire will run up it, and the harder it will be to climb. Contour maps can "let you know what you're in for."

CONTOUR LINES

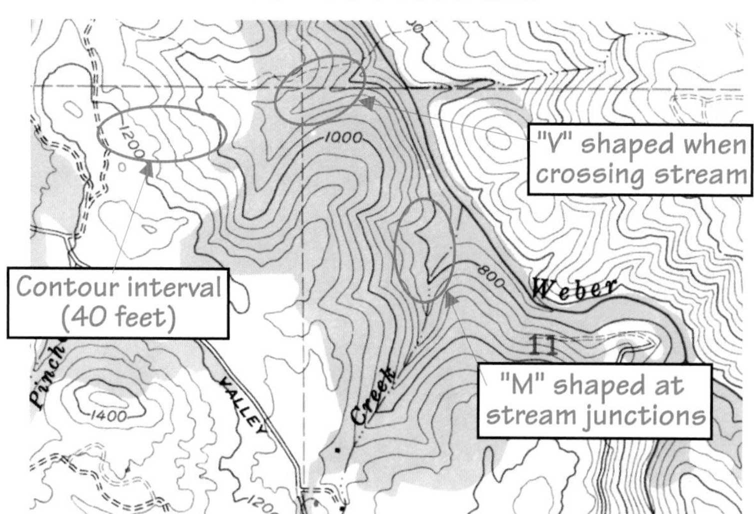

Figure A1.17 The contour line will tell you the elevation, but the shape of the contour line can also tell you a lot.

Stream Crossings - As contour lines approach streams, they turn upstream and run nearly parallel to it for a short distance. They cross streams in a "V"-shape with the "V" pointing upstream. Where a contour line crosses a stream junction, the two upside-down "V"-shapes combine to form a characteristic "M"-shape (Figure A1.17).

Forks, Crossings, and Closures - Contour lines never fork. They never cross or touch each other, except in special cases where there is a vertical or an overhanging cliff. Contour lines always "close" on themselves, but this may be on another map sheet.

Hills and Ridges - Peaks or hilltops are easily identified because the contours close in a small area, making concentric circles or ovals (Figures A1.18 to A1.21). Where two hilltops are enclosed within the same contour line, the depression between them is called a saddle. A ridge is a long hill, so it is identified by long ovals. The contours go around the end of a ridge forming a broad "U"-shape, with the bottom of the "U" pointing downslope.

Sharp V's in contour lines show streams.

Topographic maps show saddles, canyons, and chimneys that complicate firefighting.

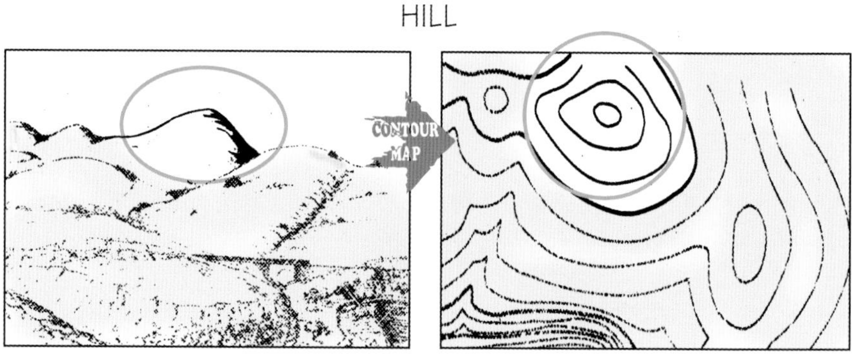

HILL

Figure A1.18 Contour lines making a closed loop or loops indicate a hill. The elevation of the more permanent hills or peaks are sometimes shown.

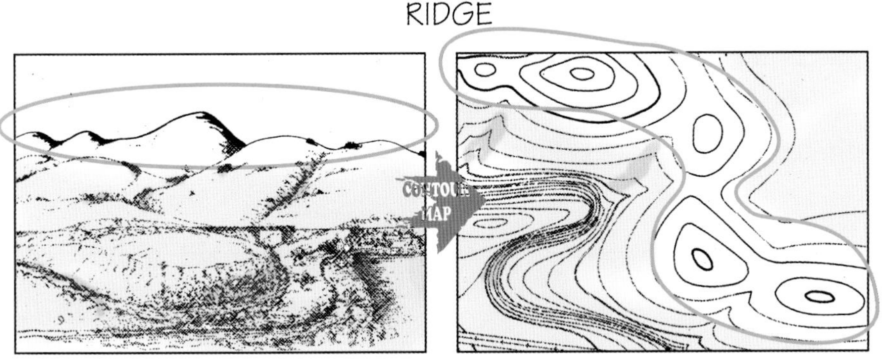

Learn how to spot good places to put fireline on a map.

Figure A1.19 Contour lines in the shape of U's with the open end pointing toward high ground indicate a ridge. A ridge may extend for many miles. It may be very winding or quite straight. It may have a uniform elevation along its top, or it may vary greatly in elevation. It may be extremely narrow or very wide.

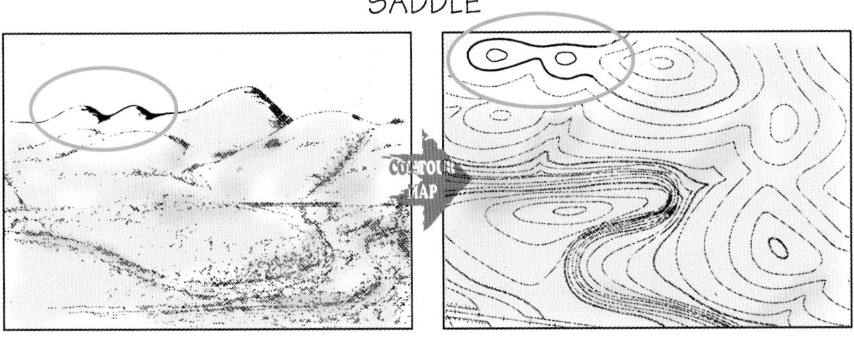

Ridges are good places for control lines.

Figure A1.20 Contour lines that show two hills enclosed within contour lines indicate a saddle. Saddles are usually low points along the crest of a ridge. You should stay out of saddles if you can. Fire moves much faster through saddles than it does on slopes.

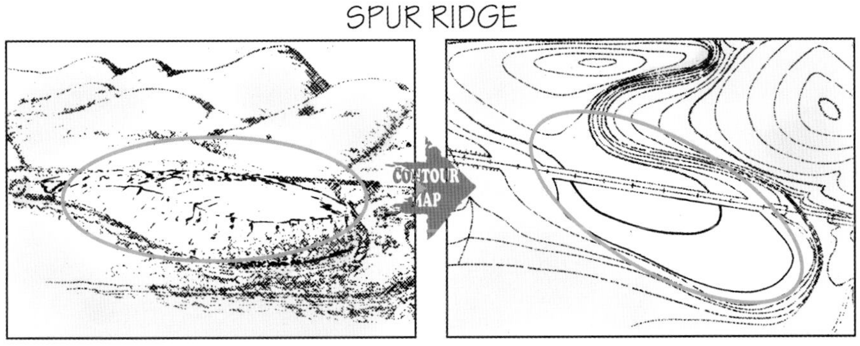

Figure A1.21 Contour lines that form a series of successive rounded U-shapes indicate a spur ridge. Unlike a ridge, a spur ridge has a continuous slope from high to low ground, or from low to high. A spur ridge is a protrusion or finger from the side of a ridge.

Stay out of saddles if you can.

Knowing what a ridge looks like can be very important because ridges are likely sites for the placement of firelines. Also, look for valleys, canyons, draws, and chimneys. All could complicate your firefighting operations (Figures A1.22 and A1.23).

Draws funnel fire up them.

DRAW

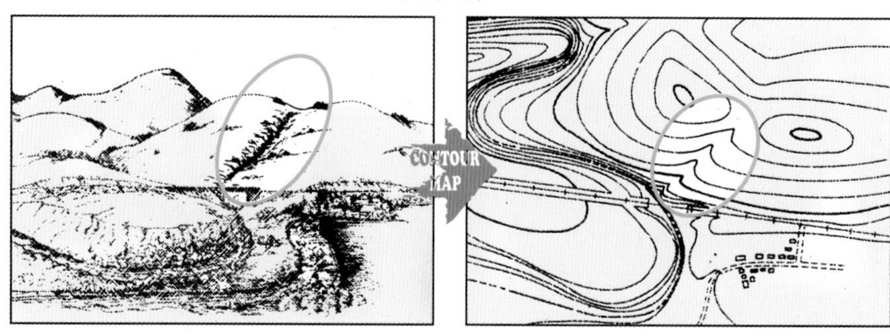

Figure A1.22 Contour lines that form a series of successive V-shapes indicate a draw. A draw is a stream course that has not developed a valley floor. Draws are topographic features that support the rapid spread of fire. Stay out of draws and where they meet the ridge line.

VALLEY

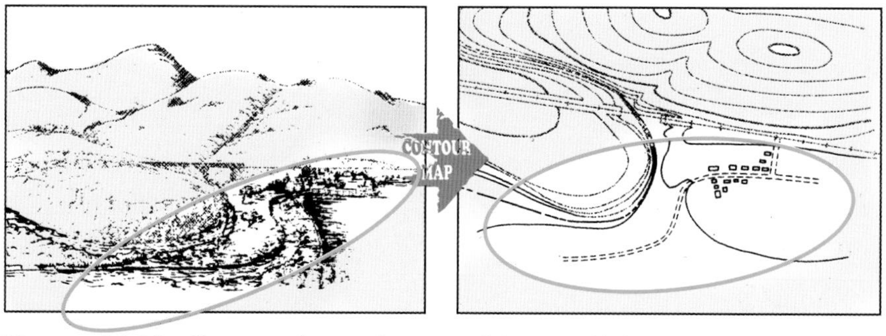

When contour lines are very close together the terrain is very steep.

Figure A1.23 Contour lines that roughly parallel a stream and represent elevations definitely lower than those farther from the stream indicate a valley. As is the case with most ridges, no single characteristic of contours shows a valley. Rather, it is characterized as the relief feature formed by the action of a stream, with enough low and reasonably level ground on one or both sides of the stream.

Cliffs - Cliffs are ground formations which are vertical or nearly vertical (Figure A1.24). Therefore, the contours for a cliff fall together or very close to each other. Where the contours are too close to each other for distinction, they may be replaced by hachures; the open ends of the hachures always point downslope. (Hachures are short lines used to depict relief. They do not represent exact elevations, but are used to show the relative slope in places where the contours do not.)

Some cliffs may erode away or undercut at the base, leaving a section jutting out at the top. This is called an overhanging cliff. In this case the contours that pass beneath the overhang are shown as broken lines. This is the only place where contour lines ever cross.

CLIFF

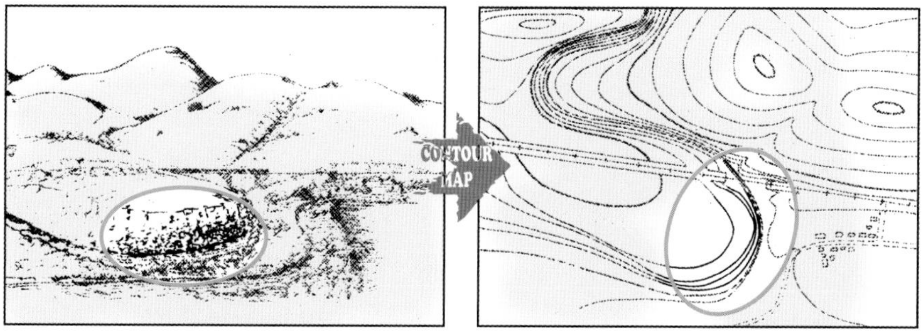

Figure A1.24 Contour lines that converge into one line indicate a cliff. Often, some of the contours converging into a cliff are broken just short of the point where they would touch, so as not to form a confusing pattern.

Overhanging cliffs are the only place contour lines ever cross, and are bad places for firefighters.

Depression Contours - Hachures are sometimes used with contours to indicate a steep cliff. They may also be used for other land forms to indicate a depression, in which case they are called depression contours (Figure A1.25). As in the case of steep cliffs, the open ends of the hachures always point downslope. Depression contours used alone help the firefighter to distinguish between a hilltop and a small basin or sinkhole.

DEPRESSION

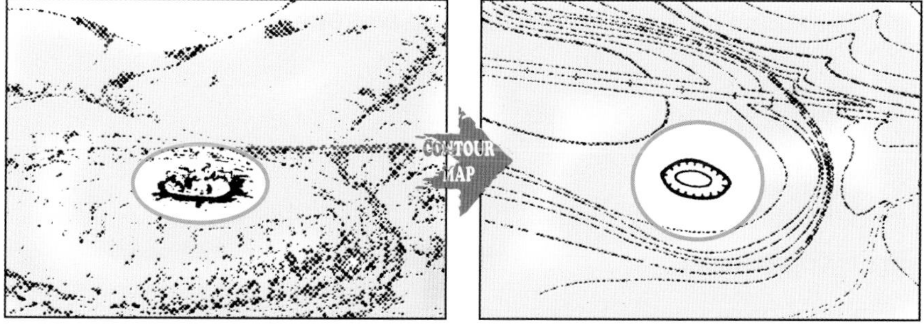

Figure A1.25 Contour lines making closed loops with ticks indicate a depression. The ticks are always on a downhill side of the contour.

A cut is a place, usually in the side of a hill, from which part of the hill has been removed to make way for a road or railroad (Figure A1.26). If it has been necessary to fill in a valley or depression for some reason, the new land is called a fill (Figure A1.27). Cuts and fills may be shown by contours alone, by depression contours, or by contours and hachures. Again, the hachures point downslope.

CUT

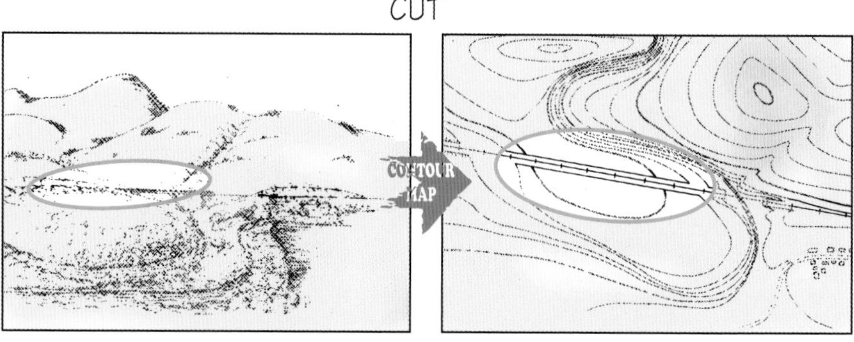

Figure A1.26 Contour lines that are straight and parallel each other adjacent to roads, railroads, and other man-made features; and pass through hills and ridges indicate a cut.

Cut banks and roadway fills usually have straight contour lines.

FILL and a BRIDGE

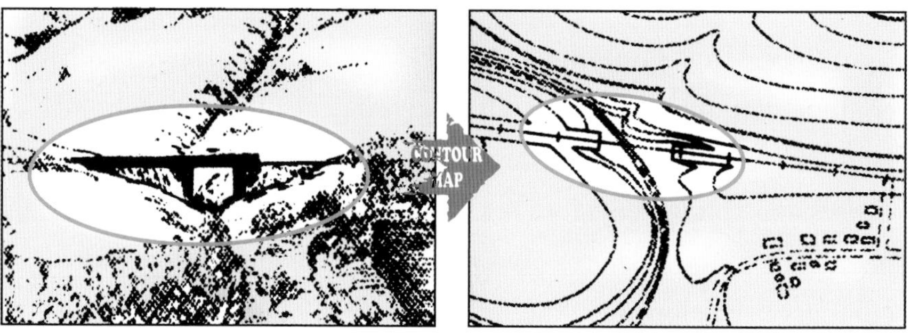

Figure A1.27 Contour lines that are straight and parallel each other adjacent to roads, railroads, and other man-made features; pass over small streams, gullies, or depressions; and show ticks along the downhill side of the highest contour indicates a fill.

Contour Maps and Symbols

The most commonly used contour maps used by the fire service are maps produced by the United States Geological Survey. These maps come in several scales, with the 7.5-minute topographic maps being the most detailed, thus the most useful. You can use these maps to plot your fire and calculate the size. You can also plot your fire behavior predictions, actual and planned firelines, etc. There are some standard colors used on most maps. Figure A1.28 lists the colors and what they represent.

USGS maps are commonly used by wildland firefighters.

Common colors used on maps

Color	What it represents
Black	Most cultural or human-made features (roads, railroads, airports, boundaries, structures, etc.)
Blue	Water features such as streams, lakes, rivers, oceans and swamps.
Green	Vegetation such as brush, woods, orchards and vineyards.
Brown	All relief features such as contour lines, special shading, etc.
Red	Main roads, built-up areas, special boundaries and features.
Magenta	Updated information. (When a map is updated, the new features are printed in magenta.)
Other colors	Other colors may be used for special purposes. They will be explained in the margin or legend.

Figure A1.28 These are some of the standard colors used on topographic maps.

Black is for human-made features.

Blue is for water features.

There are two sets of map symbols that you should become familiar with. The first are the ones placed on the printed map (Figure A1.29). Some of the features you will find are:

Magenta is for updated information.

- Human-made features - City limits, political boundaries, roads, structures, orchards, surveying benchmarks, power lines, etc.
- Water features - Lakes, streams, oceans, intermittent streams, etc.
- Vegetation features - Areas covered with forest and brush.
- Elevation features - Elevations of prominent peaks and contour lines.

TOPOGRAPHIC MAP SYMBOLS

ROADS AND RELATED FEATURES

Roads on Provisional edition maps are not classified as primary, secondary, or light duty. They are all symbolized as light duty roads.

- Primary highway
- Secondary highway
- Light duty road
- Unimproved road
- Trail
- Dual highway
- Dual highway with median strip
- Road under construction
- Underpass; overpass
- Bridge
- Drawbridge
- Tunnel

RAILROADS AND RELATED FEATURES

- Standard gauge single track; station
- Standard gauge multiple track
- Abandoned
- Under construction
- Narrow gauge single track
- Narrow gauge multiple track
- Railroad in street
- Juxtaposition
- Roundhouse and turntable

BOUNDARIES

- National
- State or territorial
- County or equivalent
- Civil township or equivalent
- Incorporated city or equivalent
- Park, reservation, or monument
- Small park

COASTAL FEATURES

- Foreshore flat
- Rock or coral reef
- Rock bare or awash
- Group of rocks bare or awash
- Exposed wreck
- Depth curve; sounding
- Breakwater, pier, jetty, or wharf
- Seawall

CONTOURS

Topographic

- Intermediate
- Index
- Supplementary
- Depression
- Cut; fill

Bathymetric

- Intermediate
- Index
- Primary
- Index Primary
- Supplementary

RIVERS, LAKES, AND CANALS

- Intermittent stream
- Intermittent river
- Disappearing stream
- Perennial stream
- Perennial river
- Small falls; small rapids
- Large falls; large rapids
- Masonry dam
- Dam with lock
- Dam carrying road
- Perennial lake; Intermittent lake or pond
- Dry lake
- Narrow wash
- Wide wash
- Canal, flume, or aqueduct with lock
- Elevated aqueduct, flume, or conduit
- Aqueduct tunnel
- Well or spring; spring or seep

SURFACE FEATURES

- Levee
- Sand or mud area, dunes, or shifting sand
- Intricate surface area
- Gravel beach or glacial moraine
- Tailings pond

LAND SURVEY SYSTEMS

U.S. Public Land Survey System

- Township or range line
- Location doubtful
- Section line
- Location doubtful
- Found section corner; found closing corner
- Witness corner; meander corner

Other land surveys

- Township or range line
- Section line
- Land grant or mining claim; monument
- Fence line

BUILDINGS AND RELATED FEATURES

- Building
- School; church
- Built-up Area
- Racetrack
- Airport
- Landing strip
- Well (other than water); windmill
- Tanks
- Covered reservoir
- Gaging station
- Landmark object (feature as labeled)
- Campground; picnic area
- Cemetery: small; large

SUBMERGED AREAS AND BOGS

- Marsh or swamp
- Submerged marsh or swamp
- Wooded marsh or swamp
- Submerged wooded marsh or swamp
- Rice field
- Land subject to inundation

MINES AND CAVES

- Quarry or open pit mine
- Gravel, sand, clay, or borrow pit
- Mine tunnel or cave entrance
- Prospect; mine shaft
- Mine dump
- Tailings

Figure A1.29 The United States Geological Survey uses a standard set of topographic map symbols.

The color of the symbols is also important. Obviously, water is blue, but red, green and magenta also mean something.

Learn and use standard symbols.

The second set of symbols are the ones used by the fire service to plot the fire, activities, and location of key facilities (Figure A1.30). These symbols are usually simple and easy to recognize. The color of a feature may be an indication of its type.

The contour map shown as Figure A1.31 has some of the more commonly used symbols highlighted for your review. When you

COMMONLY USED FIRE MAP SYMBOLS

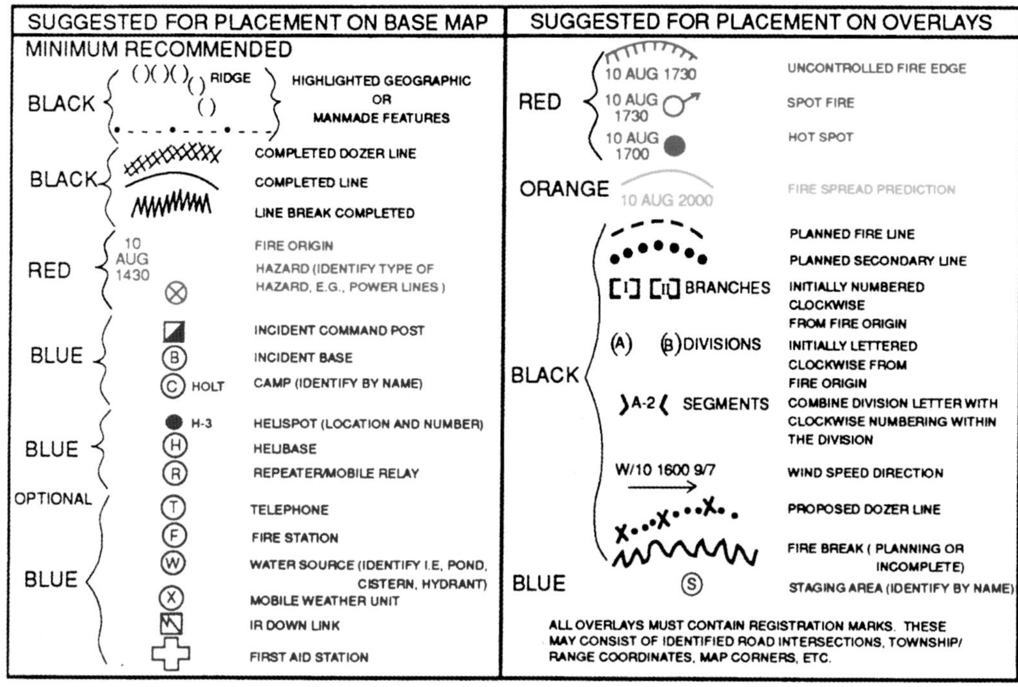

Figure A1.30 These are the map symbols that are used on fire maps in incident base. They are very simple and straight forward.

CONTOUR MAP

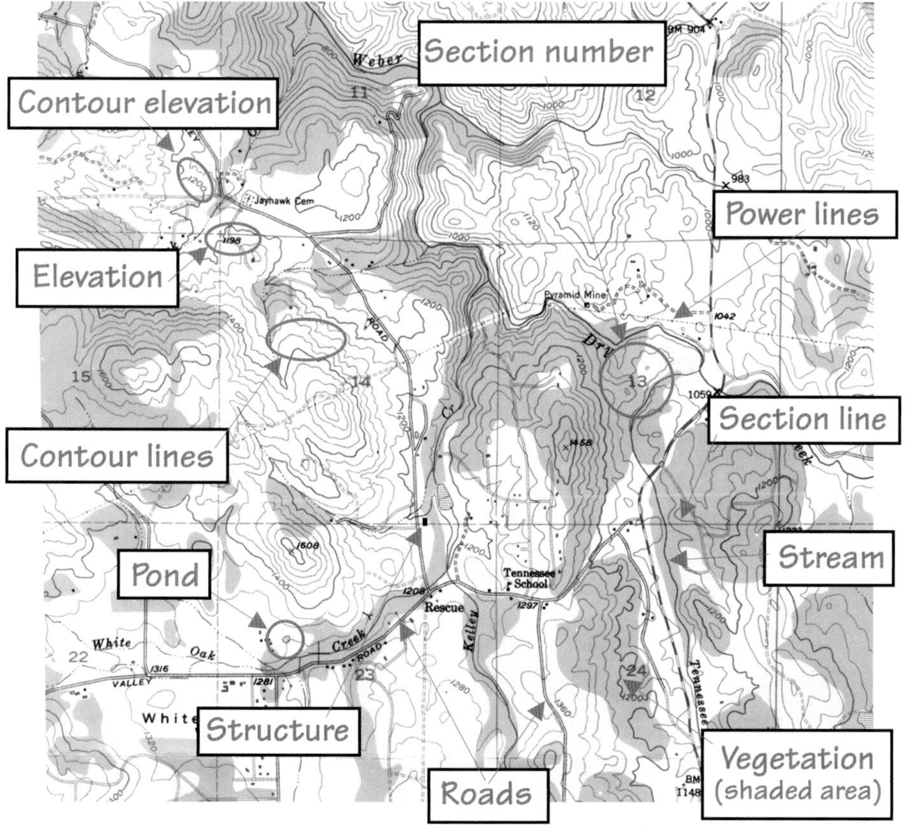

Topographic maps depict the terrain.

Figure A1.31 This is an example of a contour map produced by the U.S. Geological Survey. This is their most detailed topographic map that is commonly called a "7.5-minute quad." This map has a scale of 1:24,000 and a contour interval of 40 feet.

become familiar with a contour map and its symbols, you can start to think in "three-dimensions." You will be able to find ridge lines, where you could place control lines, and determine the general steepness of the slopes.

Slope or Grade

Slope is an aid to fire spread, but an obstacle to firefighters and fire equipment.

The rate of rise or fall of a piece of ground or roadway is known as its slope, and can be described as being steep or gentle. You may be asked how steep the slope is (Figure A1.32)? The exact amount or rate of slope (grade) is often important to know when using bulldozers and other mechanized firefighting equipment. The speed at which personnel and equipment can move is affected by the slope of the ground. Most equipment has a limit as to the steepness it can negotiate.

There are several ways of representing the amount of grade, but they are all simply a comparison of vertical distance to horizontal distance.

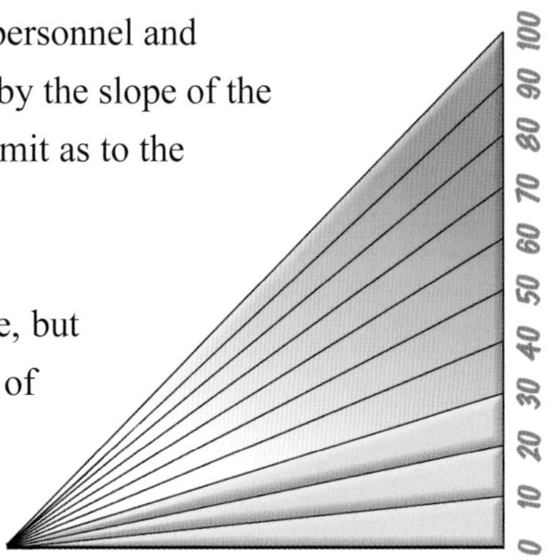

0 10 20 30 40 50 60 70 80 90 100

Figure A1.32 Slope is the rise or fall of a piece of ground. Anything over 40% is considered steep.

Slope in Percent

Slope is a relationship of vertical distance divided by horizontal distance.

Slope is most commonly expressed as a percent, the number of units of vertical distance in every hundred units of horizontal distance. If two points are 100 feet apart and one is 5 feet higher than the other, the slope between them is 5 percent. As a formula:

Percent of slope = (vertical distance / horizontal distance) x 100

The vertical distance is the difference in elevation and can be determined by the contours. The horizontal distance can be scaled off the map.

EXAMPLE: The elevation of one point is 345 feet and another is 371 feet. The horizontal distance between these two points is 340 feet. Vertical distance is 371 - 345 = 26 feet.

Percent of slope = (26 / 340) x 100 = 7.6 percent

EXAMPLE: The elevation difference between a road intersection and the top of the hill is 1,050 feet. The road distance between these two points is 1.2 miles (1.2 miles = (5,280 x 1.2) = 6,336 feet).

Percent of slope = (1,050/6,336) x 100 = 15.6% This would be considered a medium slope.

Topographic maps have a wealth of information valuable to firefighters.

You can also calculate slope using a contour map. Figure A1.33 lays out two slope calculation problems. Calculate the slope between Point A and B, and from Point B to C.

Point A to Point B

The elevation of Point A is 1,280 feet and Point B is about 1,565 feet. The distance between the two points can be estimated by comparing the length of the line on the map with the graphic scale usually found at the bottom of the map. The estimated distance between Point A and Point B is 1,985 feet. With this information you can calculate the slope between these two points:

Percent of slope = (1,565-1,280)/(1,985)x100 = 14.4%

Point B to Point C

The elevation of Point B is 1,565 feet and Point C is about 1,195 feet. The estimated distance between Point B and Point C is 2,195

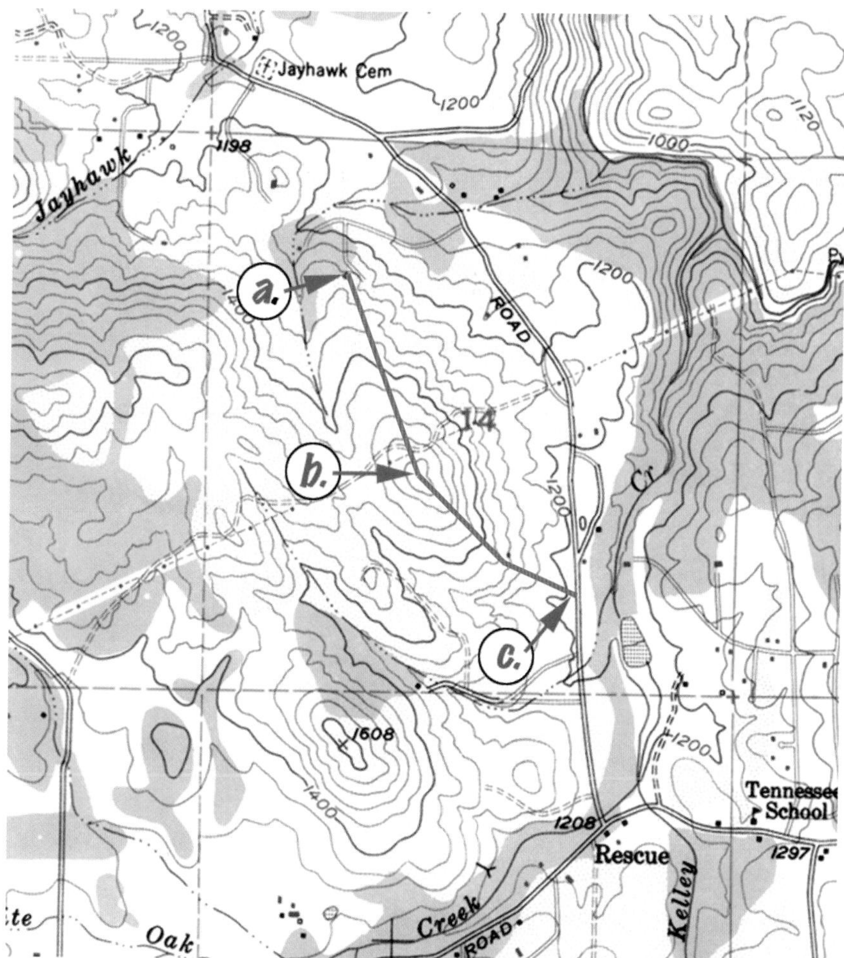

Figure A1.33 You can use a topographic map to calculate slope. Estimate the elevations at two points and the distance between them. This is enough information to calculate the slope between the two points.

feet. With this information, you can calculate the slope between these two points:

Percent of slope = (1,565-1,195)/(2,195)x100 =16.9%

Use of a Compass

If you have ever spent a lot of time in the woods, you have usually had a compass close at hand. Not that you were using it during your every walk, but because if you became lost or disoriented, it was there to help find your way home. The compass has been used for navigation for centuries, and it is still of some use on the fireline.

Compasses have been used for navigation for centuries.

There are two primary purposes of a compass. The first is that it will help you orient yourself to magnetic north, and true north if you know the proper declination.

(Something about declination – it isn't constant. If you are using a map that is over 10 years old, the declination may be wrong. Magnetic north is actually the center of a mass of floating iron ore magma located deep under the earth's crust in northern Canada. It moves around, thus magnetic north moves around.)

The second use of the compass is to aid you in moving in a straight line toward some point, which could be smoke on a ridge, or along a property line, etc.

Parts of a Compass

Before you pick up a compass and start to use it, you need to know something about its parts (Figure A1.34). There are many different compass designs or models, but here is a description of the common parts of a compass:

The only thing a compass can do is point you toward magnetic north.

- ***Travel Arrow*** – The travel arrow on the compass is like the sight on a rifle. It is a line that points in the direction you want to travel.

- ***Dial*** – is marked with the cardinal directions and the 360 degrees of a circle. You can rotate the dial to the bearing or azimuth you want to follow, or to the proper declination between true north and magnetic north.

- ***Index Pointer*** – is a mark on the compass at which you read the bearing; the index pointer and the travel arrow are parallel.

- ***Magnetic Needle*** – This is the little floating arrow that points to magnetic north or a strong magnet held in proximity. Note: If you are in an area of high concentrations of iron ore, or if you are near iron or steel objects (pocket knives, belt buckles, railroad tracks, trucks, electrical lines, etc.), the magnetic needle may not necessarily point toward northern Canada and magnetic north.

- ***Orientation Arrow*** – is located on the dial. The orientation arrow is used to align the dial and the magnet needle.

- ***Declination Scale*** – is located inside the dial and denotes magnetic declination from true north (not found on all compasses).

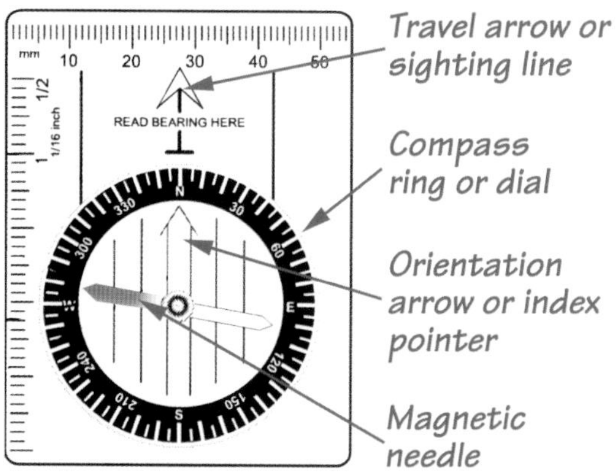

Figure A1.34 These are the parts of a compass. The only thing a compass does is point you to magnetic north. With a map, you can use this information to gain other information.

Azimuths or Bearings

An azimuth or bearing is way of describing a direction. Everyday we use terms like "go to the right" or "go straight ahead;" these are directions. If we add some reference to distance, "go to the left six feet," this tells you from the place you are now, move six feet to the left.

Azimuth and bearing are the same.

To a firefighter, an azimuth is a direction from a base line; this is normally north. So, if you are told to move six feet, on an azimuth bearing of 90 degrees, you are to move six feet to the east.

There are 360 degrees in a circle. North is usually the base line at 0 or 360 degrees, and all azimuth readings are read in degrees clockwise. There are eight cardinal directions, each with a specific azimuth reading in degrees (Figure A1.35). They are:

There are eight cardinal directions.

North at 0 or 360 degrees	*South* at 180 degrees
Northeast at 45 degrees	*Southwest* at 225 degrees
East at 90 degrees	*West* at 270 degrees
Southeast at 135 degrees	*Northwest* at 315 degrees

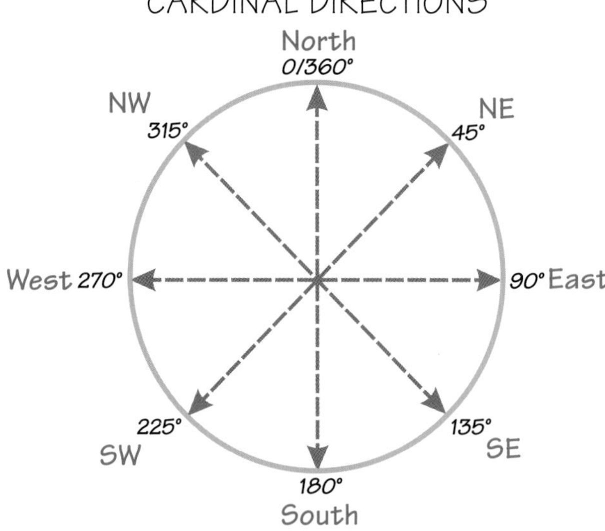

Figure A1.35 The cardinal directions are all based on north being the base line at 0 or 360 degrees.

You read the azimuth at the index pointer. If you were told a fire was at 112 degrees, one mile from your present location, you would set your compass at 112 degrees, and move in that direction. After walking a mile, you should find the fire.

When using the compass to give you the proper direction to move, you should use these techniques:

• Hold the compass level so the magnetic needle floats freely. Some compasses are held at waist height and others at eye level.

• Rotate the whole body toward the proper azimuth, watching the needle as it aligns with the orientation arrow. When they are aligned, you are at 112 degrees from north. The fire is in that direction. Pick a spot off in the distance that falls on this line and get moving!

Declination is the difference between magnetic and true north.

Declination

As mentioned earlier, north is at the top of the world, where the North Pole, is but the magnetic arrow points at magnetic north somewhere else. ***The difference is what is called declination.*** Figure A1.36 shows the latest world magnetic chart. This shows that the declination to be used in Rescue, California in 2005 is 15 degrees West. That means that true north is really 15 degrees west of where the magnetic arrow is pointing. (Note: In 1973 the declination for this same spot was 17 degrees west.) Having the proper declination calculated into your azimuth is more important the farther you are from your beginning point (Figure A1.37). The error becomes greater the farther you travel.

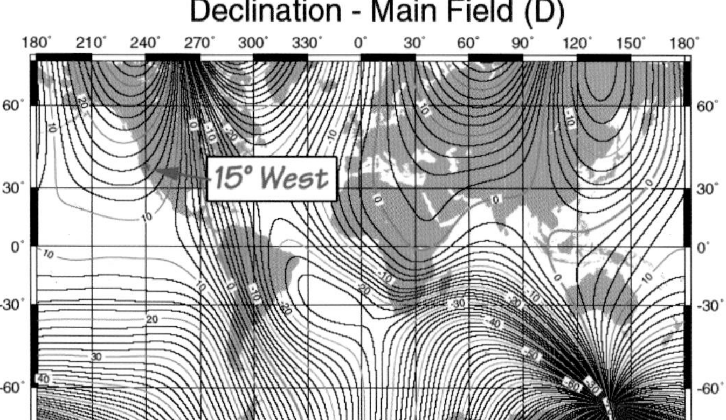

Figure A1.36 This is the world magnetic chart produced in the year 2000. It shows the declinations that would have to be used to convert a magnetic north reading to true north.

There are several ways for you to set the declination. If your compass has a built-in declination adjustment, simply make the adjustment for the area you will be working.

If not, you may want to make a mark on the compass (scratch a line or put a line on a piece of removable tape). Or, you may just make a mental note of the declination and adjust your bearing accordingly. Another way is to add the declination to the azimuth you want to follow (add for declinations that are east; subtract for declinations that are west).

So, if your desired bearing is 112 degrees, and the declination is 15 degrees west, you should set the orientation arrow at 97 degrees. When the magnetic and orientation arrows are in alignment, you will be pointed 112 degrees from north.

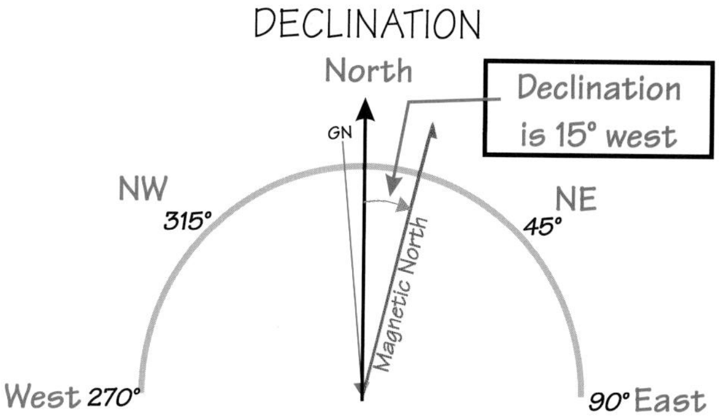

Figure A1.37 The present declination for Rescue, California is 15 degrees west. This information will allow you to convert magnetic north to true north.

As you move along the way toward your destination, there may be times that you want to check to see if you are on track. The easiest way is to use the back azimuth. The back azimuth is 180 degrees from the direction you want to go. You can calculate the back azimuth by adding 180 to the azimuth (and subtract 180 from azimuths that are over 180 degrees). See Figure A1.38.

Now turn back toward the starting point and align your compass, set with the back azimuth toward it. If you are on track, the reading to the starting point should be correct. If the compass points to the left, your are right of the line, and vice versa. This is an excellent way to check to see if you are going in the desired direction.

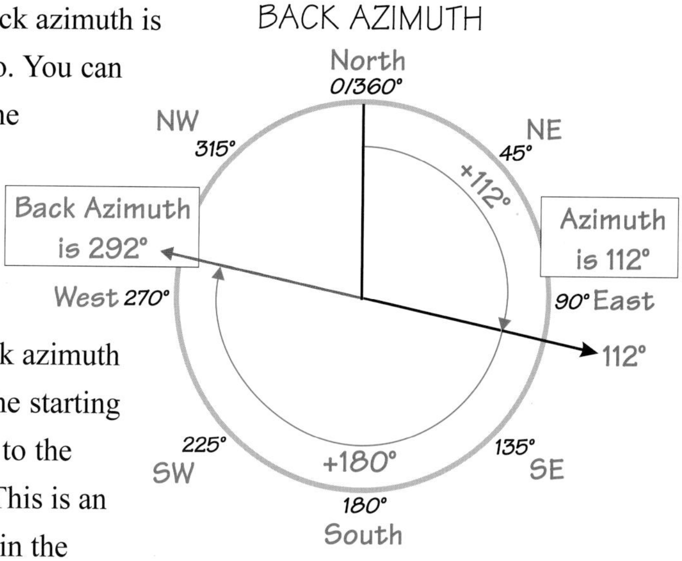

Figure A1.38 shows you how to determine the back azimuth. This bearing is used to check your alignment with your starting point.

Aligning a Map

There are several ways for you to align a map. They are:

- If you can find your location on the map, and you have some reference points (like a road, power line, or peak in the distance), align these points by rotating the map. If the road on the map and the road on the ground are in alignment, the map is aligned. North is in the direction of the top of the map.

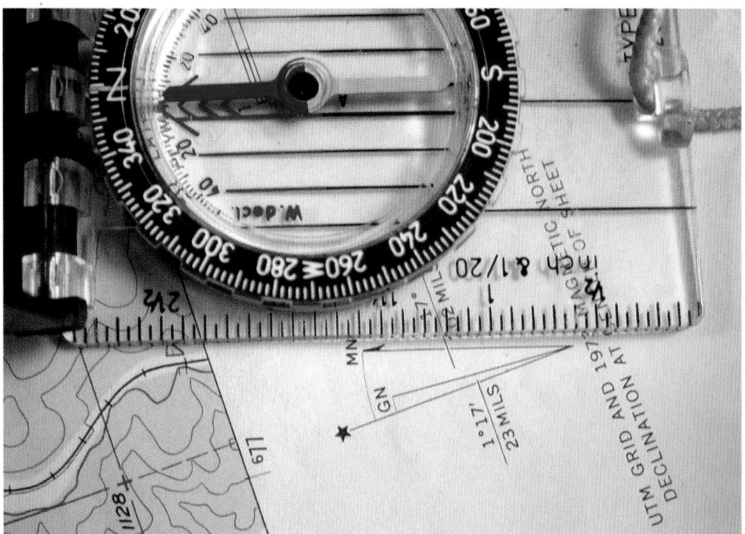

Figure A1.39 The compass can be used to align your map. Here, the compass is aligned with the magnetic north arrow printed on the map. Rotate the map and compass until they align.

Figure A1.40 Set the declination on the compass and align it with the edge of the map. Rotate the map and compass until the magnetic and index arrows are in alignment. The map is now oriented with the top of the map pointing north.

• If you have a compass, and the map has a declination reference at the bottom, simply place the compass on the map, next to the declination reference. Rotate the map and align the magnetic arrow with the declination arrow. When this is done, the map is aligned and north is in the direction of the top of the map (Figure A1.39).

• If you have a compass and a map which does not have a declination arrow, but you do know the proper declination, apply the declination to the compass and align the side of the compass with the side of the map. Rotate the map until the magnetic arrow point is aligned with the index arrow and the map is aligned. The top of the map is pointing north (Figure A1.40).

Distance to the Objective

The compass cannot really help you with distance. You can use the map to measure the distance and determine the real distance using the scale found at the bottom of the map. Then you can pace off the distance as best you can.

Use of Global Positioning System Receiver

In recent years, global positioning system (GPS) receivers have become the new "in" thing in cars, boats, aircraft, and even cell phones. GPS has filled a real need on the fireline. GPS can tell you exactly where you are; can tell you the direction and distance to a point you want to find; and, it can track you as you move around a fire, plotting its location, which can later be transferred to a map.

History of GPS

A little background is good. The global positioning system was designed and built by Rockwell International for the United States Department of Defense. The first of the 24 satellites was placed into orbit in 1978. The system was fully operational in 1994. The satellites are in an orbit that is just less than 12,000 miles and they make two complete passes around the earth every 24 hours.

GPS was originally built to fight wars.

The original purpose focused on the military and their uses. In the 1980s, the government opened up the technology for civilian use. The system was designed to provide two levels of accuracy. The military had equipment that could utilize the higher accuracy (within 10 meters), whereas other users would only be able to attain accuracies of about 100 meters. The reason; the military didn't want to give our enemies the ability to use our technology against us. In 2000, the system was opened up so that all users could utilize the higher level of accuracy.

24 satellites circle the earth every 12 hours.

How does it Work?

Each of the satellites transmits low-strength FM radio signals. These signals include the time, its location, and other information the receiver is going to need for its calculations. The receiver needs to "see" at least three satellites to function at all, and prefers four for best performance. The concept is very simple…triangulation. The location (latitude and longitude) and elevation of each satellite is known. The receiver calculates the distance from the satellite(s) to the receiver's location, using the time it takes for the signal to travel the distance times the speed of light. Once this math is done, the receiver then displays the latitude and longitude, and elevation on its display panel.

It is all about triangulation.

Once it knows its location on the ground, the GPS unit can do all sorts of other stuff. Some of the more important capabilities from a firefighter's point of view are, it can tell the user:

- Exact location, to within feet, in terms of latitude and longitude.

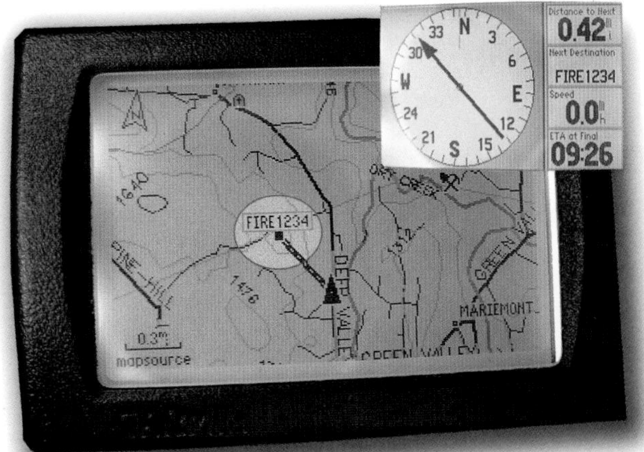

Figure A1.41 If you enter the lat/long for a fire, the GPS unit will tell you the direction and distance to the fire. These two images show you the relationship between your location and that of the fire. The image in the upper right gives you the bearing and distance to the fire.

- Exact time, to the second – Very accurate time is needed by the system to complete the calculations. Each satellite has an atomic clock on board to ensure the time is right.
- Elevation, in feet.
- Direction of travel - As you move, the receiver plots your direction of travel in degrees from north and in cardinal directions.
- Speed of travel - The receiver compares your last location to your new location and calculates your speed of travel.
- Direction and distance to another location - Since the GPS unit knows its location, it can calculate the direction and distance to another point, as long as you know the latitude and longitude of the new point. This point is called a waypoint. It could be the location of a fire, a drop off point or base camp (Figure A1.41).
- Track your route of travel - The receiver can be told to record your movement by every so many seconds, or by so many feet. You can use this to record escape routes, your way back home, or to record the fire perimeter as you walk or fly it. The travel log can then be downloaded from the GPS unit to a computer for use in mapping the fire.

GPS can help you in so many ways.

- Topographic map – Most of the later models of GPS units include the capability to store some map information in them. The maps will not have the same quality of the printed versions, but enough to make them very useful.

The compass had the ability to orient you to magnetic north, only. The GPS unit, as you can see, can do much more. But it is not without some limitations:

GPS radio signals are blocked by tall buildings and dense forests.

- Line of site – GPS relies on signals from space to work. These radio signals do not pass through buildings, trees, metal of any kind, soil, and water. So, you will have problems with the units around tall buildings, in deep canyons and in dense forests.

- Battery life – The GPS receiver uses batteries to power it, so carry an extra set or two.

- Conversion to Public Land Survey – It is difficult and time consuming to attempt to convert latitude and longitude (a point specific system) to public land survey section, township and range (an area system). Some maps do not include lat/long information in their margins.

We will not attempt to train you in the full use of a GPS unit. There are too many models and levels of sophistication. GPS is a great tool for the wildland firefighter, and its use will become increasingly valuable.

Notes:

Fire Prevention

About 1 billion acres of the United States (about one half of the US land mass) is forest, brush or grasslands. Annually, over 100,000 wildland fires start and burn about 4½-million acres. About 90 percent of these fires are caused by people and their activities, and 10 percent are caused by lightning. About 30 percent of the human-related fires are caused by arson. As you can see from these numbers, the most important factor in protecting the wildlands from fire, is an effective prevention plan that concentrates on modifying human activities and behaviors (Figure A2.1).

Figure A2.1 Most of the wildland fires are started by some human-related cause. Debris burning, machine use and arson lead the list of causes. Lightning is a leading cause in some areas, but fires started this way are "hard" to prevent.

The activities of people cause most wildland fires.

More and more people are entering the woods to work, live and recreate. This is why fire prevention targets people. Every firefighter should know something about fire prevention: why it is so important; how to discuss it with members of the public; what it takes to make a structure safe; how to conduct an inspection; and how to preserve the origin of a fire for the investigators.

Use fire "spot maps" to identify fire cause in problem areas.

Fire Prevention Components

There are three elements to a wildland fire prevention program: ***education***, ***engineering***, and ***enforcement***. The purpose of a fire prevention program is to eliminate or reduce ***risk*** (fire cause) and ***hazards*** (where and what it burns). Figure A2.2 outlines the elements that make up each component of a fire prevention program and cites examples of various prevention activities.

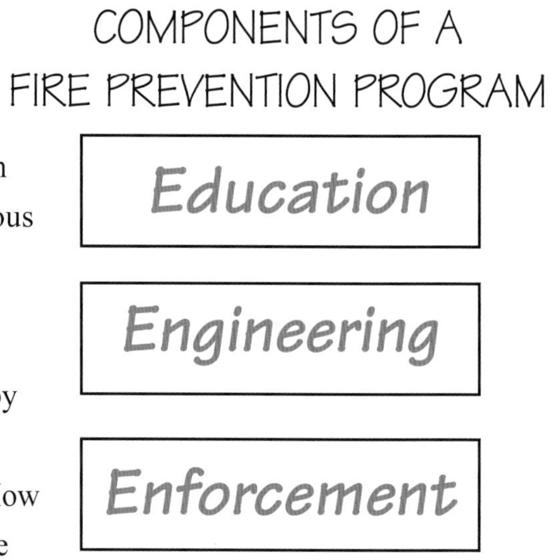

COMPONENTS OF A FIRE PREVENTION PROGRAM

Education

Engineering

Enforcement

Use law enforcement when education fails.

Fire Prevention Plan

Most fires are caused by the careless activity of man, thus they are preventable. How a fire starts and/or where the fire starts must be changed/ reduced to lower the incidence of fire occurrence.

Risk is the cause of fires.

- ***Risks*** are those "people" things (human-related activities) which cause fires to start. They can include such things as debris burning, arson, smoking, mishandling hot ashes, logging operations, sparks from a chimney or machine use, and railroad operations (Figure A2.3).

Hazard is what burns.

- ***Hazards*** are the flammable materials in which the fire starts. They include such things as grass, leaves, forest fuels, logging slash, and trash (Figure A2.4).

There are several things that can be done to prevent fire in the wildlands. The first step in developing a fire prevention plan is to determine what the specific risks (causes) and hazards (location of burnables) are. In some areas the predominant cause is debris burning or equipment use; in other areas it may be lightning or arson. Look over the fire reports for the last several years and "chart" the fire starts by cause and location.

Once you have determined the causes of the fires, and in which fuels they are starting, develop a plan to reduce these incidents. If the predominant cause is debris burning, start a campaign targeting those who are doing the burning. Point out the damage caused by these fires, and the costs of putting them out. Teach them how to burn safely, following area burn regulations.

Find out what is causing most of your fires and develop plans to reduce the numbers.

If the cause is logging or farming activities, work on reducing these causes by emphasizing the importance of educating the workers; the installation of spark arresters on machinery; or by

The three components of a fire prevention program

	Key elements	Examples
EDUCATION	Informing the public of wildland fire prevention practices; changing attitudes and behavior; and creating an awareness of fire prevention.	School and civic group programs; person-to-person contacts; FireWise education; parades, exhibits and displays; Smokey Bear posters and billboards; and public service contacts on TV and radio.
ENGINEERING	Separating heat sources from fuels; reducing or eliminating fuels when heat sources must remain; and shielding fuels from heat sources to prevent contact.	Creating and maintaining fuel breaks; clearance around structures and power lines; design and installation of spark arresters; and the development of fire codes and regulations.
ENFORCEMENT	Compliance with local, state and federal fire codes and regulations; fire cause determination; and law enforcement action.	Inspection for compliance with FireWise related laws and regulations; investigation of fire cause; red flag patrolling and closure of areas; and law enforcement and court action.

Figure A2.2 The three components of fire prevention, key elements and examples.

Figure A2.3 ***Risks*** *are the human-related things that start fires. Debris burning is one of the leading "preventable" fire causes. This was a pile of trash someone was burning and it got away. Note an attempt to cut a scratch line, a shovel and a can of gas.*

Northtree Fire International

Figure A2.4 The stuff that burns is the ***hazard****. This can include homes. Tucked under all of these trees are hundreds of homes and thousands of people.*

restricting the hours of work during periods of high fire danger. You could also modify the hazard. Have fuel breaks constructed around operational areas or along roads.

The key to preventing human-caused fires is working with people. Education will go a long way, but if the local laws prohibit these types of fires (debris burning, machine use, etc.), start prosecuting offenders. Law enforcement is the last element of prevention. Use it when everything else fails.

FireWise

The term *FireWise* refers to the concept of designing, constructing, modifying and maintaining one's home in such a way that it is less vulnerable to fire that may threaten it. The concept includes nonflammable roof and siding materials; covered attic vents; adequate clearance of flammable vegetation; as well as smoke detectors; adequate street widths and clearly posted names; adequate water for fire protection; and fire extinguishers. (Figure A2.5 outlines the key concepts of a fire prevention program.)

The presence of homes in the wildland brings more people, thus increasing the likelihood of risks and hazards. There have been homes, barns, sheds and other types of buildings scattered throughout the countryside for centuries. Only recently has this phenomenon been referred to as the "wildland/urban interface fire problem." The presence of homes dramatically changes the firefighting strategy and tactics (e.g., wildland firefighting agencies have less opportunity to do perimeter control because they have to protect more structures).

KEY CONCEPTS IN FIRE PREVENTION

Risk *is the chance of a fire starting: the causes.*

Hazard *is the fuels in which the fire will burn: type, volume, condition, arrangement and location.*

Fire danger *is the chance of a fire starting given the current/expected risks and hazards. Categories of fire danger are low, moderate, high, very high, and extreme.*

Exposures *are the resources or properties or lives that would be threatened by a fire.*

Damage Potential *is the net economic change as the result of a fire.*

A Fire Prevention Program

Fire prevention education and enforcement programs target risk; engineering targets hazards and exposures. A comprehensive fire prevention program uses all three elements (education, engineering and enforcement) to try to reduce risk, hazards, and exposures.

Figure A2.5 A complete fire prevention program includes an evaluation of risk, hazards, fire danger, exposures and damage potential.

Appropriate defensible space gives firefighters room to work, and homes a "fighting chance."

Another term that you should know is *defensible space*, the cleared area between the structure and wildland vegetation that may "save" the home if there are no firefighting personnel available to protect it, or the area where you will set up a defense of the home. Adequate defensible space and proper design elements are the key factors in protecting a home from an advancing fire (Figures A2.6 and A2.7). You must have both to have maximum advantage.

Figure A2.6 In the event of a wildland fire, this home would be lost because there is no clearance and no place for firefighters to defend it. It is a "real loser."

Figure A2.7 Defensible space doesn't mean you have to clear everything for 30 or more feet; it means you have to ensure that a wildland fire cannot just roll up unabated to the structure. This home was saved by having enough clearance.

It is well-known what is necessary to protect a home from an advancing wildland fire. It should be is constructed of nonflammable materials, and have adequate clearance to keep the fire from entering through an attic vent or by radiating through a window. You don't have to live in a steel house to be protected, but many modern materials are now available that allow homeowners to "build smart from the start," or modify their existing homes.

Fire-resistant roofing is critical to the survivability of structures.

The Flammability of the Home

The most critical part of a home is its roof covering. If a home has an untreated wood roof, the chance of it catching fire from flying firebrands is dramatically increased (Figure A2.8). If there is an accumulation of leaves and needles, the likelihood of a fire starting is increased that much more. The fire does not have to be right up to the home to threaten it from burning brands. The *FireWise* recommendations to deal with these problems are:

Figure A2.8 Not only is this a very flammable roof, they have allowed "kindling" in the form of leaves to accumulate. One tiny burning ember is all it will take to burn this home down.

- A 30-foot (minimum) "green area" cleared of flammable vegetation to reduce heat and spotting.
- Every spring, or as necessary, remove all debris (needles, leaves, twigs and branches) from the roof and gutters.
- Encourage the homes in your area to be constructed with a fire-resistant roof covering. Steel roofs are considered better than tile because they are lighter in weight.

The flammability of the siding materials is also important. Discourage the use of wood siding, but if the homeowner insists, recommend plywood, or another less porous wood surface. The best exterior covering would be stucco or another nonflammable material. If wood is the material of choice, the clearance around the home should be increased.

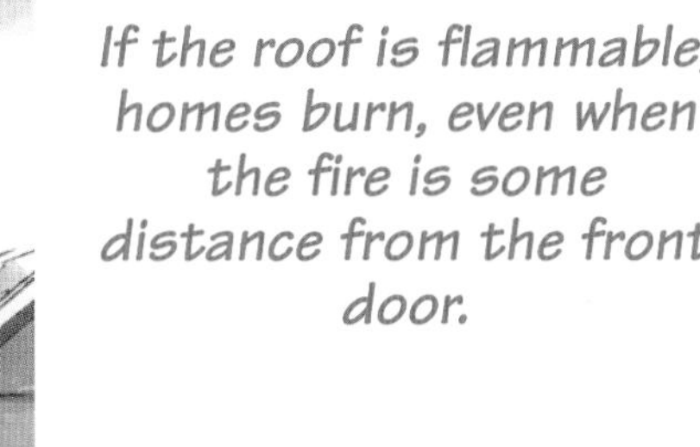

If the roof is flammable, homes burn, even when the fire is some distance from the front door.

There are several other ways a fire can enter a home. Heat can accumulate under eaves, cantilevered floors and balconies.

If there is a way, fire will find it.

Firebrands can get into the attic through unprotected attic vents. Radiant heat can pass through windows, igniting curtains and nearby furnishings.

The FireWise recommendations are:

Be sure the attic vents are screened. Plastic vent screens melt... they don't hold up when subjected to heat.

- Enclose eaves, cantilevered floors and balconies with materials rated for a one hour separation.

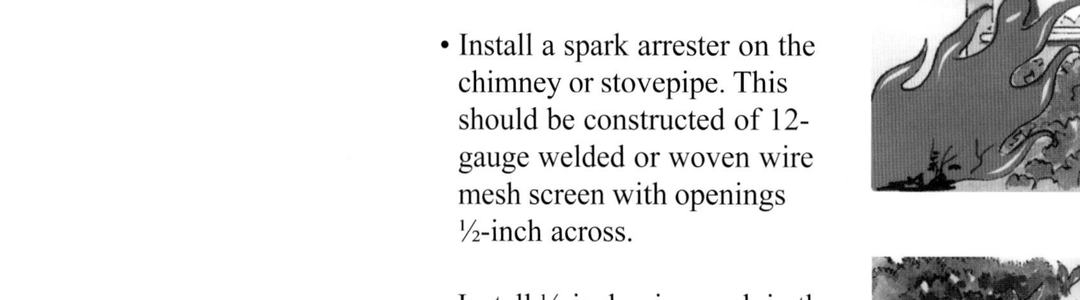

- Install double-pane windows and remove flammable fuels that are adjacent to windows.
- Install a spark arrester on the chimney or stovepipe. This should be constructed of 12-gauge welded or woven wire mesh screen with openings ½-inch across.
- Install ¼-inch wire mesh in the attic vents.
- Install shutters over windows and other openings.

No limbs within 10 feet of chimneys or stove pipes.

The Access to the Home

A home has to be adequately marked so firefighters can find it. Streets have to be marked, and the home has to have a readily visible street number. The roadway has to be wide enough to allow fire apparatus to move into an area at the same time the residents are evacuating. If a bridge is involved, it must have the load capacity to support heavy fire apparatus. There has to be adequate area to allow the

Streets, roads, and bridges must be capable of handling heavy fire apparatus.

fire apparatus to maneuver and turn around. The FireWise recommendations are:

- The streets be signed and the homes be marked with reflective numbers of such a size that they can be read from the road. The numbers should be at least 4 inches tall.

- The access roads should be a minimum of 20 feet wide and should not have a grade steeper than 15 percent.
- Bridges should have a minimum load capacity of 40,000 pounds.
- If a cul-de-sac is present, it should not be over 1,000 feet in length and should have a radius of at least 45 feet.

If fire personnel can not find your home, they can't protect it.

Location of the Home on the Property

Placement of the home may also be critical to its survivability. The steeper the terrain, the further back from the edge of the slope it should be set. If this is not possible, the vegetation must be cleared that much farther down the hill. The FireWise recommendation is:

- Place the home at least 30 feet back from a slope that is greater than 40 percent.

Other Hazards

There are several other things that may cause a problem in the event the property is threatened by a wildland fire:

- Log piles close to the structure. Log piles tend to be a great place for fire to take hold, and once on fire can hold heat for hours.

• LPG tanks are very common in rural areas. Be sure they have proper clearance, and if threatened, are kept cool. If they do start to vent and burn, let them burn out, if possible.

Fuel Modification to Provide Defensible Space

The development of defensible space is critical to the survivability of the home (Figures A2.9 and A2.10). Creating the space doesn't mean the removal of all the vegetation for 100 feet around the house. The key is the modification of the fuels so that the fire can't easily

Figure A2.9 The ideal situation is a home that is ignition resistant, located well back from a slope, with proper clearance and easy access. It can be done, in most cases.

How homeowners can make their homes defensible.

Problem	Concern	Action needed
ROOFING	The roof of a house is the first area to be "assaulted." Flying brands will hit the roof long before flames contact the exterior walls.	Roofs MUST be constructed of fire-resistant material. Roofs and rain gutters must be cleared of all flammable debris.
CLEARANCES	Flammable fuels can allow the fire to move from the burning vegetation to the structure.	Clear back a minimum of 30 feet. Limb up trees so that a ground fire cannot move into the crowns.
EXTERIOR WALLS and OPENINGS	The fire can enter the structure via the siding, windows, attic venting or other openings.	Use fire-resistant materials on exterior walls. Windows should be double-pane to reduce the passage of radiant heat; roof and attic vents should be covered with wire screen, and the area under decks should be cleared of all flammable material. Eaves should be boxed.
SLOPE	If a structure is constructed on or near a steep slope, fire coming upslope will hit it harder.	Increase clearance to compensate for the steepness of the slope. It may have to be as much as 200 feet.
ACCESS	It is sometimes very difficult to find homes in remote, wooded country because of unmarked street names and numbers.	Roads must be marked; address numbers should be visible. Roads wide enough and bridges strong enough to handle heavy fire apparatus.
LACK OF WATER	Most rural areas do not have fire hydrants.	Provide water for the fire department. Use tanks, cisterns or access to swimming pools.
INTERIOR FIRES	Adequate warning of a fire starting inside the home to save lives.	Install a smoke detector; change the batteries once a year. Have a fire extinguisher. Consider the installation of a residential sprinkler system.

Figure A2.10 Things homeowners can do to make their homes FireWise and defensible.

move through them to the home and ignite it. Brush and dense undergrowth are primary hazards to buildings. They ignite easily, burn with intense heat, and spread fire rapidly. Vegetation clearance requirements are necessary to reduce flame exposure and radiant heat, AND to give residents and firefighters a reasonable chance of protecting structures. The FireWise recommendations are:

Defensible space can be landscaped with fire-resistant plants, following FireWise recommendations.

- Maintain a firebreak, cleared of ALL dead material, for a ***minimum of 30 feet***. Dry grass should be mowed.
- Expand the area, possibly up to 100 feet, if the structure is located on sloping terrain.
- Remove limbs of any tree which extends within 10 feet of a chimney or stovepipe.

Debris Burning Standards

Burn barrels and the burning of small piles are still allowed in some rural areas. If permits are required, become familiar with the regulations and process. You may also have to be familiar with any applicable air pollution regulations.

One of the most common causes of wildland fires is debris burning.

Discourage burning during the fire season, and never allow burning when it is windy or it is forecasted to be windy. The recommended debris burning standards are:

- If using a burn barrel, ensure that it doesn't have holes in it, that a screen with a maximum of ½-inch holes is used and that there is a minimum of a 10-foot-radius firebreak, void of any flammable material, around the barrel.
- Piles should be kept small, no more than 4 feet high and 4 feet around. Ensure that there is a minimum of a 10-foot-radius firebreak, void of any flammable material, around the pile.
- Ensure that an able-bodied person always be in attendance during burning. This person should have a shovel AND a water hose, in case the fire does escape.

Debris burning can be done safely with adequate precautions and compliance with codes or regulations.

Conducting a Fire Prevention Inspection

You may be called upon to inspect various buildings and surrounding areas for compliance with local, state or federal fire codes and regulations. Before you start, you should be familiar with the laws and regulations you will be enforcing.

Be prepared and make a good first impression.

You will also have to gather up some tools. Be sure to have a map of the area, a supply of inspection forms, note pad and pencils, a measuring tape, and flashlight. You may want to have a camera to document any serious infractions. You should also have copies of all the laws and regulations you will be using as the basis of your inspection.

It is best that you enter the property in a marked fire department vehicle. This helps explain your presence. You should wear a neat, clean uniform. Introduce yourself and explain to the homeowner why you are there and what you would like to do. Be courteous and friendly. The law may give you the right to be there and perform the inspection, but don't use this authority unless absolutely necessary. ***Your mission is to gain the willing cooperation and compliance of the homeowner***.

Look for clues to the origin of the fire.

Some of the FireWise elements to check are:

- Adequate clearance around the structure. If the structure is on a slope, has the minimum clearance been increased to over 30 feet? Explain the need for "defensible space."

- Litter on the roof and in the rain gutters. Don't climb on the roof. Some roofing materials can be easily damaged. Explain that dried litter on the roof is ready to catch any flying brands that may come this way.

Know the fire regulations that apply in your area.

- A well-maintained screen over the chimney and no trees or limbs within 10 feet of the chimney.

- Screens on attic vents. Show the homeowner how easy it would be for a brand to enter the attic and start the home on fire.

Check small engines for spark arresters.

- That any firewood pile is at least 25 feet from the structure. Also check to see that no other flammable materials are close enough to the structure to pose a threat. Describe to the homeowner the effects of radiant heat and how it has the ability to take a fire into a structure through perfectly good windows.

Know the regulations for your area.

- Look for the street name and adequate house numbers. Explain how important it is to a firefighter or other emergency responder to have the roads and homes properly marked and identified.

You should also check any barbecues, incinerators or burn barrels, and LPG tanks for proper clearance. Barrels should have a properly maintained screen and a minimum of 10 feet of clearance. LPG tanks should be cleared to a minimum of 5 feet. Ask the homeowner how they dispose of hot ashes. Explain that every year hundreds of homes are lost because a homeowner places "cold" ashes in a flammable container, or dump them into a hazardous area.

If you see an internal combustion engine, check to see if it has the proper spark arrester installed. Check all potential sources of ignition.

The homeowner has the primary responsibility to protect their home.

Recommend to homeowners that they have enough garden hoses to reach all the way around the home, and that they have a ladder so they can easily and safely get onto the roof. Suggest that they keep a shovel and rake handy during the summer months, and that they keep a large bucket available for carrying water.

Remember, the key to a successful fire prevention program is to gain compliance without having to use the authority given you by the law. You are there to help people protect themselves.

Fire Cause Determination

Most often, the fire cause determination and fire investigation are conducted by a specialist. But, you may be asked to assist in the investigation in several very important ways. You may be asked, "Did you see anyone while you were responding who may know something about the cause of the fire?"

Flag the origin or evidence to protect it.

While responding to a fire you should look for anyone who may know something about the fire. Look for:

- People or vehicles that "don't fit."
- Vehicles speeding away from the fire area.
- Vehicles running without lights at night.

- Vehicles or people in unlikely places.
- Children leaving the fire area.
- Gates and fences down or damaged.
- Power lines that are down, damaged or arching. This is a firefighter safety issue and all personnel on-scene or en-route should be notified of this hazard.

Downed power lines can be deadly!

Write down descriptions of people or vehicles, licenses, direction of travel and time, or any other information that will aid in identification.

Once you arrive, you should:

- Make note of the location of the fire. Also, write down the descriptions of any people or vehicles at the fire when you arrive. This will assist the fire investigators in their task.
- If there are people there when you arrive, ask them what caused the fire and where it was when they first spotted it.

Jake says, "Be sure you do not destroy evidence; keep notes on what you see as you arrive."

- If the origin is evident, protect it. Keep everyone, including firefighters, out of the area.

Are people leaving the area as you arrive?

- Do not disturb any material that you suspect is evidence. This material is usually very fragile and can be damaged or lost. In addition, the rules of evidence control are very strict. If not handled properly, the evidence may be lost or inadmissible in a court of law.
- Look for any evidence that may lead investigators to the cause of the fire and those responsible. Are there signs of a campfire or debris burning? Are there signs that the fire was intentionally set?
- Scan the crowd or individuals that are there. Look for anyone that is looking nervous or agitated, or who is trying to help.
- Flag the origin, protecting it from people tramping it down or firefighters mopping it up and destroying evidence. Post a guard if you have to. Use flagging tape, toilet paper, or place hand tools around the origin or evidence to protect it.

Write down descriptions of people and vehicles that look or act suspicious.

Be sure to give any information you have to the fire investigators; don't wait for them to come to you.

Notes:

Introduction to Hazardous Materials Recognition

Hazardous materials are literally known by hundreds of thousands of names. Within the Department of Transportation (DOT) guidebook there are approximately two thousand-four hundred (2,400) hazardous chemicals listed. Hazardous materials can also be found everywhere in our daily lives.

Hazardous materials are used to preserve the food we eat, and are used in some medical treatments. They are used in the manufacturing of our clothes, within the fuels we use to power our cars and heat our homes, and are also found in the process used to purify our drinking water. This segment will introduce you to the recognition and identification of hazardous materials, or more commonly called within the fire industry as "Haz-Mat Recognition and Isolate (R and I)."

DANGEROUS WHEN WET 4

FLAMMABLE 3

There are thousands of chemicals that can harm you.

Placards like these are not for looks!

This section of the handbook is not intended to train you to become a specialist in hazardous materials emergencies, but rather to enable you to better assess the hazardous materials threat to people, property and systems through awareness, and mitigate potential risks and hazards if you encounter a haz-mat incident.

A hazardous material can be one substance or chemical, or a combination of several.

Before further discussion, it is important to first understand the definition of hazardous materials. Hazardous materials can be defined as "a single material or a combination of materials that may produce serious health, fire or explosive hazards." Hazardous materials can come in many different forms and types. They can be poisonous, explosive, or biologically hazardous, depending on their make-up.

General Guidelines when Responding to a Possible Haz-Mat Emergency

In the event that you are exposed to a "hazardous materials" incident, you must ask yourself the following question: "Will my involvement favorably change the outcome?" The following are general guidelines when reacting to a possible hazardous materials emergency. Some simple directions on isolating the area can be found in Figure A3.1. Review your agency's policy before placing yourself or others at risk.

Jake says, "You may only be able to isolate the area and keep the material from spreading."

If you feel that you are involved in a hazardous materials incident you should approach cautiously, wearing appropriate protective gear, and with supervisor authorization. Resist the urge to "rush in." You need to be knowledgeable to accurately identify a hazardous material. Reacting without proper planning and preparation may cause you to be part of the problem, not part of the solution:

Isolation Distances (Rule of Thumb)
(In all directions)
Minor event (1 drum, bag, etc.) = 150 ft.
Residential/Commerical area = 300 ft.
Major event (1 drum or more) = 500 ft.
Open area = 1,000 ft.
B.L.E.V.E. potential = 2,500 ft.

Figure A3.1 Hazardous materials are nothing to fool with. Clear the area...here are some simple guidelines for how far you should set the limits.

- Without entering the immediate hazard area, do what you can to isolate the area and assure the safety of yourself, your fellow workers, the surrounding community and the environment.

- Direct people away from the scene, and keep them at a safe distance.

- Allow enough room to move or remove your own equipment, or accommodate any emergency response or evacuation efforts.

- Maintain communications with your supervisor and/or dispatch.

- Above all, ***do not*** walk into or touch spilled materials! Avoid inhalation of fumes, smoke or vapors, even if no hazardous materials are known to be involved. Odorless gases or vapors may be very harmful. Your ability to identify danger and act decisively will help you minimize the potential harm and risks associated with the release of the many toxic substances known as "hazardous materials."

Jake says,
"Remember the song titled, 'Only fools rush in!?' You must never rush in at a haz-mat incident."

If something doesn't look right, approach with caution.

To ensure you have covered all of the bases, use a checklist like the one in Figure A3.2 to remind you of all the necessary steps. Follow the steps in the "D.E.C.I.D.E." process, and look for the six clues when determining the presence of hazardous materials.

HazMat INCIDENT CHECK

THINK SAFETY - Is the benefit worth the risk?

- ☐ Safe approach: upwind; from above; or from upstream.
- ☐ Isolate and deny entry.
- ☐ Involve only trained and authorized personnel.
- ☐ Ensure proper protective equipment is used.

SCENE MANAGEMENT

- ☐ Assume control of area and emergency response personnel.
- ☐ Protect life, environment and property.
- ☐ Contain and control the substance.
- ☐ Develop an ACTION PLAN.
- ☐ Attempt to identify substance.

Figure A3.2 This simple checklist may remind you of something you have not done.

The D.E.C.I.D.E. Process

Once you are aware that a hazardous situation may exist, a series of decisions must be weighed in order to act effectively and safely.

The acronym, "D.E.C.I.D.E." will help you execute the decision making process, and can be broken down into six areas of recognition, identification and awareness. **D**etect, **E**stimate, **C**hoose the response objective, **I**dentify action options, **D**o the best option, and **E**valuate your progress.

DETECT
ESTIMATE
CHOOSE
IDENTIFY
DO
EVALUATE

The first step to take in Hazardous Material Recognition and Identification is to **"DETECT"** any hazardous materials that are present. Hazardous materials are often located in or near structures, including garages, sheds, barns, trash piles, ditches, and even abandoned vehicles. You should also be aware of detecting large capacity liquid petroleum gas (LPG) tanks, which will generally be located within 100 feet of the structure. Other hazardous material concerns include drug labs.

Drug Labs

Drug labs found throughout the United States are mostly temporary, makeshift operations set up in apartments, storage facilities, motel and hotel rooms, campsites, fields, vacant buildings, and vehicles. Discarded lab equipment and chemical waste are also being found in roadside ditches. As a wildland firefighter, it is important to be able to detect drug labs in the wildland or urban interface environment.

Meth labs often catch fire.

In detecting drug labs, more commonly known as methamphetamine labs or clandestine labs, you should be able to identify the hazardous materials and/or products associated with their operation. Methamphetamine (meth) is a potent amphetamine, typically produced from over-the-counter cold medications and sold illegally in pill form, capsules, powder and chunks. Meth stimulates the central nervous system, is extremely addictive, and can cause an intense, prolonged withdrawal. Users may be violent or combative, or display paranoid behavior. Common street names for methamphetamine include: speed, crank, glass, crystal, ice, and zip. Meth can be swallowed, snorted, smoked or injected.

Estimate the likely harm.

One type of hazardous material container connected with methamphetamine or clandestine labs are LPG tanks. These tanks are generally small, and contain a very toxic gas (not liquid petroleum), called anhydrous ammonia, which is used in the drug lab process. If you detect an LPG tank with a turquoise (blue-green) discolored valve stem, tag the tank and leave it alone. Under no circumstance should you open, or allow any other person to open the valve. This turquoise-colored valve stem (oxidation of the brass valve) indicates the LPG tank contains, or has contained, anhydrous ammonia. This could also be an indicator that a drug lab exists in the general vicinity. If you detect this type of hazard, it should be reported immediately to your supervisor, incident command and local authorities for proper removal.

In addition, it is important to detect and be aware of other drug lab paraphernalia which could be stored and/or used in quantity amounts. Some of the more common materials found in meth labs include: chemicals with potential for explosion; fire; toxic release and health effects, including damage to skin and lungs; as well as long-term health effects. These dangers are in addition to those related to using the manufactured drugs.

TOXIC

6

Some of the items found in a drug lab explode.

Substitution of common ingredients in a basic recipe, based on ingredient availability, is the norm in meth cooking. The following is a list of common substitutions, in specific ingredient categories:

- ***Ephedrine*** **-** Pseudoephedrine, Efidac, Sudafed, Mini-thins.
- ***Organic solvents*** **-** Toluene, Coleman fuel, ether, acetone, white gasoline.
- ***Corrosive alkalines*** **-** Calcium, potassium or sodium hydroxide, caustic soda.

An individual who believes he or she has discovered an illegal drug lab, or the site of an abandoned lab, should immediately notify local law enforcement and should not enter the area of the suspected lab. Any individual who inadvertently enters a lab, should leave immediately without disturbing any cooking process, chemicals or equipment. Confrontation with makers of illegal drugs is very dangerous and is strongly discouraged. While dangerous, drug labs are not the only type of hazardous materials incident you may detect. Be prepared if you detect others.

A drug lab is a crime scene.

If hazardous materials have been detected, your next move should be to **"*estimate*"** the likely harm. After completing your estimation of harm, you should then determine the extent of the problem. If possible, predict the behavior of the material.

Don't destroy evidence.

Next, you should **"*choose the response objective.*"** This includes staying away from the area of concern. Keep a safe distance from the suspected hazardous material. Next, notify the authorities, and determine if all the necessary people have been called to handle the incident. Keep others out of the suspected area! Don't let anybody into the area except for authorized personnel.

Next, ***"identify"*** your action options. If you identify that the hazardous material in question is an immediate life threatening hazard, withdraw immediately from the area of the hazardous material, notify your supervisor and proper authorities, and secure the area from unauthorized entry.

Check for clues.

"Do best option": After you have withdrawn from the area, you should observe the effects of your decision. Follow up; make sure your action decisions are working, and then get feedback from residents, fellow workers, supervisor, and incident command.

"Evaluate your progress." Follow your agency's policies to determine who to call and what course of action should be taken next to remove or mitigate the hazardous material incident.

Six Clues for Detecting the Presence of Hazardous Materials

Cover all of your bases.

There are six clues that you should look for when detecting the presence of hazardous materials. These clues will help you determine the type and level of hazardous material exposure, and are prioritized to be followed in this order:

Occupancy and location.

- ***Type of occupancy and location*** - The occupancy or location of an area effected by exposure to hazardous materials needs to be addressed. Is it in a populated neighborhood, or near a busy intersection? The occupancy and location are primary considerations of a hazardous material exposure.

Shape of the container.

- ***Container shapes*** - The shape of the container used to haul hazardous materials may give you a clue as to what is inside. A gasoline tanker, for example, looks very different from a container which is used to haul non-hazardous materials such as vegetables. However, a can of gasoline may contain other chemicals besides the obvious.

Markings and color.

- ***Markings and colors*** - Some containers may have special markings or colors that will give you a clue if they are transporting a hazardous material or not.

Placards and labels.

- ***Placards and labels*** - Placards and labels will allow you to identify if a material is poisonous, flammable or explosive (Figure A3.3).

- ***Shipping papers*** - The shipping papers that must accompany any load will tell you what the material is and its hazard level.

- ***Your senses*** - Your senses may be able to tell you if a product is hazardous. Some materials such as gasoline, diesel, ammonia and chlorine all have distinct smells. Be extremely cautious! Although some toxic gasses or vapors are odorless, they may still be harmful or deadly, if inhaled.

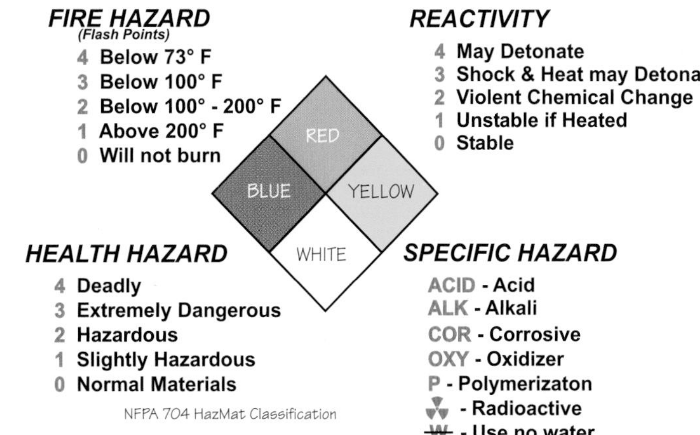

Figure A3.3 This is the standard hazardous materials placard that provide a clue on how to deal with the "stuff" within the car, truck or building.

The release of hazardous materials can have wide-ranging effects on people, property and the environment. Your ability to properly identify and react to these situations can make a positive difference on the outcome.

Shipping papers.

Being able to identify a hazardous material exposure incident and respond properly using the "D.E.C.I.D.E." process, can help you minimize the effects hazardous materials have on an area and the surrounding environment.

Your senses!

Using "The Six Clues" of hazardous materials detection will help you determine the presence of hazardous materials and set hazardous material management goals and objectives quickly, efficiently and safely.

Being aware of your role as part of an emergency response team will allow you to make the area safe for the surrounding community, your fellow workers and yourself.

You may only be able to isolate and keep from spreading.

Knowing how to react to hazardous material exposure can make all the difference. Responding appropriately to hazardous material situations can save lives.

Figure A3.4 There may be times you will be fighting fire on a military or former military site, and unexploded ordnance is always a possibility. Be very careful.

Unexploded Ordnance

There may be times when you will encounter unexploded ordnance (UXO). This can be very dangerous, and you need to recognize when UXO exposure is possible (Figure A3.4).

Some of the more common UXO that could be found on military or former military sites are:

- Small arms munitions
- Projectiles
- Grenades
- Bombs
- Rockets
- Guided missiles
- Submunitions

There may be times where fully intact munitions or fragments will be found. Consider either to be dangerous. Most of these munitions will be old, and their chemical make-up may have been altered over time.

UXO Safety and Reporting

If you see unexploded ordnance:

- Stop and do not move closer. Do not move or remove anything close to the UXO.
- Never transmit on the radio (hand-held, mobile, etc.).
- Never touch or disturb.
- Clearly mark the area.
- Keep a minimum of 500 feet away.
- Report the discovery to your supervisor.

Use and Care of Hand Tools

When it comes right down to it, firefighters with hand tools ultimately contain wildland fires. While all of those other tools discussed in Chapter 7 assist the firefighter, he or she will actually take the final actions that ensure the fire will not spring back to life.

Correct Tool for the Job

The various hand tools available to wildland firefighters have been developed over years of experience and experimentation. Several have multiple uses, while some are intended for specific tasks only. It is very important to select the correct tool for each task, as the improper use of a tool increases the risk of injury to yourself, other firefighters, and/or damage to the tool.

Use the proper tool for the task at hand.

Tool Inspection

Handles should be smooth.

Each time you pick up a hand tool, inspect it to make sure that it is in good condition and safe for you to use. Examine all the parts of the tool to make sure they are not loose, cracked or broken. Make sure the blade is sharp. Make sure the tool is properly assembled and that you have any accessories you will need to properly use the tool. Test the handle by placing the head on the ground, the handle at a 45-degree angle, and applying firm, downward pressure. Any tool that cannot be repaired or is unsafe should be removed from service and flagged.

Figure A4.1 The tool is firmly planted in this stump, the firefighter's hands are gloved, the file is held in the palm of the hand and a file handle is in place. Use a smooth stroke away from the edge.

Sharpening

Most hand tools should be sharpened by hand with a file. A manually powered wet grinding wheel can also be used, but is not very portable. Do not use a power grinder to sharpen hand tools unless you have been specifically trained for this procedure, as improper use of a grinder can ruin the tool and make it unsafe. Always follow the correct procedure and use proper safety equipment when sharpening hand tools. Use a flat, mill bastard file with a safety guard to sharpen hand tools. Wear gloves (Figures A4.1 and A4.2). Make all strokes with the file in a forward motion, holding the file at the same angle for each stroke. Use long, smooth strokes, applying even

Most hand tools are sharpened with a file.

Wear your gloves!

Figure A4.2 When done, don't run your finger across the edge to see if it is sharp. This is one injury that is completely avoidable.

pressure, and releasing the pressure on the backstroke. When sharpening axes, Pulaskis, and brush hooks, the filing direction should always be away from the eye toward the cutting edge to reduce the chance of injury. Clean the file with a wire brush or file card between uses. Protect the edge with masking tape. This will protect the edge from damage while in storage. You can also use boxes, old fire hose, sheaths, old innertubes and conveyor belt to protect the tools while in storage. When storing the tools on a vehicle, be sure the safety guards are properly secured and placed in the proper compartment. If they are to be transported by aircraft, be sure they are properly bundled.

Dull tools are dangerous!

Handle Care and Replacement

Many tool handles are made of wood, which is light and strong, easy to work, and affordable. Wood handles need to be properly maintained to extend their useful life and make them safe for the user. Handles should be smooth and free of burs, splinters and cracks. Sand rough handles smooth, and treat with a light coat of boiled linseed oil as a preservative. Never paint or varnish the tool handle. Tool heads should be tight on the handles. Use metal wedges to tighten tool handles on heads. Handles that are bent, cracked, splintered or otherwise damaged should be replaced before the tool is used again. Some tools may have handles made of metal, fiberglass, or other synthetic material that may require different care.

Inspect the tool before use.

Carrying

There is a correct way to carry each type of hand tool. Carrying a tool any other way increases your risk of injury. Learn the correct way to carry each hand tool and practice doing it right every time. As a general rule, tools should be carried on the downhill side when traveling side hill, so that *when* (not if) you fall, you will NOT land on the sharp tool. When you pass the tool to another firefighter, always pass it handle first.

Usually carry your hand tool on the downhill side.

Personal Space

When walking or working with hand tools, give yourself at least ten feet of space between you and your co-workers to reduce the

Figure A4.3 Call out, "Coming through!" before trying to pass all of these firefighters swinging tools.

chance that you will accidentally strike someone else with your tool, or that "the other guy" may strike and injure you.

If you have to move up or down the line of working firefighters, signal your passing by calling out, "Coming through!" Wait until they stop swinging and grant you right-of-way (Figure A4.3).

Maintain a safe working space (10 feet) from other people. Remove any obstructions before using the tool. Use the tool properly.

If you lay your tool down along the fireline, be sure it is a safe distance from foot traffic, but still visible to other crew members. Attach guards, when necessary, and flag the tools if you need to. The key is to leave the tool in a safe place so no one can stumble across it and be injured.

Regardless of how you will be traveling, be sure your tools are properly stowed.

Safety Tips

Inspect and test each hand tool before leaving the vehicle. Carry each tool with its appropriate guards in place to protect you from sharp edges. Carry each tool at its balance point to improve control and reduce fatigue. Use the correct tool for each job. Keep tools sharp and handles tight and smooth. Always wear gloves and any other recommended PPE when using hand tools.

Cutting Tools

Cutting tools are primarily intended for cutting wood (usually limbs, small trees and bushes), although they may be used for felling full-sized trees if you have lots of energy or manpower. They have a sharp edge to penetrate hard wood, and can be very dangerous if not maintained or used properly.

Axe

There are several types of axes available to firefighters. The most common are the single-bit axe, the double-bit axe and the fireman's axe. Each has a head with one or more cutting edges attached to a handle. Short-handled axes are called hatchets. One side of the head of a single-bit axe can be used like a hammer for driving wedges. The fireman's axe has a pointed hook opposite the cutting edge to pull or pry open doors and windows.

Carry the axe at the balance point, just behind the head.

All axes should be sharpened with a smooth, even, double-tapered edge extending back two and one half inches from the edge of the blade on each side of the head. Cover the cutting edges with tape or a plastic guard after sharpening.

Carry an axe at the balance point, just behind the head with the head facing forward and the handle parallel to the ground.

When using the axe, stand with your feet about shoulder-width apart to improve balance. Use a firm, flexible grip on the handle to allow for controlled swings. Be sure there are no overhanging branches or limbs to interfere with your swing. When cutting wood, strike the wood at an angle of 45-60 degrees to achieve best penetration. Learn how to swing the axe from both sides of your body to make notched cuts or work around obstacles.

Pulaski

The Pulaski is a combination cutting/grubbing tool invented by Ranger E.C. Pulaski of the USDA Forest Service, who is famous for saving 150 men trapped in the "Big Burn" of 1910 in Idaho. It has a flattened mattock blade opposite an axe blade, and can be used for grubbing and trenching in the soil in addition to cutting wood. It weighs about 3¾-pounds and has a 36" handle. The axe blade is sharpened like any

The Pulaski is used to cut and grub.

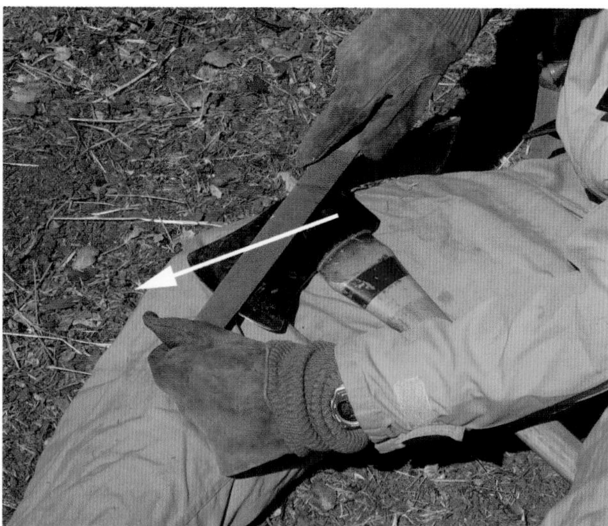

Figure A4.4 and A4.5 When sharpening an axe or Pulaski, be sure the tool is well-anchored before you begin. You don't want it to move around. Always wear gloves and use a file with a handle attached. Stroke the file away from the sharp edge.

other axe. The grubbing edge is sharpened on a single bevel, angled back $^{3}/_{8}$" from the edge. Install a guard on the cutting edges when finished sharpening (Figures A4.4 and A4.5).

Figure A4.6 The Pulaski can be used for cutting, but it is primarily used for grubbing and trenching.

Carry a Pulaski by gripping the handle at the balance point behind the head, with the handle parallel to the ground and the mattock blade up. A very versatile tool, the Pulaski can be used in any situation an axe can be used, but the mattock side also allows it to be used for grubbing, trenching or prying, as well as scraping line in grass or duff (Figure A4.6). The same rules for using an axe also apply to the Pulaski tool.

Brush Hook

The brush hook is a specialized firefighting tool with a long, curved cutting edge mounted on a straight axe handle. It is an effective tool for cutting small diameter wood material such as brush.

Because of its long, curved cutting edge, the brush hook can be difficult to control, and requires greater skill to use safely and efficiently than other cutting tools.

To sharpen a brush hook, file both sides of the long, straight section of the cutting edge at an even angle back at least one inch from the edge. Take care when working on the throat (curved part of the blade), for this must have an even, circular sharpening pattern. The bevel at the point of the curve should be sharpened back to three-fourths of an inch.

Take special precautions when sharpening a brush hook.

Carry the brush hook by gripping the handle at the balance point with the hook of the blade pointed down.

This tool is difficult to master and requires extensive practice to use safely. It requires a pulling action at the end of the swing to sever and pull apart brush stems. Strike the stem at an angle with the straight cutting edge at a point just behind the curve. Then, pull back towards you in one motion, allowing the throat to finish the cut and drag the stem off the stump. Follow the same procedures and precautions as when using an axe.

Scraping Tools

Scraping tools are designed to scrape grass, duff, litter, etc., off of the mineral soil to create a firebreak. They have a beveled edge that is sharpened, but not as sharp as a cutting tool.

Shovel

The long-handled, round-point (LHRP) shovel is one of the most common and most versatile hand tools available to the wildland firefighter. It has a convex, pointed blade about 8-9" wide and 11-12" long, tapering to a heart-shaped point. The handle is about 47" long, and the tool generally weighs about 4½ pounds.

The shovel is one of the most versatile tools on the fireline.

Sharpen the shovel from the point to within 3½" of the heel of the blade with a shallow bevel, about 3/8" deep. After sharpening, install a safety guard (Figure A4.7).

The shovel is carried gripped at the balance point behind the head, with the handle parallel to the ground and the point of the blade down.

Some of the uses of the shovel are:

- Throwing dirt: The most effective way to throw dirt to smother flames is overhand, in a fan-shaped pattern (Figures A4.8 a/b).

Figure A4.7 Stroke the file away from the edge at a 45-degree angle.

Figure A4.8 (a/b) Besides using the shovel to dig, you can use the shovel's edge to scrape fuels from a control line, and throw dirt to cool or smother burning fuels.

- Digging: Dig up mineral soil for throwing at the fire's edge, or shovel it on stationary objects to smother the fire.
- Scraping: Use the flat side of the blade to scrape grass or duff down to mineral soil, using a pivoting motion with the shovel handle braced above the left knee for leverage.
- Cutting: The sharp edge can be used for cutting small brush and weeds.
- Trenching: Dig a ditch in mineral soil on a side hill to catch rolling material.
- Carrying: Move hot materials from one place to another, such as to start a burn-out, or to get near a water source.
- Smothering: Use the back of the blade like a swatter to smother flames in duff or short grass.
- Raking: Use the side of the blade to rake duff or burned material off logs and stumps.
- Shield: The blade of the shovel can be used to protect the face from radiant heat.

The McLeod is one of the most recognized tools on the fireline.

McLeod

The McLeod is one of the most widely known and used firefighting tools in America. It is a combination scraping/raking tool head on a metal shaft attached to a straight handle measuring about 48" long. The scraping blade of the head is about 11" wide by 5" deep; the rake has wide tines about 3½" deep. It is designed to scrape grass off of dirt, and then rake the grass to the side of the firebreak.

The hoe, or scraping edge of the McLeod is filed at a 45-degree angle, with the bevel sloping away from the handle (Figure A4.9). The cutting edge should be flat and level, not concave. The tines are not sharpened. Install a safety guard on the tool after sharpening.

Figure A4.9 Anchor the tool and sharpen the edge at a 45-degree angle.

Carry the McLeod at your side, balanced parallel to the ground with the hoe edge up to avoid falling on the rake tines (puncture wounds are more dangerous than lacerations).

Use this tool in grass, brush, or timber fuel types for scraping, cutting or trenching in light fuels; and then to rake the cut debris out of the firebreak (Figures A4.10 a/b). To use the McLeod as a trenching tool, stand on the downhill side of the line, and using a scraping action, angle one corner of the blade down to create a "V" trench (Figure A4.11). In light grass, use a sideways scraping motion rather than trying to chop as with a garden hoe. To cut light brush, swing the cutting edge in a chopping motion.

Figures A4.10(a/b) The McLeod is a combination scraping/raking tool. It is used to clear fuels from a control line, and to aid in mop-up.

Figure A4.11 The McLeod is also a good trenching tool. The tool pictured here is a Thau Claw. It has three scraping edges, rather than just one.

When working in light grass or shallow duff, two people with McLeods can team-up to build a line quickly. The lead person faces the direction-of-travel with the tool behind, scraping edge down. The second person is behind the first, with the tool in front, rake tines intertwined on top of the first tool. The lead person provides horizontal travel for the tools, while the rear person exerts enough downward pressure to scrape through the material. This is a quick way to make a scratch line through light fuels in an emergency.

A variation of the McLeod is the Barron, the Thau Claw or California tool, which is smaller and lighter with a longer metal shank. The blade is 10¼" wide and 2¾" deep with seven 2½" deep rake tines. The hoe side of the blade is angled, rather than straight like a McLeod. The cutting edge is sharpened at a 65-degree angle from the blade. This tool is designed for use in grass, pine needles and open woodland. The long steel shank allows the rake to be used to carry burning material for firing-out without damaging the wood handle.

Carry the McLeod with the hoe edge up to avoid falling on the tines.

U-bolt Hoe

The U-bolt hoe is like a heavy-duty, oversized garden hoe, and is designed for scraping in heavy grass, weeds or duff. The name derives from the "U-shaped" bolt that attaches the handle to the blade, which is about 10" wide by 8" deep. Handles are usually 36" long, but may be up to 48 or 50" for greater leverage.

The cutting edge of the U-bolt hoe is sharpened the same way as a McLeod, by filing the bevel edge at a 45-degree angle, sloping away from the handle. The cutting edge should be straight and level. Place a guard on the cutting edge after sharpening. Check to ensure that the two nuts on the U-bolt are tight. Carry the U-bolt hoe at the balance point, handle parallel to the ground and with the cutting edge down.

The U-bolt hoe can be used in heavy grass and duff and in light brush. Because of its light weight and unbalanced handle, swinging it like a cutting tool can be awkward and ineffective. Use a scraping action similar to a McLeod rather than a chopping action.

Combi Tool

The Combi tool is a combination tool that can be assembled in various configurations for digging, scraping and cutting (Figure A4.12). When assembled in the 90-degree configuration,

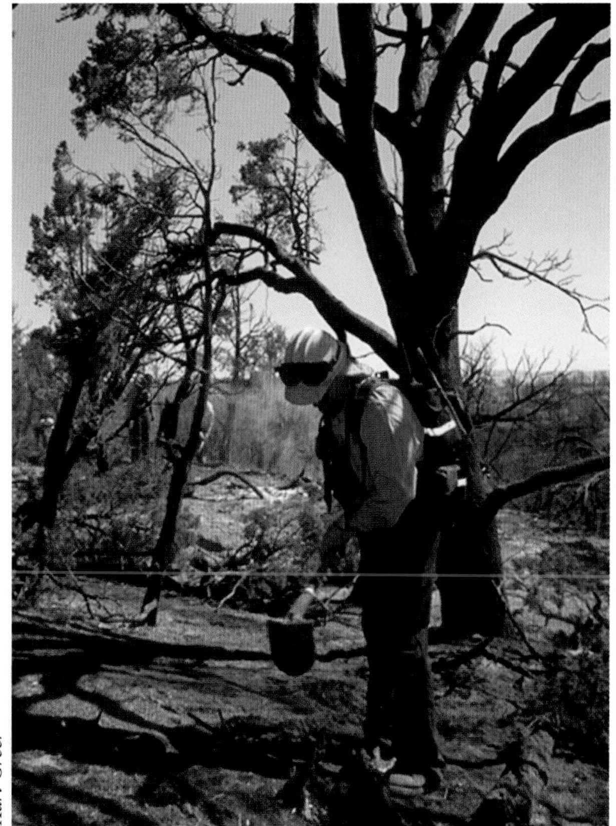

Kari Greer

Figure A4.12 This firefighter is using a combi tool during mop-up.

it works much like the hoe side of a Pulaski, and is good for building trenches in soft dirt. The pick end works well in tight, rocky areas that the hoe end of a Pulaski is too wide to reach.

Fire Rake/Broom

Brooms are good at scraping moss off of rocks.

The fire rake is used in fireline construction, mop-up, and burning-out operations in fuel types where the hoe side of a McLeod is not needed. It is well-suited to building fireline in deciduous leaf cover.

The fire broom is a heavy wire broom that can also be used for sweeping pine needles or leaves off hard ground to create a firebreak.

Felling Tools

Felling tools are used to cut down, or "fell," trees and snags, and to "buck" logs into manageable lengths. Saws are more efficient ways to cut down large trees than axes. Chain saws are very efficient, but also very dangerous. Felling timber requires special training and experience to avoid serious accidents.

Hand Saw

The most common hand saw used in wildland fire fighting is the crosscut saw, or "misery whip." These saws are usually 6-10 feet long. One person can operate shorter crosscut saws, but most require two people...and good teamwork.

Hand saws aren't used much any more.

Check the saw for kinks, cracks, pitch and rust. Keep the blade clean, using solvent as necessary to remove pitch, rust and dirt. Use a guard over the cutting edge. Only experienced saw filers should sharpen crosscut saws.

To use, attach the handles tightly to the blade and remove the guard. Stand on opposite sides of the tree with the saw at about waist height. Place feet about shoulder-width apart for good balance. Each person *pulls*, never pushes, the saw toward him/her in turn, establishing a rhythm aimed at endurance, not speed. Designate a lookout to watch for falling limbs, bark, or sparks and have an escape route pre-planned.

Chain Saw

Gasoline-powered chain saws are much more efficient felling tools than crosscut saws, but they are also very dangerous. Only specially trained firefighters should operate chain saws (Figure A4.13).

Chain saws are quick, but dangerous.

The chain saw consists of a small gas engine, a combination of handles and controls, and a bar around which a flexible chain with cutting and raking teeth is driven.

Use a saw if you are trained to cut large material.

Check the saw for proper amounts of fuel and bar oil, and proper chain tightness before each use. Use caution when fueling on/near the fireline. Always wear full protective equipment, including chaps, gloves and hearing protection when operating a chain saw. Only experienced saw filers should sharpen chains. Clean and lightly oil the chain to prevent rust. Place a guard on the chain when the saw is not in use. **NOTE:** It is NOT cool or macho to carry your chain saw thrown over your shoulder without a chain guard. Again, always use a lookout, and have a preplanned escape route when felling trees and snags.

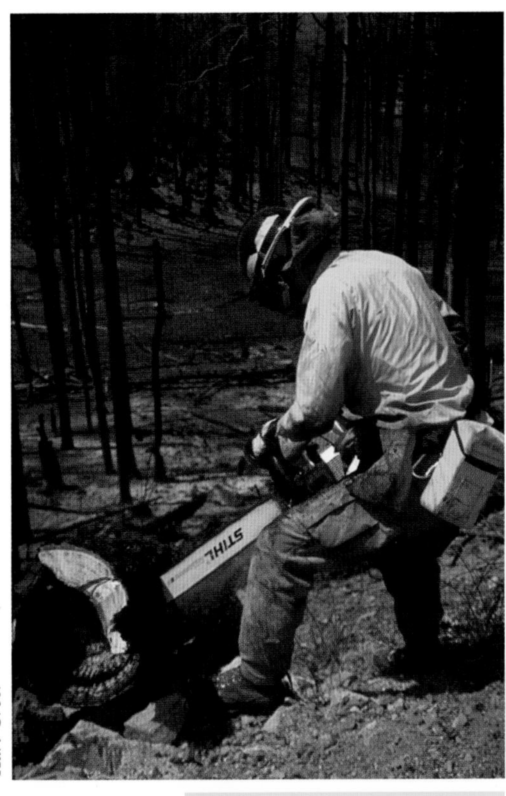

Kari Greer

Figure A4.13 Chain saws are very effective, but dangerous. Only trained personnel should be allowed to use them.

Wedge

Various combinations of splitting and felling wedges may be used to open up saw kerfs (cuts) to keep saw blades from binding, or to finish a cut. The wedge provides leverage where needed while felling, bucking or splitting wood products. Wedges are made of wood, metal or plastic, and require little maintenance. Keep wedges clean and examine them for cracks, splinters or burs before each use. If a metal wedge develops a mushroom edge, grind a bevel on the edge of the face to eliminate weak metal that can break off.

Use wedges to hold saw kerfs open so that the saw blade or bar won't bind, or to counteract the lean of a tree to be felled.

Always use plastic wedges with chain saws to avoid kickback or ruining a chain.

Sledgehammer

Wedges can make your job much easier.

The sledgehammer is a long-handled, heavy (4 to 16 pounds), double-faced hammer used primarily for driving wedges into wood in felling operations. It may also be used for breaking down doors if forced entry into a building is necessary.

Examine the sledgehammer before each use to be sure the hammer faces are free of burs and cracks, and that the handle is tightly attached to the head.

Carry the sledgehammer at the balance point, just behind the head with the handle parallel to the ground and the hammer faces vertical.

Smothering Tools

Smothering tools are used to extinguish a wildfire by depriving it of oxygen, or by blowing the air away from the flame. This is a method that works best for fires burning with short flame lengths in light fuels such as short grass or leaf litter.

Swatter

In some fuels, a swatter is very effective.

The swatter is a large flap of rubber/plastic material (like a mud flap) attached to a long, straight handle. The user literally swats the flames with the flat side of the swatter, fanning away the air, and depriving the fire of oxygen. The swatter is used with light "pattings," or by dragging it directly on and

Jake Oosthuizen

Figure A4.14. Swatters are used in a light "patting" motion, or dragged long the fire's edge.

along the fire's edge (Figure A4.14). Hard, vertical swatting can cause hot embers to spring back to life, and can spread fire.

Check the swatter to be sure the handle is smooth and free of cracks or splinters, and that the swatter is firmly attached to the handle before each use. Carry the swatter at the balance point just behind the hinge, with the handle parallel to the ground.

Gunnysack

The good, old-fashioned gunnysack (aka: crocus sack, croaker sack, grass sack, tow sack and barley sack) can be used to smother fires, and can be very effective when the conditions are right. Only use sacks made from natural burlap, not synthetic materials.

Wet the sack, if possible. Grab it by one corner, and swing it down on top of the flames in a smooth, firm—but not fast—motion. Make swings parallel to the fireline, but follow through into "the black." The idea is to blow air away from the fire without swinging so hard you scatter sparks and embers all over the place.

Wet the sack if possible.

This technique works best in short grass or leaf litter with short flame lengths and little wind. It is time consuming and hard on people, and should be used only for first attack where no better tools are available.

Backpack Pumps

Backpack pumps (aka: backpumps) are manual or powered water pumps and tanks that can be carried on a persons back, and are used to extinguish flames by direct attack with water.

Manual

The manual backpack pump consists of a rigid (metal, fiberglass) or flexible (rubber, plastic) water storage tank connected by a hose to a trombone pump and equipped with shoulder straps for carrying on the back. The rigid tanks generally have a 5-gallon capacity, and the whole unit weighs as much as 55 pounds. The flexible tanks hold 5-6 gallons and weigh 46 or more pounds.

Because it has several moving parts, the backpack pump requires a lot of maintenance. Some of the items that should be checked after each use are:

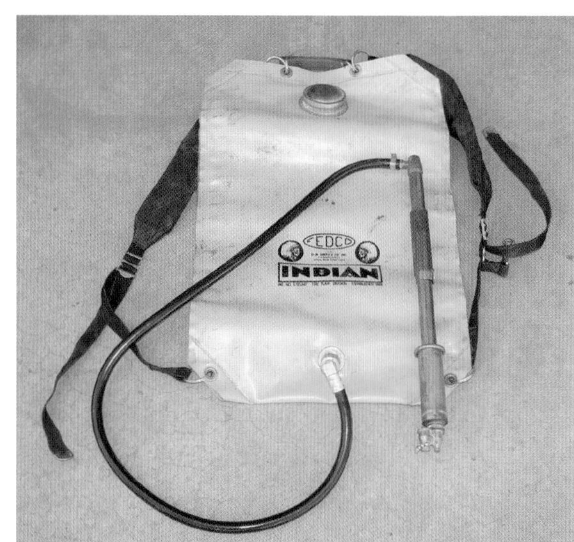

- The pump barrel and inner cylinder to make sure that neither is bent.

- The pump packing or "O-ring" for condition of seal.

A backpack pump can put out a lot of fire.

- The nozzle tip – If plugged, use a fine wire to clean it out.

- The pump itself – for any sign of corrosion or gumming-up. Clean with solvent.

- The hose – Check for kinks, cracks or leaks; check clamps for tightness.

- The tanks – Check inside for rust or corrosion; check the filter screen at the hose connection; clean out any debris. Check for leaks and repair as necessary.

- The straps – Check for cuts, cracking, or rotting.

- The air vent in the tank lid – If it is plugged, the tank can collapse; use a fine wire to clean it out.

Use your finger to produce a spray pattern.

Check the unit for function before donning, including the condition of the backpack straps and the operation of the pump. Make sure the straps are properly adjusted to help maintain your balance.

The idea of the backpack pump is simply to squirt water on the flames, but it takes some practice to develop the right technique for different situations. On a grass fire, the firefighter should direct the stream of water (either straight stream or spray, depending on the situation) parallel to the fire's edge in a sweeping motion coordinated with pumping the trombone. If a fine mist or fog is desired (useful in mop-up), a finger can be placed over the pump nozzle tip. The hand holding the nozzle end of the pump should remain stationary, with the other hand doing the pumping of the trombone. To direct a straight stream of water at a target, lock your elbow tightly against your body to steady your aim.

Deploy backpack pumps liberally when burning-out to control spot fires. Make water go farther by adding a dose of wetting agent to each tank. With experience, a single firefighter with a manual backpump should be able to contain a half-acre fire in light grass.

Backpack pumps are great for mop-up.

Power

Various types of power backpumps are available that use a small gasoline engine to power a pump that moves water from the tank to the nozzle. They are more efficient if used correctly, but are heavier than manual pumps, and can waste water if used incorrectly.

The same types of inspection and maintenance procedures apply, plus the care and feeding of the gas engine. Use caution when refueling on/near the fireline. Hearing protection is required due to the noise from the gas engine. Power backpumps can be used for the same tasks as manual backpumps.

Be sure the backpack pump is maintained and ready for action.

Firing Devices

These are tools that are used to set fire, either for backfiring, burning-out, or to ignite prescribed burns. Each type has advantages and disadvantages. Some specialized types of firing devices may require special training, permits, or licenses. When conducting a burning operation, be sure to watch for rolling material, or your operation may spread fire across the fireline. Always stay in close contact with those around you.

Drip Torch

The drip torch (frequently called the orchard torch) is designed to ignite fuel that may be only semi-dry, as well as for spreading fire over a large area. The orchard torch is easy to use, but because firing requires expertise, only experienced firefighters under proper supervision should conduct burn-outs or backfires.

Inspect the orchard torch for cracks or leaks. Carry the torch about two-thirds to three-quarters full of fuel (4 parts diesel to 1 part gasoline) to allow room for heat expansion. Too much gasoline may cause the torch to explode upon ignition.

Be very careful when mixing the fuel for the drip torch. Too much gasoline will turn it into a bomb.

To use the drip torch, remove it from the vehicle-mounting bracket and place it in a safe area (road, firebreak, etc.) away from flames or sparks. Remove the locking nut at the top of the torch; lift out the spout and igniter assembly, and remove the discharge-sealing plug. Place the spout back on the tank in the extended position with the loop facing away from the handle. Screw the locking nut down securely. Open the breather valve; tip the torch forward to collect fuel in the igniter and light with a fusee, lighter, match or other flame. Adjust the breather valve to achieve the appropriate rate of flow of burning fuel to the ground.

Be sure to "drip it" on the right side of the line.

Carry the drip torch at arms length at your side, swinging it gently back and forth with a flip of the wrist at the end of the swing to toss burning fuel forward (Figure A4.15). Use caution to keep from splashing fuel on your clothing. Stay on the correct side of the fireline, and carry the drip torch on the downhill side to prevent personal burns.

When done using the torch, set it down in a cleared, safe area and allow it to burn itself out.

Wes Schultz

Figure A4.15 The drip torch is used to ignite light fuels in backfiring and burning-out operations. Be sure you wear all of your PPE when using.

Don't try to extinguish the torch with water or dirt, or by blowing out the flame. Let the torch cool down completely before storing. Remove the lock ring, reverse the spout, reinstall the discharge-sealing plug, and close the breather valve. Clean the outside of the tank and replace the torch in the vehicle-mounting bracket.

Fusee

The fusee was originally developed as a railroad and highway emergency signaling device. Because it is light, easy to store and simple to use, it became popular as a backfiring tool. Wildland fire fusees burn hotter and longer than regular road flares.

Fusees are light and very effective.

Keep fusees dry, and inspect them for cracks or leaks before you activate them. Properly stored, fusees last for years.

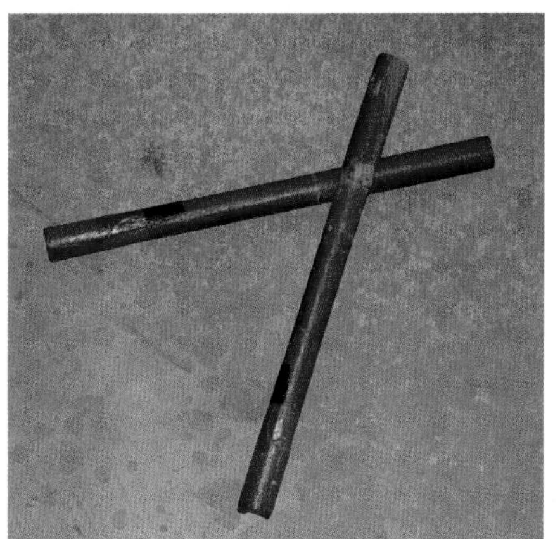

To use a fusee, remove the cap, hold the fusee pointing ***away*** from you, and strike the igniter patch on the cap across the tip of the fusee (Figure A4.16). Always keep the fusee pointed away from you. Do not hold it upright, as the burning substance may splatter onto your hand. Avoid breathing the fumes from fusees, and always wear PPE—including gloves and eye protection.

Be very careful of the hot end of a fusee.

The hottest point is at the tip of the flame, so you don't have to stick the fusee way down into the fuel to light it. To avoid excessive stooping, plug several fusees together to make a long handle, or cut a small branch to use as a handle. Fusees are easy to carry, and each firefighter should carry several. They may be

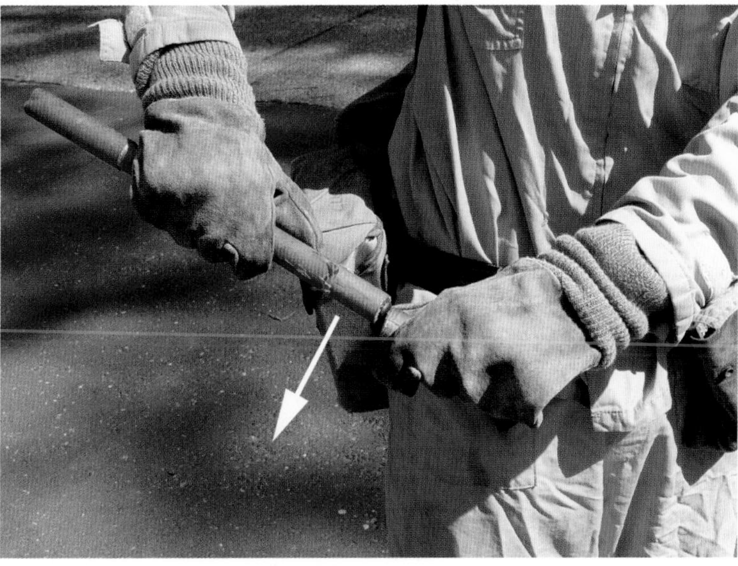

Figure A4.16 Strike the fusee away from you to avoid any possibility of sparks hitting you. This motion also points the burning end away from your body.

thrown into slash piles or brush thickets to avoid exposing personnel to danger. To extinguish a partially used fusee, stick the tip in loose dirt to smother the flame. ***Avoid breathing the smoke from burning fusees!***

Other

Other types of firing devices sometimes used in wildland firefighting include:

- Flare gun (or very pistol) – shoots an incendiary projectile up to a quarter of a mile (Figure A4.17).
- Napalm grenade – small metal can of gelled fuel with an igniter that is thrown by hand.

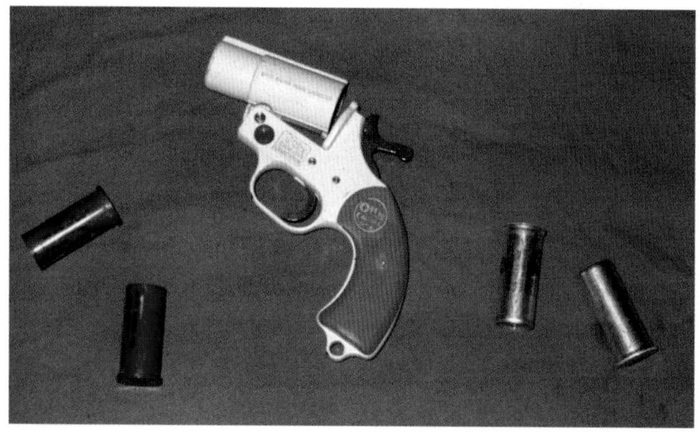

Figure A4.17 Flare guns can aid in firing areas you wouldn't want to put people. Shooting these type of guns is not for the first-year firefighter.

Figure A4.18 With a heli-torch you can set a lot of fire very quickly. Sometimes too quickly.

- Terra torch – vehicle-mounted powered flame thrower.
- Heli-torch – helicopter-mounted drip torch (Figure A4.18).
- Aerial ignition device – helicopter-mounted machine that drops "ping-pong" balls that burst into flame.
- Field expedient methods such as matches, or dragging burning ground fuels with a hand tool.

All firing devices and firing operations involve risks. To avoid danger and/or injury, you must exercise great care. Backfiring and burning-out are tactics that require training and experience to accomplish safe, efficient results. They should be directed only by very experienced firefighters acting under orders from appropriate command levels.

A good flashlight is a must on the fireline after dark.

Lights

A variety of lights are available to help firefighters work safely after dark. Each firefighter should carry a personal flashlight in their web gear for emergencies.

Flashlight

A variety of handheld flashlights are available, ranging from small disposable pen and pocket lights to bright multi-cell aluminum police-type flashlights that are nearly indestructible. Each wildland fire vehicle should be equipped with at least a couple of good flashlights. Use alkaline batteries for longer life. Store flashlights with a thin cardboard wedge between the batteries and the flashlight contact to avoid unintended discharge of batteries.

Lantern

Large, heavy-duty, battery-operated lanterns can provide adequate light to a whole work area. Each wildland fire vehicle should be equipped with at least two large lanterns. They should be mounted in brackets, not stored loose.

Be sure your headlamp is ready for action, BEFORE you leave the station.

Headlamp

Headlamps are specialized flashlights with the bulb on a cord so that it can be attached to a helmet. The older models had the battery pack carried on the belt. Various types and sizes of headlamps are in use, with the smaller, lighter units using halogen bulbs and small AA batteries becoming increasingly popular. Every person on a fire crew should have a headlamp with them when they leave the

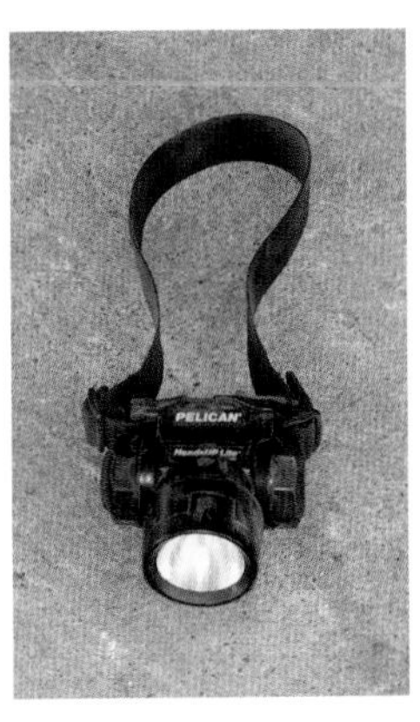

vehicle. Use a cardboard wedge between the batteries and the switch contact to prevent accidental battery discharge during storage. Carry extra batteries and bulbs on the vehicle.

Fireground Communications

Communication is critical in any firefighting operation. It is the thread that ties all of the various elements of a firefighting organization together. It is the way information and instructions are passed up and down the chain of command. If there is no communication, there is no organization.

Jake says,
"Face-to-face communications is the best form."

Communication on the fireline comes in many forms: hand signals; use of runners; mirrors; whistles or air horns; face-to-face spoken dialogue; transmitted voice; and written messages. The best and most reliable form of communication is face to face, with written notes to record what was said.

Basic Radio Theory

In this section you will not be taught to be electronics engineers, but you will be given information that may help you use your radio more effectively or understand why you are having a problem.

Radio Waves

Radio waves travel at the speed of light.

A radio converts the voice (sound waves) into an electrical charge and transmits it into space. Another radio detects the transmitted electronic signal and converts it back to sound waves. Sound waves travel at about 700 miles an hour or take 5 seconds to travel a mile. Radio waves travel at the speed of light or 186,000 miles a second.

(Note: Use this information when near a thunderstorm. Start counting the seconds when you see the flash of lightning. For every 5 seconds, you are about a mile from the strike.)

Radio waves are a form of energy called radio-frequency energy (RF). RF moves through space in waves from a positive charge to a negative charge. As RF changes from a positive charge to a negative charge and back to a positive charge, it has moved through a cycle (Figure A5.1). A cycle per second is called a *hertz.* If it does this a million times in a second, it is called one *megahertz.*

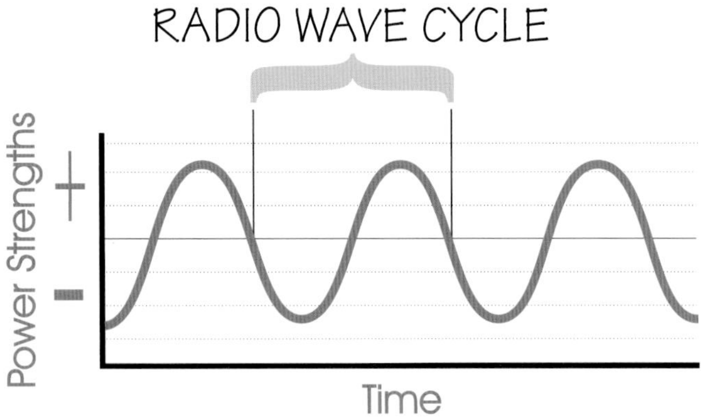

Figure A5.1 One radio wave cycle is in the form of a sine wave. One cycle per second is a hertz; one million cycles per second is one megahertz.

There are three forms of radio waves used by each of us on a daily basis: amplitude modulated (AM) waves, frequency modulated (FM) waves, and digital (Figure A5.2). AM radio waves are used for public radio broadcasting. They are long waves and can travel for miles, even up and over mountains, but are subject to noise interference and are not very efficient. AM radio transmitters need a lot of electrical power to work because they vary the power level of their RF to transmit (modulate) the voice information. That is why AM radio stations run between 5,000 and 50,000 watts of power. AM radios are a lot less expensive to build, but are highly susceptible to noise from power lines, automotive ignition systems, and even sunspots and solar flares.

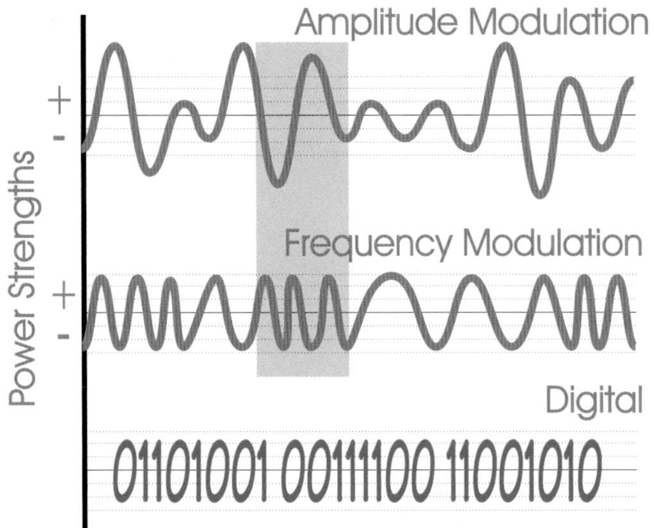

Figure A5.2 Amplitude modulation varies the strength of the plus and minus swing of energy; frequency modulation varies the number of cycles in a second; and digital signals are zeros and ones.

FM radio systems are less susceptible to noise and use less power. This is why most public service agencies use FM radio systems. FM radios keep the power levels constant and vary the frequency of the radio waves.

The newer radio systems are digital. Instead of modulating the radio signal, digital information in the form of "zeros" and "ones" are transmitted. Digital signals do not have the same limitations as AM or FM signals.

There are basically three radio bands used by the fire service. They are:

- Low Band - Equipment that used low-band frequencies were used years ago. The longer waves tended to bend a little over hills and up canyons, but lost favor because they also skipped off of the atmosphere, and interference from great distances was common.

Most forestry agencies use very high frequencies (150 to 170 MHz.).

- Very High Frequency - Systems that use very high frequencies (VHF) are still very common. Most of the state and federal forestry agencies still use this type of equipment in the 150 to 170 MHz frequency range. Radio frequencies in this range do not travel as far as low frequencies, so radio relay stations are used.

- Ultra High Frequency - Municipal fire departments use ultra high frequency (UHF) radio systems. They are much shorter range, but this is compatible with the area they usually protect. Radio waves in this range are absorbed by vegetative cover, trees, etc.

A high percentage of municipalities are moving to the 800 MHz. frequency range.

Other Frequency Ranges - There are other radio frequency bands used by the fire service, but the three already discussed are the most widely used.

You might hear some discussion about "narrow banding." This is a move by the federal government to squeeze even more frequencies from the existing allocations. Historically, there were 15 kilocycles between channels. Technology now allows for this gap to be cut to 7.5 kilocycles. The only problem is that with so many different kinds of radios used in the fire service, most of the time the systems are not compatible with each other. Lots of radios...but no way to talk to each other.

Old wide-band radios can "talk" to new narrow band radio...but it is "make-do" at best.

Federal Communications Commission

Radio frequencies are assigned for various uses, ranging from "ham" radio operations, to the military, law enforcement, forestry organizations, and fire departments. Blocks of frequencies are allocated to specific functions. The Federal Communications Commission (FCC) is responsible for the allocation of radio frequencies and the policing of their use. Because there are a limited number of frequencies available for use,

Shared use of radio frequencies is regulated by the FCC.

frequencies are sometimes shared. This means inappropriate use of these frequencies may interfere with other agencies. The FCC has the authority to cite individuals or agencies for violations and they do so on a regular basis.

Know and follow the rules of radio usage.

The Radio Equipment

There are three types of radio equipment that firefighters must be familiar with: the base station radio and mobile relay, the mobile radio, and the portable or handheld radio.

Base Stations and Mobile Relays

Base station radios are at a fixed location.

The base radio is like the one in the fire station. It is fixed to a location, not portable. It is usually capable of transmitting and receiving one or two frequencies. This type of radio usually has an alerting device attached to it. This is the equipment that dispatch uses to alert the station to a response. Some base stations use remote equipment. This means the transmitter and receiver are actually at another site, such as on top of a tall building or on the top of a mountain.

Figure A5.3 These dispatchers use these screens to show them what nets they are on...the radio equipment is some distance away in a radio vault.)

The dispatch offices are becoming very sophisticated, using computers to link the dispatcher with the radio equipment (Figure 5A.3).

Mobile relays are just what the name implies: they relay transmissions from mobile radios. Their sole purpose is to extend the range of the mobile radio over mountains and for greater distances. Most mobile relays are positioned on high points; in mountainous country on high peaks, and on flatter terrain atop tall buildings or towers. To use a mobile relay, you must set the mobile radio to transmit on a specific frequency (Figure A5.4). You may also have to select a specific tone. The frequency may be used by more than one mobile relay, but the tone would select a specific one. If a tone is used to activate a mobile relay, you may have to activate the transmitter a second or so before you speak. This will allow the tone time to "turn on the relay" before your voice is transmitted.

Some of the newer radio systems use "trunking systems" and "voting equipment." These are very sophisticated systems and very site specific. Learn your local system, so you know what it can do, and what it can't do. Some systems use what is called "private line." This is a tone system that is assigned to a specific department or agency. If "private line" is used, you not only need the proper frequency, you will need the proper tone.

OPERATION OF A MOBILE RELAY

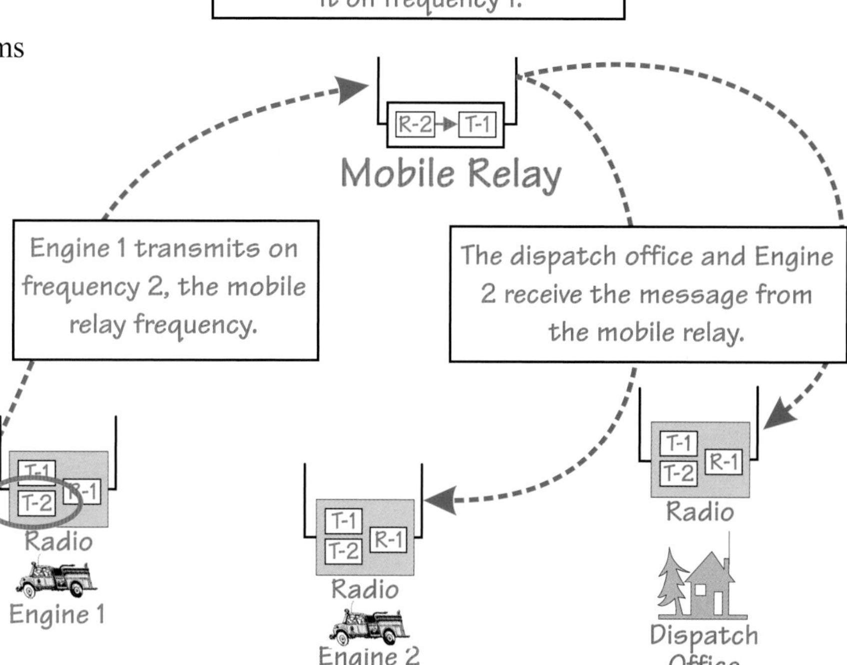
The mobile relay receives the transmission and retransmits it on frequency 1.
R-2
T-1
Mobile Relay
Engine 1 transmits on frequency 2, the mobile relay frequency.
The dispatch office and Engine 2 receive the message from the mobile relay.
T-1
T-2
R-1
Radio
Engine 1
T-1
T-2
R-1
Radio
Engine 2
T-1
T-2
R-1
Radio
Dispatch Office

Figure A5.4 Engine 1 is transmitting on "frequency 2." Since the mobile relay can receive "frequency 2," it hears the message and retransmits it on "frequency 1." Since the dispatch office and Engine 2 both receive "frequency 1," they now hear the message.

Mobile Radios

The mobile radio is the one in the engine. This radio is usually a little more sophisticated than the base radio (Figure A5.5). It will have the capability to transmit and receive on many more frequencies. This is needed to ensure adequate fireground communications. The mobile radio will probably have the ability to transmit either through a mobile relay (repeater) or simplex (car-to-car). A simplex transmission means the radio waves are transmitted on a frequency that can be received directly by another mobile radio. Consider simplex a line-of-sight mode normally used in tactical operations. Mobile relays are normally used to communicate between radios

Mobile relays extend the range of mobile or portable radios.

KENWOOD
RESCUE-PINE
KENWOOD
HD SERIES
Panasonic

Figure A5.5 This engine has two radios installed; one for AM low band and one for FM very high frequencies. Both mobile radios are multi-channel, with scan capabilities. You don't use these "babies" without some instruction and training.

separated by considerable distances, with those that are using low power, or when terrain features make line-of-sight communications difficult or impossible. Use simplex whenever possible, to minimize interference with other users (Figure A5.6).

OPERATION OF SIMPLEX (CAR-TO-CAR) COMMUNICATIONS

T-1 T-2 R-1 Radio

T-1 T-2 R-1 Radio

R-2 → T-1

Mobile Relay

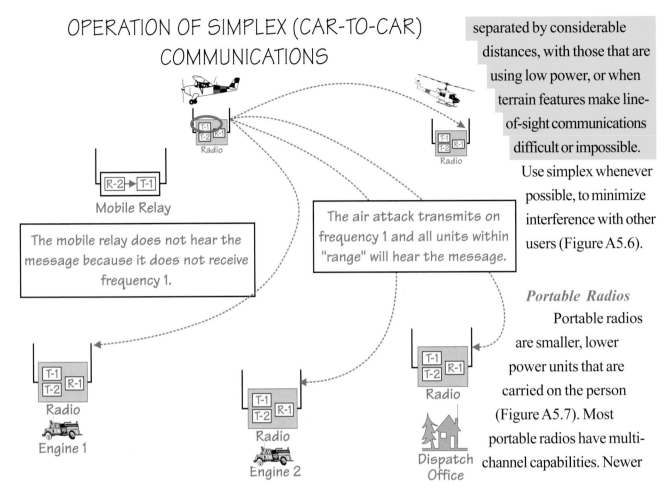

The mobile relay does not hear the message because it does not receive frequency 1.

The air attack transmits on frequency 1 and all units within "range" will hear the message.

T-1 T-2 R-1 Radio Engine 1

T-1 T-2 R-1 Radio Engine 2

T-1 T-2 R-1 Radio Dispatch Office

Figure A5.6 The air attack is transmitting on "frequency 1," the simplex or car-to-car frequency. The mobile relay only receives "frequency 2," so it does not hear the message. All units that receive "frequency" and are within range will hear the message.

Portable Radios

Portable radios are smaller, lower power units that are carried on the person (Figure A5.7). Most portable radios have multi-channel capabilities. Newer models are programmable, meaning the specific frequencies can be changed very easily. The portable radio may be able to function through a mobile relay, but is primarily used in the simplex mode. Some portable radios have the capability to scan several frequencies. This is a useful feature, but it is hard on batteries. Scan only when you need to. Always carry an extra set of batteries.

Learn how your local radio system works.

Portable radios are handy and versatile.

Figure A5.7 This handheld radio is in a battery charger and ready for use.

Use of the Radio Equipment

You don't have to know how a radio works inside, but you must have a working knowledge of the external parts of the radio to ensure it will function properly. You should know how to turn it on and set the volume and squelch, how to select the proper channel and tone, how to activate the transmitter, and where to speak.

Know how to operate all the radios in your local system.

Volume and Squelch

Setting the volume is very straightforward: set it to a comfortable level. Don't set the volume to a level that may be damaging to your ears, but set it high enough to be clearly audible over background noise.

Squelch is a control that is a little more complicated (Figure A5.8). The squelch circuitry in the radio acts as a noise limiter. Turn the squelch knob counterclockwise until you hear static. Turn it clockwise and you eliminate the static. The key is to set the squelch just to the point where the static is removed. Setting the squelch any "tighter" (more in the clockwise direction) will also block weak signals that you want to hear.

Figure A5.8 The on-off/volume control is very straight-forward. Turn the radio on and adjust to the proper volume. The squelch control "filters out" the noise...be sure it is adjusted properly, or you may not hear some weaker signals. The other controls are more radio specific.

Frequency Selection

On day-to-day operations, it is not normally necessary for you to use too many frequencies. The real challenge is when you respond out of the area to assist at a major fire. It is just as important to receive a radio frequency assignment as it is to know where the staging area is. Once given the assigned frequency, you have to determine if you have that capability. If you don't, be sure to let your dispatcher or supervisor know. Once you are on the proper channel, stay there. Don't get lost in the "sea of frequencies, nowhere to be found." If you have to change channels, tell the dispatch or your supervisor.

Know which radio frequency applies to your specific task.

Maintain your radios in proper working condition.

Antenna and Batteries

Save and reuse rechargeable radio batteries.

The antenna and the batteries are the "business ends" of the radio for the firefighter. If the antenna is loose, broken or even bent, it will not function properly. Check the antenna before use. The batteries are just as important. As mentioned before, be sure that you have a fully charged set of batteries before you leave the station AND you have a spare set of batteries with you. Your life may depend on it. Also, if you are using rechargeable batteries, don't just discard them when they are low; they are very expensive and you will need them another day.

Proper Use of Radios

It should go without saying, but fire department radios are "For Official Use Only" and even if you have official business, emergency traffic has priority. If you have an emergency message, do not hesitate to announce you have "emergency traffic" or "priority traffic."

Jake says, "Think before you push the mike button and talk. Keep the message short and to the point."

Historically, the "Ten Code" or a derivation of it was used for radio communications. There was "10-4" which meant "Okay," "10-20" which meant "what is your location?," "10-9" which meant "repeat," etc. With the advent of ICS and the need for compatible communications between assisting agencies, codes were replaced with clear text. This avoids confusion. Simply say what you mean. If you have arrived at the scene of the fire, say "at scene." If you can handle the incident with what you have, say "can handle with equipment at scene." There are some common word phrases, but your message should be phrased so as to be easily understood by anyone hearing it.

Tac nets are for simplex communications between groups of tactical resources.

There are also several common identifiers. Some of the identifiers need no explanation: "Chief," "engine," or "squad." A "tanker" is an airtanker...a "water tender" is a water truck. A "truck" is not an engine, it has a ladder or aerial platform. An engine with a pump, tank and aerial platform or ladder is called a "quint."

Jake says, "In every large fire review I have been involved in, radio communications were a mess."

As you prepare to transmit a message, think about what you need to say before you push the transmit button. By thinking about what needs to be said, there will be less "ahs" and "umms" before you say something. If you are asked to relay a message, take the time to write it down.

Be courteous, answer calls as quickly as possible, and never use profanity on the radio. Not only will your agency come down on you, it is against the law.

Be brief and to the point. The radio channel(s) may be clear now, but once fire activity gets heavy, the radio system is the first to be overloaded.

Some key points to remember when using the radio:

- Wait until the person using the radio is finished before you transmit your message.

- When initiating a call, say the station name or unit number of the person you are calling, followed by your station name or unit number. (Some agencies use personal names.)

- Hold the microphone from 2 to 4 inches away from your mouth; avoid blowing into the microphone when you speak.

- Wait a full second after you press the key before you speak.

- Have a note pad and pencil ready to record information.

- Answer your radio when your station or unit is called. There may be times when more than one unit is attempting to use the net at the same time. Wait until they are done, and make your call again.

- The person who initiates the call closes with the proper FCC station identification or some other form of acknowledgment.

Select the correct radio frequency for the situation.

Clear text means saying what you mean; no code.

Radio Troubleshooting

You are not a radio technician, so don't get your screwdriver out and attempt major surgery if your radio doesn't work. There may be some simple things you can check before shipping the radio back to the service center. They are:

- ***Location*** - You may be in an area closed-in by terrain and your radio is working perfectly, but the signal is not getting to who you want to talk to. If you are on a car-to-car frequency, you may be able to change to a frequency that uses a mobile relay. You may have to change locations.

- ***Batteries*** - If you are using a portable radio, your batteries may simply be spent. Replace them if needed. The more "features" you are using on the radio, like scan, the shorter the battery life.

Too much squelch control may block out weak signals.

• ***Net*** - You may be on the wrong net or frequency. Check and change if needed.

• ***Antenna*** - Your antenna may be loose. Check and tighten.

Use the correct identifier with your radio number.

Care and Handling

Proper care and handling can extend the life of your portable radio, and help ensure that it is in good working order:

• Protect the radio from dust, moisture, fire retardant, excessive vibration, dropping, extreme heat, etc.

• Keep radios in a protective cover.

• Do not modify or attempt repairs on any radio. Notify your supervisor if your radio is not functioning properly.

• Do not use the antenna to pull the portable from its case, or put unnecessary bending pressure on the antenna.

Radio Nets

Under the Incident Command System, there are specific radio nets (or channels or frequency assignments) for certain functions: command, tactical, support, air-to-air, and air-to-ground.

Air-to-air net is reserved for aircraft.

The command net is used to communicate between the various command and general staff functions down to the Division/Group Supervisor level. This is an exclusive channel for command and control communications. It is not for general tactical operations.

The tactical nets are used for tactical communications at the Division or Branch level. "Tac nets" may be established around agencies, geographical areas or specific functions. On large operations, there will be several tactical nets assigned. The specific radio frequencies will be listed in the Incident Action Plan on the bottom of each Division Assignment List (Figure A5.9). Use the "tac net" to

DIVISION ASSIGNMENT LIST

9. Division/Group Communications Summary

FUNCTION		FREQUENCY	SYSTEM	CHAN.	FUNCTION		FREQUENCY	SYSTEM	CHAN.
COMMAND	Local	169.100	BOISE C-2	4	SUPPORT	Local	154.295	FIREMARS	3
	Repeat	170.450	BOISE C-2	5		Repeat	154.280	FIREMARS	4
DIV/GROUP TACTICAL		159.330	CDF	3	GROUND TO AIR		170.000	BOISE	6

PREPARED BY (Resource Unit Leader)	APPROVED BY (Planning Section Chief)	DATE	TIME
Al Smith, Unit Leader	WTI	6-21-94	2100

Figure A5.9 On larger fires, the Incident Action Plan will outline the frequencies to be used.

communicate with other equipment or personnel in your Division (Figure A5.10).

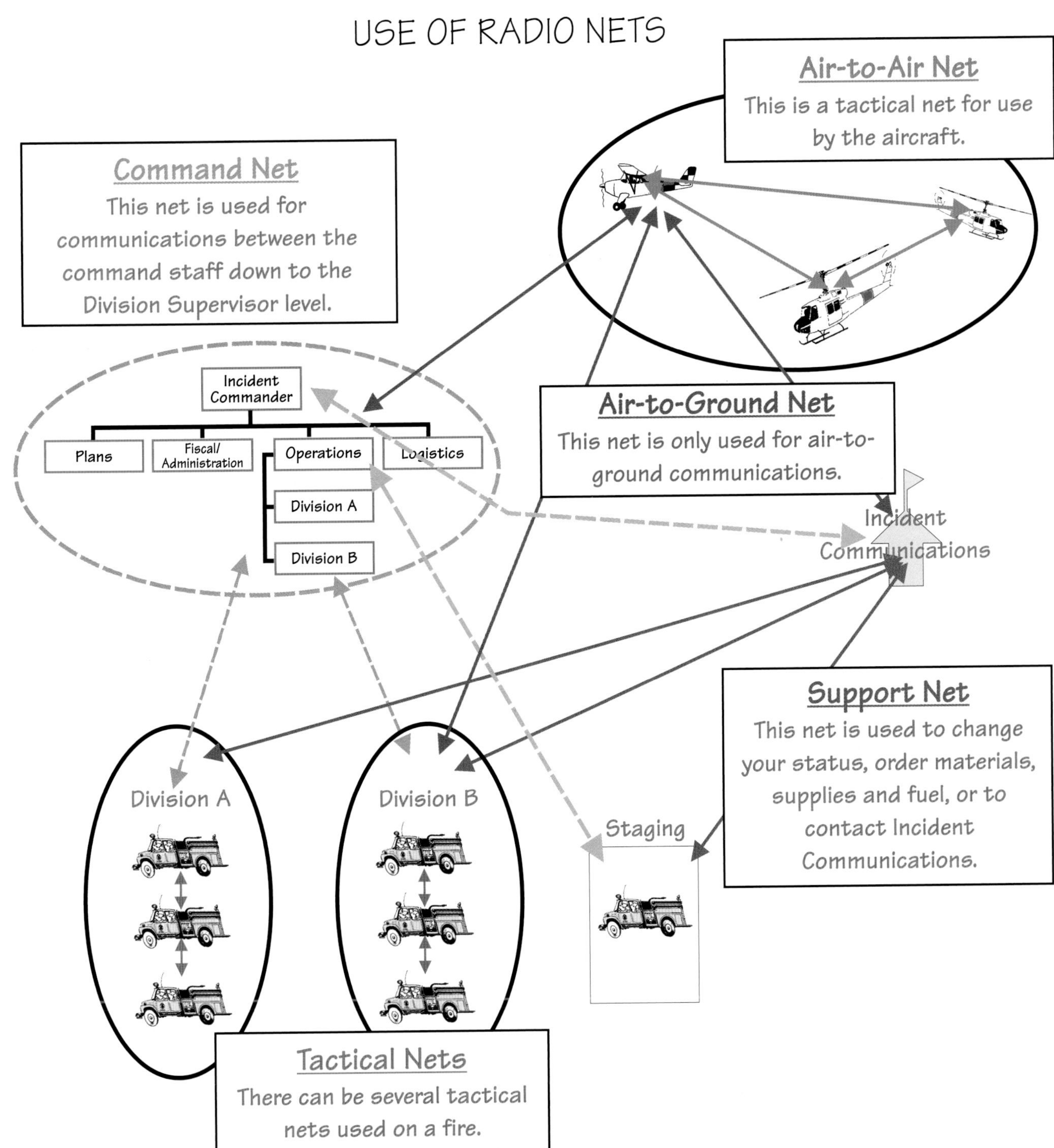

Figure A5.10 It doesn't take a large fire anymore before tactical, command and support nets are assigned and used. It is imperative that you know what frequencies are assigned. Switch your radio to that frequency and stay there.

The support net is established for the purpose of handling status changes, requests for support needs, and other non-tactical and command functions. Use the support net to communicate to "Incident Communications" change in status, like "out of service mechanical" or "in incident base, off shift." If you need fuel, water or food, you would use the support net.

The air-to-ground net is used to communicate with the incident Air Attack function. It is an exclusive net for air-to-ground communications. This net is critical during major fire operations where several tactical nets are in use. It is impossible for the Air Attack to monitor each and every "tac net." If ground forces need air support they can switch to a specific air-to-ground frequency and request assistance.

The air-to-air net is the net used for tactical air operations. It is the net used by the airtankers and helicopters to communicate between themselves and with the Air Attack Officer. This net should never by used by ground forces. The only exception may be the communications between a helitack crew and its helicopter.

NIFC radio caches are frequently used on large fires.

Mutual Aid Frequencies

The Federal Communications Commission has established several national fire mutual aid frequencies. These are normally controlled at the state level. They are to be used for communications between assisting agencies.

The federal wildland firefighting agencies also maintain a very sophisticated major emergency radio communications system. This includes not only large blocks of frequencies, but the equipment needed to use them. When assigned to a major fire, you may be assigned one of the National Interagency Fire Center (NIFC) frequencies. All of these frequencies and the equipment are controlled by NIFC in Boise, Idaho.

INCIDENT ORGANIZER
WILDLAND FIRE

Date/Time

Fire Name

Incident Number(s)

Location/Map Grid

Latitude N

Longitude W

Radio Nets

Command	Tactical	Air-to-Ground

Hazards or Special Instructions

Incident Commander(s)

Name	Date/Time	✓
IA		☐
		☐
		☐
		☐
		☐
		☐

Staging Area - is located at:

Incident Command Post - is located at:

RESOURCES SUMMARY

Resources Ordered ✓ if needed immediately.	Resource Identification	Date/ ETA	At Scene	No. of People	Location/Assignment	✓ Released
☐		( )	☐			☐( )
☐		( )	☐			☐( )
☐		( )	☐			☐( )
☐		( )	☐			☐( )
☐		( )	☐			☐( )
☐		( )	☐			☐( )
☐		( )	☐			☐( )
☐		( )	☐			☐( )
☐		( )	☐			☐( )
☐		( )	☐			☐( )
☐		( )	☐			☐( )
☐		( )	☐			☐( )
☐		( )	☐			☐( )
☐		( )	☐			☐( )
☐		( )	☐			☐( )
☐		( )	☐			☐( )
☐		( )	☐			☐( )
☐		( )	☐			☐( )
☐		( )	☐			☐( )
☐		( )	☐			☐( )
☐		( )	☐			☐( )
☐		( )	☐			☐( )

DVP Form 230.2 (revised August 2004)

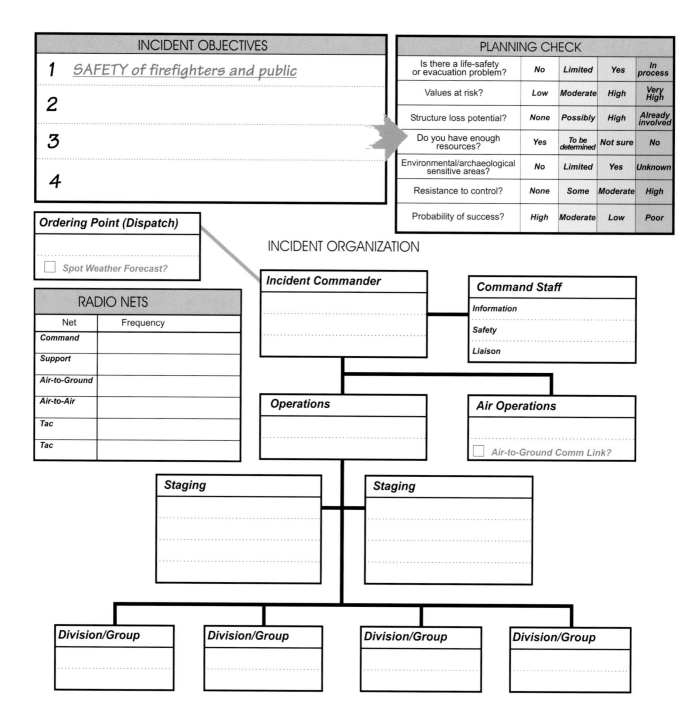

INCIDENT OBJECTIVES
1 SAFETY of firefighters and public
2
3
4

PLANNING CHECK				
Is there a life-safety or evacuation problem?	No	Limited	Yes	In process
Values at risk?	Low	Moderate	High	Very High
Structure loss potential?	None	Possibly	High	Already involved
Do you have enough resources?	Yes	To be determined	Not sure	No
Environmental/archaeological sensitive areas?	No	Limited	Yes	Unknown
Resistance to control?	None	Some	Moderate	High
Probability of success?	High	Moderate	Low	Poor

Ordering Point (Dispatch)

☐ Spot Weather Forecast?

RADIO NETS	
Net	Frequency
Command	
Support	
Air-to-Ground	
Air-to-Air	
Tac	
Tac	

INCIDENT ORGANIZATION

Incident Commander

Command Staff
- Information
- Safety
- Liaison

Operations

Air Operations

☐ Air-to-Ground Comm Link?

Staging

Staging

Division/Group

Division/Group

Division/Group

Division/Group

Step #1

SITUATION AWARENESS	YES	NO
Objectives - Are they still valid?	☐	☐
Communications - Have they been confirmed?	☐	☐
Who's In Charge? - Is this known by all?	☐	☐
Previous Incident Behavior - Any lessons?	☐	☐
Weather Forecast - What is predicted behavior?	☐	☐
Local Factors - What will their impact be?	☐	☐
Other Incidents - Will other incident impact you?	☐	☐
Keep Reassessing Your Situation!	☐	

Re-evaluate

WEATHER READINGS							
Date/Time							
Temperature	*Wet Bulb*						
	Dry Bulb						
Relative Humidity (%)							
Wind	*Direction*						
	Speed (mph)						

MAP SKETCH

N
W E
S

Prepared by: by:	Position	Date/Time

Step #2

HAZARD ASSESSMENT

☐ *Estimate potential fire behavior.*

Look up, look down, look all around. What fire behavior factors will cause you a problem?

FIRE BEHAVIOR CHECK

Relative Humidity (%)	over 45	35 to 45	25 to 35	under 25
Wind Speed (mph)	Calm	under 7	8 to 15	over 15
Slope (%)	Flat	under 25	25 to 50	over 50
Flame Length	under 2'	2' to 4'	4' to 8'	over 8'
Aspect	North	East	West	South
Spotting	None	Minor	Moderate	Extensive
Time of Day	2000 to 1000	1600 to 2000	1000 to 1200	1200 to 1600
Haines Index	1 to 2	3 to 4	5	6

☐ *Identify tactical hazards.*

Using the 18 Situations that shout "Watch Out!" Identify any tactical hazards that may impact your operation.

☐ *What other safety hazards exist?*

List any other hazards and what you will do to mitigate them.

	Mitigated?
....	☐
....	☐
....	☐
....	☐
....	☐
....	☐
....	☐
....	☐

☐ *What is the Severity vs. Probability?*

Is what you are attempting to accomplish worth the risk? Is this something you don't do very often and it has a high risk?

Do you really have to do this task this way?

High Frequency **High Risk**	**Low Frequency** **High Risk**
Low Frequency **Low Risk**	**High Frequency** **Low Risk**

Step #3

HAZARD CONTROL

☑ Reviewed
☑ Present
☒ Mitigated

18 Situations that shout "Watch Out!"

- ☐ Fire not scouted and sized up.
- ☐ In country not seen in daylight.
- ☐ Safety zones and escape routes not identified.
- ☐ Unfamiliar with weather and local factors influencing fire behavior.
- ☐ Uniformed on strategy, tactics, and hazards.
- ☐ Instructions and assignments not clear.
- ☐ No communication link with crew members/supervisor.
- ☐ Constructing line without safe anchor point.
- ☐ Constructing fireline downhill with fire below.
- ☐ Attempting frontal assault on fire.
- ☐ Unburned fuel between you and fire.
- ☐ Cannot see main fire, not in contact with anyone who can.
- ☐ On a hillside where rolling material can ignite fuel below.
- ☐ Weather is getting hotter and drier.
- ☐ Wind increases and/or changes direction.
- ☐ Getting frequent spot fires across line.
- ☐ Terrain and fuels make escape to safety zones difficult.
- ☐ Taking nap near fireline.

☐ *Have you complied with LACES?*

LACES
Safety must never be compromised.
☐ **L**ookouts
☐ **A**wareness
☐ Predicted Weather
☐ Fire Behavior
☐ Plan
☐ **C**ommunications
☐ **E**scape Routes
☐ **S**afety Zones

☐ *Are you building fireline downhill with fire below you?*

If you are, refer to the special precautions that are required.

☐ *Have you anchored all of your fireline?*

If you have fireline that is not anchored, you must develop a plan that ties all of your lines to good anchor points.

When in doubt STAND-EM DOWN in a safe location.

Step #4

DECISION POINTS

☐ *Controls in place for identified hazards?*

Yes ☐ No ☐ If, NO, reassess your situation.

☐ *Are selected tactics based on expected fire behavior?*

Yes ☐ No ☐ If, NO, reassess your situation.

☐ *Have instructions been given and understood?*

Yes ☐ No ☐ If, NO, reassess your situation.

INITIATE ACTION

PLANNING

A properly prepared and safely executed plan is vital to the success of any operation. Take the time to complete the ***SITUATION/RESOURCES STATUS MATRIX***. You will need a current map, with the projected fire perimeter.

Think in operational periods, basing your estimated needs on your projections, and a set of objectives. Think ahead! Think Safety!

Step #5

EVALUATIONS

☐ *What is the state of your personnel?*

Yes	No	
☐	☐	Low experience with local factors?
☐	☐	Distracted from primary tasks?
☐	☐	Fatigue or stress reaction?
☐	☐	Hazardous attitudes?
☐	☐	Full use of PPE and Wheel Chocks?

☐ *What is the situation?*

☐	☐	Is it changing?
☐	☐	Are strategy and tactics working?

Re-evaluate or re-process as needed!

Comply?

10 Standard Firefighting Orders

- ☐ Keep informed on fire weather conditions and forecasts.
- ☐ Know what your fire is doing at all times.
- ☐ Base all actions on current and expected fire behavior of the fire.
- ☐ Identify escape routes and safety zones, and make them known.
- ☐ Post lookout when there is possible danger.
- ☐ Be alert. Keep calm. Think clearly. Act decisively.
- ☐ Maintain prompt communications with your forces, your supervisor and adjoining forces.
- ☐ Give clear instructions and insure they are understood.
- ☐ Maintain control of your forces at all times
- ☐ Fight fire aggressively, having provided for safety first.

SITUATION/RESOURCES STATUS MATRIX

	CURRENT STATUS	PROJECTED STATUS
SITUATION	*Record current situation status in this block, including a map, major tactical issues, etc.*	*Project future situation status in this block, including a map, major tactical objectives, etc.*
RESOURCES	*Record current resource status in this block, including what is there and enroute.*	*Project future resource needs in this block, including in which operational period they will be needed.*

Date/ Time	SUMMARY OF ACTIONS (ICS 214)
( )	
( )	
( )	
( )	
( )	
( )	
( )	
( )	
( )	
( )	
( )	
( )	
( )	
( )	
( )	
( )	
( )	
( )	
( )	
( )	
( )	
( )	
( )	
( )	
( )	
( )	
( )	
( )	
( )	
( )	
( )	
( )	
( )	
( )	
( )	
( )	
( )	
( )	
( )	

Document:
- Important Decisions
- Significant Events
- Briefings
- Reports on Conditions

TELEPHONE NUMBERS	
Person/Function	Telephone / Cell Number

Glossary

A

Adapter: A hose-coupling device for connecting hose threads of the same size.

Aerial Ignition Device (AID): Inclusive term applied to equipment designed to ignite wildland fuels from an aircraft.

Aerial torch: An ignition device suspended under a helicopter, capable of dispensing ignited fuel to the ground for assistance in burning-out or backfiring.

Agency dispatcher: Dispatch organization for the agency with primary jurisdictional responsibility for the incident.

Airtanker: Any fixed-wing aircraft capable of transport and delivery of fire suppressant or retardant materials. There are four ICS size classes of airtanker.

Allocated resources: Resources dispatched to an incident that have not yet checked in.

Anchor point: An advantageous location, usually a barrier to fire spread from which to start constructing fireline. Used to minimize the chance of being flanked by the fire while the line is being constructed.

Apparatus: A fire engine or other firefighting piece of equipment, or grouping of such equipment.

Area Command: An organization established to 1) oversee the management of multiple incidents that are being handled by an ICS organization; or 2) to oversee the management of a very large incident that has multiple Incident Management Teams assigned. Area Command has the responsibility to set overall strategies and prorities, allocate critical resources based on priorities, ensure that incidents are properly managed, and ensure that objectives are met and strategies followed.

Arson: The setting of fires to defraud, or for other illegal or malicious purposes.

Area ignition: The ignition of a number of individual fires throughout an area either simultaneously or in quick succession, and so spaced that they soon influence and support each other to produce fast, hot spread of fire throughout the area.

Aspect: The direction a slope faces the sun. Expressed in cardinal direction. Same as exposure.

Assigned resources: Resources checked in and assigned work tasks on an incident.

Assisting agency: An agency directly contributing suppression, rescue, support, or service resources to another agency.

Attack a fire: Limit the spread of fire by cooling, smothering, or removing or otherwise treating the fuel around its perimeter.

Attack line: Line of hose on an engine or water tender, used to fight or attack the fire directly.

Attack time: Date, hour, and minute the person who does the first suppression work on the fire starts to do it.

Available fuels: The portion of the total fuel that actually burns.

Available resources: Resources assigned to an incident and available for an assignment.

B

Back-burn: Used in some localities to specify fire set to spread against the wind in prescribed burning. Also called backing fire.

Backfire: 1) Fire set along the inner edge of a fire control line to stop a spreading wildfire by reducing the fuel or changing the direction of force of the fire's convection column. The term applies best where skilled techniques are required for successful execution. Using such fire to consume unburned fuel inside the fireline to speed up line holding and mop-up is usually called "burning-out" or "clean burning." 2) A prescribed fire set to burn against the wind.

Barrier: Any obstruction to the spread of fire. Typically, an area or strip devoid of flammable fuel.

Berm: Outside or downhill side of a ditch or trench.

Black line: Fuel between the fireline and the fire that has been burned-out. Line is not complete until fuel is burned-out between fireline and fire.

Blowup: Sudden increase in fire intensity or rate of spread sufficient to preclude direct control or to upset existing control plans. Often accompanied by violent convection and may have other characteristics of a firestorm.

Borate bomber: An archaic term referring to an airtanker. Borate was one of the first chemicals added to water as the thickener, which was used in the '50s.

Branch: The organizational level having functional/geographic responsibility for major segments of incident operations. The Branch level is organizationally between the Operations Section Chief and the Division/Group Supervisors.

Breakout: A fire edge that crosses a control line or natural barrier intended to confine the fire and the resultant fire. Also called a slopover.

Broadcast burning: Intentional burning of a specific area within well-defined boundaries.

Brush: Shrubs and stands of short scrubby trees that do not reach merchantable size.

Buildup: 1) Cumulative effects of drying (during a preceding period) on the current fire danger. 2) Acceleration of a fire with time. 3) Increase in strength of a fire control organization.

Burning conditions: The state of the combined factors of environment that affect fire in a given fuel association.

Burning index: A number related to the contribution that fire behavior makes to the amount or effort needed to contain a fire in a particular fuel type within a rating area. An index for describing fire danger.

Burning-out: Setting fire inside a control line to consume fuel between the edge of the fire and the control line.

Burning period: That part of each 24-hour period when fires will spread most rapidly.

C

Camp: A geographical site, within the general incident area, separate from the Incident Base, equipped and staffed to provide sleeping, food, water, and sanitary services to incident personnel.

Campfire: As used to classify the cause of a forest fire, a fire that was started for cooking or for providing light and warmth and that spreads sufficiently from its source to require action by a fire control agency. Usually exclusive of fires started by railroad or lumbering employees in connection with their work.

Cat face: A defect on the surface of a tree resulting from a wound in which healing has not reestablished the normal cross section. See fire scar, fire wound.

Cat line: Line constructed using bulldozers or tractor-plows. In lighter fuels, other types of mechanized equipment, such as graders and scrapers, can be used.

Causes of fires: For statistical purposes forest fires are grouped into broad cause classes. The eight general causes defined in U.S. Forest Service practice are: campfire, debris burning fire, incendiary fire, lightning fire, lumbering fire, railroad fire, smoker fire, miscellaneous fire.

Center firing: A technique of broadcast burning in which fires are set in the center of the area to create a strong draft. Additional fires are then set progressively nearer the outer control lines as indraft builds up to draw them in toward the center.

Chance: Conditions suitable for a given kind of operation; for example: gravity chance, water so located that it can be delivered by gravity to a desired point; pump chance, the quantity and location of water with respect to a fire that is suitable for power pumps.

Check in: Locations where assigned resources check in at an incident.

Check line: A temporary line constructed at right angles to the control line and used to hold a backfire in check as a means of regulating the heat (or intensity) of the backfire.

Class of fire (as to kind of fire): Class A fires are those that are burning solid fuels, fires are including wood and vegetation; Class B fires are those that involve flammable liquids; Class C fires are those that involve electrical equipment; and Class D fires involve burning metals.

Class of fire (as to size of wildland fires): Class A is a fire that is 1/4 acre or less; Class B is a fire that is more than 1/4 acre, but less than 10 acres; Class C is a fire of 10 acres or more, but less than 100 acres; Class D is a fire of 100 acres or more, but less than 300 acres; and Class E is a fire that is 300 acres or more.

Clear text: The use of plain English in radio communication transmissions. No ten codes or agency specific codes are used when using clear text.

Closed area: An area in which specified activities or entry is temporarily restricted to reduce risk of human-caused fires.

Closure: Legal restriction, but not necessarily elimination, of specified activities such as smoking, camping, or entry that might cause fires in a given area.

Cold trailing: A method of controlling a partly dead fire edge by carefully inspecting and feeling with the hand to detect any fire, digging out every live spot, and trenching any live edge.

Cold line: Fireline that has been controlled. The fire has been mopped up for a safe distance inside the line and can be considered safe to leave.

Command: The act of directing, ordering and/or controlling resources by virtue of explicit legal, agency, or delegated authority.

Command staff: The Command Staff consists of the Information Officer, Safety Officer, and Liaison Officer. They report directly to the Incident Commander.

Compacts: Formal working agreements among states to obtain mutual aid for existing disasters.

Complex: Two or more individual incidents located in the same general proximity which are assigned to a single Incident Commander or Unified Command.

Condition of vegetation: Stage of growth, or degree of flammability, of vegetation that forms part of a fuel complex. Herbaceous stage is at times used when referring to herbaceous vegetation alone. In grass areas minimum qualitative distinctions for stages of annual growth are usually green, curing, and dry or cured.

Conduction: Heat transfer by direct contact or through an intervening heat-conducting medium.

Conflagration: A raging, destructible fire. Often used to describe a fire that has a fast-moving fire front; different from a fire storm.

Contain a fire: To take suppression action, as needed, which can reasonably be expected to check the fire's spread under prevailing conditions.

Containment: When a fire is encircled by a fireline, but not under control.

Control (a fire): To complete firelines around a fire, any spot fires, and any interior islands; burn-out any unburned area adjacent to the fireline; and cool down all hotspots that are an immediate threat to the control line.

Control burning: The planned application of fire with intent to confine it to predetermined area.

Control force: Personnel and equipment used to control a fire.

Control line: An inclusive term for all constructed or natural fire barriers and treated fire edge used to control the fire.

Control time: The time the fire is controlled, with the expectation that it will not escape; different from containment.

Cooperating agency: An agency supplying assistance other than direct suppression, rescue, support, or service functions to the incident control effort; e.g., Red Cross, law enforcement agency, telephone company, etc.

Counter fire: Fire set between main fire and backfire to hasten spread of backfire. Also called draft fire.

Cover type: The designation of a vegetation complex described by dominant species, age, and form.

Coyote tactics: A progressive line construction technique involving self-sufficient crews which build fireline until the end of shift, remaining at or near the point while off shift, and begin building fireline again the next shift where they left off the previous shift.

Creeping: Fire burning with a low flame and spreading slowly.

Crown fire: A fire that advances from top to top of trees or shrubs, more or less independently of the surface fire.

Crown out: Fire burning principally as a surface fire that intermittently ignites the crowns of trees or shrubs. A fire that rises from ground level into the tree crowns and advances from treetop to treetop.

D

Danger class: A segment of a fire danger scale identified by a qualitative or numerical term.

Danger index: A relative number indicating the severity of forest fire danger as determined from burning conditions and other variable factors of fire danger.

Debris burning: A fire spreading from any fire originally set for the purpose of clearing land; or for rubbish, garbage, range, stubble, or meadow burning.

Deputy: A fully qualified individual who, in the absence of his/her superior, could be delegated the authority to manage a functional operation or perform a specific task. Deputies can be assigned to the Incident Commander, General Staff Heads and Branch Directors.

Detection: The act or system of discovering and locating fires.

Direct attack: See Direct method.

Direct line: Any treatment of burning fuel by wetting, smothering or chemically quenching the fire or by physically separating the burning from the unburned fuel.

Direct method: A method of suppression that treats the fire as a whole, or all its burning edge, by wetting, cooling, smothering, or chemically quenching the fire or by mechanically separating the fire from unburned fuel.

Discovery: Determination that a fire exists. In contrast to detection, location of a fire is not required.

Dispatch: The implementation of a command decision to move a resource or resources from one place to another.

Dispatch center: A facility from which resources are directly assigned to an incident. Sometimes called a command center.

Dispatcher: A person who receives reports of discovery and status of fires, confirms their location, takes action promptly to provide the people and equipment likely to be needed for control in the first attack, and sends them to the proper place.

Division: A unit established to divide an incident into geographical areas of operation.

Dogleg: A sharp direction change in a fireline that usually develops into a location where a fire can cross the line due to heat concentration.

Dozer: Any tracked vehicle with a blade for exposing mineral soil, with transportation and personnel for its operation.

Dozer line: Fireline constructed by a bulldozer.

Drift smoke: Smoke that has drifted from its point of origin and has lost any original billow form.

Drought index: A number representing net effect of evaporation, transpiration, and precipitation in producing cumulative moisture depletion in deep duff or upper soil layers. The Palmer Drought Index is the most widely used.

Drop Point: A pre-identified location where personnel, equipment, and supplies are to be delivered or picked-up.

Dry lightning storm: A lightning storm with negligible precipitation reaching the ground.

Duff: The partly decomposed organic material of the forest floor beneath the litter of freshly fallen twigs, needles, and leaves.

E

Edge firing: A technique of broadcast burning in which fires are set along the edges of an area and allowed to spread to the center.

Emergency medical technician (EMT): A health care professional with special skills and knowledge in prehospital emergency medicine.

Engine: Any ground vehicle providing specified levels of pumping, water, hose capacity, but with less than the specified level of personnel.

Escape route: A route of travel known to all that leads away from a point of danger, generally to a safety zone. It should be preplanned.

Escaped fire: A fire which has exceeded initial attack capabilities.

Exposure: Property that may be endangered by a fire in another structure or by a wildfire.

Extended Attack: Situation in which a fire cannot be controlled by initial attack resources within a reasonable period of time. The fire usually can be controlled by additonal resources within 24 hours after commencing suppression action.

Extra-period fire: A fire not controlled by 10 a.m. of the day following discovery.

F

False alarm: A reported smoke or fire requiring no suppression; for example, brush burning under control, mill smoke, false smoke, etc.

False smoke: Any phenomenon likely to be mistaken for smoke, such as gray cliffs, sheep driveway or road dust, or fog.

Feeling for fire: Examining burned material after fire is apparently out and feeling with bare hands to find any live coals.

Fine fuel moisture: The probable moisture content of fast-drying fuels which have a time lag constant of one hour or less, such as grass, leaves, and small twigs.

Fingers: Narrow points of fire, caused by a shift of wind or change in topography. They develop behind the head extending from the flanks. They may become second heads.

Fire analysis: Process of reviewing the fire control action on a given unit or the specific action taken on a given fire in order to identify reasons for both good and poor results and make recommendations for ways of doing a more effective job.

Fire behavior: The manner in which a fire reacts to the variables of fuel, weather, and topography.

Firebreak: A natural or constructed barrier utilized to stop or check fires, or to provide a control line from which to work.

Fire concentration: A situation in which numerous fires are burning in a locality. Also used at times to denote the rate of fire occurrence per unit of area.

Fire control: All activities to protect wildland from fire. Includes prevention, presuppression, and suppression.

Fire control equipment: All tools, machinery, special devices, and vehicles used in fire control, but excluding structures.

Fire control improvements: The structures primarily used for fire control, such as lookout towers, fireguard cabins, telephone lines and roads to lookout stations, etc.

Fire control planning: The systematic technological and administrative management process of designing organization, facilities, and procedures to protect wildland from fire.

Fire cooperator: A fire-trained local person or agency who has agreed in advance to perform specified fire control services.

Fire damage: The loss, expressed in money or other units, caused by fire. Includes all indirect losses, such as reduction in future values produced by the forest area, as well as direct losses of cover, improvements, wildlife, etc., killed or consumed by fire.

Fire danger: Resultant of both constant and variable fire danger factors, which affect the ignition, spread, and difficulty of control of fires and the damage they cause.

Fire danger rating: A fire management system that integrates the effects of selected fire danger factors into one or more qualitative or numerical indices of current protection needs.

Fire edge: The boundary of a fire at a given moment.

Fire effects: The physical, biological, and ecological impact of fire on the environment.

Fireline: The part of a control line that is scraped or dug to mineral soil. Sometimes called fire trail.

Fire management: All activities required for the protection of burnable forest values from fire, and the use of fire to meet land management goals and objectives.

Fire plow: A heavy-duty plowshare or disc plow usually pulled by a tractor to construct a fireline.

Fireproof: 1) Not burnable. 2) To treat an area, hazard, road, etc., so as to reduce the danger that fires will start or spread; e.g., to fireproof a roadside or campground.

Fire-progress map: A map maintained on a large fire to show at given times the location of the fire, deployment of suppression forces and progress of suppression.

Fire retardant: Any substance except plain water that by chemical or physical action reduces flammability of fuels or slows their rate of combustion.

Fire scar: 1) A healing or healed injury or wound, caused or accentuated by fire, on a woody plant. 2) The scar made on a landscape by fire.

Fire shelter: An aluminized, heat reflective, personal, protective pup tent. Required to be worn by most wildland agencies.

Fire season: The period or periods of the year during which fires are likely to occur, spread, and do sufficient damage to warrant organized fire control.

Firestorm: Violent convection caused by a large continuous area of intense fire. Often characterized by destructively violent surface indrafts near and beyond the perimeter, and sometimes by tornado-like whirls.

Fire suppression organization: 1) The management structure designed to enable carrying out line and staff duties of the fire boss with increases in size and complexity of the suppression job. 2) All supervisory and facilitating personnel assigned to fire suppression duty under the direction of a fire boss.

Fire tool cache: A supply of fire tools and equipment assembled in planned quantities or standard units at a strategic point for fire suppression use.

Fire trap: 1) An accumulation of highly flammable fuels. 2) A situation in which it is highly dangerous to fight a fire.

Fire weather forecast: A weather prediction specially prepared for use in forest fire control.

Fire weather station: A meteorological station specially equipped to measure weather elements that have an important effect on fire control.

Firing wound: Fresh or healing injuries of the cambium of a woody plant caused by fire.

Firing-out: Also called burning-out. The act of setting fire to fuels between the control line and the main fire in burning-out operation.

First attack: The first suppression work on a fire.

Flammability: The relative ease with which fuels ignite and burn, regardless of the quantity of the fuels.

Flanking action: Attacking a fire by working along the flanks either simultaneously or successively from a less active or anchor point, and trying to connect the two lines at the head.

Flank fire: A fire set along a control line parallel to the wind and allowed to spread at right angles to it.

Flanks: The sides of the fire. Usually they are not burning as hot as the head. The left flank is the left side looking toward the head from the origin or base of the fire. The right flank is the one on the right side of the fire.

Flare-up: Any sudden acceleration of fire spread or intensification of the fire. Unlike a blowup, a flareup is of relatively short duration and does not radically change existing control plans.

Flash fuels: Fuels such as grass, leaves, dropped pine needles, fern, tree moss, and some kinds of slash which ignite readily and are consumed rapidly when dry. Also called fine fuels.

Flashover: Rapid combustion and/or explosion of unburned gases trapped at some distance from the main fire front. Usually occurs only in poorly ventilated topography.

Foam: A fire-extinguishing chemical that forms bubbles when mixed with water. It adheres to the fuel and reduces combustion by cooling, moistening, and excluding oxygen.

Follow-up: The act of augmenting the first people who go to a fire by sending additional people or equipment to facilitate suppression. Sometimes called reinforcement.

Free-burning: The condition of a fire or part of a fire that has not been checked by natural barriers or by control measures.

Friction loss: Resistance to flow of liquids (usually water) through hose and appliance.

Fuel break: A wide strip or block of land on which the native vegetation has been permanently modified so that fires burning into it can be more readily extinguished. It may or may not have a fireline constructed in it prior to fire occurrence.

Fuel break system: A series of modified strips or blocks tied together to form continuous strategically located fuel breaks around land units.

Fuel moisture content: The quantity of moisture in fuel expressed as a percentage of the weight when thoroughly dried at 212 degrees Fahrenheit.

Fuel-moisture-indicator stick: A specially prepared stick or set of sticks of known dry weight continuously exposed to the weather and periodically weighed to determine changes in moisture content as an indication of moisture changes in forest fuels.

Fuel tender: Any vehicle capable of supplying fuel to ground or airborne equipment.

Fuel type: An identifiable association of fuel elements of distinctive species form, size, arrangement, or other characteristics that will cause a predictable rate of fire spread or difficulty of control under specified weather conditions.

Fuel type classifications: The division of wildland areas into fire hazard classes.

G

Gels: A chemical that is added to water to thicken it, so it will be able to absorb more heat.

General staff: The group of incident management personnel composed of an Operations Section Chief, a Planning Section Chief, a Logistics Section Chief, and a Finance Chief.

Going fire: A fire on which suppression action has not reached an extensive mop-up stage.

Group: Groups are normally established to divide the incident into functional areas of operation. Groups are composed of resources assembled to perform a special function not necessarily within a single geographic division.

Ground fire: Fire that consumes the organic material beneath the surface ground litter, such as peat fire.

Gutter trench: A ditch dug on a slope below a fire, designed to catch rolling burning material.

H

Hand crew: A number of individuals that have been organized and trained and are supervised principally for operational assignments on an incident.

Hand line: Line constructed using hand tools.

Hazard: A fuel complex defined by kind, arrangement, volume, condition, and location that forms a special threat of ignition or of suppression difficulty.

Hazard reduction: Any treatment of a hazard that reduces the threat of ignition and spread of fire.

Head: Pressure due to elevation of water. Equals 0.433 pound per square inch (PSI) of elevation. Back pressure.

Head of the fire: A "running edge" of the fire, usually spreading with the greatest speed. It is driven by the wind or topography. It is not uncommon to have two or more heads on a fire.

Head fire: A fire spreading or set to spread with the wind.

Heavy equipment transport: Any ground vehicle capable of transporting a dozer or tractor.

Heavy fuels: Fuels of large diameter such as snags, logs, and large limbwood, which ignite and are consumed more slowly than flash fuels. Also called coarse fuels.

Held line: All worked control line that still contains the fire when mop-up is completed. Excludes lost line, natural barriers not backfired, and unused secondary lines.

Helibase: The main location, within the general incident area, for parking, fueling, maintenance, and loading of helicopters. It is usually located at or near the incident base.

Helibase crew: A crew of individuals who may be assigned to support helicopter operations.

Helicopter tender: A ground service vehicle capable of supplying fuel and support equipment to helicopters.

Helijumper: A firefighter equipped and trained to jump from a helicopter to fight fire in areas where helicopters cannot land.

Helispot: A temporary landing spot for helicopters.

Helitack: Fire suppression using helicopters and trained airborne teams to achieve control of wildfires.

Helitack foreman: A supervisory firefighter trained in the tactical and logistical use of helicopters for fire suppression.

Helitanker: A helicopter equipped with a fixed tank or a suspended bucket-type container that is used for aerial delivery of water or retardants.

Holdover fire: A fire that remains dormant for a considerable time. Also called hangover fire or sleeper fire.

Hose lay: Arrangement of connected lengths of fire hose and accessories on the ground beginning at the first pumping unit and ending at the point of water delivery.

Hot line: Line that still has active fire along it.

Hotshot crew: A highly trained firefighting crew used primarily in hand line construction.

Hotspot: A particularly active part of a fire.

Hotspotting: Checking the spread of fire at points of rapid spread or special threat. Is usually the initial step in prompt control with emphasis on first priorities.

I

Incendiary fire: A fire willfully set by anyone to burn vegetation or property not owned or controlled by that person, and without consent of the owner.

Incident: An occurrence or event, either human caused or a natural phenomenon, that requires action by emergency service personnel to prevent or minimize loss of life or damage to property and/or natural resources.

Incident action plan: Contains objectives reflecting the overall incident strategy and specific control actions for the next operational period.

Incident base: That location at which the primary logistics functions are coordinated and administered. There is only one base per incident.

Incident command post (ICP): The location at which the primary command functions are executed. Usually collocated with the incident base.

Incident command system (ICS): The combination of facilities, equipment, personnel, procedures and communications operating with a common organizational structure, with responsibility for the management of assigned resources to effectively accomplish stated objectives pertaining to an incident.

Independent action: Suppression action by other than the regular fire control organization or cooperators.

Indirect line: A method of suppression in which the control line is located along natural fuelbreaks, favorable breaks in topography, or at considerable distance from the fire and intervening fuel is burned-out.

Infrared (IR): A heat detection system used for fire detection, mapping, and hotspot identification.

Initial attack: (Initial action) The control efforts taken by resources which are the first to arrive at the incident.

Islands: Patches of unburned fuels inside the fire's perimeter. Some form of suppression action must be taken on islands that are close to the line.

J

Jump spot: A selected landing area for smokejumpers or helijumpers.

Jurisdictional agency: The agency having jurisdiction and responsibility for a specific geographical area.

K

Knock down: To reduce the flame or heat on the more vigorously burning parts of a fire edge.

L

Lead plane: Aircraft flown to make trial runs over the fire and used to direct the tactical deployment of air tankers.

Leapfrog method: A system of organizing workers in fire suppression in which each crew member is assigned a specific task such as clearing or digging fireline on a specific section of the control line, and when that task is completed, passes other workers in moving to a new assignment.

Light burning: Periodic broadcast burning to prevent accumulation of fuels in quantities that would cause excessive damage or difficult suppression in the event of accidental fires.

Lightning fire: A fire caused directly or indirectly by lightning.

Litter: The top layer of the forest floor, composed of loose debris of dead sticks, branches, twigs, and recently fallen leaves or needles, little altered in structure by decomposition.

Live burning: Progressive burning of green slash as it is cut.

Lookout: 1) A person designated to detect and report fires from a vantage point. 2) A location from which fires can be detected and reported. 3) A fire crew member assigned to observe the fire and warn the crew when there is danger of becoming trapped.

M

Message Center: The message center is part of the Incident Communications Center and is co-located or placed adjacent to it. It receives, records, and routes information about resources pertaining to the incident, resource status, and administration and tactical traffic.

Mobilization center: An off-incident location at which emergency service personnel and equipment are temporarily located pending assignment, release, or reassignment.

Modular airborne firefighting system (MAFFS): A pressurized self-contained retardant system for use in Lockheed C-130 military aircraft.

Mop-up, dry: A method in which burning materials are extinguished without the use of water.

Mop-up, wet: A method in which burning materials are extinguished with the aid of water, or in combination with water and soil.

Multi-agency coordination (MAC): A generalized term which describes the functions and activities of representatives or involved agencies in a geographic area who come together to make key decisions regarding the prioritizing of incidents and to share the use of critical resources. A MAC organization is not part of the ICS and is not involved in incident strategy or tactics.

N

National interagency incident management System (NIIMS): Consists of five major subsystems which collectively provide a total systems approach to all-risk incident management. The subsystems are the Incident Command System; Training; Qualifications and Certification; Supporting Technologies; and Publications Management.

National wildfire coordinating group (NWCG): A group of people formed under the direction of the Secretaries of the Interior and Agriculture and composed of Representatives of FS, BLM, NPS, FWS, and Association of State Foresters. The group's purpose is to improve the coordination and effectiveness of wildland fire activities and provide a forum to discuss, recommend appropriate action, and resolve issues and problems of substantive nature. NWCG is the certifying body for all courses in the National Fire Curriculum.

Normal fire season: 1) A season when weather, fire danger, and number and distribution of fires are about average. 2) Period of the year that normally comprises the fire season.

O

One foot in the black: Constructing fireline next to the fire. Usually safest method of attacking a fire of low or moderate intensity in light fuels.

Open line: Refers to open fire front, where no line has been constructed.

Operational period: The period of time scheduled for execution of a given set of operation actions as specified in the Incident Action Plan.

Origin: Location where the fire started.

Orthophoto maps: Aerial photographs corrected to scale such that geographic measurements may be taken directly from the prints. They may contain graphically emphasized geographic features and may be provided with overlays of such features as: water systems, facility location, etc.

Out-of-service resources: Resources assigned to an incident but unable to respond for mechanical, rest, or personnel reasons.

P

Patrol: 1) To travel a given route to prevent, detect, and suppress fires. 2) To go back and forth watchfully over a length of fireline during or after its construction to prevent slopovers and to control spot fires. 3) A person or group who carry out patrol actions.

Perimeter: The total length of the outside edge of the burning or burned area.

Planning meeting: A meeting, held as needed throughout the duration of an incident, to select strategies and tactics for incident control operations and for service and support planning.

Pockets: Deep indentations of unburned fuel along the fire's perimeter. Normally, fireline will be constructed across pockets and they are then burned-out.

Parallel method of attack: A method of suppression in which fireline is constructed approximately parallel to, and just far enough from the fire edge to enable personnel and equipment to work effectively. The line length may be shortened by cutting across unburned pockets. The unburned strip between the fire's edge and the fireline is burned-out as the fireline is being constructed.

Prescribed burning: Controlled application of fire to wildland fuels in either their natural or modified state, under specified environmental conditions which allow the fire to be confined to a predetermined area and at the same time to produce the intensity of heat and rate of spread required to attain planned resource management objectives.

Presuppression: Activities in advance of fire occurrence to ensure effective suppression action. Includes recruiting and training, planning the organization, maintaining fire equipment and fire control improvements, and procuring equipment and supplies.

Progressive hose lay: A hose lay that is laid as it is used to suppress the fire. Lateral hose lines are connected to the main hose line at regular intervals to assist in the fire suppression effort and mop-up.

Protection boundary: The exterior boundary of an area within which a given agency has assumed a degree of responsibility for emergency operations. It may include lands protected under agreement or contract.

R

Radio cache: A cache may consist of a number of portable radios, a base station, and in some cases, a mobile repeater, all stored in a predetermined location for dispatch to an incident.

Rate of spread: The relative activity of a fire in extending its size. It is expressed as rate of increase of the total perimeter of the fire, as rate of forward spread of the fire front, or as rate of increase in area, depending on the intended use of the information.

Rear of the fire: Usually opposite from the head, closest to the origin and nearest the source of the wind. The rear edge of the fire is usually burning slower than other sectors of the fire. Sometimes this is called the *heel* or *base* of the fire.

Reburn: 1) Subsequent burning of an area in which fire has previously burned, but has left flammable fuel that ignites when burning conditions are more favorable. 2) An area that has reburned.

Relative humidity: The ratio of the amount of moisture in a given volume of space to the amount that volume would contain if it were saturated. The ratio of the actual vapor pressure to the saturated vapor pressure.

Resistance to control: The relative difficulty of constructing and holding a fireline as affected by resistance to line construction and by fire behavior. Also called difficulty of control.

Resources: All personnel and major items of equipment available, or potentially available, for assignment to emergencies. Resources are described by kind and type.

Retardant line: Usually constructed by an airtanker or helicopter. Treat retardant line just like wet line. Follow-up with some ground action.

Risk: 1) The chance of fire starting as determined by the presence and activity of causative agents. 2) A causative agent. 3) A number related to the potential number of firebrands to which a given area will be exposed during the rating day.

Running fire: Behavior of a fire spreading rapidly with a well-defined head.

S

Safety zone: An area used for escape in the event the line is outflanked or in case a spot fire causes fuels outside the control lines to render the line unsafe. In firing operations, crews progress so as to maintain a safety island close at hand, allowing the fuels inside the control line to be consumed before going ahead.

Scratch line: Hasty, narrow line cut in the fuels to temporarily stop the spread of the fire. It can be widened later and become the final control line.

Secondary line: Any fireline that is constructed at a distance from the fire perimeter concurrently with or after a line already constructed on or near to the fire perimeter of the fire.

Segment: A geographical area in which a task force/strike team leader or single resource boss is assigned authority and responsiblity for the coordination of resouces and implementation of planning tactics. A segment may be a portion of a division; an area inside or outside the perimeter of an incident; or a fire or group of fires within a complex. Segments are identified with Arabic numbers.

Set: 1) An individual incendiary fire. 2) The point or points of origin of an incendiary fire. 3) Material left to ignite an incendiary fire at a later time.

Simple hose lay: A hose lay consisting of consecutive coupled lengths of hose without laterals. The lay is extended from the water source or pump to the nozzle. The hose lay is filled with water only after it is put in place.

Slash: Debris left after logging, pruning, thinning or brush cutting. It includes logs, chunks, bark, branches, stumps, and broken understory trees and brush.

Slopovers: Where the fire crosses the fireline. The fire and slopover are adjacent to each other with no unburned fuel between them.

Smokejumper: A firefighter who travels to fires by aircraft and parachutes to the fire.

Snag: A dead standing tree.

Span of control: The supervisor ratio of from three to seven individuals with five being established as an optimum.

Spot fire: Fire burning outside the perimeter of the main fire caused by flying sparks or embers.

Spotting: Behavior of a fire producing sparks or embers that are carried by the wind and start new fires beyond the zone of direct ignition by the main fire.

Staging area: A temporary on-incident location, managed by the operations section, where incident personnel and equipment are assigned on a three-minute availability status.

Strike team: Specified combinations of the same kind and type of resources, with common communications and a leader.

Strip burning: 1) Burning by means of strip firing. 2) In hazard reduction, burning narrow strips of fuel and leaving the rest of an area untreated by fire.

Strip firing: Setting fire to more than one strip of fuel and providing for the strips to burn together. Frequently done in burning-out against a wind where inner strips are fired first to create drafts which pull flames and sparks away from the control line.

Suppressant: An agent that extinguishes the flaming and glowing phases of combustion by direct application to the burning fuel. (Water is a suppressant agent.)

Surface fire: Fire that burns surface litter, other loose debris on the forest floor, and small vegetation.

Swamper: A firefighter that leads a bulldozer.

T

Tactics: Deploying and directing resources on an incident to accomplish the objectives designated by the overall strategy.

Task force: Any combination of single resources, within the span of control, assembled for a particular tactical need, with common communications and a leader.

Tractor-plow: Any vehicle with a plow for exposing mineral soil, with transportation and personnel for its operation. Tractor-plows are used primarily in the southern United States.

Trench or Trenched line: A small ditch often constructed below a fire on sloping ground (undercut or underslung line) to catch rolling material.

U

Undercut line: Line that is constructed on a hillside when there is the possibility of burning materials rolling down and crossing the fireline. Undercut line incorporates a trench into its construction. It can also be called trenched line.

Unified command: A command structure which provides for all agencies or individuals who have jurisdictional responsibilities, either geographical or functional, to jointly manage an incident through a common set of objectives.

W

Water tender: Any ground vehicle capable of transporting specified quantities of water.

Wet line: Line that has been constructed using water or foam. Wet line is constructed to extinguish the flame front or to be used to burn from. Except in VERY light fuel, a wet line should not be considered the final control line. The final control line should be cut through the fuel to mineral soil.

Wetting agent: A chemical that reduces the surface tension of water and causes it to spread and penetrate more effectively.

Wet water: Water with added chemicals, called wetting agents, that increase its spreading and penetrating properties.

Wildfire: Any fire occurring on wildland except a fire under prescription.

Wildland: An area in which development is essentially nonexistent, except for roads, railroads, power lines, and similar transportation facilities.

Wildland/Urban Interface: The area where structures and other human development meet or intermingle with natural wildland vegetation.

Index

A

Abilene Paridox 372–375
Access 456
  to structures 456
Action plan 304–308
Aerial Fuels 116
After Action Fire Reviews 56
Air Attack
  use of 268–269
Air attack officer 258, 268–269
Air Operations 321, 333, 383
  air attack officer 190
Air Stability
  related to extreme fire behavior 152
Aircraft 257
  tactical support 257
  use of 257–258, 269–272
Airtankers 257, 258–260, 297, 374, 513
  drop height 260
  retardant drop
    being hit 262
Anchor point 26, 182, 270, 383, 513
Antiquities Act of 1906 231
Archaeological Resources Protection Act of 1979 231
Area Command 355, 513
Area Ignition 162
Aspect 60, 61, 105, 122, 513
  eastern 106
  northern 106
  southern 106, 367
  western 107
Atmosphere
  elements 58
  stability 76, 86, 88
    Haines index 91
    relationship to fire behavior 132, 156
  troposphere 58
  unstable 88
Atmospheric Pressure 59
Attack strategies
  direct 212, 213
  hotspotting 220
  indirect 212, 217
  mobile 240–242
    Envelopment Attack with an Engine 216, 242
    Flanking Attack with an Engine 215, 242
    Pincer Attack with an Engine 215, 242
    Tandem Attack with an Engine 215, 242
  parallel 216
Axe 477

B

Backfiring 192, 194, 513
Backpack Pumps 487
Backpumps 253
Barometric Pressure 60
Barriers 183
  existing 183
  natural 183
Basics
  getting back to 386
Battling Winds 94, 151
Big Iron 287
Black, The 169
Blowup 514
Boots 8
Branches 514
Brass fittings 244
Brush Hook 478
Bulldozers 234, 278–288, 296, 383
  classification 279
  damage they can do 282
  factors that influence the performance 278–279
  how to use 272–288, 279–282
  management of 284–285
  piles 228
  safety 285
Burning-out 192, 514

C

CAFS 173
Camps 342
Canyons 108, 148
Carbon Monoxide 5
Cause
  determination 462
Chain of Command 23
Chain Saw 485
Chimneys 108
Chinook Winds 74
Class of Fire
  size**515**
  type **515**
Clothing
  safety
    second layer 7
Clouds 58, 61, 96–102, 149
  identification 96
  relationship to elevation 96
  relationship to relative humidity 86
  relationship to tempertures 86
Cold Fronts 149
Cold Trailing 221, 515
Combi Tool 483
Command 515
Command Staff 515
Common Denominators
  fatal fires 29
Communications
  need for strong links 23, 33
Compactness
  fuels 128
    fire intensity 138
Compass
  parts of 440
  use of 440–444
Compensation/Claims Unit 337
Complexes 354
Conduction 64, 515
Contour Lines 428–434
Control Lines 229
Convection 64, 144, 161
Coriolis Force 68
Cost Unit 338
Crown Fire 146, 157, 516
  active 158
  independent 158
  passive 157
Crown Spacing 159
Cultural Differences
  hand crews 276
Cutting Tools 476–479

D

Debris Burning 516
Declination 442
Defensible space 454
Demobilization Unit 335
Dew 64
Dew Point 84
relationship to relative humidity 84
Direct Attack 213, 516
Dispatch Information 202–211
Divisions 331, 516
Documentation Unit 334
Downdrafts 74, 166
Drip Torch 255, 489
Drug Labs 468
Dust Devils 76, 90

E

Earth 61
rotation 67
surface
color and texture 62
moisture 63
Ejector 251
Elevation 61, 104, 111, 123
relationship to fire behavior 132
En route to fire 203
Endangered Species Act 232
Engines
ICS typing 240
not all created equal 381
types and sizes 238–240
use of 238–240
Environmentally Sensitive Strategies and Tactics 233
Escape Routes 21, 33, 221
Evacuation 295
Evacuations 302–322
Experience
needed 373
Explosives
Fireline 190
Extended Attack 517

F

Facilities Unit 336
Fatal Fires 145
Fatigue 3, 273
Federal Communications Commission 497
Felling Tools 484–486
Finance/Administration
finance/administration section chief 336
Fire
a chemical reaction 112
cause determination 462–463
elements 112
parts of 181
small vs. big 375–376
use of 190, 192–199
backfiring 194
burning-out 191, 192
general rules 197
prescribed 369–375
Fire Behavior
103, 131, 294, 299, 302, 517
aspect 132
characteristics 142, 204
charts 143
current conditions 19
direction of spread 141
fire environment 132
vertical dimension 140
fuel moisture 133
Haines index 91
mid-flame winds 66
night 380–381
predictions
relative humidity 83, 85
related to topography 132
slope 132
vegetation 63
vertical dimension 139
winds
fatal fires 66
Fire Behavior, Extreme
contributing factors 154–166
fuel characteristics 145
indicators
firewhirls 153
fuel temperature 147
smoke column 152
spot fires 154
terrain 148
trees torching 153
patterns 164
plume-driven fires 165
wind-driven 164
Firebrands 135, 136, 160, 294
relationship to wind 136, 294
Firebreaks 460
Fireline 40, 178–190
Construction 188–190
construction 179, 179–199, 182–183, 184
general rule 188–190
downhill 27
Snags 40
hazards 38–50
sharp tools 42
mopping-up
feeling out 227
patrolling 230
placement and construction 182, 316
Wet 179
width 185
fuel 186
slope or topography 187
weather conditions 186, 187
Fire Characteristics
extreme fire behavior 145
Fire Danger
Haines index 91
Fire Environment 132–166
open or closed 141
Fire Intensity
compactness and arrangement of fuels 138
fuel loading 138
relationship to fuel moisture 138
relationship to rate of spread 138
slope and wind 138
surface fires 144
Fire Orders 15, 15–24, 18
Fire prevention 449–463, 450
conducting an inspection 461–462
FireWise 452–460
planning 450
Fire Shelter 10
Fire suppression
Principles 113
Firewhirls 76, 90, 153
FireWise 452–460
Firing Devices 489–493
Drip Torch 489
Fusee 491
Firing tools 254–256
drip torch 255
fusee 255–256
Firing-out 518
general rules 192, 320
Flashlight 493
Foam
Class A 167, 170, 313, 519
environmental concerns 172
Safety 172
use 171, 174
mixing systems 172
Foehn Winds 73, 94, 151
Chinook winds 74
Santa Ana winds 74
Food and Nutrition 6
Food Unit 336
Forest
cover
shade 63

Friction loss 247
Fronts
cold front 68
winds 72
frontal lifting 99
occluded front 69
troughs and ridges 69
warm front 69
weather 61
Fuel Loading 118
Fuel Moisture 120, 145
dead fuel moisture 120
indicators
extreme fire behavior 147
live fuel moisture 120
relationship to fire behavior 133
relationship to fire intensity 138
relationship to ignition 135
Fuel Temperature 120
Fuels 20, 103, 112, 294, 299
aerial 116
characteristics 115
extreme fire behavior 145
compactness 128
grasses 126
ground 116
heavy or slow-burning 115, 116
horizontal continuance 128
ladder 118, 129
light or fast-burning 115, 169
loading
fire intensity 138
modification 458
reburn 130
size and shape 117
surface 116
temperture
ignition 135
related to extreme fire behavior 145
type change
relationship to rate of spread 138
vertical arrangement 129, 159
Fusee 255, 491

G

Gels
Firefighting 170
Use of 176, 313
General Staff 330
Global Positioning System
how it works 421, 445–447
using it 444
Gloves 9
Goggles 9
Graham, Gordon 362
Grasses 126
Gravity
spotting 136
Gravity Sock 251
Ground Fuels 116
Ground Support Unit 336
Groups 331

H

Haines Index 91
Hand Crews 519
cultural differences 276
effectiveness 273–274, 297
inmate crews 276
military crews 277
operational uses 272–288, 275
production rates 274–275
Hand Line 179
trenching 229
Hand Saw 484
Hand tools 252–253, 473–494
axe 477
backpumps 253
brush hook 478
carrying 475
combi tool 483
cutting and scraping tools 252–253, 476–479
felling tools 484–486
hand saws 484
inspection 474
ladders 253
McLeod 481
Pulaski 477
scraping 479
sharpening 474
shovel 479
u-bolt hoe 483
Hazardous Materials
drug labs 468–470
guidelines 466–467
introduction 465–472
unexploded ordnance 472
Hazards
fire prevention factor 450
retardant drops 45
sharp tools 42
Headlamp 493
Heat
balance 61
solar 60
Sources of 113
transfer
conductive 64
convection 64
radiation 64
Heat stress 3
Heat stroke 3
Heat Wave 320
structure protection 227
Helibase 342
Helicopters 263–288, 297, 374
engine company support 267–272
Helitack crews 265
working around 266–267
Helispot 343
Helitack crews 265
Helmet 9
Hillside
rolling material 28
Historical Preservation 230
Antiquities Act of 1906 230
Archaeological Resources Protection Act of 1979 231
National Historic Preservation Act of 1966 231
Homeowners
FireWise 308
Hood or Shroud 9
Hose 243
progressive hose lays 249–251
reel lines 248
simple hose lays 249
Hose Lay evolutions 317–322
Hose Lays 383
Hotspotting 220
Human Factors 50–56
Hydraulics 245–288
friction loss 247
head pressure 246
lift 245
nozzle pressure 248

I

Ignition 136
compactness of fuels 135
fuel moisture content 135
relationship to fuel temperature 135
relationship to size and shape of fuels 134
Incident Action Plan 344, 352
Incident Base 341
Incident Briefing 343
Incident Command Post 338
Incident Command System 323
area command 355
command
briefing 343

incident action plan 344
liaison officer 329
safety officer 329
command staff 515
complexes 354
components 325–348
engine typing 240
finance/administration
compensation/claims unit 337
cost unit 338
procurement unit 337
time unit 337
general staff 330
how to use
expanding situation 349–360
small incident 348
incident facilities 338–360
camps 342
helibase 342
helispot 343
incident base 341
staging areas 340
information officer 329
logistics
facilities unit 336
food unit 336
ground support unit 336
logistics section chief 335
medical unit 336
service branch 336
supply unit 336
support branch 336
multiagency coordination 356
operations
air operations 333
divisions 331, **516**
groups 331
operations section chief 330–360, 352
single resource 345
staging areas 333
strike team 346
task force 345
organizational functions 327
command 328
planning 343
demobilization unit 335
documentation unit 334
planning section chief 334
resource unit 334
situation unit 334
technical specialists 326, 335
unified command 351
Incident Facilities (ICS) 338–360
incident command post 338
Indirect Attack 217
Indrafts 166
Information Officer 329
Inmate Crews 276
Interface Condition 290
Intermix Condition 291
Interzone 289
Inversion Layers 85, 91
marine 94
nighttime 92
relationship to fire behavior 93
relationship to smoke 95
subsidence 94

L

LACES 22, 32–38, 374
Ladder Fuels 129
Ladders 253
Lag Time 61
Lapse Rate 87
average lapse rate 88
dry lapse rate 87
moist lapse rate 87
Latitude and longitude 61, 420–421
Lenticular Clouds 78, 98, 149
Let Burn Policy *370–371*
Liaison Officer 329
Life Safety
protecting people from fire 209
Live Fuels
moisture content 158
Local Winds 69
Logistics Section Chief 335
Lookouts 22, 32, 34

M

Major Incident Management 350–360
area command 355
complexes 354
multiagency coordinations 356
unified command 351
Map Reading 417, 417–447
azimuths and bearings 441–442
contour lines 428–434
declination 442
elevation and relief 427
latitude and longitude 420–421
magnetic north 424, 424–425
public land survey 421–424
scale and distance 425
slope and grade 438–447
topographic map 418
Maps
coordinates 419
latitude and longitude 420
need 211
symbols 434
type 417–419
Marine Layers 94
Matches 256
McLeod 481
Media 297, 382
Medical Unit 336
Mid-flame Lengths 80
Military crews 277
Minimum Impact Suppression Tactics 233–235
Mopping-up 222–224, 383
dry 222
feeling out 227
wet 224
MultiAgency Coordination (MAC) 356
Mutual Aid
frequencies 506

N

National Historic Preservation Act of 1966 231
National Incident Management System 358–360
Natural Barriers 208
Night
fighting fire at *380–381*
Nozzle Pressure 248
Nozzles 168, 174, 245
applying water 168, 169
aspirating 174
fog 168
straight stream 168

O

Objectives
incident *371*
Occluded Condition 291
Oceans 82
Operations Section Chief 330, 352
Orders
ten standard *366*
Organizer
incident *365, 511-515*
Origin
fire
cause determination 463
Orographic Lifting 99
Oxygen 167

P

Parallel Attack 216
Patrolling
fireline 230
Personal Protective Equipment (PPE) 6–15
Physical Fitness 1–6
fatigue 3
work and rest 3
Planning Section Chief 334
Plume-Driven Fires 165
Polar Caps 61
Portable Pumps 254
Power Lines 46. ***See also* Hazards**
Power tools 254
chain saws 254
portable pumps 254
Precipitation 123
related to fire behavior 132
Pressure
barometric 60
high and low 67
Procurement Unit 337
Protection
property 209
resources 209
Public land survey 203, 421–424
Pulaski 477

R

Radiation 114
heat transfer 64
incoming 61
outgoing 61
solar 60
Radio
batteries 502
equipment 498
base stations 498
how to use 502, 503
mobile radio 499–500
mobile relays 498
portable radios 500
nets 504–506, 506
air to ground 506
command 504
mutual aid frequencies 506
tactical nets 504
theory
radio frequency energy 496
radio waves 495–497
Rate of Spread 138
relationship to extreme fire behavior 149
relationship to fire intensity 138
relationship to fuel type change 138
relationship to slope 108, 138
relationship to spotting 139
relationship to wind speed and direction 138
Reburn 130
Refusing an Assignment 374–375
Relative Humidity 81–86
impact on fire behavior 84, 132
relationship to dew point 84
relationship to wind 86
Resource Unit 334
Respirator 14
Retardant Drops 45, 179
being hit 262
Retardants and suppressants 260–261, 263–265
environmental concerns 234
Ridges 108
Risks 450
balancing with values 361–362
Rocks
rolling 42
Roofing 455
fire-resistant 455
Rules of Engagement 368–370
Rural condition 291

S

Saddles 148
Safety 300
awareness 32
Six-Minutes for Safety Program 362
Safety Officer
role 329
Safety Zone 20, 25, 29,33, 37
Santa Ana Wind 74, 151
degrading 94
SCBA 13
Scraping Tools 479
Self-Contained Breathing Apparatus 13
Service Branch 336
Shade 122
Shovel 479
Signs
street 457
Single Resource 345
Situation Avoidance 382–386
Situation Unit 334
Situational Awareness 53–56, 363–369
ten standard firefighing orders 18
Situations
"watch out" 25–29
Six-Minutes for Safety Program 362
Size Up 25, 204
Skidders and "Skidgines" 286
Sledgehammer 486
Slope 61, 123, 159
calculate 438
relationship to fire behavior 148
relationship to fire intensity 138
relationship to rate of spread 108, 138
winds 70
Smoke 5
column 151
inversion layers 95
Smothering Tools 486–487
Snags 40
removal 229
Solar Heating
winds 76
Solar Radiation 60, 62, 148
lag time 61
reflected 61
South Canyon Fire 372
Spotting 135, 137, 154, 160, 162, 204
probability 160, 162
relationship to extreme fire behavior 149
relationship to rate of spread 139
spot fires 194
Spread
direction 141
Stability of Air
extreme fire behavior 145
Haines index 91
vertical dimension 140
Staging Areas 296, 333, 340–360
Strategy and Tactics 26, 201
environmentally sensitive
minimum impact suppression tactics 233
offensive/defensive action 212
Strike team 326, 346, 349
Structure Protection 207, 290
defending the structure 317
heat wave 227
triage 298–308
Stump Holes 227
Sun 60, 61
Supply Unit 336
Support Branch 336
Surface Fuels 116

T

Tactics 26, 201
- defensive 212
- initial attack 212
- offensive 212

Task Force 326, 345
Team Work 55
Teams
- too much reliance 374

Technical Specialists 335
Temperature
- relationship with dew point 84
- relationship with fire behavior 132
- surface 91
  - factors 61
  - inversion layer 91
- surface heating
  - thermal lifting 99

Ten Standard Wildland Firefighting Safety Orders 15–24
Terrain 61, 85
- relationship to relative humidity 85
- relationship to solar radiation 85

Thermal Belts 91, 95
- relationship to fire behavior 95

Thermal Lifting 99
Think Big 378–379
Thunderstorm 96
- development 75, 98
- downdrafts 74
- relationship to atmospheric stability 87
- relationship to fire behavior 97, 101, 150
- stages 100
  - cumulus stage 100
  - dissipating stage 102
  - mature stage 101
- winds 75

Time of Day 20
Time Unit 337
Topographic map 418
Topography 20, 104, 206, 294
- effects on wind 77
- relationship to fire behavior 103

Torching 137, 153
Tractor-plows 286
Training 17, 384–387
Trenching 229
Trigger Points 366–369
Troposphere 58
Troughs and Ridges 69
Turnout gear 13

U

U-bolt Hoe 483
Unexploded Ordnance 472
Unified Command 326, 351

V

Valley Winds 71
Vegetation 62
- relationship to relative humidity 85

Vehicle Hazards 44
Vortices 163

W

Water
- drinking, importance of 13
- Use of 167, 225
- Wetting Agent 170

Water Vapor 60, 63, 81
- from lakes, streams, etc. 82
- from transpiration of plants 82

Weather 19, 57, 365
- changes 57
- clouds 58
- cold front 64
- conditions
  - en route to fire 206
- fronts 61
- strategic 18
- tactical 17
- warm front 64
- winds 64
  - influences fire spread 64

Wedge 485
Wetting Agents 179
Widow Makers 40
Wildland/Urban Interface 207, 289–322
- common ignition points 314
- dealing with civilians 308
- defensible space 315
- evacuation 295
- hazards 48
- number of structures being threatened 294
- other firefighting resources 296–297
- site preparations 315–317
- size-up 293–296
- special strategies and tactics 292–322, 308
- structure fire protection 290
- structure protection 290
  - triage 298
- watch outs 382

Wind 64, 87, 123, 159
- aloft
  - vertical dimension of a fire 140
- battling 94
- fire intensity 138
- firebrands 136
- foehn 73
- general winds 65
- gravity winds 73
- indicators
  - extreme fire behavior 149, 156
  - lenticular clouds 78, 98
- local winds 66, 69
- midflame winds 66, 80
- problem winds 72
  - foehn winds 73
- related to extreme fire behavior 145
- relationship to fire behavior 65, 66, 94, 132
- relationship to pressure gradient 68
- relationship to relative humidity 86
- relationship to topography 77, 79
- speed and direction 79
  - relationship to rate of spread 138
- slope winds 71
- surface winds 66
- valley winds 71

Wind-Driven Fires 164